2nd International Conference on Smart City Innovation 2019

IOP Conference Series: Earth and Environmental Science
Volume 396

Semarang, Indonesia
9 October 2019

Editor:

Maya Damayanti

ISBN: 978-1-7138-0263-1
ISSN: 1755-1307

Printed from e-media with permission by:

Curran Associates, Inc.
57 Morehouse Lane
Red Hook, NY 12571

Some format issues inherent in the e-media version may also appear in this print version.

Institute of Physics
Dirac House, Temple Back
Bristol BS1 6BE UK

Phone: 44 1 17 929 7481
Fax: 44 1 17 920 0979

techtracking@iop.org

Additional copies of this publication are available from:

Curran Associates, Inc.
57 Morehouse Lane
Red Hook, NY 12571 USA
Phone: 845-758-0400
Fax: 845-758-2633
Email: curran@proceedings.com
Web: www.proceedings.com

TABLE OF CONTENTS

ENERGY AND ENVIRONMENT

ENVIRONMENT

INFRASTRUCTURE

IOP Conf. Series: Earth and Environmental Science **396** (2019) 011001 doi:10.1088/1755-1315/396/1/011001

Preface

We are pleased to present the proceedings of the 2nd International Conference on Smart City and Innovation that was held in Semarang, Indonesia, October 9, 2019 organized by Diponegoro University. This event is funded by the United States Agency for International Development (USAID) and the result of cooperation among six universities in Indonesia, namely the University of Indonesia, Diponegoro University, Padjajaran University, Sriwijaya University, Udayana University, and Sumbawa University of Technology that are involved in the project of Sustainable Higher Education Research Alliances (SHERA).

The SHERA-SMART CITY (Scientific Modeling, Application, Research, and Training for City-centered Innovation and Technology) aims to enhance the quality and quantity of science and technology research in Indonesia's higher institutions. Hence, this conference is organized to encourage the researchers to share their scientific works in response to urban challenges, including the adoption of sustainable energy for urban built environment and urban systems, as well as the provision of reliable urban mobility and sustainable water management.

The proceeding presents the work of the academics around the world who research various topics of smart city discourse particularly related to the topics of energy-environment and infrastructure. This conference had received considerable attention from academics. A total of 132 abstracts and 83 full papers were submitted and followed a blind peer review process. There are 41 papers selected to be published in this proceeding. We like to take this opportunity to thank the reviewers who donated their time to ensure the quality of the papers.

The Editor and the 2nd International Conference on Smart City and Innovation Committee expect that readers of this proceeding will find the research presented in this proceeding are interesting and encouraging future collaboration among the researchers.

Best wishes,

Maya Damayanti, Editor

IOP Conf. Series: Earth and Environmental Science **396** (2019) 011002 doi:10.1088/1755-1315/396/1/011002

Peer review statement

All papers published in this volume of *IOP Conference Series: Earth and Environmental Science* have been peer reviewed through processes administered by the proceedings Editors. Reviews were conducted by expert referees to the professional and scientific standards expected of a proceedings journal published by IOP Publishing.

Sustainability of Using Low-Rank Coal as Energy Source through The Upgrading Brown Coal (UBC) Process by Adding Waste Cooking Oil

B D Afrah[1], M I Riady[2], N Thereza[3], H Widhaningtyas[1], M Siregar[1]

[1]Chemical Engineering Department, Engineering Faculty, Universitas Sriwijaya
[2]Mechanical Engineering Department, Engineering Faculty, Universitas Sriwijaya
[3]Electrical Engineering Department, Engineering Faculty, Universitas Sriwijaya

E-mail: bazlina.afrah@ft.unsri.ac.id

Abstract. The use of coal which is dominated by hard coal is projected as the main energy source (± 25%) in 2050 according to the Indonesian Government Regulation No. 79 of 2014. The use of low-rank coal as a substitute for hard coal through the Upgrading Brown Coal (UBC) process is supported by the Indonesian Government Regulation No. 77 of 2014 concerning the obligation to increase the added value of mining products. The use of UBC products can reduce the amount of coal consumption by up to 50% of the use of low-rank coal The UBC process is still being developed, as in this study, a mixture of low-rank coal and waste cooking oil is used as a coating material to cover the coal pores after the watering process. The variation in the mixing ratio of low-rank coal and waste cooking oil is 1:1; 1:2; and 1:3, with a variation of the heating temperature and mixing mixture of 100°C; 125°C; 150°C; 175°C; and 200°C. The analysis showed that the best conditions are found at a ratio of 1: 3 with a temperature of 175°C, wherein the results of the inherent moisture showed a decrease of 21.95% with a calorific value of 7462.82 cal/g. The high calorific value obtained was proven by the flame that occurred for 14 minutes for 50 g of the UBC product sample in the best variable with ash content of 2.92 %, and volatile matter by 64.77%.

Keywords: Upgrading Brown Coal, Waste Cooking Oil, Environmentally Friendly Energy

1. Introduction

Based on Government Regulation of The Republic of Indonesia No. 79 of 2014 concerning National Energy Policy, some achievement of national energy policy targets is needed to meet the supply and utilization of energy as a target for the provision and utilization of primary energy and final energy. This energy policy is aimed at achieving an optimal energy mix, namely the increase in the role of renewable energy, as well as natural gas, followed by the target of reducing the use of oil and coal as an energy source. Based on the data below, the role of coal as an energy source in 2050 is targeted to reach 25%, replacing the role of oil which is targeted to decrease to 20%, which was previously the main energy source [1]. The policy is based on the high coal production that has existed so far. This can be seen from the export demand for mining commodities, especially coal demand, from abroad increasing from year to year. The volume of coal exports increased, from 287 million tons (2011) to 343.5 million tons (2016) [2]. The coal demand is also met from imported coal, which is projected to increase from 4 million tons in 2016 to 11 million tons or 2% in 2050 [3].

Coal production in Indonesia is expected to continue to increase along with the addition of coal contracts and an increase in the need for national coal. This is because coal is still a cheap and very necessary type of fossil fuel to meet the needs of steam power plants as a basic burden and the needs of several industries, such as cement, textile, steel, and others. National coal production is targeted to continue to increase to reach 648 million tons in 2050 which will certainly deplete national coal reserves, especially the much-needed coal with calorie content above 5,100 kcal/kg [3]. Based on coal reserves data as of January 1, 2016, the proven reserves of medium to high-quality coal (> 5,100 kcal/kg) reached 9.9 billion tons, while the total coal reserves reached 14.3 billion tons [4].

Types of medium quality coal to the top (hard coal) are needed by many power plants and industries. Therefore, it is estimated that coal reserves and total medium to high-quality coal reserves will be exhausted in 2038 and 2048 respectively [2]. This condition needs attention from the government by promoting coal exploration, encouraging the use of low-quality coal (<5,100 kcal/kg) as fuel for coal power plants at the mouth of the mine, increasing efficient use of technology, and encouraging the use of coal upgrading. One of the uses of low-quality coal is the utilization of the Upgrading Brown Coal (UBC) process. Previous research mentioned that the net annual cash flow (ACF) has a positive value, this UBC commercial plant is considered to be economically feasible [5].

This condition needs attention from the government by promoting brown coal exploration. The limitations of old coal energy sources in Indonesia, encourage the use of low-quality coal (<5,100 kcal/kg) as an export resource and fuel for the coal power plant at the mouth of the mine, increase efficient use of technology, and encourage the use of coal upgrading [4]. From previous research, it was found that the use of UBC product use can reduce the amount of coal consumption by 50% of the use of low-quality coal. This is due to the quantity of high humidity and the low heating value of low-quality coal (brown coal) which causes the amount of fuel needed by the boiler to be 3 to 4 times more than the use of high-quality coal (hard coal) to produce the same amount of electricity [6].

Based on Government Regulation of The Republic of Indonesia No. 77 of 2014 concerning the Implementation of Mineral and Coal Mining Business Activities that require the process of adding value to low-quality coal (brown coal). This regulation that demands coal processing for added value includes quality improvement (upgrading), briquetting, cokes making, liquefaction, gasification, and coal slurry/coal water mixture [7].

The process of adding value with the Upgrading Brown Coal (UBC) method is a solution to improve the quality of low rank coal before it is used or converted in other forms of energy so that the energy and heating values obtained are more important and can minimize environmental pollution. Brown coal with low characteristics is difficult to compete with and has an adverse impact on the environment. Low-quality coal consists of many pores and the moisture trapped between the pores is removed by evaporation to be further mixed with asphalt. Asphalt function to fill the pores of coal after the evaporation process and prevent the re-absorption of water content into coal [8]. In this research, the use of used cooking oil as waste oil is expected to be a substitute for asphalt which is added as a mixture in the UBC process to fill coal pores after the dewatering process. The use of waste cooking oil will also save costs so that the process of upgrading coal can be maximized.

2. Experimental methods

The use of research methodology was experimental by taking low-rank coal sample at the mine site of PT Bukit Asam while waste cooking oil was from the seafood restaurant in South Sumatera, and tested in the laboratory of the Research Center of Universitas Sriwijaya. The stages of the UBC process are carried out as follows [7]:

a. Coal preparation: low –rank coal was crushed and then sieve to size up to 32 mesh.
b. Slurry dewatering: low-rank coal was mixed with waste cooking oil in different composition: 2:1, 1:1, 1:2, and 1:3 (w/w) total of the mixed ingredients is 300 g. The mixture was mixed and heated with the various temperature 100 °C, 125 °C, 150 °C, 175 °C, and 200°C for about 1 hour in magnetic stirrer. This stage is used to evaporated the water.
c. Coal-oil separation: The mixture was dried in the oven about 1 hour. The slurry at every composition were filtered to separated coal and waste cooking oil.
d. Coal briquetting: The mixture was pressed to separate coal at 7000-10.000 psi. The briquette was leaved at open space for 24 hours.

Figure 1. Upgrading Brown Coal (UBC) process block diagram

The product analysis stage of UBC product such as inherent moisture, gross calorific value, ash content, and volatile matter would be examined at Laboratorium Dinas Pertambangan, South Sumatera. The next step is to see the correlation between the inherent moisture content value and the calorific value of the UBC product. This correlation is done to see whether the addition of waste cooking oil in the UBC process which has the main purpose to reduce water content can increase the calorific value of the UBC product or not.

3. Results and discussion

Raw material from PT. Bukit Asam is coal which is not sold commercially because of the quality carried by the standards set by the company. Therefore, even though the heating value is above 5000 cal/gr, this coal still categorized as low-rank coal because only used for the company's internal needs. Table 1 shows the results of the raw material analysis in more detail.

Table 1. Raw material analysis

Inherent Moisture (%)	Calorific Value (cal/gr)	Ash Content	Volatile Matter
8.2	6242	4.6	42.32

In this experiment, focus the study is the effect of the composition of briquettes given different heating temperatures with different compositions. The UBC product produced in this experiment has a denser texture due to the low-rank coal section coated with waste cooking oil.

3.1. Moisture Content

The decrease in moisture content is the main target of this experiment considering that the basis of the UBC process is the slurry dewatering process so that it can be a supporting factor in increasing the quality of low-rank coal [7]. The temperature of the heating and variations in the mixture between coal and oil affect the moisture content in UBC products. The value of the moisture content of UBC products is presented in Figure 2.

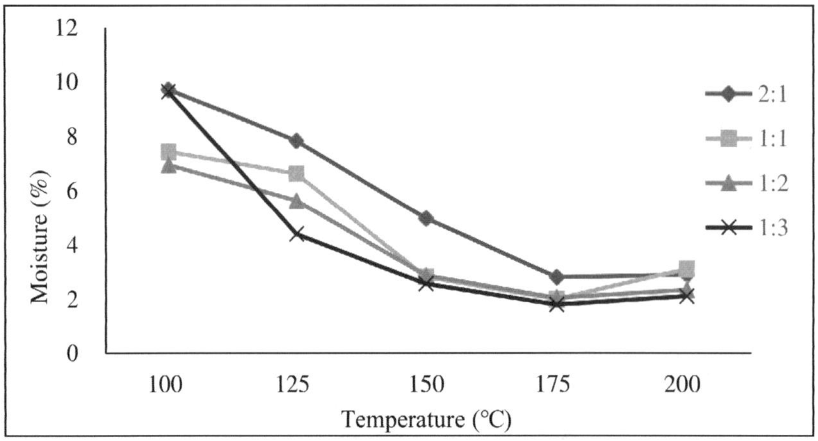

Figure 2. The effect of temperature and mixed ratio on moisture content

Each heating temperature in various mixed variations shows different results. From the graph, it can be seen that the decrease in moisture stops when the heating temperature reaches 200°C. Heating to a temperature of 200°C causes an increase in the moisture content of the previous variable because oil is defect or degraded by temperature which causes defect to the oil structure [9]. This defect causes the presence of oil on briquettes to not have much influence on the quality of briquettes. The UBC product in this condition has not been well coated by oil and allows water to re-enter after the pores in the product UBC open with the dewatering process in the early steps of the UBC process.

If viewed from moisture content, the best heating temperature is 175°C in each composition of coal and wasted cooking oil shows the smallest value. In addition, the graph shows that the greater the composition of the oil from the mixture, the lower the moisture content. The composition of the oil content in the briquettes determines the quality of coal, especially the value of inherent moisture. The presence of oil in coal affects the attractiveness (cohesive force) between coal materials. Increasing the cohesive force will strengthen the pores so that is become smaller and prevent the entry of impurities. The cohesiveness of coal will increase if the moisture content in coal decreases [10]. The presence of pores can cause a decrease in the quality of briquettes because many impurities that can enter fill the pores. The more oil that fills the pores, the fewer impurities that enter the pores of the briquettes. From the explanation above, the lowest moisture content was obtained at 175°C and 1: 3, when compared to low-rank coal, decreased to reach 1.8% of the overall composition of UBC.

3.2. Calorific Value

Calorific value influenced by a decrease in moisture in UBC product, it can be seen in the lowest moisture content in the composition of 1:3 briquettes with a temperature of 17°C there is an increase in calorific value up to 7552.55 cal/g (figure 3). Calorific value is also affected by water content because of the lower the water impurity, the higher the calorific value. These results, when compared with the low-rank coal before going through the addition of waste cooking oil, have increased the calorific value of more than 1000 cal/g.

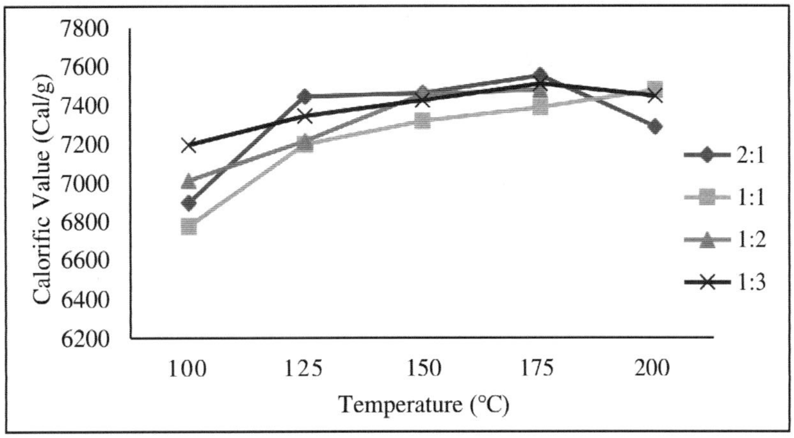

Figure 3. The effect of temperature and ration on calorific value

The increase in the lowest calorific value occurs in the ratio of coal and oil 2:1 at a temperature of 100°C which is 6,777.4 cal/g. The water at that temperature has not been completely evaporated due to the reason (the water has not been evaporated at 100°C). In addition, the amount of coal that is more than oil causes not all low-rank coal pores that have undergone a dewatering process to be filled with used cooking oil.

3.3. *Ash Content*

The ash content of coal in each variation in composition and temperature is shown in figure 4.

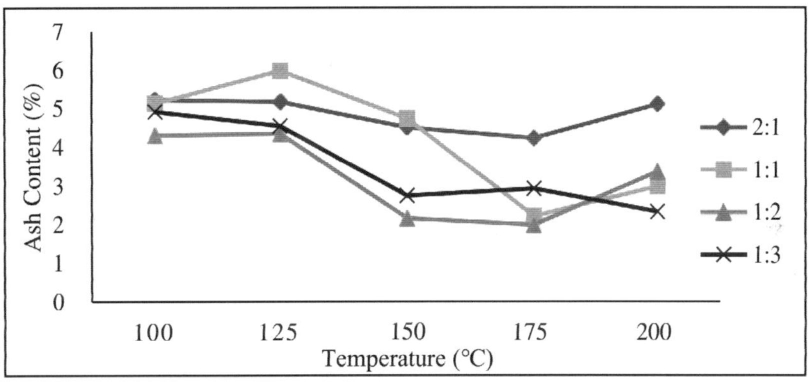

Figure 4. Ash content on UBC products

The ash content is an undesirable impurity in the process of making UBC products and in the coal itself. Most of the ash content in each composition decreases with increasing temperature and affects the calorific value of the UBC product [11,12]. The higher the ash value will reduce the calorific value because the calories burned become decreased when there is impurity ash. The high calorific value was obtained (conditions 175°C and 1: 3) were proven for 14 minutes for 50 g of the UBC product samples in the best variable with ash content of 2.92%.

3.4. *Volatile Matter*

The level of the fly substance in briquettes is one of the impurities that can trigger the initial combustion. Gas content such as H_2, CO, CO_2, CH_4, and steam affect the ease of briquettes to ignite and affect the amount of smoke produced [9]. The greater the level of volatile matter, the easier the UBC products to light up with side effects that will produce a lot of smoke. The substitution of waste cooking oil in coal can increase volatile matter levels in UBC products, resulting in an increase of up to 65.77% from 42.32%. The results of volatile matter analysis are presented in the graph below.

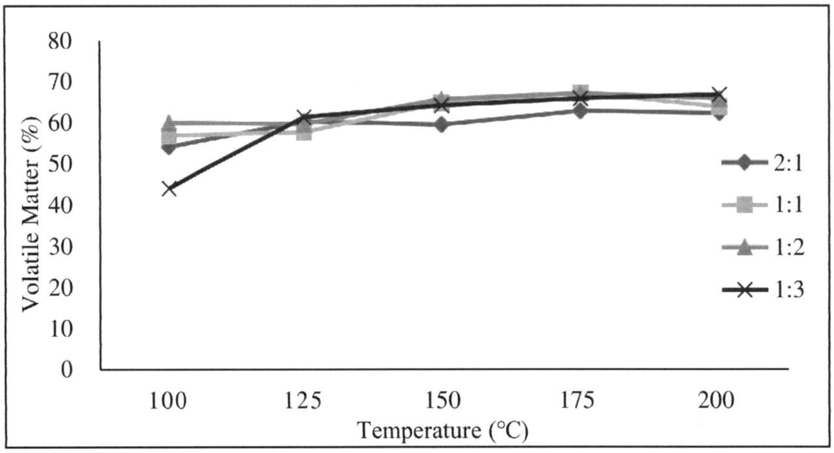

Figure 5. Volatile Matter in varying mixed ratio and temperature

3.5. Inherent Moisture Content – Calorific Value Correlation

The attachment between inherent moisture (IM) and calorific value (CV) is displayed by comparing the composition between the low-rank coal composition and used cooking oil to show the effectiveness of adding wasted cooking oil in the UBC process.

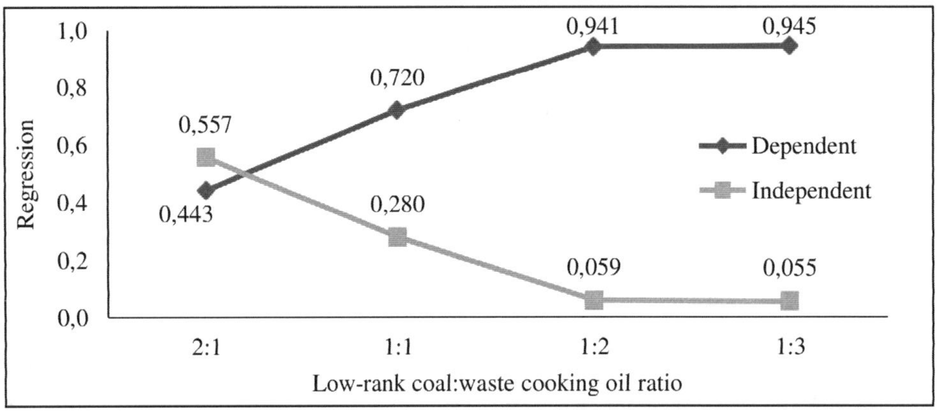

Figure 6. Inherent Moisture Content – Calorific Value Correlation

In figure 6 the graph shows that the greater the composition of oil contained in UBC products will bind the material so that the coal material will be increasingly bound. UBC product with a composition of 2:1 shows that oil does not have a big enough role because the attachment between CV and IM reaches 44.30%, but with the addition of binding oil between these qualities the greater, this is shown in the ratio 1:3, the relationship between the two is bound to 94.5% and are not bound by 0.055%. This indicates that the presence of oil plays an active role in increasing the CV contained in the UBC product.

4. Conclusion

UBC process can manufactured upgraded coals with better combustion characteristics compared to those of raw coals. The process of improving the quality of low-rank coal decreases the water content of UBC products, this is evidenced by the decrease in water reaching 1.8% at the composition of 1:3 and 175°C. In the composition of the lowest moisture content, the highest calorific value is more than >1000 cal/g reached 7552.22 cal/g. The quality of UBC products increases with increasing temperature with the best temperature of 175°C and decreases in quality at 200°C due to damage caused by the degradation of wasted cooking oil. The composition of the mixture affects the cohesion of the UBC

IOP Conf. Series: Earth and Environmental Science **396** (2019) 012001 doi:10.1088/1755-1315/396/1/012001

products marked by the more oil added, it will eliminate the impurities, is indicated by the composition 1:3 which has the best value. The presence of ash content and the volatile matter has an effect on the calorific value, the lower the ash content and the higher the level of volatile matter, the higher the heating value. The attachment between the value of the inherent moisture content and the calorific value decreases as the composition of wasted cooking oil increases, as evidenced by the value of the binding (regression) in the composition of 1:3 of 94.48%.

5. Acknowledgment

This article was presented at the 2nd International Conference on Smart City INNOVATION (ICSCI) 2019, jointly held by Universitas Indonesia and Universitas Diponegoro. ICSCI conferences have been supported by the United States Agency for International Development (USAID) through the Sustainable Higher Education Research Alliance (SHERA) Program for Universitas Indonesia's Scientific Modelling, Application, Research and Training for City-centered Innovation and Tcchnology (SMART CITY) Projcct, Grant#AID-497-A-1600004, Sub Grant#IIE00000078-UI-1. Thc authors wish thank to USAID SHERA, Lembaga Penelitian dan Pengabdian kepada Masyarakat (LPPPM) Universitas Sriwijaya, Pusat Penelitian Lingkungan Hidup (PPLH) Universitas Sriwijaya and Dinas Pertambangan Laboratory of South Sumatera Province.

6. References

[1] Indonesian Government 2014 *Indonesian Government Regulation no.77/2014 on National Energy Policy* (Indonesia)

[2] Atteridge A, Aung M T and Nugroho A 2018 *Contemporary Coal Dynamics in Indonesia* (Stockholm Environment Institute.)

[3] Yudiartono, Anindhita F, Sugiyono A, Abdul Wahid L O and Adiarso 2018 *Outlook Energi Indonesia 2018: Energi Berkelanjutan untuk Transportasi Darat*

[4] D. and I. C. of E. and M. Resources 2012 *Indonesian Energy Outlook Review* (Indonesia: Ministry of Energy and Mineral Resources of the Republic of Indonesia)

[5] Afraha B D, Sajjakulnukita B and Bustanb M D 2014 Life Cycle Cost Analysis of Upgrading Brown Coal (UBC) Process in Indonesia *Int. J.* **5** 153–7

[6] Afrah B D, Sajjakulnukit B and Bustan M D 2017 Product Competitiveness of Upgrading Brown Coal (UBC) Process in Indonesia *Int. J. Adv. Sci. Eng. Inf. Technol.* **7** 1289

[7] Kinoshita S, Yamamoto S, Deguchi T and Shigehisa T 2010 Demonstration of Upgraded Brown Coal (UBC) Process by 600 Tonnes/Day Plant *Kobelco Technol. Rev* **29** 93–8

[8] Center J C E 2007 Clean Coal Technologies in Japan *Technol. Innov. Coal Ind.*

[9] Thoha M Y and Fajrin D E 2010 Pembuatan Briket Arang Dari Daun Jati Dengan Sagu Aren Sebagai Pengikat *J. Tek. Kim.* **17**

[10] Zhang H, Wan Z, Ma D, Zhang B and Zhou P 2017 Coupled Effects of Moisture Content and Inherent Clay Minerals on the Cohesive Strength of Remodelled Coal *Energies* **10** 1234

[11] Diaz R O, Montañez A and Cuentas J 2016 Coal Tar, Material Used in Soil Improvement For Use in Road Engineering *J. Phys. Conf. Ser.* **687** 12045

[12] Said M and Fanani Z 2017 Peningkatan Kualitas Batubara Terhadap Sub-Bituminus Dengan Metode Froth Flotation *J. Penelit. Sains* **11** 587–93

Plasma Gasification With Municipal Solid Waste As A Method Of Energy Self Sustained For Better Urban Built Environment: Modeling and Simulation

P A Sesotyo[1] M Nur[1,2], J E Suseno[2]
[1]Master of Energy, Diponegoro University, Semarang – Indonesia
[2]Departement of Physics, Diponegoro University, Semarang – Indonesia

E-mail: psesotyo@gmail.com

Abstract. The plasma gasification method is known as a new technology to produce hydrogen from MSW as a fuel for power generation. Recently, it has become a crucial issue due to the rapid increasing amount of waste, along with the increasingly national population in Indonesia. The goal of waste management is as a material recovery (MR), as an energy recovery (ER) and minimizing the environment impact. The plasma gasification modeling was developed based on plasma torch technology to estimate the syngas composition and energy potential. The process simulation is optimized for mixed steam and air as a gasifying agent. This study is to obtain the essential parameter to have the better content of H_2 in the syngas, better plant efficiency and lowest CO_2 emission for developing urban built environment with lesser greenhouse gases (GHG). Result shows that, the ratio of Steam to Waste: 0.08 %, the air mass flow rate: $1.11*10^{-3}$ kg/s and the steam mass flow rate: $01.11*10^{-4}$ kg/s gain the highest molar fraction of H_2 : 32.09%, Syngas Yield : 1.28 Nm^3/kg, Carbon Conversion Efficiency : 5.83%, Cold Gasification Efficiency : 77.96% and the CO_2 emission : $2.23*10^{-4}$ kg/s.

Keyword: Plasma Gasification; Urban Build Environment; H_2 molar fraction; Gasifying Agent; CO_2 emission.

1. Introduction

The disposal of waste, especially in the urban built environment, such as municipal solid waste, in the open landfill site, expose a problem of contamination from hazardous materials and emission of greenhouse gases (GHG). Substituting the consumption of fossil fuels with renewable fuel, is one of the solution for overcoming the greenhouse gases [1]. The problem of hazardous material landfills or increasingly coverage area of municipal solid waste landfills can be resolved with waste management system, as the newly method is plasma gasification.

Plasma gasification is a method of waste disposal by introducing the high temperature plasma to decompose the waste into its decomposed chemical compound. In the decompose process, the thermal plasma is applied, generated by passing high electric current through gasifying agent. At high temperature, electrons dissociate from the gasifying agent molecules and an ionized gas / plasma stream is obtained [2]. The plasma stream breaks all the chemical bonds of the feedstock material, resulting the amount of highly active radicals, electrons, ions and excited molecules in the plasma gasification reactor.

The plasma gasification method has multiple benefits, such as a fast reaction time, due to high temperature and high energy density. Within compact reactors, a high material flow rate can be applied due to the fast reaction time. Desired result is as high as the molar fraction of H_2, in the composition of the syngas. No combustion takes place inside its reactors, resulting in lower CO_2 emission, compared to the waste incineration. The heat sources of plasma gasification can be controlled to regulate the temperature inside the reactor, independent from the fluctuation of its feedstock flow rate, quality and mass flow of its gasifying agent. The disadvantage is that the required energy to generate the high temperature plasma is considered extensive, which might can be compensate by the energy potential generated from the plant, which is H_2, by using Internal Combustion Engine (ICE) or Fuel Cells system.

Like gasoline or natural gas, H_2 is a fuel that must be handled properly. H_2 is colorless, odorless, tasteless, non-toxic and corrosive. Hydrogen has a high energy content by weight, but not by volume, which is a particular challenge for storage. In order to store sufficient quantities of hydrogen gas, it's compressed and stored at high pressure. Liquid hydrogen has different characteristics and different potential hazards than gaseous hydrogen, so different control measures are used to ensure safety. An understanding of the properties of hydrogen, either gaseous or liquid, is critical for the proper design of a plasma gasification facility [3].

One of the success stories for the firstly installed integrated plasma gasification and fuel cell system, is in Cheongsong, South Korea. It has the capacity of 15 ton/day of municipal solid waste and can produce 50 kW of electricity from Fuel Cell (5kW Fuel Cell unit x 10 unit). Waste to energy technology with zero waste can't be achieved if the public sector without government support for adequate policies and give opportunities to the public sector to contribute in the implementation of the policies [4].

This study follows the work of Minutillo et al [5], where the developed EPJ (equiplasmajet) was the modelling of thermal equilibrium of plasma gasification and using the municipal solid waste characteristics from the works of Khuriatie et al [6]. The carbon conversion efficiency, the CO2 emission aside the quality of syngas and reactor efficiencies also hasn't been explained in previous research papers. The purpose of this study is to obtain the essential parameter to have the better content of H_2 in the syngas, better plant efficiency and lowest CO_2 emission by using mix steam and air as a gasifying agent and its variation to the feedstock mass flow to be considered, which refer to the Favas et al [7].

2. Material and methodology

2.1 Material

Municipal solid waste from Jatibarang Landfill is used as a plasma gasification feedstock. Its characteristics are defined as Proximate Analaysis, Ultimate Analysis and Calorivic Value, can seen in Table 1.

Table. 1. Waste Characteristic of MSW [6]

Ultimate Analysis (wt. % db)	
Ash	9.51
C	43.71
H	7.74
N	1.95
Cl	0
S	0.40
O	36.69
Proximate Analysis (wt. %)	
Fixed Carbon	12.82
Volatile Matter	77.66
Ash	9.51
Water Content	20
Calorivic Value (kJ/kg)	
HHV	18530.4
LHV	16013.6

2.2 Method

The method of simulation for this study, rely on the sequence workflow of the developed model, with the feedstock properties as in the Tabel 2 to be inputted in the as the ingredients. The boundary conditions of plasma gasification process modeling are process parameters along with the decomposing formula and chemical reaction in the thermochemical equilibrium stoichiometry. Molar fraction of desired syngas compound with its process temperature, density and mass flow rate is the output parameters from the simulation. Validating the modified model output parameters from the reference model [5], with the similar feedstock properties and process parameters resulting in the little variance. The observed process parameters will work as the prepared scenario, in order to compare among those results. The syngas molar fraction as well as the other process parameter in the simulation results will be used to calculate the performance parameters such as Syngas Yield, Syngas LHV, Carbon Conversion Efficiency and Cold Gasification Efficiency. Analyst will be done to those performance parameters obtained to observe the best scenario of H_2 generation along with the lowest CO_2 emission.

2.3 Modelling

Within this study, the plasma gasification has been investigated and analyzed through numerical model which properly developed by applying the thermochemical equilibrium approach as mentioned in the references. Fig. 1 shows the flowsheet of plasma gasification model, called modified EquiPlasmaJet (EPJ), firstly developed by Minutillo, namely EquiPlasmaJet, with the little modification by the introduction of RStoic for the chemical stoichiometric approach to determine the molar fraction of each syngas compound compositions [5].

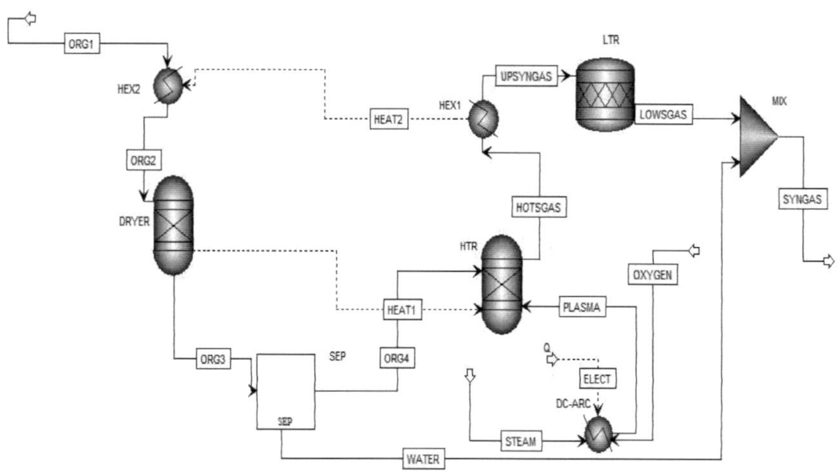

Fig. 1. Gasification Process model (modified EPJ model).

Table 2, shows the short description of the block used in the model. Since only the organic fraction of the MSW is gasified, the modified EPJ model will abandon the inorganic fraction, which will be vitrified, and not in the focus of this study.

IOP Conf. Series: Earth and Environmental Science **396** (2019) 012002 doi:10.1088/1755-1315/396/1/012002

Table. 2. Process Block Description of Modified EquiPlasmaJet

Block Name	Block Type	Description
DRYER	RYield	Non-Stoichiometric reactor based on known Yield Element Distribution from Calculator Feature
HTR	RGibbs	Rigorous Hydrate Reactor and multiphase equilibrium based on Gibbs Free Energy Minimization without known chemical reaction
LTR	RStoic	Stoichiometric reactor with known chemical reaction from its reactant and product
HEX1 & HEX2	Heater	Simple Thermal heat exchanger with only 1 material stream
SEP	Separator	Water separation from Feedstock due to the high temperature
DC-ARC	Heater	Simple Electric Thermal Conversion within DC environment
MIX	Mixer	Material Stream Mixer from two or more streams.

A DRYER block is located before the HTR block as a means of decompose the feedstock into an organic fraction. Inside dryer block, material yield distribution is specified by the help of the FORTRAN calculator according to the proximate analysis and ultimate analysis. The result of calculator is used to estimate the decomposition of material into molecular elements. The excess heat from the gasification process is associated with the HTR block is balanced as a "heat stream" (HEAT1) that connect the DRYER with the HTR block.

As the temperature inside the reactor will vary between certain zones, the temperature divided into two reaction zones for the convention of the modelling. There are two reactors, which will represent the two different temperature zones. The first reactor is the HTR (high temperature reactor), in which thermochemical equilibrium is gain by a non-stoichiometric method in high temperature environment and LTR (low temperature reactor), in which the thermochemical equilibrium is gain by a stoichiometric method in low temperature environment. Inside the HTR, the equilibrium composition is gain by direct minimization of the Gibbs free energy for a given set of expected products without the specific known chemical reaction. The HTR reactor operating temperature on the average of 2500 degC, is the main zone of the plasma gasification reactor, where the plasma jet directly impacted the treated MSW. Inside the LTR reactor, the operating temperature on the average of 1250 degC, where the plasma gasification process is completed with the known chemical reaction, and the organic decomposed element is converted into syngas.

Plasma torch device, modeled by a DC-ARC block which supplied the heat required to generate the hot plasma gas. The plasma stream and estimated power consumption of the plasma torch is calculated by the thermal power transferred into the mixed of steam and oxygen as gasifying agent in the DC-ARC heat exchanger with the assumption of its efficiency, which is ratio between the heat energy transferred and the consumption of energy.

Since the feedstock, which is municipal solid waste is gravitationally flow downward of the gasification reactor, it is preheated by the hot syngas that flow upward of the reactor. The moisture content from the solid waste evaporates due to the hot syngas and leaves the reactor along with the flow of syngas. The block model HEX2 for solid waste (ORG1) and HEX1 for syngas (HOTGAS). And the separation block model (SEP) and the mixer block model (MIX) have been selected in the reactor model.

2.4 Model validation

The result of the EPJ model which developed by Minutillo et al 2009 is used as the validation reference for the modified EPJ model which used by this study. The validation will use the same feedstock characteristics, the same type of gasifying agent and the other process parameters. Below is the feedstock characteristics from Minutillo et al, which used the RDF (refuse derived fuel)

Table. 3. RDF characteristics [5]

Proximate (% mol)	
Moisture	20
FC	10.23
VM	75.96
Ash	13.81
Ultimate (% mol)	
Ash	13.81
C	48.23
H	6.37
N	1.22
CI	1.13
S	0.76
O	28.48
MSW HHV (MJ/kg)	17.8

A result comparison between the molar fractions of syngas composition by modified EPJ model (see Fig. 1) and those obtained by Minutillo et al are summarized in Table 4, showing tolerable results. The results in Table 4 are using the air as plasma gas and RDF as the feedstock material.

Table. 4. Validation Result

Syngas Composition (%mol)	Simulation from modified model	Minutillo, Perna and Bona, 2009	error (%)	RMSE
N2	27.43	26.97	1.69	
CO	32.28	33.79	4.48	
CO2	0.00	0.00	0.00	0.85
H2	20.01	21.04	4.89	
CH4	5.73	5.97	3.96	

The variation between the reference [5] and the simulation result from the modified model, especially on the molar fraction of CO, H_2 and CH_4 are less than 10%, considered tolerable.

2.5 Gasification Reaction

The chemical reactions which occur in the gasification reactor can be summarized as follows [8]:
$C + H_{\%}O \leftrightarrow CO + H_{\%}$ *(Heterogenous Water Gas)*
$C + 2H_{\%} \longrightarrow CH_{+}$ *(Methanation of Carbon)*
$CO + H_{\%}O \leftrightarrow CO_{\%} + H_{\%}$ *(Water Gas Shift)*
$C + O_{\%} \longrightarrow CO_{\%}$ *(Carbon Combustion)*
$H_{\%} + 0.5O_{\%} \longrightarrow H_{\%}O$ *(Hydrogen Partial Combustion)*

Those chemical reactions are fits in with the chemical composition of the feedstock (see Table. 4) and the product results from the simulation software (see Table. 5)

2.6 Boundary condition assumptions and variable

Boundary condition assumption are as follow:

- Steady state process
- The gasification process is considered isobaric and adiabatic.
- The HCoalgen and DCoalligt property models were used to estimate biomass formation enthalpy, specific heat capacity in constant pressure and chemical density based on the proximate and ultimate analysis.
- The chemical element & compound used are : H_2, O_2, N_2, CO, CO_2, CH_4, H_2O, C, Cl, S, H_2S, S and COS.

IOP Conf. Series: Earth and Environmental Science **396** (2019) 012002 doi:10.1088/1755-1315/396/1/012002

- Ash is considered as a non-reactive and non-conventional solid.
- Modeling approach use is thermochemical equilibrium non-stoichiometric Boundary condition variable are as follows:
- Steam Gasifying Agent Temperature : 120 °C [9]
- Torch Plasma Temperature : 4000 °C [5]
- Ambient Temperature: 25 °C
- Plasma Generation Efficiency : 90 % [5]
- Steam Generation Power Consumption: 4.27 MW per 1 kg/s
- Gasification Pressure: 1 atm

3. Results and Discussion

3.1 Simulation result

The modified EPJ model has been employed to predict the syngas composition, syngas yield, syngas LHV, Carbon Conversion Efficiency, Cold Gasification Efficiency and the CO_2 emission. To define the optimal performance parameter of the Plasma Gasification Process, five difference scenarios can be seen on Table 5.

Table. 5. Scenario for the simulation

Steam & Air case	Steam to waste ratio [10]	Feedstock massflow [10] (kg/s)	Steam massflow (kg/s)	Air mass flow (kg/s)	Air actual	ER [10]	Air Stoic
1	**0.080**	**1.39.E-03**	1.11.E-04	1.11.E-03	0.801	**0.180**	4.450
2	0.225	**1.39.E-03**	3.13.E-04	1.53.E-03	1.101	0.248	4.450
3	0.370	**1.39.E-03**	5.14.E-04	1.95.E-03	1.402	0.315	4.450
4	0.515	**1.39.E-03**	7.15.E-04	2.36.E-03	1.702	0.383	4.450
5	**0.660**	**1.39.E-03**	9.17.E-04	2.78.E-03	2.003	**0.450**	4.450

The performance parameter formula as our concern in this study, will be as follows:

$LHV_{234567} = HHV - 10.79Z_{>?} + 12.62Z_{AB} + 35.81Z_{A>E}$ (1)

Where Z is the % volume of mentioned syngas components and LHV in (MJ/Nm^3)

$Yield = \frac{K}{L}$ (2)

Syngas Yield, with □ as the volumetric flow rate (Nm^3/s) and ṁ as the mass flow rate (kg/s)

$CGE = \frac{O{>}K_{PQRSTP}U_{PQRSTP}}{O{>}K_{VWX}L_{VWX}Z[_{\backslash JTP\wedge T}} * 100\%$ (3)

CGE as the Cold Gasification Efficiency, LHV_{Syngas} in (MJ/Nm^3), F_{Syngas} as the volumetric flow rate of syngas (Nm^3/s), LHV_{MSW} as the low heating value of the feedstock in (MJ/kg), $\dot{m}_{MSW}$ as the feedstock mass flow rate (kg/s) and P_{plasma} as the power consume to supply the plasma torch.

$X_b = c\frac{d\%e(AB\%ZAB\%\%ZA{>}{+}\%ZAB2\%)}{\%\%.{+}{*}A\%}h * 100\%$ (4)

X_c as the Carbon Conversion Efficiency, the, Y is the Syngas Yield (Nm^3/kg), and C, CO, CO_2, CH_4, COS is the mol fraction of those chemical compound existed in the syngas composition.

The following figure show the trends as per the configuration 1 to configuration 5

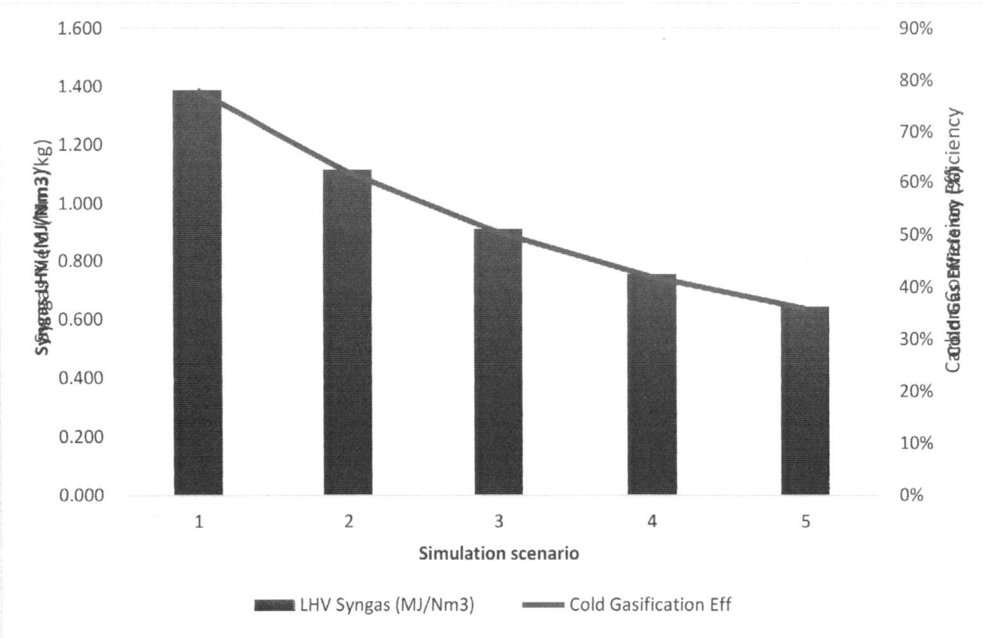

Fig. 2. Trends on Syngas LHV and Cold Gasification Eff.

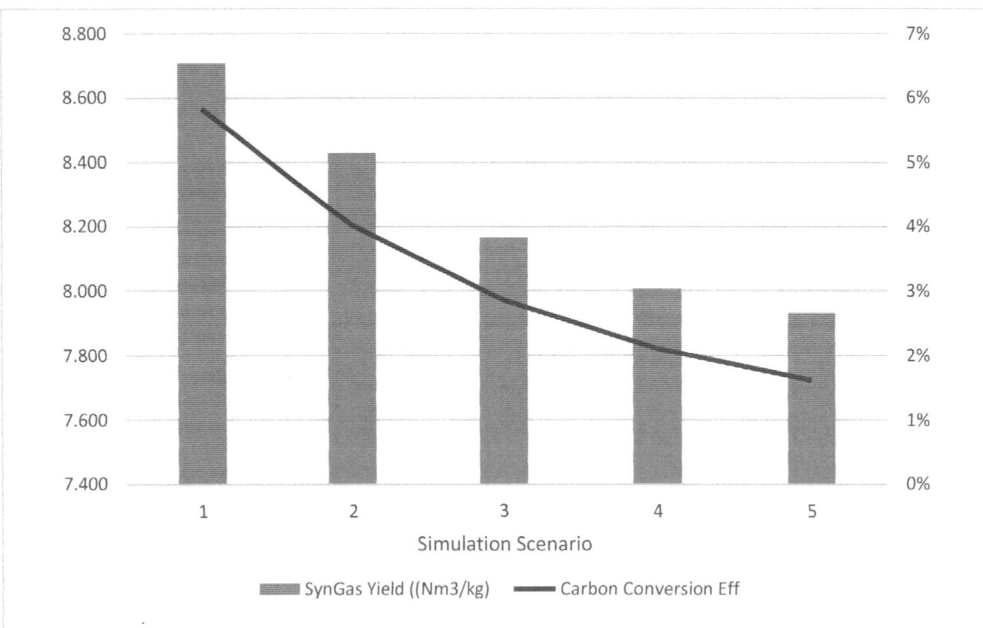

Fig. 3 Trends on Syngas Yield and Carbon Conversion Eff.

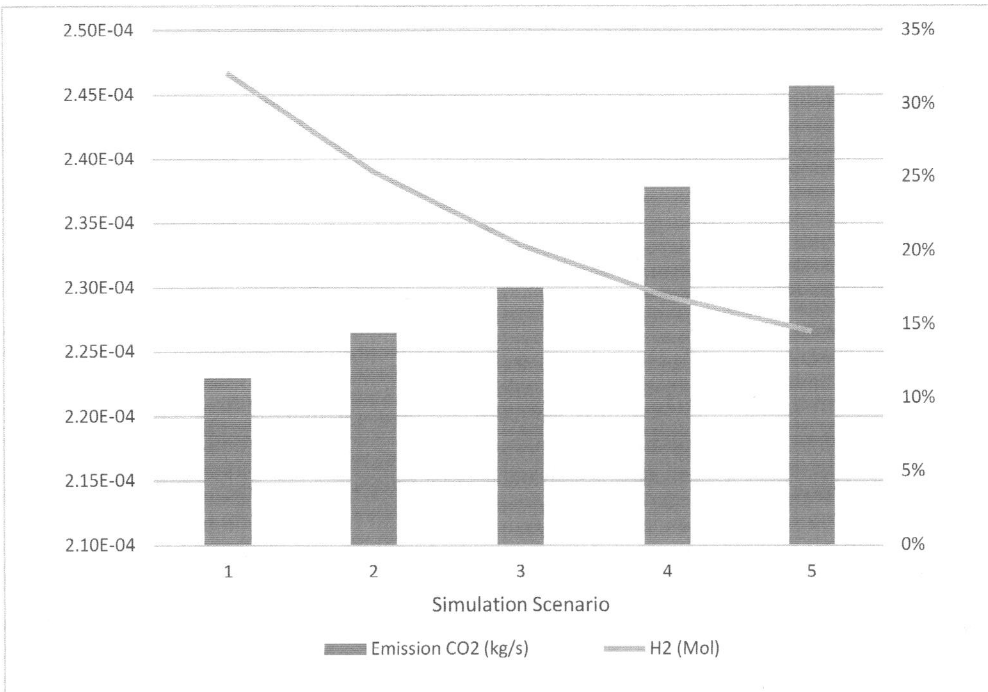

Fig. 4. Trends on molar fraction of H_2 and CO_2 emission

3.2 Simulation result analysis

The decrease of CO and molar fraction of H_2 are influenced by the temperature values, since a greater temperature values is required by the endothermic chemical reaction, causing lower rate of the Water Gas Shift reaction and automatically reduce these chemical elements, since water gas shift is an endothermic reaction. The reason to use mix steam-air as the gasifying agent is to enrich the molar fraction of H_2 in the syngas [7] and has the advantage by introducing more hydrogen atom into the system, thereby increasing the quality of the syngas, instead of using normal air. Disadvantage of using mix steam-air is the greater cost of power consumed. As expected, the use of mix steam-air, resulting in higher partial pressure of water vapor inside the reactor, which required by the water gas shift reaction, leading to the increase molar fraction of H_2.

Combining the three of process parameter mentioned previously, steam to waste ratio (SWR), MSW mass flow, and equivalent ratio, resulting the variation of the molar fraction of H_2 in syngas, also the other process parameter such as syngas volumetric flow, syngas density, the syngas temperature and CO_2 emission. Among the five configurations as per Table 5, the 1st configuration give the highest H_2 molar fraction, which is 32,09%, also give the lowest CO_2 emission, which is 2.23 $*10^{-4}$ kg/s, highest CGE, which is 77.96%, highest X_c, which is 5.83%, highest syngas LHV, which is 1.387 MJ/Nm^3 and highest syngas yield, which is 1.281 Nm^3/kg. The 1st configuration, shows the smallest steam flow rate and the smallest SWR and the 5th configuration, shows the highest steam flow and the highest SWR. The injection of mix steam-air as gasifying agent has to be in proportional with the MSW mass flow rate.

The lower the SWR, the better result of molar fraction of H_2 thus there is an option to save energy from generating steam, save the water as a raw material to generate steam and reduce the CO_2 emission (see Fig. 4). The more mix steam-air introduce to the reactor, the carbon conversion efficiency becoming worse and syngas yield also decreasing (see Fig. 3). Therefore, higher concentration of water is found in the syngas and automatically decreasing the gasification efficiency. The less fraction of carbon and the more oxygen fraction in the ultimate analysis of feedstock, implied a low LHV of the feedstock, resulting in the more plasma power consume to reach the required gasifier temperature. Plasma power requirement also determined by the type of gasifying agent used. The syngas LHV and Cold Gas efficiency should decrease with the increasing SWR (see Fig. 2).

4. Conclusions

The highest reactor efficiency, the highest quality of syngas and the lowest CO_2 emission when the SWR is on 0.080 and the MSW mass flow rate is on $1.39*10^{-3}$ kg/s and equivalent ratio is on 0.180. The results indicate that low SWR is more favorable for the production of H_2 than the high SWR, because the syngas has lesser moisture, which also increase the gasification efficiency. When the SWR is only 0.080, the molar fraction of H_2 is 32.09%, the reactor efficiency (CGE) is 77.96%, X_c is 5.83%, the syngas LHV is 1.387 MJ/Nm^3, syngas Yield is 1.281 Nm^3/kg and CO_2 emission is only 2.23 $*10^{-4}$ kg/s. These data are crucial to describe scenarios concering the method to be applied to increase the production of H_2 while minimizing the impact of CO_2 emission to the environment.

5. Acknowledgment

This article was presented at the 2nd International Conference on Smart City INNOVATION (ICSCI) 2019, jointly held by Universitas Indonesia and Universitas Diponegoro. ICSCI conferences have been supported by the United States Agency for International Development (USAID) through the Sustainable Higher Education Research Alliance (SHERA) Project for Universitas Indonesia's Scientific Modelling, Application, Research, and Training for City-centered Innovation and Technology (SMART CITY) Center for Collaborative Research, administered through Grant #AID-497-A-1600004, Sub Grant #IIE-00000078-UI-1.

6. References

[1] Valmundsson A S and Janajreh I 2011 Plasma Gasification Process Modeling and Energy Recovery From Solid Waste *{ASME} 2011 5th International Conference on Energy Sustainability, Parts A, B, and C* (ASMEDC)

[2] Nur M, Bonifaci N and Denat A 2014 Ionic Wind Phenomenon and Charge Carrier Mobility in Very High Density Argon Corona Discharge Plasma *J. Phys. Conf. Ser.* **495** 12041

[3] Pacific Northwest National Laboratory 2015 *Hydrogen Compared with Other Fuels | Hydrogen Tools* (H2 Tools)

[4] Byun Y, Cho M, Hwang S-M and Chung J 2012 Thermal Plasma Gasification of Municipal Solid Waste (MSW) *Gasification for Practical Applications* (InTech)

[5] Minutillo M, Perna A and Bona D Di 2009 Modelling and Performance Analysis of An Integrated Plasma Gasification Combined Cycle (IPGCC) Power Plant *Energy Convers. Manag.* **50** 2837–42

[6] Khuriati A, Purwanto P, Huboyo H S, Suryono S and Putro A B 2018 Application of Aspen Plus for Municipal Solid Waste Plasma Gasification Simulation: Case Study of Jatibarang Landfill in Semarang Indonesia *J. Phys. Conf. Ser.* **1025** 12006

[7] Favas J, Monteiro E and Rouboa A 2017 Hydrogen Production Using Plasma Gasification with Steam Injection *Int. J. Hydrogen Energy* **42** 10997–1005

[8] Haugen H H, Halvorsen B M and Eikeland M S 2015 Simulation of Gasification of Livestock Manure with Aspen Plus *Proceedings of the 56th Conference on Simulation and Modelling (SIMS 56), October, 7-9, 2015, Linköping University, Sweden* (Linköping University Electronic Press)

[9] Lv P, Yuan Z, Ma L, Wu C, Chen Y and Zhu J 2007 Hydrogen-rich Gas Production From Biomass Air and Oxygen/Steam Gasification in A Downdraft Gasifier *Renew. Energy* **32** 2173–85

[10] Gil J, Corella J, Aznar M P and Caballero M A 1999 Biomass Gasification in Atmospheric and Bubbling Fluidized Bed: Effect of The Type of Gasifying Agent on The Product Distribution *Biomass and Bioenergy* **17** 389–403

Purification of Used Cooking Oil Using Activated Carbon Adsorbent from Durian Peel

S Miskah[1], T Aprianti[2], M Agustien[3], Y Utama[4] and M Said[1]

[1]Department of Chemical Engineering, Sriwijaya University, South Sumatera, Indonesia
[2]Department of Chemical Engineering, The University of Western Australia, Perth, Australia
[3]Department of Civil Engineering, Sriwijaya University, South Sumatera, Indonesia
[4]Department of Information Systems and Computer Science, Sriwijaya University, South Sumatera, Indonesia

E-mail: sitimiskah@gmail.com

Abstract. Repeatedly use of cooking oil can cause increasing free fatty acids and peroxide value contained. Therefore, continuous exposure to used cooking oil has known resulting hazardous impacts on the environment and also to human health. Hence, it is necessary to purify used cooking oil before it is discharged to the environment. Durian peel contains high cellulose ranging from 50% - 60%, which is eligible to be used as carbon adsorbent's raw material for used cooking oil purification. This study produced a kind of carbon adsorbent from durian peel through a carbonization process that lasted for 2 hours at 500°C combustion temperature with a chemical activation using H_2SO_4 1 N. This study was carried out with the adsorbent weight of 6 grams and various contact times of 30, 60, 90, 120, and 150 minutes. The most favorable results obtained after contact time was 0.0637% of free fatty acid, 0.41 meq O_2/kg of peroxide value, and 0.9022 gr/ml of density, respectively, at 150 minutes contact time.

Keywords: Durian Peel, Activated Carbon Adsorbent, Cooking Oil Purification

1. Introduction

In Indonesia, durian (*Durio zibhetinus murr*) is a kind of seasonal fruit that is very well known as "the king of fruits" has a very distinctive aroma, thorny peel, abundant availability and is highly demanded by the public during its season. With its enormous amount, the environmental problem arises, that is the accumulation of durian peels. Proportionally, durian peel contains lignocellulose consisting of cellulose of 50%, 60%, lignin of 5%, and starch of 5% [1], Lignocellulose found in any raw materials can be used to make activated carbon. The utilization of activated carbon as an effective and affordable adsorbent has been investigated for many years. The previous study shows that activated carbon as a ceramic adsorbent is highly useful to remove heavy metals pollutant contained in acid mine drainage [2]. Besides, ceramic composite adsorbent of zeolite-activated carbon can be an alternative to treat wastewater containing heavy metals derived from the pulp and paper industry [3].

On the other hand, waste cooking oil handling is still considered insufficient. The use of cooking oil more than surely will darken its color and releases a kind of odor. Besides, the number of peroxides and free fatty acids also become higher. The changes in the chemical properties of cooking oil are due to hydrolysis, polymerization, and oxidation, these are caused by heating cooking oil at a high temperature accompanied by air contact [4]. These conditions are resulting in hazardous impacts on human health. Therefore, it is necessary to purify used cooking oil before it is discharged to the environment.

There are various efforts to purify used cooking oil, one of those is by using adsorption process. There are several studies regarding the adsorption process of used cooking oil using H_2SO_4 activators with various raw materials. A previous study used *ketapang* activated carbon as an adsorbent for used cooking oil purification using H_2SO_4 activator with the most favorable results on free fatty acids of 0.78% [5]. A previous study also examined the quality of used cooking oil after using carbon adsorbent from bagasse with H_2SO_4 as the activator. It obtained the most favorable peroxide value of 6.4295 meq/kg [6]. Also, another study utilized corn stalks for used cooking oil treatment using two different activators of HNO_3 and H_2SO_4, the best results obtained were 0.0548% for FFA value and 14.78 meq/kg for peroxide value [7]. There are many kinds of activated carbon from various raw materials have been used for adsorption process of used cooking oil purification, but unfortunately, there are some other results that do not meet Indonesian standard quality (SNI 7709: 2012) [8]. This research is aimed to investigate the quality of used cooking oil using activated carbon adsorbent from durian peel. The result is expected to produce an environmentally friendly used cooking oil using a more economical process.

The proximate analysis of durian peel shows that its water content is 14.5%, ash is 0.4%, flying substances is 64.4%, fixed carbon is 20.7%, and heating value is 13.8 mj/kg [9]. The carbon content in durian peel meets the requirement to be used as the raw material for activated carbon. The repeated use of cooking oil for frying process accompanied by heating with high temperatures can cause damages to its chemical contents due to changes in the chemical composition regarding the oxidation of components in the cooking oil. One of the characteristics of the oxidation reactions is the color of the cooking oil turns darker as cooking oil is used more often. Also, FFA value is to measure the oxidation and hydrolysis reactions in the cooking oil, whereas the amount of water contained in the cooking oil is because of its hydrolysis process. Repeatedly used cooking oil can also cause foam formation, increased viscosity, and density due to unwanted components from the results of hydrolysis, oxidation, and polymerization reactions in the cooking oil [10].

2. Material and Method

2.1. Making Activated Carbon

A batch of durian peel is washed using deionized water to remove its impurities, then cut into small pieces and dried for three days with the help of sunlight. Then the durian peel is carbonized in the furnace for 2 hours with a combustion temperature of 500 °C. The carbon resulted from durian peel is then smoothed with a grinder to refine its size. Afterward, durian peel carbon is activated using a 1 N H_2SO_4 solution. The activation process lasts for 24 hours. Activated carbon is washed with deionized water until pH 7 and then filtered. The activated carbon is dried to remove its water content at 110 °C for 2 hours. Lastly, the activated carbon is mashed and sifted using a sieve measuring 0.328 mm.

2.2. Purification Process

500 grams of used cooking oil is weighed and then mixed with deionized water with the composition of (1:1). The mixture of oil and water is heated to reduce its volume to half of the initial. Next, the oil layer is taken and deposited for 1 hour using a separating funnel. After that, the oil layer is filtered to separate the remaining impurities. Used cooking oil of 450 gr is heated at 35 °C for odor removal process, 18 ml of 16% NaOH was added to used cooking oil and stirred for 10 minutes with a heating temperature of 40 °C. The mixture is left for 10 minutes until it cools down, the mixture of oil and NaOH will produce soap, and it has to be removed from the oil. 200 mL of used cooking oil is mixed and stirred with 6 gr of activated carbon at room temperature, the contact times between used cooking oil and activated carbon are varied of 30, 60, 90, 120, and 150 minutes. Finally, purified cooking oil is filtered with filter paper for analysis.

3. Results and Discussion

Table 1. Free fatty acids at various contact time

Used Cooking Oil (gr)	Adsorbent Weight (gr)	Contact Time (min)	Free Fatty Acids (%)
450	6	30	0.1513
		60	0.1311
		90	0.1085
		120	0.0839
		150	0.0637

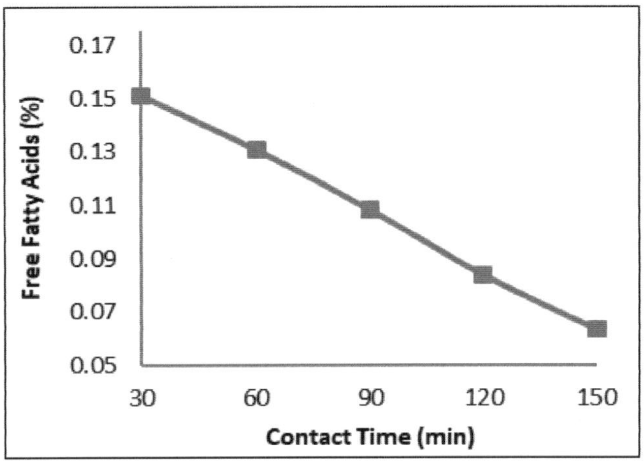

Figure 1. Effect of Contact time on Free Fatty Acids Percentage

3.1. Free Fatty Acids

Free fatty acids are produced in any of the steps of the process by hydrolytic reactions. Free fatty acids contained in palm oil are normally < 1%, the higher the level of free fatty acids can reduce the quality of oil, the value of free fatty acids > 1% causes oil becomes rancid, and it can be tasted on the surface of the tongue [11]. Besides, the free fatty acid is known as the cause of metabolically unhealthy obesity, and type diabetes to human and likely contributes to end-stage kidney disease irrespective of the underlying kidney disease [12].

Based on figure 1 and table 1, the values of free fatty acids tend to get lower along with longer contact time. The longer the contact time, the smaller the value of free fatty acids, this is because the long contact time causes more intense contact between the surface of the adsorbent and the used cooking oil. In addition, the more mass of adsorbent is used, the more surface of the adsorbent is produced so that the adsorbent can adsorb more free fatty acids. It was stated that the better the adsorption process is because, the more surface area where the adsorption process takes place and the longer contact time causes the freer fatty acids bound by the adsorbent's surface, and the reactions took place were both oil hydrolysis and unsaturated fatty acid oxidation.

The carbon is activated using H_2SO_4 1 N, where the carbon is immersed in H_2SO_4 solution to increase its pores length and width, which are covered by materials or organic metal oxide components. Thus, these carbon pores play a role in adsorbing impurities so that it can reduce the free fatty acids contained in used cooking oil. Based on the analysis results, the free fatty acids contaminant is significantly reduced after the adsorption process. This occurs due to the more prolonged the use of the adsorbent, the higher the level of adsorption. The adsorbent adsorbed impurities into the carbon pores during the adsorption process, and the activation process is resulting in increases the rate of adsorption.

Table 2. Peroxide value at various contact time

Used Cooking Oil (gr)	Adsorbent Weight (gr)	Contact Time (min)	Peroxide value (meq O_2/kg)
		30	2.28
		60	2.16
450	6	90	1.80
		120	1.72
		150	0.41

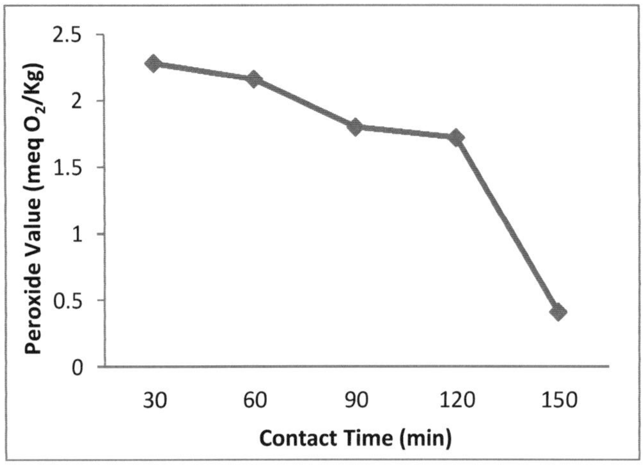

Figure 2. Effect of Contact time on Peroxide value

3.2. Peroxide value

A chloroform-acetic acid mixture is used to dissolve the used cooking oil, and this mixture is subjected to an excess of iodide via a saturated solution of potassium iodide. The peroxides present oxidize the iodide to iodine, and the iodine is then titrated to a colorimetric endpoint using sodium thiosulfate with starch as an indicator. The amount of iodine produced is directly proportional to the peroxide value. In general fresh oils, the peroxide value is >10 mEq/Kg while peroxide values in the 30-40 mEq/Kg range are generally associated with a rancid taste. Therefore, peroxide value in oil is also used to determine the damage degree to oil. Table 2 and figure 2 above show that the graph represents peroxide value gradually decreases with longer contact time. The decrease in peroxide value due to the termination of the double bonds in oil. Besides that, peroxide contains oxygen as a polar compound so that it is more comfortable to bound to the adsorbent [13]. Thus, the lower the peroxide value, the smaller the damage to the oil.

The carbon used in this study is activated with H_2SO_4 1 N so that it has a large surface area and pores to be able to bind and adsorb more peroxide compounds in the oil. Also, the adsorption process does not use heating, so that the decrease in peroxide value is not inhibited [14] if the adsorption process is accompanied by heating, it actually can cause the reformation the peroxide compounds. Increasing peroxide values occurs because of the oxidation process of unsaturated fatty acids in oil with the aid of oxygen.

From the analysis of peroxide value, it can be seen that the adsorption process uses activated carbon from durian peel resulting in peroxide values that meet Indonesian Standard (SNI 7709: 2012) [8] which is <10 meq O_2/kg with 30 minutes contact time. Thus, the carbon-based adsorption process is considered effective for the adsorption process.

Table 3. Free fatty acids at various contact time

Used Cooking Oil (gr)	Adsorbent Weight (gr)	Contact Time (min)	Density (gr/ml)
		30	0.9085
		60	0.9073
450	6	90	0.9070
		120	0.9062
		150	0.9022

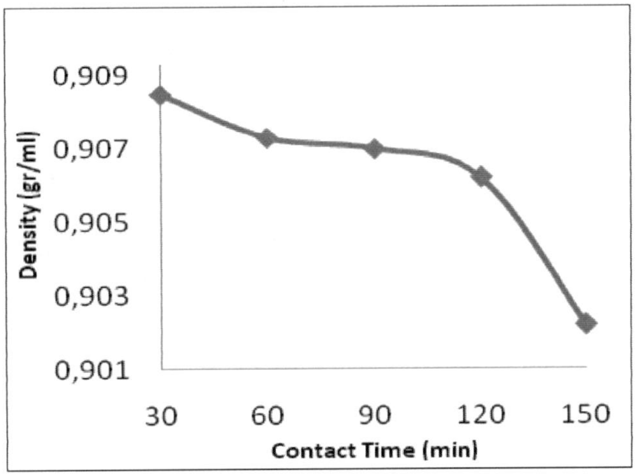

Figure 3. Effect of Contact time on Density

3.3. Density

Based on figure 3, it can be seen that the density of used cooking oil decreases quite sharply. The decrease is because the adsorbent has adsorbed significant amount of the impurities contained in used cooking oil, therefore the molecular bonds in the oil become reduced causing the removal of odors and dark coloring in the oil [15], determination of density is influenced by the water content and the level of impurities attached to oil. From the density analysis of used cooking oil after being adsorbed using activated carbon shows that used cooking oil for all samples does not meet the Indonesian Standard (SNI 7709: 2012) [8] which is 0.900 gr/ml at the mass of adsorbent 6 gram with 150 minutes contact time.

4. Conclusions

The results of the analysis of the value of free fatty acids, peroxide values, and the best density using the adsorbent without activating, respectively, are at the contact time of 150 minutes and the mass of 6 grams of adsorbent is 0.0637%, 0.41 meq O_2/kg, and 0.9022 gr/ml. The results of the analysis of the value of free fatty acids, peroxide values, and the best density of the oil using the activated adsorbent are at the contact time of 150 minutes, and the mass of 6 grams of adsorbent is 0.0637%, 3.04 meq O_2/kg, and 0.9052 gr/ml.

5. Acknowledgments
This article was presented at the 2nd International Conference on Smart City INNOVATION (ICSCI) 2019, jointly held by Universitas Indonesia and Universitas Diponegoro. ICSCI conferences have been supported by the United States Agency for International Development (USAID) through the Sustainable Higher Education Research Alliance (SHERA) Project for Universitas Indonesia's Scientific Modelling, Application, Research, and Training for City-centered Innovation and Technology (SMART CITY) Center for Collaborative Research, administered through Grant #AID-497-A-1600004, Sub Grant #IIE-00000078-UI-1

6. References

[1] Yusuf B and others 2016 Pemanfaatan Arang Aktif dari Kulit Durian (Durio Zibethinus L.) sebagai Adsorben Ion Logam Kadmium (II) *J. Kim. Mulawarman* **13**

[2] Aprianti T, Afrah B D and Agustina T E 2017 Acid Mine Drainage Treatment using Activated Carbon Ceramic Adsorbent in Adsorption Column *Int. J. Adv. Sci. Eng. Inf. Technol.* **7** 1241

[3] Aprianti T, Miskah S, Selpiana, Komala R and Hatina S 2018 Heavy Metal Ions Adsorption From Pulp and Paper Industry Wastewater using Zeolite/Activated Carbon-ceramic Composite Adsorbent (Author(s))

[4] Kusumastuti K 2004 Kinerja Zeloit dalam Memperbaiki Mutu Minyak Goreng Bekas *J. Teknol. dan Ind. Pangan* **15** 141–4

[5] Megiyo M, Aldila H, Afriani F, Mahardika R G and Enggiwanto S 2017 Sintesis Karbon Aktif Tempurung Ketapang (Terminalia Catappa) sebagai Adsorben Minyak Jelantah *Pros. SNFA (Seminar Nas. Fis. dan Apl.* **2** 137

[6] Triyanto A and others 2013 *Peningkatan Kualitas Minyak Goreng Bekas Menggunakan Arang Ampas Tebu Teraktivasi dan Penetralan dengan NaHSO3* (Universitas Negeri Semarang)

[7] Yulianti E, Mahmudah R and Royana I 2018 Pemanfaatan Biosorben Batang Jagung Teraktivasi Asam Nitrat dan Asam Sulfat untuk Penurunan Angka Peroksida - Asam Lemak Bebas Minyak Goreng Bekas *ALCHEMY* **5** 9

[8] Badan Standardisasi Nasional 2012 SNI 7709-2012 : Definisi Minyak Goreng Sawit Perlu Koreksi

[9] Ismadji S 2012 Kulit Durian Sebagai Bahan Baku Pembuatan Bio-oil: Sumber Energi Terbarukan *Surabaya Jur. Tek. Kim. Unika Widya Mandala*

[10] Ketaren S 1986 Pengantar Teknologi Minyak dan Lemak Pangan

[11] Sopianti D S, Herlina H and Saputra H T 2017 Penetapan Kadar Asam Lemak Bebas pada Minyak Goreng *J. Katalisator* **2** 100

[12] Sieber J and Jehle A W 2014 Free Fatty Acids and Their Metabolism Affect Function and Survival of Podocytes *Front. Endocrinol. (Lausanne).* **5**

[13] Tyas S D C and Tjahjani S 2011 The Utilization of Pyrophilite Before and After Activation as an Adsorbent on Decreased Level Process *J. People Environ.* **18** 184–90

[14] Rahayu L H and Purnavita S 2018 Pengaruh Suhu dan Waktu Adsorpsi Terhadap Sifat Kimia-Fisika Minyak Goreng Bekas Hasil Pemurnian Menggunakan Adsorben Ampas Pati Aren dan Bentonit *J. Ilm. Momentum* **10**

[15] Handoko D S P 2009 The Improvement of Waste Cooking Oil Quality using H5-NZA Adsorbent in Fluid Fixed Bed Reactor *J. Ilmu Dasar* **10** 121–32

Local Wisdom in Malay Kampung Semarang as Representatives of Smart Environment

W Kurniawati[1], Mussadun[1], D Suwandono[1], T Z Islamey[1]

[1]Department of Urban and Regional Planning, Faculty of Engineering, Diponegoro University

E-mail: wakhidahkurniawati3@gmail.com

Abstract. One of the smart city indicators is a smart environment. Smart environment is a setting which can be a catalyst for its citizens, support them to be aware of a local engagement in managing their existing natural resources. The local engagement is not only shaped by formal strategies, but also by local wisdom. Local wisdom is a community understanding that is used as a reference in daily life activities. Based on its history, the Malay Kampong (Kampung Melayu) Semarang which was formed in 1743 has strong local wisdom as a basis for arranging the district. Settlement patterns based on ethnicity and culture, building patterns that are sensitive to the floods and tidal ranges, multicultural-based district, make this region able to survive hundreds of years to maintain its existence and can be a representation of the smart multicultural environment in Semarang. The current environmental degradation in this area is the basis for the research. Is local wisdom still maintained here? Or is there a shifting meaning of local wisdom at this time? This study aims to explore the role of local wisdom of Malay Kampung Semarang as a representation of a smart environment. This study combines quantitative and qualitative approaches. To achieve this goal, it is necessary to identify the influence of local wisdom in Malay Kampung Semarang on flood resilience and tidal. The analysis of the smart environment model is based on the interaction of wisdom and influence on the district arrangement. The result of the study shows that local wisdom is an effort to improve the living place to be more sustainable in the event of flooding and tidal. Efforts made to elevate the floor of the building, height of buildings, elevate the street, build houses with a stage model, and make the second floor of the house represent a smart environment in the form of sustainable urban planning in the Malay Kampung Semarang.

Keywords: Local Wisdom, Local Engagement, Smart Environment

1. Introduction

Floods and tidal have become a threat of disaster for the city of Semarang. Handling floods and tidal has become a priority as stated in mission III of Semarang Mid-term Regional Development Plan 2016-2021. With the aim of the mission is to realize a strong, productive and sustainable city development. Moreover, the Land subsidence in some parts of the Semarang city and the location on the edge of the northern coast of Central Java make the city of Semarang facing flood and tidal.

Strategic areas are usually prone to changes in land use, for example in cities along the post highway, or located in disaster-prone areas such as in Banda Aceh, Padang, Kotagede, north coast of Jakarta and Semarang. In "The geography of urban disaster" it is mentioned that colonial cities are as disaster-prone cities. Cultural heritage areas (urban heritage) are usually the core areas forming city structures, layers and the history of the city (based on the theory of Urban Tissue, Athens Charter 1943, New Urbanism Charter 1993). Flood and tidal disasters have an impact on cultural heritage sites. The existence of floods and tidal can lead to coastal erosion, seawater intrusion into artifacts and locations, permanent immersion in lowland areas, population migration, disruption of community activities, and result in the loss of rituals and social interaction of the community.

Semarang City has decided to implement smart city as a priority in the Mid-Term Regional Development Plan (RPJMD). One of the smart city applications is smart environment. Smart environment is a setting that can catalyze its citizens, support them to be aware of local engagement in managing their existing natural resources. The local engagement is not only shaped by formal strategies, but also by local wisdom. Local wisdom is a community understanding that is used as a reference in daily life activities. According to Rukmana (2010) Planning in Indonesia as a developing country, needs greater and more explicit reliance on local wisdom rather than end-rationality [1].

Malay Kampung is located in North Semarang District. This Kampung was established when Malay and Banjar traders landed in Semarang. Malay Kampung has been inhabited since in 1743 [2,3]. Since most people who inhabit the area are Malays, it is called Malay Kampung. At that time there were places in the old kampung to land ships and boats carrying merchandise, so it was not surprising that there was a section called Malay Darat. Its strategic location attracts people to stay there too. So that currently not only Malay people who live in the Malay Kampung but have become a multicultural region.

Currently, the Malay Kampung area is a cultural heritage area in Semarang. The uniqueness of strong local wisdom as a basis for regulating the area, settlement patterns based on ethnicity and culture, building patterns that are sensitive to flooding and tidal reach, and now have been strengthened by the plan of Malay Kampung Environmental Management. However, like other northern coastal settlements, the region has many problems, such as the problem of continuous flooding and the unhealthy living conditions. In addition, many old buildings were damaged and drowned.

Various efforts were made by the community to maintain their existence and become a representative of the smart multicultural environment in Semarang. Environmental degradation is the basis of this study. The issue arises is local wisdom still maintained here? Or is there a shift in local wisdom currently? The stages of analysis were carried out to achieve the objectives, namely: 1. Identify local wisdom as disaster resilience of flood and tidal. 2. Analyze smart environmental models that occur based on the interaction of local wisdom and their influence on district regulation.

2. Literature Review

2.1 Smart City Concept

Smart City is defined as a city that is able to use human resource, social capital, and modern telecommunications infrastructure (Information and Communication Technology) to realize sustainable economic growth and high quality of life, with source management wise power through community-based government participation [4]. Meanwhile, According to Cohen (2013) Smart city is an approach that enhances integration and efficiency in conducting city operations by improving the quality of life and economy of its population [5]. Cohen emphasizes the environmental aspects by collaborating with ICT to save on the use of resources, energy, improve service quality and live the population and support innovation and economics that are environmentally friendly. Smart City is a leading city in the economy, human resources, government, mobility, environment, and community life, which is built entirely smart, independent and has awareness from the people. There are six main key indicators in the form of implementing smart cities are Smart economy, Smart people, Smart governance, Smart mobility, Smart environment, Smart living [6]:

The above main indicators are explained by Cohen (2013) through this following diagram (Figure 1):

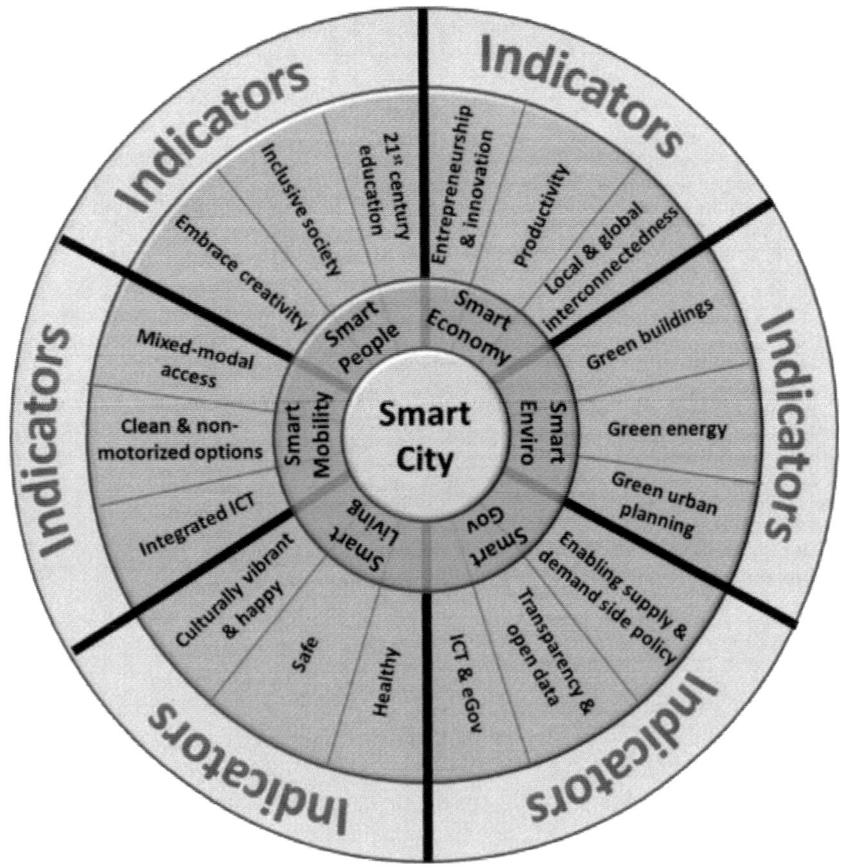

Figure 1. Smart City Wheels Cohen [7]

Smart city is one of the planning concepts used to solve urban problems that are getting worse with the increasing of population levels. Excessive use of natural resources and low quality of human resources. Declining environmental quality due to continuous development is carried out without considering the environment. The concept of smart city planning prioritizes public services and the use of technology to overcome problems that occur in cities. Using these six main indicators to overcome the city becomes more complex. The problem of each developing city is in the main indicator.

2.2 *Smart Environment*

Smart environment is one of the main indicators of implementation in smart city. According to Hollands in Rachmawati & Pertiwi (2017), implementing Smart environment can be focused on natural environment and ecological implication on urban growth and development [8]. Therefore, smart environment can be viewed from attractive natural conditions (climate, availability of green open space, etc.), pollution, resource management, and also efforts towards environmental [6].

Creating a smart environment can be supported by a smart application that regulates policies in environmental management[9]. For example for measuring, controlling and monitoring natural conditions, renovating buildings and facilities, monitoring the use of fossil and renewable energy smartly. In the Cohen Smart City Boyd diagram, the implementation of Smart Environment to be sustainable can be applied in the scope of Green Building, Green Energy, and Green Urban Planning. This is the indicators table developed by Cohen (2014) in the Yogyakarta City Road Map towards Smart City by the PSPPR UGM Team (2016) (Table 1):

Table 1. Indicator Development of Smart Environment [10]

Dimension	Working Area	Indicators
Smart Environment	Smart Building	Sustainability-certified Buildings
		Smart Homes
	Resources Management	Energy
		Carbon Footprint
		Air Quality
		Waste Generation
		Water Consumption
	Sustainable Urban Planning	Climate Resilience Planning
		Density
		Green Space per capita

There are many indicators are used to create a smart environment. In this study smart environment of the Malay Kampung can be identified based on existing indicators and later seen how local wisdom in a multicultural environment in the Malay Kampung faces natural conditions that cause floods and tidals in the Malay Kampung area.

3. Methods

This study aims to identify local wisdom in Malay Kampung as a representative form of smart environment. Based on these objectives, this study used a combination of quantitative and qualitative method approaches. The quantitative method was used to identify the local wisdom of the Malay Kampung through facing flood and tidal disasters through the facts in the field and their influence in structuring buildings and the environment in the Malay Kampung. Data from the questionnaire containing questions about the condition of the village before, after and when the disaster occurred. Then qualitative method was used to identify the smart environment model that is carried out by the community as a form of interaction of local wisdom and its influence in its regional settings. The analysis presented is based on the results of in-depth interviews with several communities that still survive in the Malay Kampung area. Therefore, with the existence of kampung condition data and statements from Malay Kampung community, it is known that the smart environment that has been cultivated by Malay Kampung community.

4. Local Wisdom in Malay kampung

Malay Kampung (Figure 2) is one of the old kampung in Dadapsari sub-district and a small part of Kuningan sub-district, North Semarang Regency. Malay Kampung was originally the entrance and landing point of merchant ships belonging to Arabs, Malays and Indians. Traders carry out trade activities and then live and settle in the area because of its strategic location. From there, the Malay Kampung residential area emerged around 1743.

In the beginning, Malay Kampung was dominated by ethnic Malays, but in its development more migrants lived and settled so that the people in the Malay Kampung were multi-ethnic communities. Not only from ethnic Malays, but also there are ethnic Banjar, Chinese, Arab, Cirebon, Javanese, etc. which then with the existence of ethnic diversity affects the kampung toponym in it such as Kampung Pulo Petekan, Kampung Baru, Peranakan Kampung, Banjar Kampung, Cerbonan Kampung, Kampung Geni, Pencikan Kampung, Kalicilik Kampung, and Kampung Darat. The toponymical of the halls in Malay Kampung is also a local wisdom [11]. Each name of the hall in the Malay Kampung has its own history and it becomes very unique and thick with local nuances.

Malay Kampung is directly limited by Kakap Street in the north, Dorang street in the south, Semarang River in the east, and Petek street in the South. Most of the land in Malay Kampung is used for residential activities and trading activities. The resident activity spread in each area which is marked by the number of kampung which become a place for settling the community. While the majority of trading activities are in the corridor of Layur Street, but even so trade activities can also be found in the middle of residential areas. Most of the people in Malay Kampung work as laborers and traders so that they are categorized into middle to lower community.

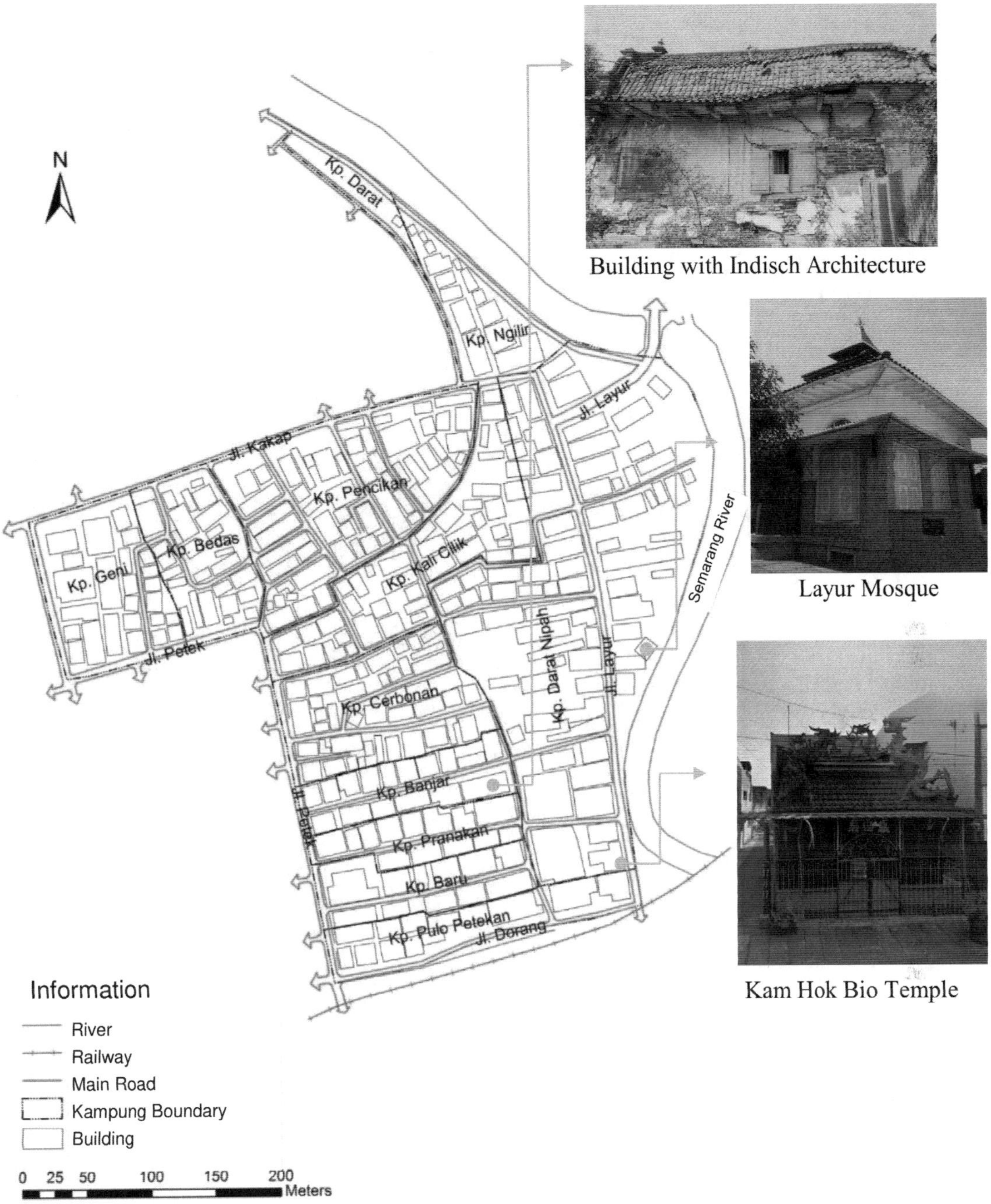

Figure 2. Kampung Delineation in Malay Kampung Area [12]

The area in Malay Kampung is vulnerable to flooding and tidal. In general, the height of the puddle is 40-60 cm with duration of approximately 2-4 days. The flood is usually caused by the overflow of water in Semarang River which is due to rain descending with a high enough intensity, the Semarang River is no longer able to accommodate the volume of water so that it would overflow to the residential area. In addition, it is worsened by poor drainage conditions so that it will ease the increase in existing water. However, in some area like in Layur Street, Kakap Street and Petek Street are not only flooded but also often tidal floods. Usually tidal occurs every day with a short period, starting in the afternoon to night or day to night. This happens because as previously explained that Malay

Kampung is below the sea level so that it becomes prone to inundation when sea water begins to install. From this description, it can be seen that throughout the Malay Kampung region is the main zone of flooding, which is the most vulnerable and potentially affected by annual flooding and tidal.

The occurrence of floods and tidal that occur in Malay Kampung not only impact on physical aspects, but also the social and economic conditions of the community. The floods and tidal affect the resilience and sustainability of these three aspects in everyday life. Waterlogged residential conditions will certainly disrupt the social and economic activities of the community. Existing climate change results in a form of climate resilience from existing historical buildings. The following is a form of resilience carried out by the Malay Kampung community in facing floods and tidal (Table 2).

Table 2 Forms of Disaster Resilience in Malay Kampung [12] (rewritten)

Disaster Phenomenon	Location	Response	Adaptation efforts
Annual flood with height of 40-60 cm and the duration is 3-4 days	Kp. Darat Nipah Kp. Pulo Petekan Kp. Baru Kp. Banjar Kp. Peranakan Kp. Cerbenon	Unable to survive, not a few who move to another place (evacuate) and tend to stop working for a while	• Elevated the street • Backfill the building
Annual flood with height of 40-60 cm and the duration is 2-3 days	Kp. Geni Kp. Bedas Kp. Pencikan Kp. Kalicilik Kp.Ngilir	Stay and move as usual	• Elevated the street • Backfill the building • Provided the water pump
Annual flood with height of 40-60 cm and the duration is 2-3 days and tidal with daily frequency, inundation duration for approximately half a day	Layur Street Kakap Street Petek Street	There is still activity as usual	• Elevated the street • Backfill the building • Provided the water pump • Construction of talud on the Semarang river

The physical form of adaptation of the Malay Kampung community towards flood and tidal disasters does not change the local wisdom. The Malay Kampung area survives by transforming house buildings (Figure 3). Worsening buildings, lifting roads and various ways are done by people to survive. In terms of non-physical, the current economy of the people is not just a trader, some people work as laborers and others. In terms of social culture, Malay Kampung are still a kampung with multi ethnicity. The grouping of each ethnic group still survives today, even though there are many newcomers. So, the local wisdom of the Malay Kampung still survives its existence.

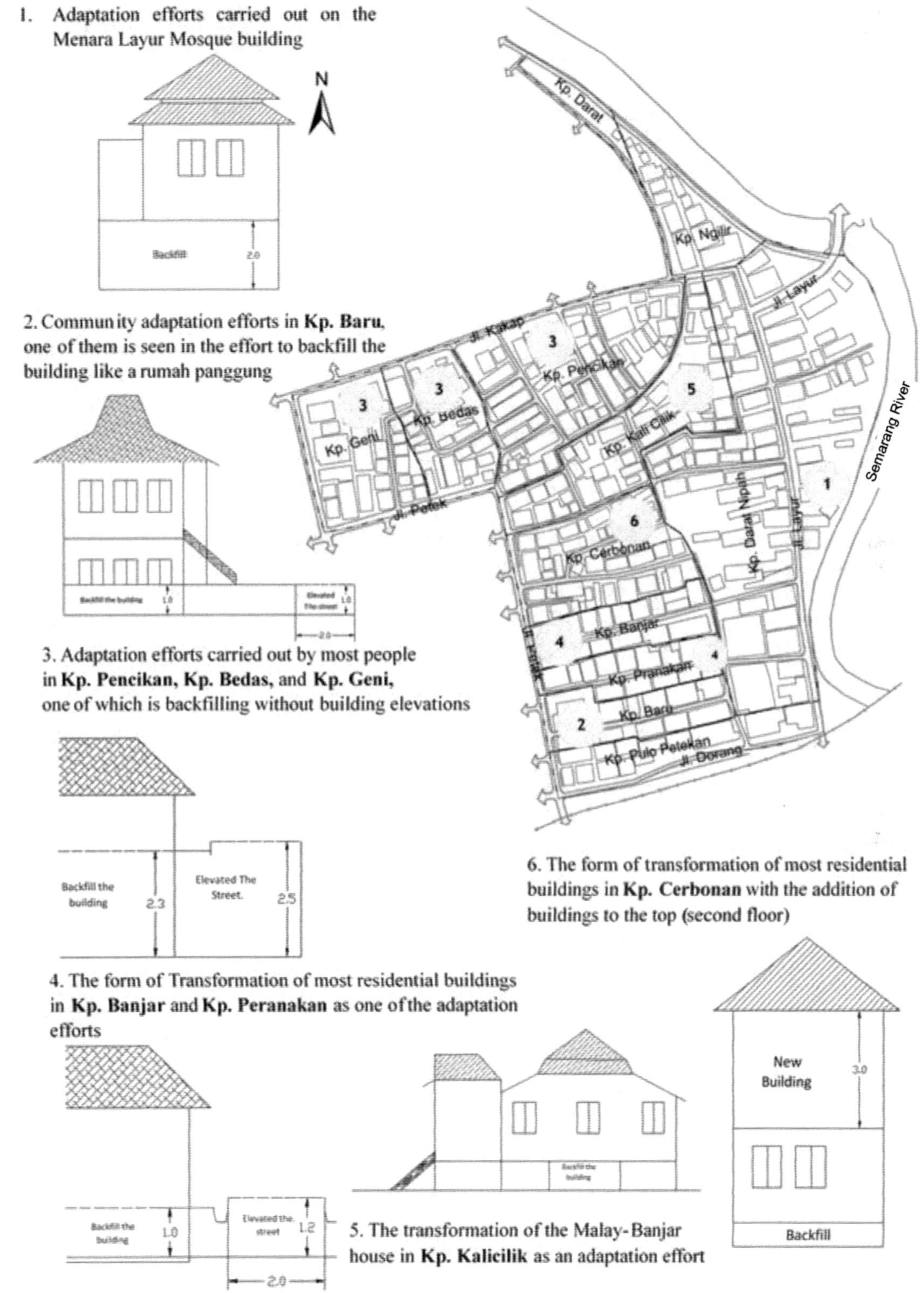

Figure 3. Building Transformation in Malay Kampung [13]

5. Smart Environment Representative

Smart environment according to Cohen is divided into several indicators that can realize the objectives of the smart environment concept. Creating a smart environment means creating an environment that can provide comfort, sustainability of resources, beauty of physical and non-beautification, for the community and the public. Basically the purpose of smart environment is for a sustainable city.

Malay Kampung, with its floods and tidal problems, can survive to this day and still maintain its existence as a multiethnic old kampung. Some ethnic cultural groups still survive in the Malay Kampung area. Although there are many newcomers, the location where people live is still in groups based on their ethnicity.

Community in Malay Kampung has various efforts for the sustainability and comfort staying there. The existence of cultural heritage in Malay Kampung can still be maintained today by the local community. Various efforts to elevate roads, height of buildings, increase the floor of the building is one of the local wisdom of the Malay Kampung community in an effort to survive the disaster. Smart environment that is carried out by Malay community is clearly seen in the smart building section. The Malay Kampung community strives to change their homes to avoid disaster while still maintaining their cultural characteristics. So it can be said that local wisdom is one of the smart environment support in an area.

Explanation of the smart environment section in Table 3, indicators of resource management in Malay Kampung are same with the other kampung, so there is no unique feature for Malay Kampung. Indicators of sustainable urban planning in terms of climate resilience plans, planning systems carried out by Malay Kampung in the form of incremental planning are not yet comprehensive planning. So that the efforts made by the community are only temporary and if occurrence of a disaster again it will be planned for handling again. Malay Kampung is also an area with a high density so that the availability of open space has not yet reached the per capita requirement standard.

Table 3 Smart Environment in Malay Kampung

Dimension	Working Area	Indicators	Condition of Malay Kampung
Smart Environment	Smart Building	Sustainability-certified Buildings	Indigenous people who stay have long lived and still maintain their culture
		Smart Homes	Local wisdom in there with efforts to height of building to avoid disaster
	Resources Management	Energy Carbon Footprint Air Quality Waste Generation Water Consumption	Management system like in general so there is no different from other kampung
	Sustainable Urban Planning	Climate Resilience Planning	Local wisdom is still incremental planning and not comprehensive planning
		Density	High Density
		Green Space per capita	Per capita needs are not enough because of high density

6. Conclusion

Local wisdom of Malay Kampong has its own adaptation to the problems it has. Malay village adaptations are representative of Smart environment dominate in Smart building. That caused by various efforts to elevate roads, height of buildings, increase the floor of the building to survive the disaster. The efforts made did not eliminate the existence of Multi ethnic cultures in there. This was proven by existence of ethnic cultures still persists to this day. Therefore, it can be said that local wisdom is a factor in the realization of the smart environment, although in Malay Kampung its only be present in the field of smart building as an effort to survive the disaster.

7. Acknowledgement

This research was financially supported by The Faculty of Engineering, Diponegoro University, Indonesia through Strategic Research Grant 2019. This research is funded by the Directorate General of Research and Development, Ministry of Research, Technology and Higher Education through Institute for research and Community Services Diponegoro University for year 2019 with contract number 257-66/UN7.P4.3/PP/2019. This article was presented at the 2nd International Conference on Smart City INNOVATION (ICSCI) 2019, jointly held by Universitas Indonesia and Universitas Diponegoro. ICSCI conferences have been supported by the United States Agency for International Development (USAID) through the Sustainable Higher Education Research Alliance (SHERA) Project for Universitas Indonesia's Scientific Modelling, Application, Research, and Training for City-centered Innovation and Technology (SMART CITY) Center for Collaborative Research, administered through Grant #AID-4=97-A-1600004, Sub Grant #IIE-00000078-UI-1.

8. References

[1] Rukmana D 2010 Urban Planning and Local Wisdom: The Shift toward Postmodernism and a New Urban Theory from Third World Cities *Indones. Urban Stud.*
[2] Kurniawati W 2002 *Penataan Kawasan Lama Sebagai Kawasan Wisata Budaya, Studi Kasus: Kampung Lama Sepanjang Kali Semarang* (ITB)
[3] Wardhani A G 2000 *Karakteristik Permukiman Kampung Melayu di Semarang, Kajian Arsitektural dengan Pendekatan Urban History, Morfologi, dan Tipologi* (Universitas Parahyangan)
[4] Caragliu A, Bo C Del and Nijkamp P 2009 Smart Cities in Europe *3rd Central European Fonference in Regional Science* pp 45–59
[5] Cohen B 2013 What Exactly a Smart City
[6] Giffinger R 2007 *Ranking of European Medium-Sized Cities*
[7] Virtudes A, Abbara A and Sa J 2017 Dubai : A Pioneer Smart City in the Arabian Territory *IOP Conf. Ser. Mater. Sci. Eng.* **245**
[8] Rachmawati T and Pertiwi P D 2017 Smart Environment Program , Smart Way to Smart City *Policy Gov. Rev.* **1** 26–36
[9] The Government Summit 2015 *Smart Cities : Regional Perspectives*
[10] Tim PSPPR UGM 2016 *Road Map Kota Yogyakarta Menuju Smart City*
[11] Madiasworo T 2009 Revitalisasi Nilai-Nilai Kearifan Lokal Kampung Melayu Semarang Dalam Pembangunan Berkelanjutan *Local Wisdom* **1** 10–8
[12] Kurniawati W and Astuti K D 2013 Bentuk Ketahanan Iklim Kawasan Bersejarah di Kampung Melayu Semarang *Ruang* **1**
[13] Kurniawati W and Astuti K D 2013 *Bentuk Ketahanan Iklim Kawasan Bersejarah di Kampung Melayu Semarang* (Semarang)

IOP Conf. Series: Earth and Environmental Science **396** (2019) 012005 doi:10.1088/1755-1315/396/1/012005

Online Taxi bike To Create Informal Public Spaces in Urban Areas

W Kurniawati[1], T Z Islamey[1]

[1]Department of Urban and Regional Planning, Diponegoro University, Semarang, Central Java, Indonesia

E-mail: wakhidahkurniawati13@gmail.com

Abstract. The changes associated with digitalization, are rapidly messing up the old pattern, leading to the creation of a new system. The expansion of technological applications such as Uber or *Gojek*, leads to the development of a unique transportation model thereby, creating informal parking spaces for online taxi bike drivers. These spaces are self-formed, random, not formally planned, and located at strategic locations to suit the needs of users. In order to explore the phenomenon of informal public space, this research conducts the following: 1) Traces online taxi bike; 2) Identifies its potentials and problems; 3) Analyses the spatial pattern; and 4) Figures out the pattern and its impact. Data was obtained by direct observation in Diponegoro University, Tembalang, Semarang. This research therefore aims to understand the existing conditions faced by the new paradigms in Indonesia. The result is a spatial interaction pattern of informal space due to online taxi bike activities, such as T-junction, and under trees spot, with digital technology used to scrub the problems associated with driving forces.
Keywords: Online taxi bike, Informal public space, Disruption era

1. Introduction

Presently, the city's public spaces are experience transition and disruption due to technology. This was critically influenced in 2004, when artist Jaume Plensa created a giant portrait of Chicago residents spraying water from the Crown Fountain in Millennium Park which proved its ability to change the urban landscape[1]. Similarly, in 2015, a new digital technology system which connected data, developed cities by making it more efficient, and attractive. It also changed the way humans behave and interact with each other in public space[2].

The relationship created by the use of digital technology is categorized into human-to-human, Human-to-Al, Machine-to-machine, machine-to-Al [3], thereby, leading to the creation of an Online transportation system [4]. Furthermore, the relationships formed between humans are facilitated by the presence of online transportation applications which provides benefits such as faster response time [5].

Urban public space is a special space, unique, and distinctive area [6]. when technology begins to influence urban open space makes it different [7]. Changes with the implementation of new forces such as online taxi bike which requires a waiting time of 30-45 minutes [8], therefore, users need a comfortable public space to wait for orders. Furthermore, the online taxi bike adaptation is influenced by various conditions such as weather, location, and adequate parking space. The issues that arises due to unplanned or unmanageable urban space is known as "informality" [9]. These online transportation systems need to properly analyst the various factors capable of triggering the formation of new and unplanned open spaces.

In the urban growth and development, the phenomenon of informality has become a significant element of the city [9,10], with several unplanned open spaces located at The Diponegoro University. This research describes the phenomenon associated with the formation of these new and unplanned open spaces caused by the existence of online taxi bike on Diponegoro University.

2. Open Public Space and Online Taxi Bike

Public space is formed due to the need for a place to converge [11]. It is either built or naturally created in the environment to which the community has free entrance [12,13] and accessible to all members of the society [14]. Basically, it is a container which accommodates certain activities of humans both individually and in groups [15]. Therefore, Public spaces are used for daily transportation, short-term stays, and recreational centers [13].

Classic public space has the ability to gather taxi bikes, with its definition developed from the existence and influence of many factors such as digital technology. This was previously investigated by Hampton, Keith N., Goulet, Lauren Sessions, & Albanesius, Garrett (2015) The above-mentioned studies proves that the Use of Mobile-phone is associated with an increased likelihood with 3% and 10% of users [16]. Meanwhile, the public space most used to privatize users from surrounding conditions (to rest and enjoy their own time) [17]. So that the use of mobile-phone is not an important thing in Public Space.

This current digital era is associated with the formation of online transportation services. According to Simonyi, Fazekas, & Gáspár (2014), online transportation provides fairly organized services [18]. It is convenience, easy , fast, and inexpensive for customers as it is accessible through smartphones [8]. Mobile-phone play a dominant role in online transportation as it connects users and drivers, thereby, increasing its activities[19].

In online transportation, the mechanism occurs between the driver and the user starts from booking, delivery, payment and service assessment by the passenger [8]. When this phase is completed, the driver waits for incoming orders which are returned by the system (the provider). The average waiting time for incoming orders is around 30-45 minutes, and during this period, drivers need a place to park and wait comfortably. This is one current example of the use of mobile phones to locate open spaces.

3. Method direct observation

This research uses a direct observation usually refers to observing a behavior and knowing exactly what is happening of online taxi bike in Diponegoro University. In order to determine the phenomenon of informal public space, this research conducts several steps as follows: 1) Traces location of online taxi bike by direct observation; 2) Identifies the potential and problems; 3) Analyses the spatial interaction pattern in this kind of space; and 4) Figures out the pattern and its impact to public space. Data was obtained from online taxi bike drivers regarding their reasons for using open space informal as their waiting destination.

4. Analysis and Results

This research needs to understand the existing conditions faced by the new paradigms forces in public spaces at the Diponegoro University Campus. The existence of online taxi bike has triggered protests from existing public transportation because it is extinct (see figure 1). In addition, the existence of online taxi bike also gave rise to a paradigm shift in the use of public space in the Diponegoro University campus. Based on distribution of online taxi bike is not only located on the main streets in Diponegoro University, but also in places that are considered strategic (see figure 2). This shows that online taxis will occupy strategic public spaces according to them.

In each faculty building of Diponegoro University usually has provided an open space that can be utilized by students and the general public, especially when not working hours. Here below are formal public spaces in the Diponegoro University (see figure 3).

Reference: http://www.kampusundip.com/2017/04/sopir-angkot-undip-protes-brt-gojek.html

Figure 1: A protest from public transportation against online taxi in the form of paper was affixed to the body of a public transport car.

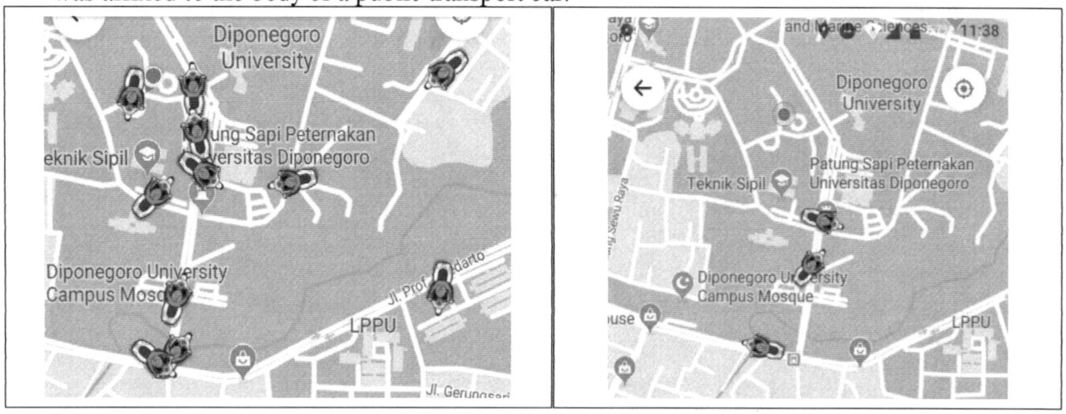

Figure 2: Distribution of online taxi bike in and around the DPWK FT Undip Campus Sample taken: left: 8 July 2019, time 15:06, Right: 12 July 2019, time 11:38.

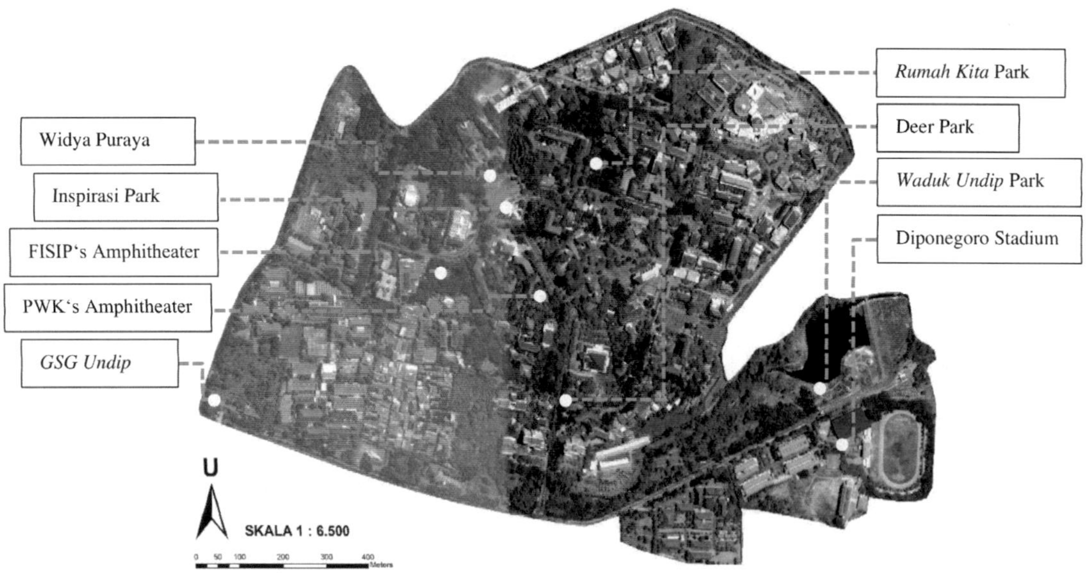

Figure 3: The location point of formal public space in the Diponegoro University.

IOP Conf. Series: Earth and Environmental Science **396** (2019) 012005 doi:10.1088/1755-1315/396/1/012005

The Location of formal public space in Diponegoro University are as follows:

1. *Widya Puraya*
2. Inspiration Park
3. Amphitheater of Social Science and Political Science Faculty
4. Amphitheater of Urban and Regional Planning Department
5. *GSG Undip*
6. *Rumah Kita* Park
7. Deer Park
8. *Waduk Undip* Park
9. Diponegoro Stadium

The massive activity of online taxi bike tends to stimulate the emergence of unplanned public spaces which are used as a waiting place for bookings. This location points are not permanent and are located in the following places Diponegoro University:

1. Roadside in front of Amphitheater of the Social Science and Political Science Faculty (FISIP)
2. Corner of the entrance to the Medical Faculty (FK)
3. Side of entrance to the Fisheries and Marine Science faculty (FPIK)
4. Roadside in front of *Rumah Kita* Parks – Science and Mathematics Faculty (FSM/MIPA)
5. Roadside on the back of the Economics and Business Faculty (FEB)
6. Along the roadside in front of the Humanities Faculty (FIB)
7. T-junction corner of the Humanities Faculty (FIB)
8. Stalls in front of the Medical Faculty (FK)
9. Roundabout Food stalls besides BRI Bank
10. In front of the *Polines* Mosque on the *Baskoro* street

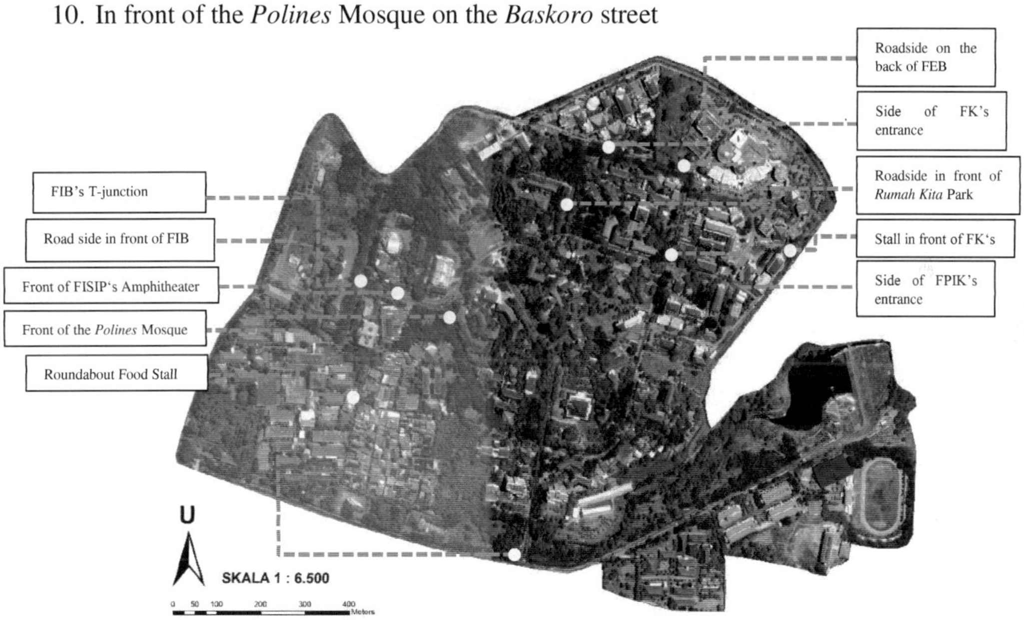

Figure 4: The location of informal public spaces for online taxi bike in Diponegoro University.

4.1 Identifying the potential and problems of informal public space

Below is a table consisting of the characteristics, this are potential and problems associated with public space in Diponegoro University.

No	Location	Characteristic	Photo	Potential & Problems
1	Roadside in front of Amphitheater of the Social Science and Political Science Faculty	• Locations: under the tree, side of the road. • Time : 07.30-16.00 • The driver spends time sitting on the motorbike. • Form of Space: Not formally Designed • Amenities: shade tree		• Potentials: located at the knot of the road and quite strategic and shady. • Problems: being in junction is dangerous for other motorists.
2	Corner of entrance to the Medical Faculty	• Locations: under the tree, and sidewalk. • Times: 07.30-16.00 • The driver spends time sitting on the motorbike and talks. • Form of Space: Not formally designed. • Amenities: shade tree		• Potentials: very shady location • Problems: being in junction of the Medical Faculty entrance access, thereby, hampering traffic, Parking on road side
3	Side of entrance to the Fisheries and Marine Science Faculty	• Locations: under the tree, side of the road and sidewalk. • Times: 09.00-16.00. • The driver spends time sitting on the motorbike, talks and have lunch or drink. • Form of Space: Sidewalk. • Amenities: shade tree, Stall (Type of Movable Stalls).		• Potentials: located at the knot of the road, strategic places and there are movable stalls. • Problems: being in junction of the Fisheries and Marine Science Faculty entrance, thereby, hampering traffic, parking on the road side causing barriers.
4	Roadside in front of *Rumah Kita* Park – Science and Mathematics Faculty	• Locations: Under the tree, side of the road. • Times: 08.00-16.00. • The driver spends time sitting on the motorbike. • Form of Space: Not formally designed. • Amenities: shade tree		• Potentials: - • Problems: on the side of the road.

5	Roadside on the back of the Economics and Business Faculty	• Locations: Under the tree, Side of the road, Side Walk. • Times: 07.30-17.00. • The driver spends time sitting on the motorbike, sidewalk, and talks. • Form of Space: Not formally designed. • Amenities: shade tree	 	• Potentials: very shady location • Problems: being in junction of the Economics and Business Faculty entrance, thereby, hampering circulation in this area.
6	Along the roadside in front of the Humanities Faculty	• Locations: Under the tree, Side of the road, Sidewalk. • Times: 07.30-17.00. • The driver spends time sitting on the motorbike, and talks. • Form of Space: Not formally designed. • Amenities: shade tree	 	• Potentials: in a crowded place • Problems: on the road side and causes barriers.
7	T-junction corner of the Humanities Faculty	• Locations: under the tree, side of the road, sidewalk. • Time: 07.30-16.00. • The driver spends time sitting on the motorbike, and talks. • Form of Space: Not formally designed. • Amenities: shade tree		• Potentials: in a shady place. • Problems: being in junction is dangerous for other motorists.
8	Food Stalls in front of the Medical Faculty	• Locations : Food Stalls • Times: 07.30-16.00. • The driver spends time sitting on the chair, and talks. • Form of Space: Building which functioned as stalls • Amenities: shade tree, Food Stall, Electricity.		• Potentials: located in a place that is quite crowded because it is near the campus and villages, there are stalls and chairs. • Problems: Parking on the road side causing barriers.

9	Roundabout Food stalls besides BRI Bank	• Locations: Food Stalls • Time: 07.30-16.00. • The driver spends time sitting on the chair, rest and talks. • Form of Space: Building which functions as Food stalls. • Amenities: shade tree, Chair, Food Stall, electricity.		• Potentials: located at the knot of the road, strategic places with stalls and chairs. • Problems: parking at the edge of alleys and interferes with its access.
10	In front of the *Polines* Mosque on the *Baskoro* street	• Locations: Food Stalls • Times: 07.30-16.00. • The driver spends time sitting on the chair, rest and talks. • Form of Space: Stalls with tents • Amenities: shade tree, Chair, Food Stall, electricity.		• Potentials: located in a place quite crowded because it is near the campus and villages. • Problems: parking on the edge of the road which is narrow enough to hamper traffic.

Reference: Direct Observation, August 2018

4.2 Spatial Pattern of Public Space by Online Taxi Bike

The behaviors of online taxi bike drivers during waiting time has created a new informal open space. There in a coherent statement with Alessandro Aurigi (2017) that there is an augmented space now due to the influence of technology[20]. Based on observations, the location chosen by online taxi bikes is strategic, due to its closeness to customers, food stalls and shelters. In addition, these locations are equipped with chairs, food stalls, and electricity. User behavior is an essential factor for creating successful public spaces [21]. Therefore, the need for drivers to rest, charge their mobile phones, eat and drink, is a user behavior which is the reason for the formation of some informal public space at Diponegoro University.

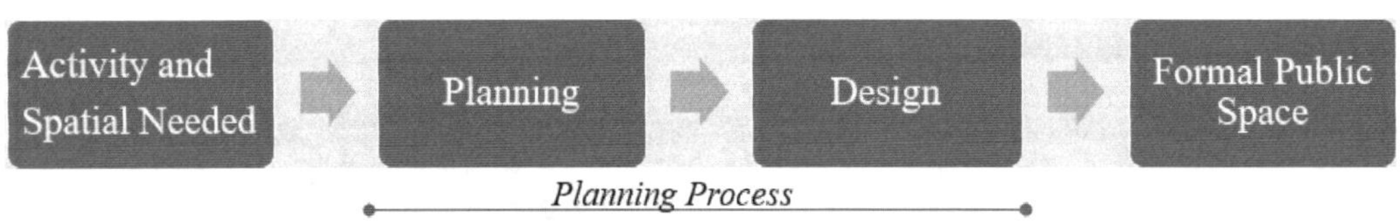

Figure 5: Forming of Formal Public Space

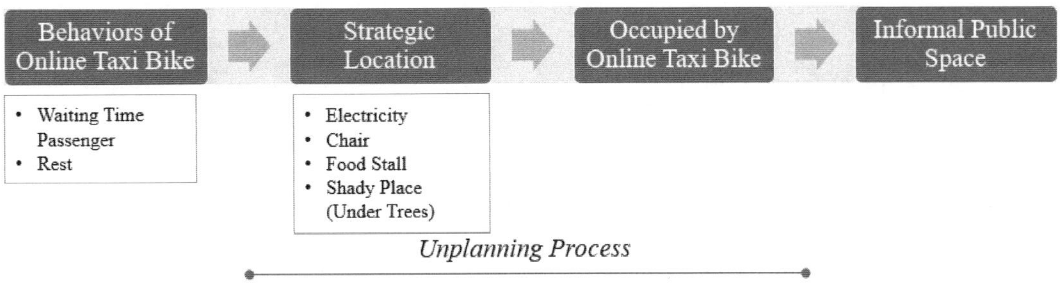

Figure 6: Spatial Pattern of Informal Public Space by Online Taxi Bike

There is a difference between formal and informal public space in their planning process. Formal public space was created by planning process (Figure 5), whereas informal public space was established by different planning process (Figure 6). In this case it is seen that the spatial pattern formed by online taxi bikes are located close to the Diponegoro University campus, availability of behaviors needs for drivers and that place has occupied by them is the reason online taxi bike create an informal public space. The formation of a public space without careful planning can cause various problems in the future [22]. Therefore, it is necessary to conduct regulatory and planning efforts in an open space for the existence of online taxi bikes.

4.3 The impact of online taxi bike on public space

The impact of online taxi bike on public space is the formation of new open spaces outside planned open spaces. The formation of public space which is usually through a planning process and design process that is in accordance with the plan needs space but in this case the online taxi bike forms a new informal space based on the occupied by them.

The existence of an online taxi bike forms a new public space in strategic located with amenities has a shady place like under trees, chair, food stall, electricity and quite crowded place near the campus. These locations which are not integrated with the formal public spaces in Diponegoro University. That space are usually on the edge of the road with shade trees, intersections, vendors, and food stalls. In addition, the choice of these online taxi bike locations causes problems such as traffic accidents due to the increasingly narrow roads (side barriers), congestion, and parking.

5. Conclusion

In conclusion, the public spaces formed by the presence of online taxi bikes are unplanned and located in strategic places such as T-junction, crossroads, and food stalls. Furthermore, the Informal public open spaces due to online taxi bikes at Diponegoro University are scattered at several roadside points such as the campus buildings at the faculties of Social and Political Science, Medicine, Fisheries and Marine, Science and Mathematics, Economics and Business, and other places with shaded trees. The existence of these unplanned public spaces, raises problems associated with of road obstacles in the form of arbitrary vehicle parking and the emergence of street vendors.

6. Acknowledgement

This article was presented at the 2nd International Conference on Smart City INNOVATION (ICSCI) 2019, jointly held by Universitas Indonesia and Universitas Diponegoro. ICSCI conferences have been supported by the United States Agency for International Development (USAID) through the Sustainable Higher Education Research Alliance (SHERA) Project for Universitas Indonesia's Scientific Modelling, Application, Research, and Training for City-centered Innovation and Technology (SMART CITY) Center for Collaborative Research, administered through Grant #AID-497-A-1600004, Sub Grant #IIE-00000078-UI-1.

7. References

[1] Society for Experiential Graphic Design 2015 15 Digital Technology Projects that Transformed Public Spaces Forever *SEGD*
[2] Taylor-Foster J Has Technology Diminished Our Understanding Of Public Space?
[3] Vogelsang M 2010 *Digitalization in Open Economies* (Dordrecht: Springer Science & Business)
[4] Berutu H C 2017 Go-jek, Revolusi Digital dalam Bisnis Transportasi *Futur. Pencil*
[5] Islamey T Z and Rakhmatulloh A R 2018 *Kajian Utilitas Pemilihan Moda Transportasi Daring di Kawasan Permukiman Kota Semarang* (Universitas Diponegoro)
[6] Bodnar J 2015 Reclaiming public space *Urban Stud.* **52** 2090–104
[7] Nesta 2015 Digital Technology in Parks and Open Spaces: Busting Myths and Exploring Opportunities
[8] Dewi N K and Rakhmatulloh A R 2018 *Analisis Penyediaan dan Pelayanan Transportasi Daring di Kota Semarang* (Universitas Diponegoro)
[9] Lutzoni L 2016 In-formalised urban space design. Rethinking the relationship between formal and informal *City, Territ. Archit.* **3**
[10] Lefebvre H 1991 *The Production of Space* (Cambridge: Blackwell)
[11] Prihutami D 2008 *Ruang Publik Kota Yang Berhasil* (Fakultas Teknik Universitas Indonesia)
[12] Carmona M, Magalhães C de and Hammond L 2008 *Public Space: The Management Dimension* (Routledge)
[13] Kurniawati W 2018 *Urban Equity In Public Space* (TU Darmstadt)
[14] Madanipour A 2010 *Whose Public Space? International case studies in urban design and development* (Routledge)
[15] Hakim R and Utomo H 2003 *Komponen Perancangan Arsitektur Lansekap* (Jakarta: Bumi Aksara)
[16] Hampton K N, Goulet L S and Albanesius G 2015 Change in the Social Life of Urban Public Spaces : The Rise of Mobile Phones and Women , and the Decline of Aloneness Over Thirty Years *Urban Stud.* **52** 1489–504
[17] Hampton K N and Gupta N 2008 Community and socialinteraction in the wirelesscity: wi-fi use in public andsemi-public spaces *New Media Soc.* **10** 831–50
[18] Simonyi E, Fazekas Z and Gáspár P 2014 Smartphone Application for Assessing Various Aspects of Urban Public Transport *Transportation Research Procedia* vol 3 (Elsevier B.V.) pp 185–94
[19] He F and Shen Z M 2015 Modeling taxi services with smartphone-based e-hailing applications *Transp. Res. PART C* **58** 93–106
[20] Zammit A and Kenna T 2017 *Enhancing Places Through Technology* (Valletta: CyberParks)
[21] Jayasundara N and Botheju P 2016 Spatial Patterns and Human Behaviours : A Study of Urban Public Space in Kandy *Proceedings of the International Conference on 'Cities, People and Places'- ICCPP* (Colombo)
[22] Sebastian Y 2013 *Time Is More Valuable Than Money Dampak Transportasi pada Hidup Kita* (Jakarta: PT Gramedia Pustaka Utama)

Organism Associated with *Cymodocea Serulata* in Different Habitats near Urban Coastal Area

N L Watiniasih[1,2], I W Nuarsa[2], I M Merdana[3], I N Budiarsa[4], A Dharma[4], I N G Antara[4], M W Poborini[1]

[1]Departement of Biology, Faculty of Mathematic and Natural Science, Udayana University, Bali, Indonesia
[2]Departement of Marine Sciences, Faculty of Marine and Fisheries, Udayana University, Bali, Indonesia
[3]Departement of Physiology and Pharmacy, Faculty of Veterinary Medicine,Udayana University, Bali, Indonesia
[4]Mechanical Engineering,Udayana University, Bali, Indonesia

E-mail: luhwatiniasih@unud.ac.id

Abstract. The development of tourism in Bali is very rapid, where most area are in direct contact with residential areas. Seagrass that grows in coastal area is under pressure due to tourism activities. This study aims to determine the state of urban coastal areas with an indicator of organisms associated with seagrass *Cymodocea serulata*. The samples of *C. serulata* leaves were collected from Sanur Beach, Denpasar and Samuh Beach, Nusa Dua, Badung, Bali. These two areas are tourism areas with high tourist visits, and are located near the residential areas. Sanur Beach is different from Samuh Beach, where the seagrass habitats at Sanur Beach are also used as a mooring place for motorized boats, crossing Penida strait from Sanur to Nuas Penida and Nusa Lembongan. The results showed that periphyton that lived in association with *C. serulata* were higher at Samuh beach (12 species) compared to Sanur beach (3 species). However, the total bacteria associated with seagrass leaves was higher at Sanur beach (58. 0 x 104 CFU / mL) compared to that of Samuh beach (51.3 x 104 CFU / mL). Three species of periphyton with the high number of individuals found on *C. serulata* leaves at Samuh beach were *Ardissonia fulgens*, *Synedra* sp, and *Thalassionema nitzschioides*, while 3 species were found at the Sanur beach were *Thalassiothrix* sp, *Oscillatoria* sp. and *Paracalanus palvus*. Four potential fungal species were found associated with seagrass leaves, 3 of them identified as *Aspergillus eff. niger*, *A. clavatus* and *Penicillium purpurogenum*, while 1 isolate has not been identified. The results of this study found that habitat differences and water designation affected the diversity of organisms associated with seagrass beds of *C. serulata*.

Keywords: urban area, tourism, *Cymodocea serulata*, periphyton

1. Introduction

The abundance and diversity of organisms depends on environmental condition. The sensitivity of the communities to the alteration of the environment and their tolerance to environmental pressure has been considered as a measure of ecosystem health [1,2]. The quality degradation of the environment can be caused by natural and anthropogenic activities. Human activities such tourism industries have important contribution to the quality of environment. Increasing tourism industries along the coastal area are, direct or indirectly contributed to the quality to the surrounding environment, exposed the fragile ecosystems to environmental degradation. Uncontrolled tourism development in the coastal brings negative impact to the area due to overcrowding, poor sewage disposal, boat-generated waste, beach erosion and destruction of wildlife habitats [3].

Sanur beach at Sanur and Samuh beach at Nusa Dua are famous as tourist destination at Bali Island. Sanur beach is differ to the Samuh beach, which Sanur beach, besides as tourist destination, also the area was functioned as harbour of boats powered by large machines to cross Penida Strait from Sanur to Nusa Penida island. Samuh beach, on the other hand, only functioning as tourist destination and harboured by few traditional fishing boat. Oil spill from boats is inevitable affecting the abundance and diversity of macro- and micro invertebrate living in those two areas. It has been found that oil spills contain polycyclic aromatic hydrocarbons (PAHs) are widely spread in marine ecosystems, correspond to acute pollution and other human activities [4]. Sanur and Samuh beaches are the habitat of seagrasses, and 7 specciesof seagrasses were identified at Sanut beach, one of them was *Cymodocea serulata* [5]. This species was also found at Samuh beach.

Seagrasses are habitats for coastal living organisms such as fish and other organisms associated with them. The leaves of the seagrasses is a unique habitat for microorganism, with specific association among diverse aquatic microorganisms are capable of colonizing surfaces of various kinds, leading to the formation of biofilms [6]. This unique association provides microorganisms a better access to nutritional resources, enhanced organism interactions, and stability. Periphyton is the aquatic biota that develops on submerged substrates when nutrients and light are available, formed assemblages, attached on any wet substrate [7]. Alteration of coastal environment due to anthropogenic activities on the coastal area and its affect to the diversity of organism attached to seagrasses in tropical area has not much studied. The limitation of information about the diversity of organism that play roles to the alteration of environment, particularly in tropical with the impact of human activities was the aim of this study.

2. Methods

2.1. Sample collection

Samples of leaves of *Cymodocea serulata* contained organism were collected from 2 sites: Sanur Beach and Samuh Beach. Samur beach located at Sanur, one of tourist destination and boat harbour to Nusa Penida Island, a small island located at the east of Sanur. Samuh beach is another tourist destination, located at Nusa Dua, but this site was not used as a bat harbour. Samples were preserved in 70% alcohol until processing.

2.2. Sample processing and analysis

The leaves samples of *C. serulata* that have been collected were brought to the Laboratory of Aquatic Resource Management, Faculty of Marine and Fishery, Udayana University. Leaves were grazed using a scalpel to collect the organism attached to the leaves. Then, they observed and identified under light microscope. Identifications of organism were based on "Encyclopedia of Marine Gastropods by Alain Robin, 2008". Bacteria and fungi identification were conducted at Laboratory on Microbiology, Faculty of Mathematics and Science, Udayana University. Bacteria isolation was done using plating methods following Nester et al. [8]. Preparation of samples for fungi identification was following Pfaller [9]. Segments from leaves of *C. serulata* were plated on potato dextrose agar medium amended with an antibiotic (chloramphenicol 150 mg l-1) in Petri dishes and incubated in a light chamber for four weeks at 26°C; the light regimen was 12 h of light followed by 12 h of darkness. The Petri dishes were observed periodically, and fungi that grew out from the tissues were isolated and identified based on spore morphology.

3. Result and Discussion

3.1. Results

3.1.1. Periphyton diversity.

The result shows that the diversity of periphyton at Sanur beach were much lower than that of Samuh beach. There were only 3 species periphyton found at Sanur beach with the average number ranged from 0.33 to 1.33 individual per collection (Figure 1).

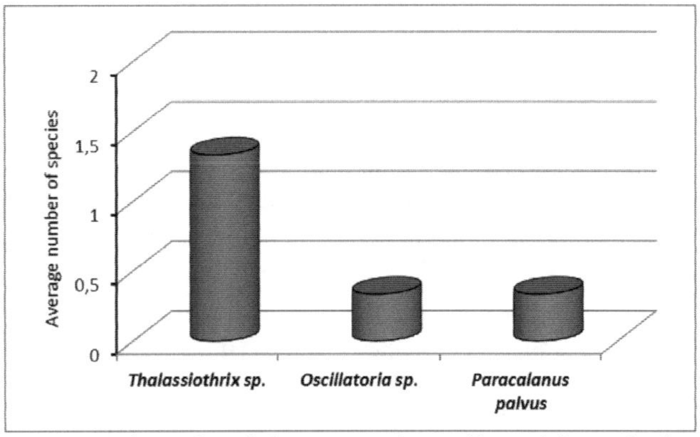

Figure 1. The average number of periphyton species collected from the leaves of *C. serulata* at Sanur beach.

3.1.2. Fungi and bacteria diversity.

At Samuh beach, on the other hand, 13 species were identified (Figure 2). The most common species found at Samuh beach was Ardissonia fulgens, with the average number of 48.33. Two other species that were commonly found at Samuh beach were Synedra sp with the average number of 29.67 and Thalassionema nitzschioides with the average number of 29.33 (Figure 2a). The other species have the average number range from 0.33 to 2.33 (Figure 2b).

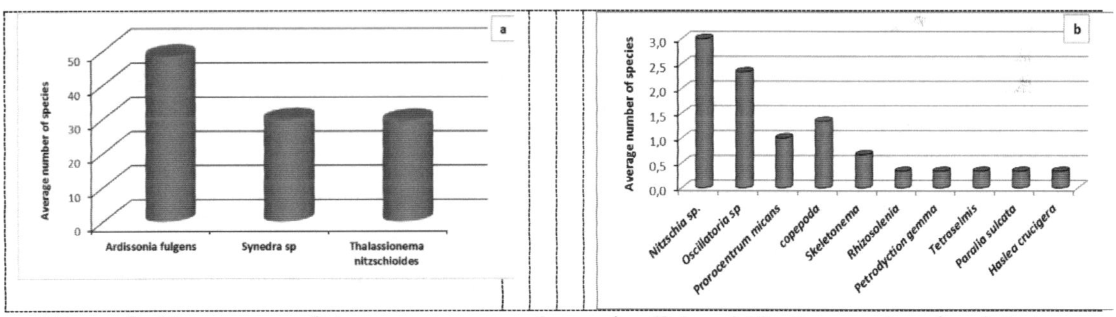

Figure 2. The average of three most commonly species (2a) and 10 other species of periphyton collected from the eaves of *C. serulata* at Samuh beach

The species of fungi isolated from the leaves of *C. serulata* were slightly differ between sites (Sanur and Samuh beaches). Five species of fungi were isolated from Samuh beach, while 4 from Sanur beach. *Apergillus niger* was isolated from both sites at all collections. *Penicillium purpurogenum* and *Aspergillus clavatus* were also collected from both sites, but only *P. Purpurogenum* was collected from all collection from Samuh beach. *Aspergillus* sp 1 and *Aspergillus* sp 2 were not found on the leaves of *C. serulata* collected from Sanur beach, but present from the samples from Samuh beach. On the other hand, the Isolate 1 was only present at the sample from Sanur beach (Table 1).

Table 1. A Species of fungi identified from the leaves of *Cymodoce serulata* collected from Sanur and Samuh beaches.

		Sanur Beach				Samuh Beach			
No.	Species/collection number	1	2	3	4	1	4	3	4
1	*Aspergillus niger*	+	+	+	+	+	+	+	+
2	*Aspergillus* sp. 1	-	-	-	-	+	+	+	+
3	*Aspergillus* sp. 2	-	-	-	-	+	+	+	-
4	*Penicillium purpurogenum*	+	+	-	+	+	+	+	+
5	*Aspergillus clavatus*	+	+	-	-	+	+	-	+
6	Isolate 1	+	+	+	-	-	-	-	-

Two species of bacteria were isolated from the leaves of *C. serulata*, namely *Sterptococcus* sp and *Bacillus* sp. The volume of both species of bacteria were much higher at Sanur beach (58 x 104 CFU/mL) compared to Samuh beach (51.3 x 104 CFU/mL).

3.2. Discusion

There is an increased awareness that marine environments are important roles in global climate changes. The structure and function of natural marine ecosystem is a concern for managing the impacts of human activities. Marine microorganism are believed to contribute to present of particulate organic matter in the ocean and drive global biogeochemical cycles [2,10]. The abundance and diversity of organisms and sensitivity of community to the alteration of the environment and their tolerance to environmental pressure has been considered as a measure of ecosystem health [1,2].

Periphyton that were found in less disturbed area at Samuh beach was higher compared to Sanur beach, indicating the pressure of environmental affecting the abundance of periphyton. Sanur beach that received more disturbance support much less periphyton population than Samuh beach. *Oscillatoria* sp. was the only species of periphyton found at both sites, indicating this species are adapted to a wide range of environmental condition. Complex communities of autotrophic and heterotrophic microorganisms such as microalgae, bacteria, fungi, protozoa, metazoa and epiphytes form periphyton biofilms [11]. The organisms in biofilms create their own microhabitats.

Microbial parameters and processes have been identified to have abilities to detect changes in enviromental quality [2]. The contribution of heterotrophic, autotrophic and detritus within the organic pools is reflection of the function of microbial assembles to the biomasses, particularly the C/N ratios [12]. Prokaryotic assembles was found as indicator of the ecosystem health by measuring its diversity related to ecological quality [12,13]. Although vibrio bacteria are usually detectable and sometimes abundant in coastal and brackish waters, especially in eutrophic environments, they are usually subdominant and opportunistic bacteria [6]. However, in this study, vibrio bacteria have not been detected, instead only two genera of bacteria were identified, namely *Streptococcus* sp and *Bacillus* sp. It has been found that some of *Bacillus* sp act together forming quorum, known as 'quorum sensing' to respond to different environmental signals, such as found in *Acidithiobacillus ferrooxidans* in colonizing and using different solid sulfur- and ironcontaining minerals [6].

Microbial surface associations, including bacteria and fungi are vital for microbial process and mechanism, providing more surface population, intraspecies and interspecies communication and interaction, and regulatory balance between cooperation and competition [6]. Bacteriodetes, group of gram negative bacteria in the form of basil was commonly found in marine environment [6]. This study also found bacteria that are associated with *C. serulata* at both Sanur and Samuh beaches were Bassilus sp. Streptomyces was another genus of bacteria that were found associated with the plants. Prokaryotes organisms have been found dynamically interact with carbon pools, mediating carbon and nutrient cycles of the sea [2], by reacting to the environmental changes, therefore suspended matter [14,15]. Diverse aquatic microorganisms are capable of colonizing surfaces of various kinds, leading to the formation of biofilms and to the development of specialized processes within these structures [16,17].

Marine environment is highly affected by the impact of anthropogenic activities. Increasing in development of tourism industries alongside the beaches have a huge impact to the marine environment. Boats powered by machines are increasingly used to transport people from one island to the other. Sanur beach, Bali, as one famous tourist destination is also a place to harbour boats powered by large machines to cross the Penida strait to Penida and Lembongan Island. Oil spills from the boat machines is inevitable. Oil spills contain polycyclic aromatic hydrocarbons (PAHs) are widely spread in marine ecosystems, correspond to acute pollution and other human activities [4]. Human activities have increase the level of PAHs in the environment, particularly in marine ecosystem raising serious environmental issues and human health concerns Some PAHs and their metabolites are considered highly toxic mutagens and carcinogens [4].

4. Conclusion

The diversity of periphyton and fungi were different between Sanur and Samuh beaches. It has been found that the diversity of periphyton and fungi were found higher at Samuh beach, where the beach was not harboured by the boat. The diversity of bacteria was not differ between sites. *Oscillatoria* sp. was the only species of periphyton found at both sites.

5. Acknowlegment

This article was presented at the 2nd International Conference on Smart City INNOVATION (ICSCI) 2019, jointly held by Universitas Indonesia and Universitas Diponegoro. ICSCI conferences have been supported by the United States Agency for International Development (USAID) through the Sustainable Higher Education Research Alliance (SHERA) Project for Universitas Indonesia's Scientific Modelling, Application, Research, and Training for City-centered Innovation and Technology (SMART CITY) Center for Collaborative Research, administered through Grant #AID-497-A-1600004, Sub Grant #IIE-00000078-UI-1

6. References

[1] Borja A, Elliott M, Andersen J H, Cardoso A C, Carstensen J, Ferreira J G, Heiskanen A-S, Marques J C, Neto J M, Teixeira H, Uusitalo L, Uyarra M C and Zampoukas N 2013 Good Environmental Status of marine ecosystems: What is it and how do we know when we have attained it? *Mar. Pollut. Bull.* **76** 16–27

[2] Caruso G, Azzaro M, Caroppo C, Decembrini F, Monticelli L S, Leonardi M, Maimone G, Zaccone R and Ferla R La 2016 Microbial community and its potential as descriptor of environmental status. *ICES J. Mar. Sci. J. du Cons.* **73** 2174–7

[3] Zahedi S 2008 Tourism impact on coastal environment *Environmental Problems in Coastal Regions {VII}* ({WIT} Press)

[4] Duran R and Cravo-Laureau C 2016 Role of environmental factors and microorganisms in determining the fate of polycyclic aromatic hydrocarbons in the marine environment ed C Brussaard *FEMS Microbiol. Rev.* **40** 814–30

[5] Arthana I W 2004 Jenis dan kerapatan padang lamun di pantai Sanur Bali *Bumi Lestari J. Environ.* **5**

[6] Dang H and Lovell C R 2015 Microbial Surface Colonization and Biofilm Development in Marine Environments *Microbiol. Mol. Biol. Rev.* **80** 91–138

[7] Levy A, Milstein A, Neori A, Harpaz S, Shpigel M and Guttman L 2017 Marine periphyton biofilters in mariculture effluents: Nutrient uptake and biomass development *Aquaculture* **473** 513–20

[8] Nester E W 2007 *Microbiology: a Human Prespective* (Boston : McGraw-Hill)

[9] Pfaller M 2015 Application of Culture-Independent Rapid Diagnostic Tests in the Management of Invasive Candidiasis and Cryptococcosis *J. Fungi* **1** 217–51

[10] Pomeroy L, leB. Williams P, Azam F and Hobbie J 2007 The Microbial Loop *Oceanography* **20**

28–33
[11] Zippel B, Rijstenbil J and Neu T R 2007 A flow-lane incubator for studying freshwater and marine phototrophic biofilms *J. Microbiol. Methods* **70** 336–45
[12] Caruso G, Ferla R La, Azzaro M, Zoppini A, Marino G, Petochi T, Corinaldesi C, Leonardi M, Zaccone R, Umani S F, Caroppo C, Monticelli L, Azzaro F, Decembrini F, Maimone G, Cavallo R A, Stabili L, Todorova N H, Karamfilov V K, Rastelli E, Cappello S, Acquaviva M I, Narracci M, Angelis R De, Negro P Del, Latini M and Danovaro R 2015 Microbial assemblages for environmental quality assessment: Knowledge, gaps and usefulness in the European Marine Strategy Framework Directive *Crit. Rev. Microbiol.* **42** 883–904
[13] Cochrane S K J, Connor D W, Nilsson P, Mitchell I, Reker J, Franco J, Valavanis V, Moncheva S, Ekebom J, Nygaard K and others 2010 Marine Strategy Framework Guidance on the interpretation and application of Descriptor 1: Biological diversity *Rep. by task Gr.* **1** 1–114
[14] Caruso G, Decembrini F, Caruso R, Zappalà G, Bergamasco A and Leonardi M 2011 Are microbial enzyme activities sensitive indicators of the trophic state of marine ecosystems *Pollut. Monit. Hauppauge Nov. Sci. Publ.* 195–210
[15] Cunha A, Almeida A, Coelho F, Gomes N C M, Oliveira V and Santos A L 2010 Bacterial extracellular enzymatic activity in globally changing aquatic ecosystems *Curr. Res. Technol. Educ. Top. Appl. Microbiol. Microb. Biotechnol.* **1** 124–35
[16] Costerton J W, Lewandowski Z, Caldwell D E, Korber D R and Lappin-Scott H M 1995 Microbial Biofilms *Annu. Rev. Microbiol.* **49** 711–45
[17] Petrova O E and Sauer K 2012 Sticky Situations: Key Components That Control Bacterial Surface Attachment *J. Bacteriol.* **194** 2413–25

The Potential of Urban Organic Waste Utilization as Neo Carbon Food

S L Putri[1], C V Marbun[1], G L Utama[1,2,3]

[1]Faculty of Urban organic Technology, Padjadjaran University, West Java, Indonesia
[2]Centre for Environmental Innovation and Technology, School of Postgraduate, Padjadjaran University, West Java, Indonesia
[3]Centre for Environment and Sustainability Science, Padjadjaran University, West Java, Indonesia

E-mail: suhartinilp@gmail.com

Abstract. Majority of the urban society has a problem with their organic waste. One alternative urban waste utilization as food production can be obtained by producing Single Cell Protein (SCP) through Neo-Carbon Food (NCF). NCF is a single cell production based on microorganisms using CO_2 emissions and urban organic waste as the main raw material for hydrogen oxidizing bacterial growth. NCF are potential because contain high protein, requiring less land and water, reducing pollutions and the GHG emissions. So, NCF can decrease GHG emmisions to overcome climate change and also as sustainable food.

Keywords: GHG, neo-carbon food, single cell protein, sustainable food, urban organic waste

1. Introduction

Nowadays, nearly 55% of the World population lives in urban areas, and it is expected that the number will increase up to 66% by 2030 [1]. As the city grows, new problems arise (e.g., traffic congestion, waste management, pollution, and GHG emissions. A greenhouse gas (GHG) is any gas in the atmosphere which absorbs and re- emits heat, and thereby keeps the planet's atmosphere warmer than it otherwise would be. The main GHGs in the Earth's atmosphere are water vapour, carbon dioxide (CO_2), methane (CH_4), nitrous oxide (N_2O) and ozone [2]. CO_2 gas is the largest contributor of GHG emissions, which is 77% [3]. One of the causes of GHG emission is waste production.

The production of wastes especially the urban organic waste remains a major source of concern as it has always been since pre historic period. The urban organic waste produced as a result of various agricultural operations. It includes manures and other wastes from farms, harvest waste, fertilizer run-off from fields, and pesticides [4]. Though the waste enters the environment it will contribute a significant to overall greenhouse gas (GHG) emissions, This waste is also an energy source and protein source because this urban organic waste contains carbon, hydrogen and nitrogen [5]. Organic wastes such as fruit and vegetables have biological and chemical potential in producing bioethanol [6,7], this bioethanol can be used as alternative liquid fuel for energy source [8].

Urban organic waste can also be used as a raw material for making Neo-Carbon Food. One of the urban organic wastes that can be used to make Neo-Carbon Food is tofu waste. Generally, solid waste tofu can be used for animal feed. Meanwhile, tofu waste in the form of liquid is discharged into the water, resulting in a detrimental effect on water quality, which results in a foul odor in the river or place around the tofu liquid waste disposal. The presence of liquid waste can give a negative value to an industrial activity. Tofu liquid waste can also provide a positive value if it can maximize the potential that exists in industrial wastewater as well as handling with the right technology [9].

The potential for the development of Neo-Carbon Food in developing countries like Indonesia is very possible to make by making several modifications in the manufacturing process. For example, like tofu waste which can be used as an energy source as well as a nitrogen source. The energy used for its production uses biogas from tofu solid waste, as well as tofu liquid waste as its nitrogen source. So, Neo-Carbon Food can decrease GHG emissions to overcome climate change and also as sustainable food.

2. Neo Carbon Food Concept

Neo-Carbon Food comes from the word "Neo" which means renewable and "Carbon" which is an element with an atomic number 12. Neo-Carbon Food is the result of biosynthesis that produces a single cell protein. In the manufacture of SCPs, in addition to microorganisms with strong ripple power, the composition of basic materials, as well as the process technology used greatly determines the quality of the products produced. The raw material for the manufacture of SCP must contain water, source N, source C and minerals [10]. Neo-Carbon Food is a future protein production that uses electricity and carbon dioxide as the main ingredients as a source of C for bacteria. This research was developed by VTT Technical Research Center of Finland Ltd. and Lappeenranta University of Technology LUT. This process is carried out using microbes. This technology allows safe food production without pesticides or agricultural land [11]. Neo-Carbon Food is processed through biosynthesis by electrolysis of water with the help of electricity so that the CO_2 occurs in the fusion reaction to C and O_2 and H_2O to H_2 and O_2.

Microbes that are used as SCP producers must have criteria that are not pathogenic, have good nutritional value, can be used as food or feed and do not contain toxic compounds [12]. Many types of microbes can produce SCP, one of which is a group of hydrogen oxidizing bacteria (eg. *Cupriavidus necator, Rhodococcus opacus,* and *Hydrogenobacter thermophilus*) capable of autotrophic growth by using hydrogen as a donor electron and oxygen as an electron acceptor to improve carbon dioxide to build their biomass. This bacterial biomass has a high protein content (50-83%), and can be used as a nutrient in humans and animals [13]. Overall growth and biomass production factors, the productivity of hydrogen oxidizing bacteria is also much faster and higher than the productivity of *Cyanobacterium* [6,7]. But this hydrogen oxidizing bacteria is still less frequent than *Cyanobacterium*.

There are other beneficial reasons besides the speed of growth to use oxidizing hydrogen bacteria for biomass production. This bacterium has a high protein content and one product of the oxidation of hydrogen is water. Hydrogen oxidation occurs, when oxidizing bacteria use hydrogen to grow substances. The bacterial protein content varies between 60-70%. There have been several experiments, in which bacterial biomass is fed to farms and furry animals. Such experiments were carried out at the biophysical institute at the Russian science academy. The result is that the use of bacterial biomass as a feed is beneficial for 25-50% of the animal protein diet. The result depends on the species and age of the animal [14]. This experiment shows that there is the potential to use these bacteria as feed use [15]. It is this high protein that bases why it is used to oxidize hydrogen to produce single cell proteins (SCP), because the need for protein in the global community is increasing.

One species of hydrogen oxidizing bacteria is the *Cupriavidus necator*. This species is commonly used in studies that focus on cell growth [16]. This microbial stoichiometry is used as a reference for calculating the needs of carbon dioxide, nitrogen, and hydrogen per kg of biomass produced. Stoichiometry for *Cupriavidus necator* is as follows [17]:

$$21.36\ H_2 + 6.21\ O_2 + 4.09\ CO_2 + 0.76\ NH_3 \rightarrow C_{4.09}H_{7.13}O_{1.99}N_{0.76} + 18.7\ H_2O$$

One kilogram of *Cupriavidus necator* requires about 1850 grams of carbon dioxide, 220 grams of hydrogen, 1020 grams of oxygen is needed [18]. Hydrogen and oxygen can be supplied to bacteria by conducting electrolysis directly in the bioreactor, where bacteria are bred. In situ water electrolysis allows the efficient use of hydrogen and oxygen by bacteria, because the transfer of the gas mass is better to the aqueous solution, than in the case of an external gas supply. Also combustible mixtures in the reactor headroom can be avoided to some extent. In addition to H_2 and O_2, CO_2 and nutrients needed for growth are supplied to reactors such as nitrogen, vitamins and minerals [13].

One of the most important growth conditions is pH. The pH used can even affect the product formed. For example, if pH is lowered, alcohol production may increase [19]. Optimal pH varies, when using different organisms, for *Cupriavidus necator* the optimal pH is in the range 6.7-6.9 with optimal temperature of 30^{0}C [20]. This bacterium belongs to the mesophilic bacteria because it grows at a temperature of 30-40°C [21].

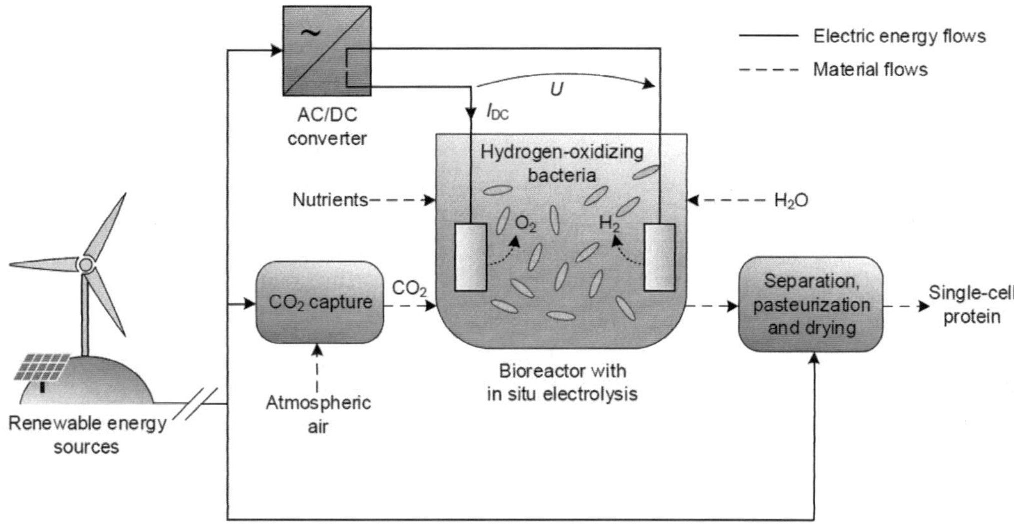

Figure 1. Scheme of Neo-Carbon Food Production [13]

An electricity-to-biomass efficiency of 54% has been achieved by applying in situ water electrolysis, which would correspond to approximately 10% solar-to-biomass efficiency (assuming 18% efficiency for photovoltaic). For comparison, the annual average efficiency of crops do not typically exceed 1%, and efficiencies during the growing seasons can reach only 3.5–4.3%. By applying direct air capture (DAC) of CO_2 by adsorption/desorption process with in situ water electrolysis, bacterial protein can be produced basically only from electricity, air water and some nutrients. The scheme of this concept is illustrated in Figure 1. [13]. The reason why DAC is used as part of the biomass production process, is that it offers a solution to produce high purity carbon dioxide with possibility to reduce the concentration of CO_2 from atmosphere without specific location. Other reason is that, it can be used sustainable, if renewable energy source is used to power the process [22]. Besides using DAC we can use CCS (Carbon Capture and Storage) as the alternative [23].

When calculating the impacts and energy consumption, the assumption is made so that, all the electrical energy used to produce biomass is produced from renewable sources. The sources are wind and solar energies. The process is assumed to function so, that all the gasses produced for biomass production are used right away. The gasses needed for biomass production can be stored, if there is so much oversupply that the process cannot handle all the material produced, but storage needs energy and materials and thus increases the energy need, GWP and land use [24].

The electricity can be produced by renewable sources and thus this application is used in the process. It is estimated that globally around 4 % of hydrogen is produced via water electrolysis. The advantages of using water electrolysis are highly dynamic modes operation, compact stack design, high unit power densities, high capacities, satisfying gas purity levels, wide partial load ranges with high enough efficiencies and low investment costs. The technology is already in use, but there is still room for improvements. Currently a lot of research is done to make this technology more efficient [25]. When using conventional electricity, the costs needed are more expensive than using electricity from renewable energy.

When planning to produce biomass from microbes in aquatic solution, the culture harvesting and drying are important unit process. There exists a couple of technologies for culture harvesting. Some usable technologies for harvesting are chemical flocculation, bio-flocculation, electro-flocculation, flotation, filtration and centrifugation. However, these harvesting technologies are not usually sufficient alone due to remaining liquid after these processes. Quite often drying process after culture harvesting is needed. Some technologies for drying are solar drying, convective drying, spray drying and freeze-drying [26].

The concept is still in research stage, but in the future, it may provide a solution to environmental impacts of conventional agriculture, such as greenhouse gases, fresh water use, land use, and pollution caused by fertilizers and pesticides. Optimization of electrode materials, cultivation medium and conditions, bioreactor construction, and process modelling and control are under research by the authors [13].

3. Neo-Carbon Food Potential as Sustainable Food

The use of microorganisms to produce single cell proteins have very fast growth rate; their biomass doubling time is typically 20-120 minutes, while e.g. soybean has doubling time 1–2 weeks. Due to high conversion efficiency and fast growth, bacterial protein production requires significantly less water and land area compared to the conventional protein sources [13].

The need of the land will also increase to grow food commodities. Land needs for this growth also affect the need for fertilizers and pesticides. Global fertilizer consumption is expected to increase by 50% in 2050 to maintain the required increase in food production capacity [27]. The use of Neo-Carbon Food can reduce the need for fertilizers and pesticides because the process of processing Neo-Carbon Food does not require pesticides and fertilizers, but what is needed is nutrition, vitamins and minerals for the growth of microorganisms that are far less than conventional protein production.

The demand for water is ought to continue rising in the future. The agriculture is the main reason why water is consumed so much. The irritation uses approximately 70% of the global water consumption and thus the lack of water affects greatly to food production [28]. Because food production plays such a huge role on water consumption in the world, we should find solutions for more sustainable water use in that sector. Especially, when there is a talk about upcoming water crisis [29]. In addition, the energy needed to produce proteins use conventional way is greater than using microorganisms [22–27].

Neo-Carbon Food obtained from the results of Nygren, et al. [13] which is single cell protein powder containing 54% protein, 1% fat, 18% carbohydrate, 8% inorganic ingredients and 19% other ingredients [11]. This research has entered the pilot plant stage by using a 20 L reactor and producing 10 grams per hour. This protein content is quite high when compared to meat with 20% protein, 3.5% milk, 25% cheese, soybeans 35-40% [28,29]. The raw material used for the production of Neo-Carbon Food also uses CO_2 which is one of the gases that contributes the most in GHG emissions [3].

4. Potential Development of Neo-Carbon Food in Indonesia

Research on Neo-Carbon Food in Indonesia is still rare. However, there have been many studies on microorganism-based single cell proteins (SCP), because this single cell protein can be used as a substitute for animal and vegetable proteins. Making these single cell proteins (SCP) requires a bioreactor, substrate and microorganism, so SCP requires some energy to produce it.

Renewable energy in Indonesia is relatively small. Nygren et al. [13] using energy from solar and wind panels, in Indonesia such energy sources are still rarely used so the use of biogas can be used as an alternative energy source. Besides being able to reduce waste, it can also produce energy and reduce GHG emissions. At present the source of biogas raw material is available abundantly and has not been fully utilized [30]. The use of biogas in Indonesia as an alternative energy source allows it to be applied, along with the increasing price of fuel oil and it's scarcity [31].

The energy needed to produce 1 kg of Neo-Carbon Food is around 13,523 kWh [31–35]. Based on the basic electricity rates from PLN in 2019, the electricity costs for the industry are Rp. 97-1115 / kWh. So for 1 kg of Neo-Carbon Food, it requires production costs (from electricity sources) of Rp. 13,482-

15,078 when using electricity from PLN. When using electricity from renewable energy, the cost of electricity used can be lower.

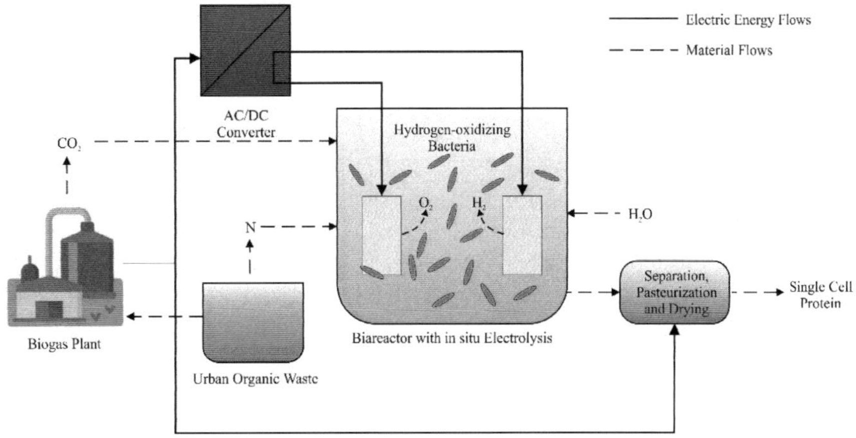

Figure 2. Scheme of Neo-Carbon Food Production in Indonesia

4.1. Urban Organic Waste

Urban organic waste is biodegradable part of households refuse, market waste, yard waste and animal and human waste. When waste is not managed properly it may cause serious health and environmental risks. The overall goal of municipal refuse management is to improve and safeguard the public health and welfare, reduce waste generation and increase resource recovery and re-use, and protect environmental qualities [36].

There are many resources from urban organic waste that can be used to produced the Single Cell Protein (SCP). For example, the horticulture waste such as vegetables and fruits from households refuse or market waste is a rich source of potentially valuable to produce the SCP. These waste contains water-soluble carbohydrates such as glucose, and water-insoluble ones such as pectin and cellulose. Carbohydrates can be used as a substrate for the SCP to grow because SCP can aerobically grow on lignin and sugar compounds [37]. Meanwhile total nitrogen level in this waste reach 2,22%. This nitrogen could be the source to produce the SCP. Another major issue associated with urban waste mismanagement is the greenhouse gas emission, which comprehends methane and nitrous oxide [38]. These gasses can be used as an energy source in biogas plant (Figure 2).

4.2. Tofu Waste

One of the organic wastes that can produce biogas is tofu waste. The tofu industry produces 20 million cubic meters of liquid waste per year and produces around 1 million tons of CO_2 equivalent. 80 percent of the tofu industry is on Java, thus emissions issued by tofu factories in Java reach 0.8 million tons of CO_2 equivalent [39].

Tofu industry waste is generally divided into two forms of waste, solid waste and liquid waste. The solid waste of the tofu processing plant is in the form of soybean cleaning (rock, soil, soybean skin, and other solid objects attached to soybeans) and the remaining soybean crude, while tofu liquid waste is produced from water used for soybean cleaning and during the processing of tofu. Tofu liquid waste is rich in protein, fat, and carbohydrates and high organic compounds. These organic compounds will produce methane (CH_4), carbon dioxide (CO_2), other gases, and water [39].

Solid waste from tofu can be used as an energy source in the bioreactor. Methane gas from tofu solid waste is used as an energy source. Bioreactor or often referred to as a bio digester, which is a dome-shaped gas-tight building that functions to capture methane gas from organic matter. The biogas reactor consists of a digester and a sink. This reactor can be filled continuously with wastewater from industrial tofu production. This reactor includes all the needs to produce methane through anaerobic processes [39].

The raw material for producing SCP must contain water, source N, source C and minerals [10]. Nitrogen can be obtained from tofu waste [40]. This tofu liquid waste which is usually immediately disposed after processing will undergo a process carried out by microorganisms (spontaneously) during disposal. This happens by hydrolyzing organic substances, such as proteins, carbohydrates, and fats which are still high in content [41], so tofu liquid waste can be used as a source of nitrogen for hydrogen-oxidizing microbes.

If all aspects; energy source (electricity), source C, source N obtained from the nature can be applied properly, then Neo-Carbon Food is economical. Neo-Carbon Food contains 54% protein, while meat only contains 20% protein [28,29]. Therefore, Neo-Carbon Food can be an alternative source of protein because of its higher protein content than meat.

This technology can be simplified and implemented in the village. The use of urban organic waste from local villages into biogas and also N sources for NCF can be done. Besides being able to empower local people, it is also source of protein. Even villages that are difficult to reach can also obtain high-protein food sources with independent production without go far to the city so they can produce new food sources through a waste to resource approach.

Based on the explanation above, the technology of Neo-Carbon Food production in Indonesia is still very possible to do by modifying several things such as the energy used and the source of N, vitamins and minerals needed by these SCP-producing. However, further research must be done to determine the optimization of Neo-Carbon Food production so that the results are more optimal and the process used is more effective.

5. Concluding Remark

Neo-Carbon Food could be implemented by using urban organic waste. A Holistic process in the utilization can take advantage of biogas as side process for the energy source. It may provide a solution for environmental impacts of organic waste, such as greenhouse gases, fresh water use, land use, and pollution. So, Neo-Carbon Food can decrease GHG emissions to overcome climate change and also as sustainable food.

6. Acknowlegment

This article was presented at the 2nd International Conference on Smart City INNOVATION (ICSCI) 2019, jointly held by Universitas Indonesia and Universitas Diponegoro. ICSCI conferences have been supported by the United States Agency for International Development (USAID) through the Sustainable Higher Education Research Alliance (SHERA) Project for Universitas Indonesia's Scientific Modelling, Application, Research, and Training for City-centered Innovation and Technology (SMART CITY) Center for Collaborative Research, administered through Grant #AID-497-A-1600004, Sub Grant #IIE-00000078-UI-1

7. References

[1] Nations U 2016 The World's Cities in 2016
[2] Brander M 2012 Greenhouse Gases What Do All These Terms Mean? 2–4
[3] Fuhs G 2009 World Greenhouse Gas Emissions: 2005
[4] Green A 2019 Agricultural Waste and Pollution *Waste* 531–51
[5] Porteous A 1992 *Dictionary of Environmental Science and Technology* (New York: John Wiley and Sons)
[6] Utama G L, Kurnani T B A, Sunardi - and Balia R L 2016 The Isolation and Identification of Stress Tolerance Ethanol-fermenting Yeasts from Mozzarella Cheese Whey *Int. J. Adv. Sci. Eng. Inf. Technol.* **6** 252–7
[7] Utama G L, Kurnani T B A, Sunardi and Balla R L 2017 Reducing Cheese-Making by-Product Disposal Through Ethanol Fermentation and The Utilization of Distillery Waste For Fertilizer *Int. J. GEOMATE* **13** 103–7
[8] Zabochnicka-Swiatek M and Slawik L 2010 (PDF) Bioethanol-Production and Utilization *Arch. Combust.* **30** 237–46
[9] Wijaya B 2008 *Potensi Limbah Cair Tahu untuk Produksi Biogas sebagai Energi Alternative*

(Yogyakarta: UGM)

[10] Tannembaun S R 1971 Single Cell Protein Food for Future *J. Food Technol.* **24** 1962

[11] Neo Carbon Energy 2015 Emission-Free Future Now Available

[12] Adedayo M R, Ajiboye E A, Akintunde J K and Odaibo A 2011 Single Cell Proteins: As Nutritional Enhancer *Adv. Appl. Sci. Res.* **2** 396–409

[13] Nygren L, Ruuskanen V, Lindh T, Ahola J, Pitkänen J-P and Vainikka P 2017 Bioreactor with In Situ Water Electrolysis for Protein Production (Copenhagen: ICE 2017)

[14] Volova T G and Barashkov V A 2010 Characteristics of Proteins Synthesized by Hydrogen-Oxidizing Microorganisms *Appl. Biochem. Microbiol.* **46** 574–9

[15] Volova T G and et al. 2013 Cell Growth and Accumulation of Polyhydroxyalkanoates from CO2 and H2 of a Hydrogen-Oxidizing Bacterium, Cupriavidus eutrophus B-10646 *Bioresour. Technol.* **146** 215–22

[16] Yu J 2014 Bio-Based Products from Solar Energy and Carbon Dioxide *Trends Biotechnol.* **32** 5–10

[17] Matassa S, Boon N and Verstraete W 2015 Resource Recovery from Used Water: The Manufacturing Abilities of Hydrogen-Oxidizing Bacteria *Water Res.* **68** 467–78

[18] Fagerbakke K M, Heldal M and Norland S 1996 Content of Carbon, Nitrogen, Oxygen, Sulfur and Phosphorus in Native Aquatic and Cultured Bacteria *Aquat. Microb. Ecol.* **10** 15–27

[19] Daniell J, Köpke M and Simpson S D 2012 Commercial Biomass Syngas Fermentation *Energies* **5** 5372–417

[20] García I L and et al. 2013 Evaluation of by-Products from the Biodiesel Industry as Fermentation Feedstock for Poly(3-hydroxybutyrate-co-3-hydroxyvalerate) Production by Cupriavidus necator *Bioresour. Technol.* **130** 16–22

[21] Munasinghe P C and Khanal S K 2010 Biomass-derived Syngas Fermentation into Biofuels: Opportunities and Challenges *Bioresour. Technol.* **10** 5013–22

[22] Choi S, Drese J H, Eisenberger P M and Jones C W 2011 Application of Amine-Tethered Solid Sorbents for Direct CO 2 Capture from the Ambient Air *Environ. Sci. Technol.* **45** 2420–7

[23] Rusdiani R R, Boedisantoso R and Hanif M 2016 Optimalisasi Teknologi Fotobioreaktor Mikroalga sebagai Dasar Perencanaan Strategi Mitigasi Gas CO2 *J. Tek. ITS* **5** 188–92

[24] Viebahn P and et al. 2007 Comparison of Carbon Capture and Storage with Renewable Energy Technologies Regarding Structural, Economic, and Ecological Aspects in Germany *Int. J. Greenh. Gas Control* **1** 121–33

[25] Lehner M, Tichler R, Steinmüller H and Koppe M 2014 *Power-to-Gas: Technology and Business Models* (New York: Springer International Publishing)

[26] Chen C L, Chang J S and Lee D J 2015 Dewatering and Drying Methods for Microalgae *Dry. Technol.* **33** 443–54

[27] Sutton M A and Bleeker A 2013 The Shape of Nitrogen to Come *Nature* **494** 435–7

[28] Wada Y and et al. 2016 Modeling global water use for the 21st century: The Water Futures and Solutions (WFaS) initiative and its approaches *Geosci. Model Dev.* **9** 175–222

[29] Hanjra M A and Qureshi M E 2010 Global Water Crisis and Future Food Security in an Era of Climate Change *Food Policy* **35** 365–77

[30] Flickinger M C, Heijnen J J and Kleerebezem R 2010 Bioenergetics of Microbial Growth *Encyclopedia of Industrial Biotechnology* (Delft: John Wiley & Sons Inc.)

[31] Hülsen T, Batstone D J and Keller J 2014 Phototrophic bacteria for nutrient recovery from domestic wastewater *Water Res.* **50** 18–26

[32] Ponsano E, Pinto M, Garcia-Neto M and Lacava P 2005 Performance and Color of Broilers Fed Diets Containing Rhodocyclus Gelatinosus Biomass *Rev. Bras. Ciência Avícola* **6** 237–42

[33] Becker E W 2007 Micro-algae as a source of protein *Biotechnol. Adv.* **25** 207–10

[34] Day L 2013 Proteins from Land Plants - Potential Resources for Human Nutrition and Food Security *Trends Food Sci. Technol.* **32** 25–42

[35] Nijdam D, Rood T and Westhoek H 2012 The price of protein: Review of land use and carbon footprints from life cycle assessments of animal food products and their substitutes *Food Policy* **37** 760–70

[36] Cointreau S J 1982 Environmental Management of Urban Solid Wastes in Developing Countries: A Project Guide *Urban Dev. Dept, World Bank*
[37] Mahan K M and et al. 2018 Production of single cell protein from agro - waste using Rhodococcus opacus *J. Ind. Microbiol. Biotechnol.*
[38] Reay D S, Hewitt C N, Smith K A and Grace J 2007 *Greenhouse gas sinks*
[39] Setiawan A and Rusdjijati R 2014 Peningkatan Kualitas Biogas Limbah Cair Tahu dengan Metode Taguchi *Pros. SNATIF* **1** 35–40
[40] Maryana L, Anam S and Nugrahani A W 2016 Produksi Protein Sel Tunggal dari Kultur Rhizopus oryzae dengan Medium Limbah Cair Tahu *Galen. J. Pharm.* **2** 132–7
[41] Yudhistira B, Andriani M and Utami R 2018 Karakterisasi: Limbah Cair Industi Tahu dengan Koagulan yang Berbeda (Asam Asetat dan Kalium Sulfat) *Caraka Tani J. Sustain. Agric.* **31** 137–45

IOP Conf. Series: Earth and Environmental Science **396** (2019) 012008 doi:10.1088/1755-1315/396/1/012008

Preliminary study on hydrogen storage for fuel of fuel cell using Fe_3Al metal hydride system

D Rohendi[1,2*], A Rachmat[1,2], N Syarif[1,2], M Said[3], I Amelia[2]

[1]*Chemistry Department, Faculty of Mathematic and Natural Sciences, Universitas Sriwijaya, Indralaya, Ogan Ilir, South Sumatera, Indonesia.*
[2] *Center of Research Excellent in Fuel Cell and Hydrogen, Universitas Sriwijaya, Palembang, South Sumatera, Indonesia.*

E-mail: rohendi19@unsri.ac.id

Abstract. Fe_3Al alloy for hydrogen storage was successfully synthesized from Fe (70 wt%) and Al (30 wt%) through ball-mill method in 5 hours milling and Ball to Powder Ratio (BPR) 10:1 to produced mechanical alloy. The results was confirmed by XRD diffraction pattern at 2θ = 44.791° (10) which is characteristic peak of Fe_3Al. The material was applied on hydrogen storage to form metal hydride in which at various temperature and pressure. The Fe_3Al alloy performance for hydrogen storage for fuel cell application was indicated from the capacity of hydrogen adsorption and the ability to supply hydrogen for fuel cell. Characterization by using SEM-EDX on adsorption used alloy showed that Fe_3Al Alloy has homogeneous surface and no cracks revealed. XRD pattern of alloy after being used to adsorb hydrogen showed a typical pattern at 2θ = 44.667° (11). Based on the results it can be concluded that hydrogen storage capacity increases with increasing temperature and pressure, and the average adsorption capacity were 58%. Hydrogen storage by the metal hydride method has met the storage criteria established by the US Department of Energy (DOE) as clean and efficient urban energy for the future.

Keywords: metal hydride, mechanical alloying, fe_3al, adsorption.

1. Introduction

Energy is a basic need of the world today for all residential, commercial, industrial and transportation activities [1] but the energy used today, most use fossil fuels that cannot be renewed. High population growth impacts on increased energy consumption. This causes fossil fuels to thin out and cannot meet human needs in the future [2], [3] so that sustainable alternative energy is needed to replace the role of fossil fuels [2]. On the other hand fossil fuels also have a negative impact on the environment caused by exhaust emissions in vehicles in the form of CO [4].

Hydrogen as an element whose abundance is very high in nature and simplest in the periodic table, with density (in the form of gas) of 0.0899 kg/Nm^3, which is 15 times lighter than air. Hydrogen is a fuel with a wide range of flammability, both in the air, from 4-75%, and in oxygen, from 4-95%. Hydrogen is also a high energy densitiy (per mass unit) with Higher Heating Value (HHV) of 3.54

kWh/Nm3 (39.42 kWh/kg), which is 2.5 times higher than methane and 3 times higher than gasoline [5]. These properties make hydrogen have the potential as a very promising alternative fuel source for the future and become one of the right solutions in overcoming fuel availability problems [6].

However, one of the limitations of exploiting hydrogen is the need for economically safe, safe and comfortable storage media because hydrogen is highly reactive to direct air contact [7]. The US Department of Energy (DOE) in 2017 sets out five main factors (such as gravimetric and volumetric capacity, operating temperature, system life cycle and filling time) that need to be considered as efficient hydrogen storage [8]. The three main hydrogen storage methods such as compressed hydrogen gas (CGH_2 and CcH_2), liquid hydrogen and solid storage (MH_2, such as metal hydrides) which have attracted many researchers since they can be operated at low temperature and pressure [9]. The MH_2 system is considered promising with the hydrogen molecule chemisorption tightly bound into the material lattice and by physical adsorption on the material surface [8] metal hydride has a high hydrogen (H_2) storage density compared to hydrogen or liquid hydrogen [9], [10]. The other methods of hydrogen storage are in formic acid and methanol [11], carbon nanostructure [12] and lithium decorated zeolite [13].

Fe and Al metals are very abundant metals on earth and the price is affordable. Fe and Al metals are atoms which have valence electrons 3+. This condition is very interesting to study the ability to bind hydrogen atoms to form metal hydrides. Based on these characteristics, a study of the hydrogen gas storage and release process was carried out using the metal hydride method using Fe-Al metal alloys.

2. Methods

Fe-Al alloy is synthesized using mechanical alloying method as much as 2.5 grams of mixture of Fe (iron) (99.99% purity) and Aluminum (99.99% purity) with a percentage ratio of Fe:Al weight is 70:30. The tool used in Mechanical Alloying is high energy milling type 3D elliptical milling machine (HEM- E3D). The mechanical alloying process is carried out with BPR (Ball to Powder Ratio), which is 10: 1 and using the PCA (Process Control Agent) of methanol. The mechanical alloying process is carried out by milling for 5 hours [14]. After Fe_3Al alloy was obtained, characterization was then performed to analyze the structure, and morphology of the particle using X-Ray Diffraction (XRD) and SEM-EDX. Storing hydrogen were performed by weighing the empty tube that has been vacuumed, and then the tube is filled with Fe_3Al alloy and then vacuumed again and weighed. Hydrogen gas is flowed into a vial containing Fe_3Al alloy at a temperature of 30°C, 40°C, 50°C. After the hydrogen flow process, the gas is released. The procedure is also done for pressure variations 2 and 3 bars.

3. Result and discussion

3.1 Manufacture of Fe-Al alloy and results of the analysis using XRD

The manufacture of Fe-Al Alloy using mechanical alloying method with HEM-E3D type ball milling tool that utilizes collisions between balls. The mechanical alloying process is influenced by several parameters such as BPR ratio, ball size milling time. BPR is a ratio of the weight of the ball and the weight of the material adjusted to the valume jar at least <50% or <60%. In the manufacture of Fe-Al alloys a ratio of 10:1 is used. The size of the ball used also gives a significant effect because the smaller the size of the ball, the greater the surface area that is 5mm in diameter for Fe (10-40 mesh) and alumunium metal (<1 mm) is used. While the milling time it self-based on the structure of the material to be used such as Fe and Al metal alloys takes <30 hours [15].

The making of metal alloys was carried out with a mixture of Fe 70 wt% and Al metal 30 wt% at room temperature and 5 hour milling time. In the process of mechanical milling, 2 mL of methanol is added as a process controlling agent (PCA) in the liquid phase which is adsorbed on the metal surface which causes a reduction in energy to control the milling process to run properly [16]. The results of milling Fe-Al metal alloys were obtained in the form of fine black powder as seen in Figure 1.

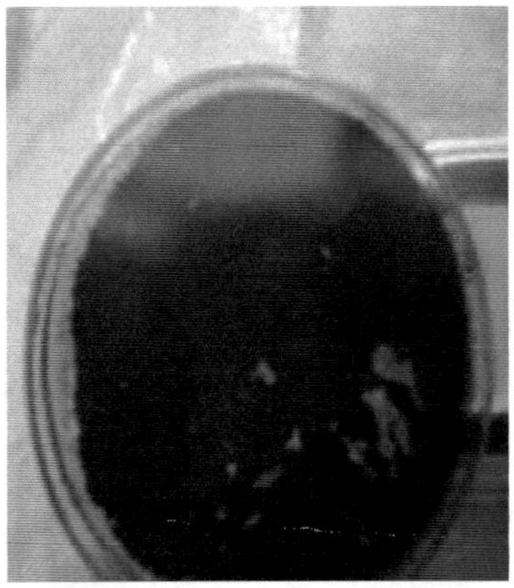

Figure 1. The Fe-Al metal milling results

The next process after milling is heating sample using oven vacuum at 0.4 bar and 70°C for 2 hours to evaporate methanol. Mass of alloy after heating was 35.3788 g. The results of metal alloys were characterized by XRD analysis as seen in Figure 2.

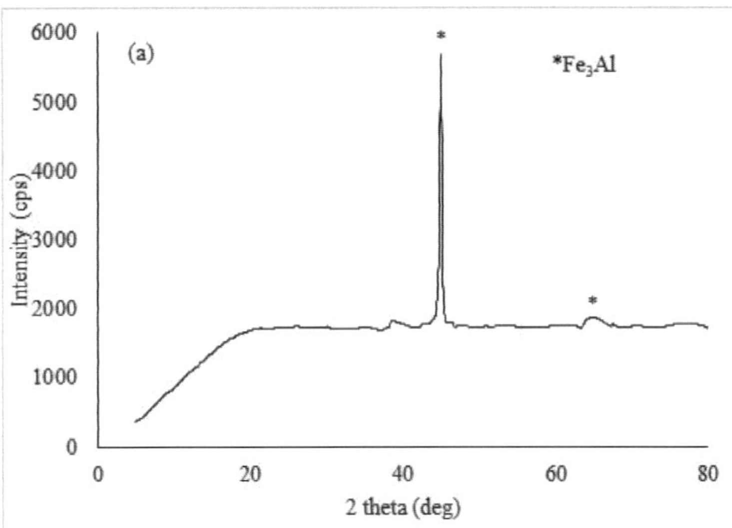

Figure 2. X-ray diffraction pattern of Fe-Al metal alloy powder with a milling time of 5 hours.

X-ray diffraction patterns for Fe-Al alloys in Figure 2 have a similar pattern with the report before [17]. This value is also in accordance with the value JCPDS No. 06-0695 (Fe3Al) [18], [19] and similar values were also shown in the study of Wei et al (2017) [20], the diffraction peaks that are characteristic of Fe_3Al alloys are at diffraction angles around 37 (42-3), 45 (110), 66 (200), 83 (211). However, in the synthesized alloy at this time, the diffractogram peak only appears at diffraction angles around 37, 45 and 66 for a 5 hour milling time having a sharp peak with high intensity with a crystal size of 31.549 nm.

3.2 The SEM-EDX analysis on Fe_3Al alloy after hydrogen adsorption

According to the Scanning Electron Microscope (SEM) and Energy Dispersive X-Ray (EDX) test results for Fe_3Al metal alloys after adsorption of hydrogen, can be seen that iron and aluminum have been evenly mixed. In Figure 3, there are other atoms such as carbon atoms and oxygen where they come from the remaining methanol used as PCA in the milling process.

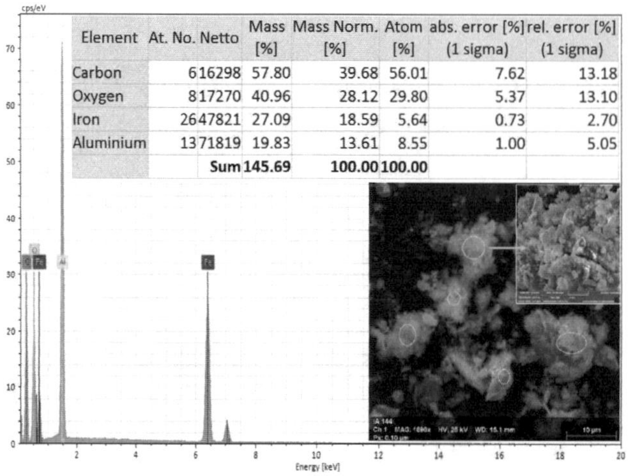

Element	At. No.	Netto	Mass [%]	Mass Norm. [%]	Atom [%]	abs. error [%] (1 sigma)	rel. error [%] (1 sigma)
Carbon		616298	57.80	39.68	56.01	7.62	13.18
Oxygen		817270	40.96	28.12	29.80	5.37	13.10
Iron		2647821	27.09	18.59	5.64	0.73	2.70
Aluminium		1371819	19.83	13.61	8.55	1.00	5.05
		Sum	**145.69**	**100.00**	**100.00**		

Figure 3. SEM and EDX analysis results on Fe_3Al alloy after the process of adsorption of hydrogen

The percentage of iron and aluminum atoms looks lower than carbon and oxygen atoms. This shows that most of the Fe and Al have changed to form Alloy. The percentage of Fe and Al mass detected from EDX did not match the actual percentage of Fe and Al (70:30). This is because the data on EDX only measures the spot from the surface. The higher percentage of carbon and oxygen indicates the presence of non-volatile methanol which partially covers the surface of the Fe_3Al alloy. Figure 3 shows a magnification of 20.20 Kx having a pore as large as it can be concluded that Fe_3Al alloy has a pore size of < 0.2 μm. The Fe_3Al alloy has a rough and thick surface that is not uniform. In Figure 3 there were also no cracks or cracks on the surface area of the Fe_3Al alloy. This indicates a strong bond between Fe-Al metals.

3.3 Qualitative analysis of adsorption of hydrogen from Fe-Al alloy and XRD characterization results for metal hydride

Initially hydrogen gas is flowed into the reactor tube containing Fe-Al alloy for 60 seconds with variations in temperatures of 30°C, 40°C, 50°C and variations in pressure 1, 2, 3 bars. Hydriding a metal is carryout by direct dissociative chemisorption and physisorption. Solid state storage of hydrogen fuel occurs absorption in the gap metals and metal alloys or adsorption at surface areas metal [10] so then hydrogen adsorption capacity was measured. In process of hydrogen adsorption in the Fe-Al alloy, hydriding process it turns into metal hydride (FeH, FeH_2, FeH_3 or AlH_3) [21]. A typical reversible interaction between metal/alloy and hydrogen can be written:

$$M + \frac{x}{2} H_2 \longrightarrow MH_x + Q$$

Where M is a metal, an intermetallic alloy, or a solid solution, x is a positive whole number, MHx is the formed hydride and Q is the heat of reaction [22]. The condition and shape of metal hydride before and after hydrogen absorption is characterized using XRD as shown in Figure 4. In Figure 4 it can be seen that the typical peak after hydrogen gas adsorption process is almost the same as the diffraction pattern before the adsorption of hydrogen gas.

Before the adsorption process obtained a typical peak in the diffraction pattern 44.791 (10) and after the process of adsorption of hydrogen gas obtained a typical peak on the diffraction pattern 44.667 (11) with a crystal size of 31.209 nm. This is because the bond between metals and hydrogen is not a chemical bond which is only bound as a physical bond so that the bond is easily released and the shift in the diffraction pattern is unreadable.

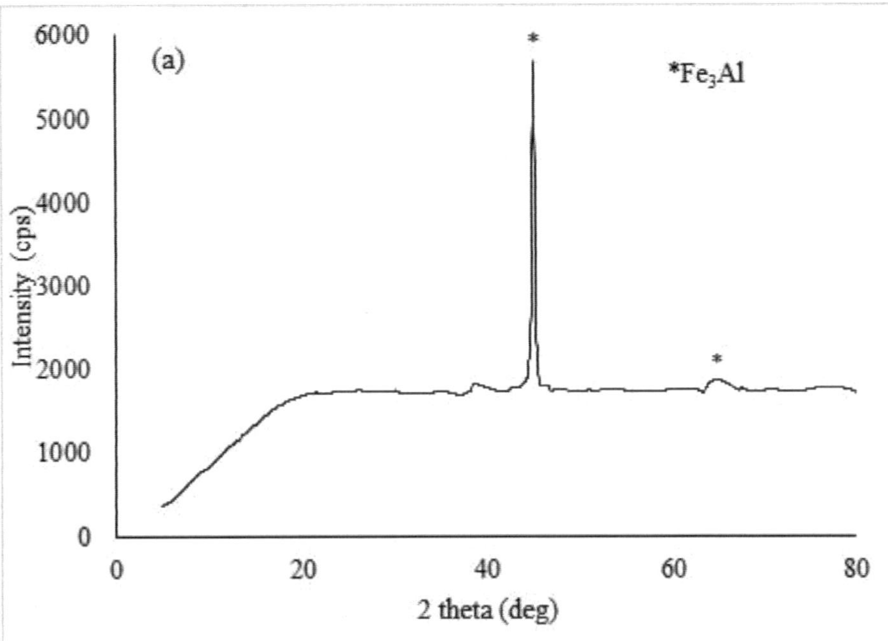

(a)

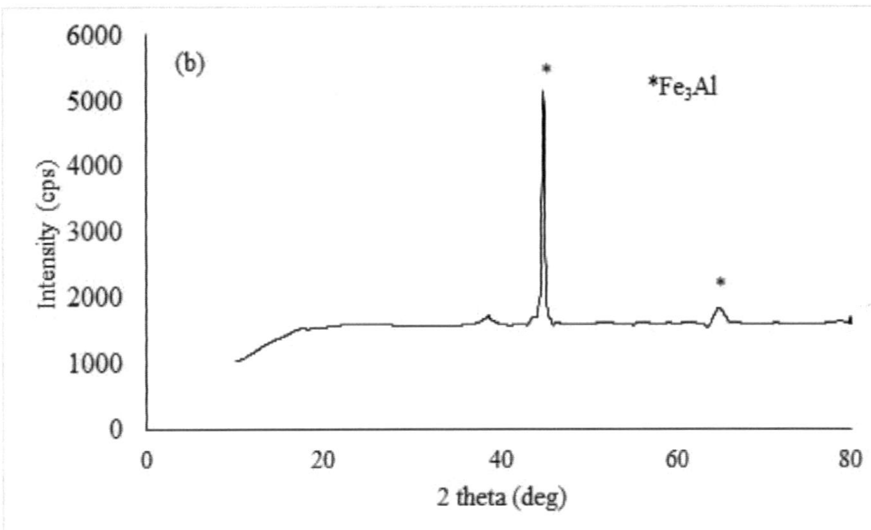

(b)

Figure 4. X-ray diffraction pattern in Fe_3Al alloy (a) before (b) after the process of adsorption of hydrogen gas

3.4 Determination of hydrogen storage capacity from Fe-Al/H_2 metal hydride

Determination of hydrogen adsorption capacity in the Fe-Al alloy system is carried out at various temperatures and pressures. Based on Table 1, it can be seen that in general, hydrogen adsorption capacity increases with increasing temperature and pressure. However, the adsorption of hydrogen in the initial conditions with a temperature of 30°C and pressure of 1 bar has a greater capacity. This is because the alloy material is still new so that hydrogen is absorbed maximally. With repeated absorption and release of hydrogen, hydrogen absorption capacity decreases because there is hydrogen retained in alloy materials. At the higher the temperature and pressure the storage capacity increases, because at the higher temperature and pressure can open the pores so that more hydrogen gas can be stored and released again.

IOP Conf. Series: Earth and Environmental Science **396** (2019) 012008 doi:10.1088/1755-1315/396/1/012008

Table 1. Results of measurement of hydrogen gas storage capacity on effects of temperature and pressure.

No	Pressure (Bar)	Temperature (°C)	Adsorption capacity
1.	1	30	60%
2.	1	40	75%
3.	1	50	56%
4.	2	30	52%
5.	2	40	55%
6.	2	50	60%
7.	3	30	38%
8.	3	40	52%
9.	3	50	78%

The relationship between temperature and pressure on hydrogen adsorption capacity of Fe-Al alloy is shown in Figure 5 and 6. From Figure 5 and 6 can generally be seen that the average storage capacity increases with increasing temperature and pressure with the average storage is 58 wt%.

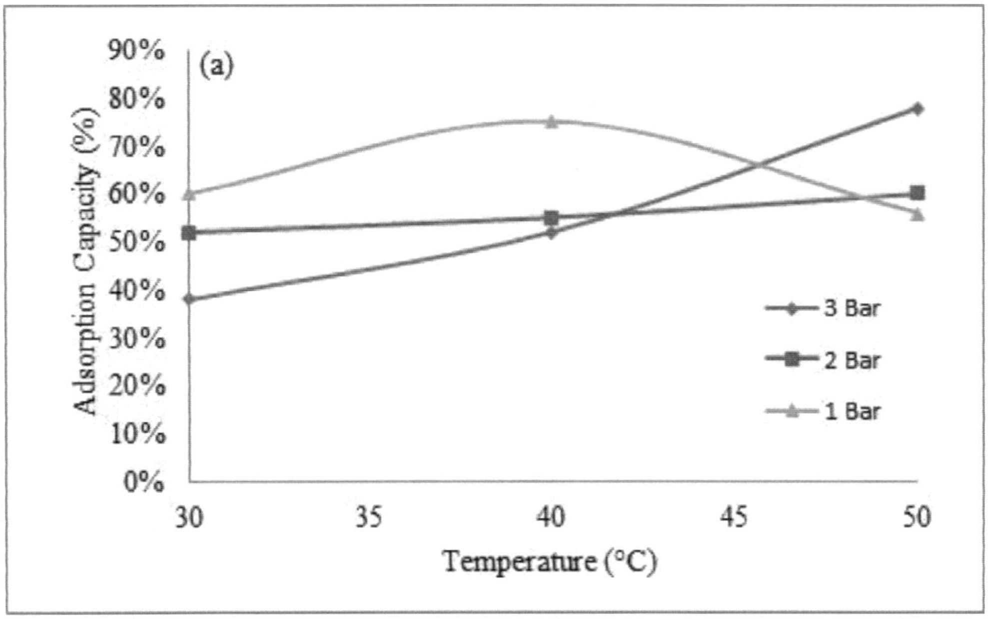

Figure 5. Graph of influence of temperature on adsorption capacity of hydrogen

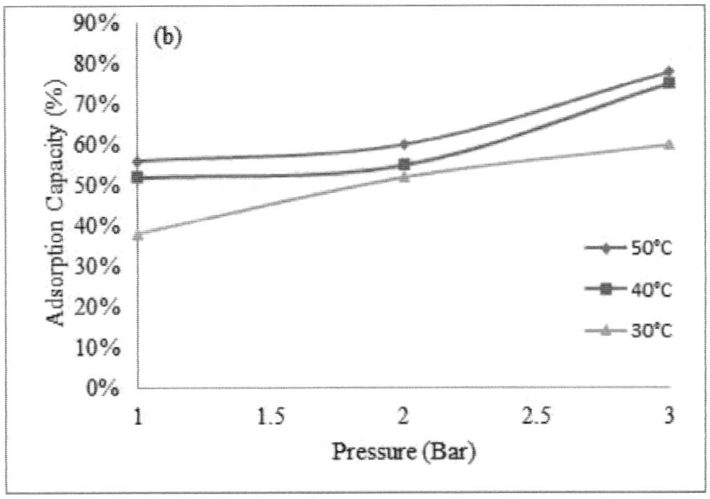

Figure 6. Graph of influence of pressure on adsorption capacity of hydrogen.

4. Conclusions

a. The Fe_3Al alloy was successfully synthesized with an alloy of Fe 70% b/b and Al 30% b/b with a milling time of 5 hours and Ball to Powder Ratio (BPR) 10: 1 and a flame time of 3 minutes and 1 minute rest time. Based on the results of XRD characterization it is known that the typical diffraction pattern at 44.79° shows a typical peak for Fe_3Al.

b. Performance of Fe_3Al alloy as hydrogen storage and release can be seen from the hydrogen storage capacity. In general, hydrogen storage capacity increases with increasing temperature and pressure. Based on the results of the study, the average Fe_3Al alloy has a hydrogen adsorption capacity of 58 wt%.

5. Acknowledgements

This article was presented at the 2nd International Conference on Smart City INNOVATION (ICSCI) 2019, jointly held by Universitas Indonesia and Universitas Diponegoro. ICSCI conferences have been supported by the United States Agency for International Development (USAID) through the Sustainable Higher Education Research Alliance (SHERA) Project for Universitas Indonesia's Scientific Modelling, Application, Research, and Training for City-centered Innovation and Technology (SMART CITY) Center for Collaborative Research, administered through Grant #AID-497-A-1600004, Sub Grant #IIE-00000078-UI-1

6. References

[1] Kaur M and Pal K 2019 Review on hydrogen storage materials and methods from an electrochemical viewpoint *J. Energy Storage* **23** 234–49

[2] Yadav T P, Yadav R M and Singh D P 2012 Mechanical Milling: a Top Down Approach for the Synthesis of Nanomaterials and Nanocomposites *Nanosci. Nanotechnol.* **2** 22–48

[3] Zhang Y, Campana P E, Lundblad A and Yan J 2017 Comparative study of hydrogen storage and battery storage in grid connected photovoltaic system: Storage sizing and rule-based operation *Appl. Energy* **201** 397–411

[4] Mubin A N A, Bahrom M H, Azri M, Ibrahim Z, Rahim N A and Raihan S R S 2017 Analysis performance of proton exchange membrane fuel cell (PEMFC) *IOP Conf. Ser. Mater. Sci. Eng.* **210** 12052

[5] K. R, R. M and S. L The first comprehensive survey of solar hydrogen generation: Reviews new and established science using solar energy and renewable sources Examines evolving technological and economic trends *springer* **49** 6221

[6] Rosen M A 2010 Advances in hydrogen production by thermochemical water decomposition: A review *Energy* **35** 1068–76

[7] Stojić D L, Kumrić S V, Grozdić T D, Koteski V J and Cekić B \DJ. 2009 Hydrogen production and storage-Investigation of Hf-based intermetallics *J. Power Sources* **193** 165–9

[8] Yanxing Z, Maoqiong G, Yuan Z, Xueqiang D and Jun S 2019 Thermodynamics analysis of hydrogen storage based on compressed gaseous hydrogen, liquid hydrogen and cryo-compressed hydrogen *Int. J. Hydrogen Energy* **44** 16833–40

[9] Andersson J and Grönkvist S 2019 Large-scale storage of hydrogen *Int. J. Hydrogen Energy* **44** 11901–19
[10] Sakintuna B, Lamaridarkrim F and Hirscher M 2007 Metal hydride materials for solid hydrogen storage: A review *Int. J. Hydrogen Energy* **32** 1121–40
[11] Onishi N, Laurenczy G, Beller M and Himeda Y 2018 Recent progress for reversible homogeneous catalytic hydrogen storage in formic acid and in methanol *Coord. Chem. Rev.* **373** 317–32
[12] Liu W, Setijadi E, Crema L, Bartali R, Laidani N, Aguey-Zinsou K F and Speranza G 2018 Carbon nanostructures/Mg hybrid materials for hydrogen storage *Diam. Relat. Mater.* **82** 19–24
[13] Isidro-Ortega F J, Pacheco-Sánchez J H and Desales-Guzmán L A 2017 Hydrogen storage on lithium decorated zeolite templated carbon, DFT study *Int. J. Hydrogen Energy* **42** 30704–17
[14] Wang J, Xing J, Cao L, Su W and Gao Y 2010 Dry sliding wear behavior of Fe3Al alloys prepared by mechanical alloying and plasma activated sintering *Wear* **268** 473–80
[15] Cardellini F, Contini V, Gupta R, Mazzone G, Montone A, Perin A and Principi G 1998 Microstructural evolution of Al-Fe powder mixtures during high-energy ball milling *J. Mater. Sci.* **33** 2519–27
[16] Shervani S, Mukherjee P, Gupta A, Mishra G, Illath K, Ajithkumar T G, Sivakumar S, Sen P, Balani K and Subramaniam A 2017 Multi-mode hydrogen storage in nanocontainers *Int. J. Hydrogen Energy* **42** 24256–62
[17] Erningpraja L and Drajat B 2006 Biodiesel berbahan baku minyak kelapa sawit *War. Penelit. dan Pengemb. Pertan.* **28** 1–3
[18] Yajiang L I, Juan W, Yonglan Z and Holly X 2002 Fine structures in Fe3Al alloy layer of a new hot dip aluminized steel *Bull. Mater. Sci.* **25** 635–9
[19] Qi N, Hu M, Wang Z, Lu Z and Xie C 2013 Synthesis of Al/Fe3Al core-shell intermetallic nanoparticles by chemical liquid deposition method *Adv. Powder Technol.* **24** 926–31
[20] Wei T Y, Lim K L, Tseng Y S and Chan S L I 2017 A review on the characterization of hydrogen in hydrogen storage materials *Renew. Sustain. Energy Rev.* **79** 1122–33
[21] Dewa M D K, Wiryolukito S and Suwarno H 2015 Hydrogen Absorption Capacity of Fe-Ti-Al Alloy Prepared by High Energy Ball Milling *Energy Procedia* **68** 318–25
[22] Abe J O, Popoola A P I, Ajenifuja E and Popoola O M 2019 Hydrogen energy, economy and storage: Review and recommendation *Int. J. Hydrogen Energy* **44** 15072–86

The Production of Hydrogen from Aluminum Waste by Aluminum-Water Methods at Various Conditions

D Rohendi[1,2*], A Rachmat[1,2], N Syarif[1,2], M Said[3], I Rihsansah[1]

[1] Chemistry Department, Faculty of Mathematics and Natural Sciences, Universitas Sriwijaya, Indralaya, Ogan Ilir, South Sumatera, Indonesia
[2] Center of Research Excellent in Fuel Cell and Hydrogen, Palembang, South Sumatera, Indonesia
[3] Chemical Engineering, Faculty of Engineering, Universitas Sriwijaya, Indralaya, Ogan Ilir South Sumatera, Indonesia

E-mail: rohendi19@unsri.ac.id

Abstract. The production of hydrogen gas from aluminum waste by Aluminum-water methods at various condition has been successfully performed. The hydrogen production is a very important for support the fuel cell application Hydrogen was produced both in the acidic and base conditions at various temperature and using lithium as the activator. The urgencies in this research were effective and cheaper methods. The highest hydrogen production from reaction 0.3 g aluminum with 30 mL hydrochloric acid 2 M at 50 °C in 10 minutes was 348 mL. Meanwhile, the best condition for aluminum water methods for 0.3 g of Aluminum as reactants, were 3 mL water, 5% w/w lithium content and 100 mesh aluminum particle size. Furthermore, using KOH as a catalyst, gave hydrogen product 670 mL with the rate at 67 mL/min.

Keywords: aluminum-water, aluminum waste, hydrogen, activator, lithium.

1. Introduction

Proton Exchange Membrane Fuel Cell (PEMFC) is electrochemical equipment that can convert the chemical energy in hydrogen as fuel and oxygen as oxidant into electrical energy directly [1]. PEMFC is the fuel cell which has been developed for various benefits, such as: high efficiency, low temperature operation and diverse applications [2,3].

The use of hydrogen gas as a fuel in PEMFC make PEMFC as a producer of clean energy and environmentally friendly. Fuel Cell, especially PEMFC convert hydrogen in the anode and oxygen in the cathode to produce water and electricity with zero-emission. In addition, the use of hydrogen will reduce dependence on fossil fuels [4]. Selection of hydrogen as a fuel has many advantages, among others: hydrogen has the highest energy density in all fuels (122 kJ per kg, about 2.75 times greater than other hydrocarbon fuels), hydrogen is believed to be an effective substitute for gasoline as 9.5 kg of hydrogen is sufficient enough to replace 25 kg of gasoline and hydrogen is the lightest of all elements in the world, and hydrogen is the most abundant chemical element comprising about 75% of the earth's elemental mass [5].

Hydrogen is found in abundance in water (H_2O), gas and oil, coal and biomass. However, the availability of hydrogen is not found to be free in nature. Hydrogen gas must be extracted from compounds in nature. This causes hydrogen gas to become a relatively expensive commodity. Hydrogen can be produced in various ways including water electrolysis [6], steam reforming [7], hydrolysis of NaBH4 [8], converting biomass [9–11] and aluminum-water [12–14].

The production of hydrogen from aluminum is one of the interesting production methods. The Methods include production hydrogen with aluminum water with lithium as activator or using KOH as catalyst.

The principle of aluminum water reaction is reaction of aluminum with water to produce oxide/hydroxide aluminum and hydrogen [15]. In this research, the source of aluminum is from aluminum waste. Reaction of aluminum with water is slow due to the formation of a thin oxide layer and coefficient which covered the surface of aluminum, the reaction of aluminum water is needed additive or catalyst. One of the additives is lithium that has a function as an activator.

The important matter of this research is how to supply the hydrogen as the source of new energy carrier for residential or office in the city. The Hydrogen energy is the clean and clear energy for fuel of fuel cell [16].

2. Methods

A parametric experimental study has been conducted in order to find the influence of different operating factors and conditions on the activated aluminum water reaction, measuring the reaction rate and efficiency. Effects of water/aluminum mass ratio, fraction of activator used, water temperature and pressure.

Batch type experiments were conducted in a glass reactor. The aluminum powder + Lithium activator form ball milling system was put in the reactor first, then N_2 was flowed to the reactor and then water was added and a spontaneous reaction started. Since hydrogen has low solubility in water the amount of hydrogen produced from the reaction was measured by water displacement. The hydrogen produced from the reaction was channeled to a glass container filled with water driving the water out from the container to a cup placed. In some experiments the measurements were conducted with a digital mass flow meter. The temperature of the reaction was measured by a digital thermometer.

The production of hydrogen gas using the aluminum water method begins with cleaning aluminum waste as raw material using KOH. Furthermore, aluminum is mixed with lithium in various comparisons and milled in a ball mill system. The mixture is then filtered using different filter sizes (20, 40 and 100 mesh).

The subsequent hydrogen gas production is carried out in reactors with an aluminum mass of 0.3 g in water volume conditions (1; 1.5; 2; 2.5 and 3 mL), aluminum grain sizes (20, 40 and 100 mesh) and lithium activator masses (3; 4; 5; 6 and 7% Al weight) varied. As a comparison material, aluminum is produced by reacting HCl and KOH at various temperatures.

3. Result and Discussion

3.1. Production of H_2 through reaction of aluminum with HCl 2 M

Aluminum will react spontaneously with HCl solution according to equation 1:

$$2\,Al_{(s)} + 6HCl_{(aq)} \rightarrow 2AlCl_{3(aq)} + 3H_{2(g)} \qquad (1)$$

The volume of hydrogen gas produced at various temperatures is shown in Figure 1.

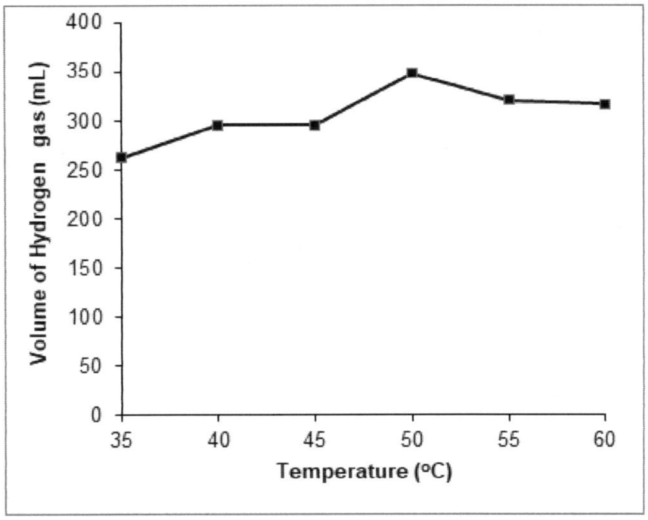

Figure 1. Volume of H_2 from reaction with HCl at various of temperature

The reaction of 0.3 g Al (0.01 mol) with HCl excess will produce 0.0167 mol H_2 stoichiometricaly. At STP conditions, will produce 373.3 mL. Based on the ideal gas equation, volume is directly proportional to the temperature. In this experiment, the highest hydrogen production is 348 mL at 50°C. The higher temperature, the volume of the generated hydrogen gas is increasing until 50°C and than decrease. This is evident from Figure 2.

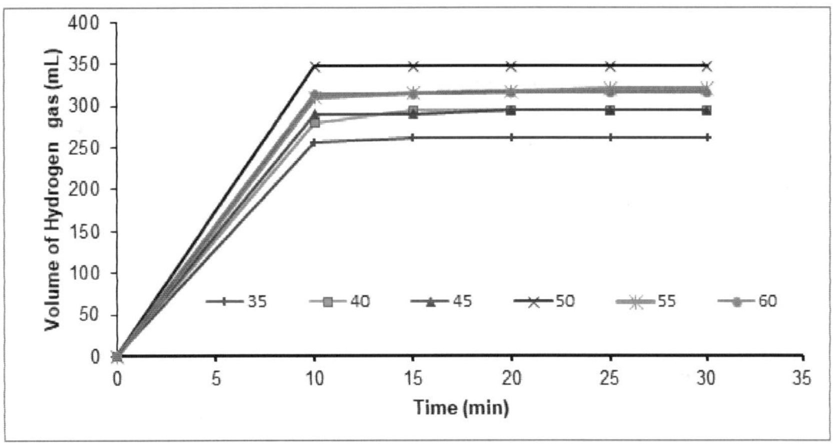

Figure 2. Volume of H_2 from reaction with HCl at various temperature as function of time

Based on Figure 2, hydrogen volumes rose sharply in the first 10 minutes, and thereafter relatively constant. This indicates that the reaction of H_2 formation tends to occur in a span of 10 minutes of the first. It can also be seen from the reaction rate as shown in Figure 3. Figure 3 shows that the reaction rate of H_2 production in the first 10 minutes is the highest rate, and the rate of production of hydrogen at 50°C is the highest rate.

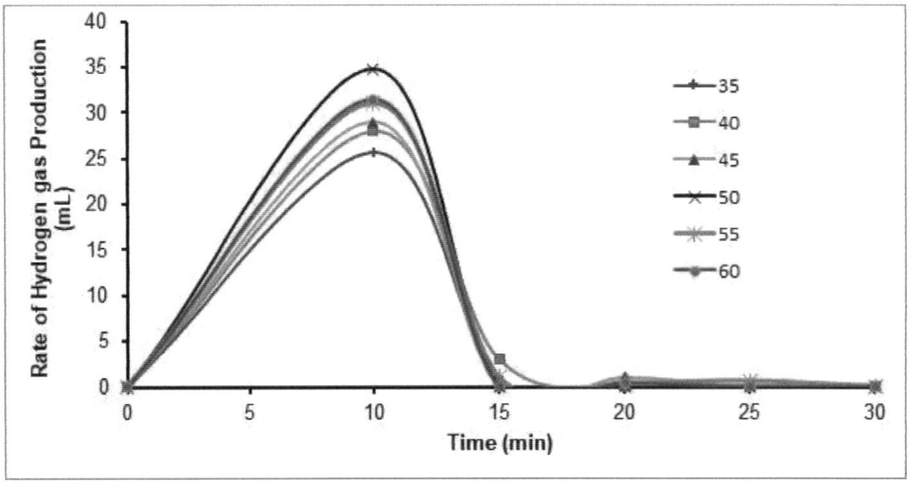

Figure 3. The rate of hydrogen production from reaction aluminum with HCl 2 M

3.2. Production of H_2 through aluminum-water reaction

Aluminum-water method is a method of hydrogen production by using aluminum treated with water. Because the existing coated aluminum oxide, aluminum reacts slowly with water. Therefore, in this study used lithium as an activator and the second method is used also KOH as a catalyst.

The principle of the aluminum water reaction is the reaction of aluminum with water at various trigger conditions / promoter to produce H_2. Stoichiometric reaction of hydrogen gas production with aluminum water method is as follows:

$$2/3Al + 2H_2O \rightarrow 2/3Al(OH)_3 + H_{2(g)} \quad (2)$$

In thermodynamics, reaction of aluminum with water will take place spontaneously. However, the fact that aluminum does not react at room temperature or high temperatures. This is because aluminum is coated by a thin layer of aluminum oxide coherent and stable. Therefore, the presence of promoters intended to damage the aluminum oxide layer and subsequently reacts with water spontaneously.

3.2.1. Production of H_2 with lithium as activator

Hydrogen production using aluminum activator is one method used by utilizing the activity of lithium in the water. In this regard, lithium serves to disrupt/destroy the oxide layer of aluminum so that the aluminum can react quickly. In addition, the presence of lithium diffuses into aluminum structure so as to modify the structure of the oxide / hydroxide of aluminum which causes aluminum surfaces are not protected anymore [17].

The production of H_2 using lithium depends on several conditions, among others, the volume of water used, the percentage litihium and temperature. In this study, factors other than the water volume, the content of lithium and temperature, also studied the effect of different types and sizes of samples of aluminum.

a. Effect of volume of water

The volume of water as a reactant affects the production of H_2. If the volume of water does not reach equilibrium stoichiometry, the volume of H_2 produced will not be maximized. Conversely, if too much volume of water will hydrate alumina (Al_2O_3) resulting from the merger of $Al(OH)_3$ becomes bemit (AlOOH) and further hydrolyzed into bayerit ($Al(OH)_3$) [18]. The presence of bayerit make the reaction of H_2 production is inhibited. This condition can be seen in Figure 4.

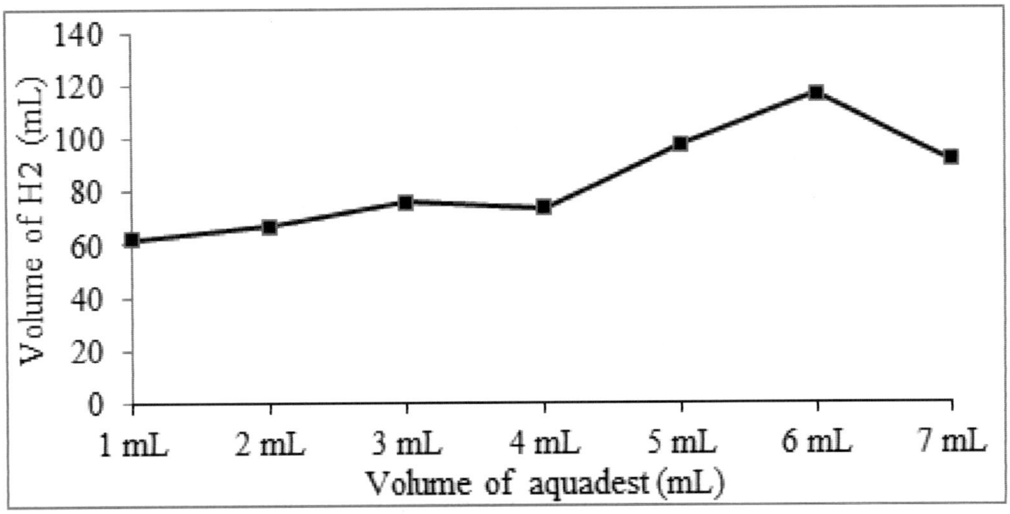

Figure 4. Production of H_2 at various volume of aquadest

b. The effect of kinds and sizes of sample

Type the source of aluminum and aluminum grain size is one that is to be assessed in this study. The influence of the type of source and size of the aluminum particles is presented in Figures 5 and 6.

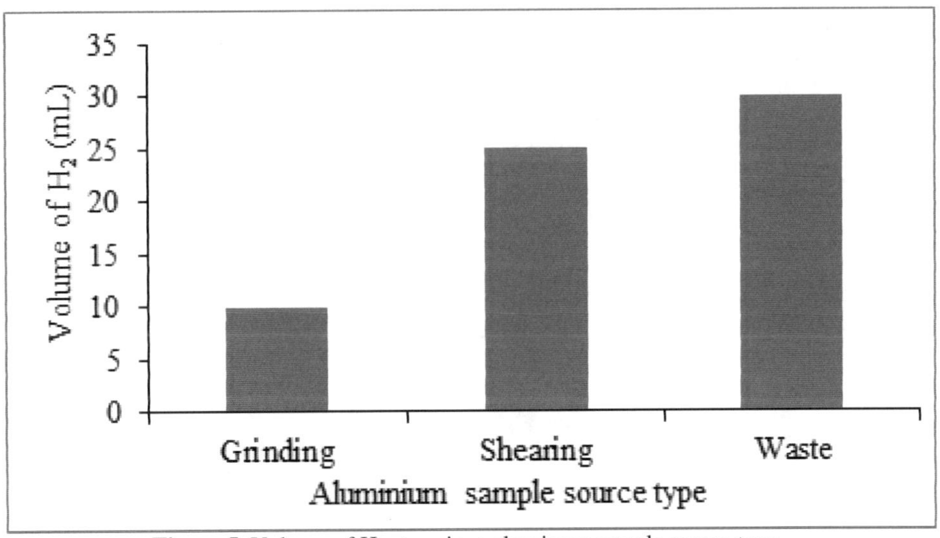

Figure 5. Volume of H_2 at various aluminum sample source type

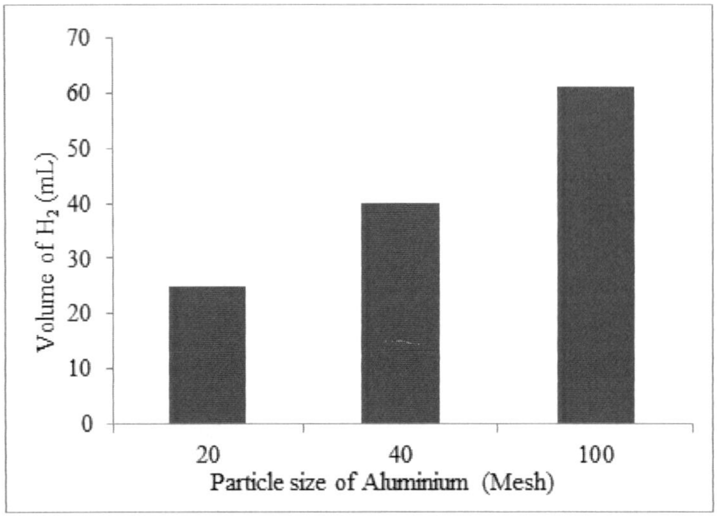

Figure 6. Volume of H_2 at various particle size of aluminum

Aluminum sample types of waste cutting aluminum based image 5 produce more H_2 than the aluminum using grinding and cutting results using scissors. It is associated with a grain size of aluminum from aluminum waste cutting results are more subtle than the result of shearing or cutting with grinding. The smaller the particle size, the greater the surface area and can increase the production of H_2. It is also evident from Figure 6. Aluminum with a grain size of 100 mesh produce more hydrogen than aluminum with a grain size of 20 and 40 mesh.

c. The effect of lithium content

The content of lithium as an activator effect on the productivity of H_2. Based on the Figure 7 and 8 can be seen that the production of H_2 increases with increasing of lithium content up to 5 wt% and then decreases again. According to some literature, effective lithium content is from 1 to 5 wt% [12,17].

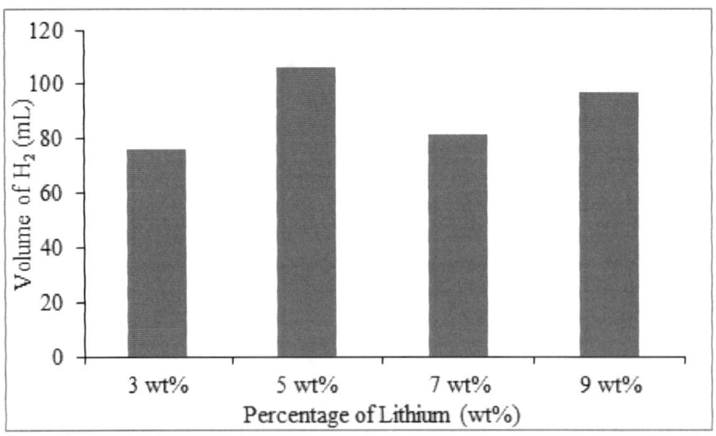

Figure 7. Production of H_2 at various of lithium content

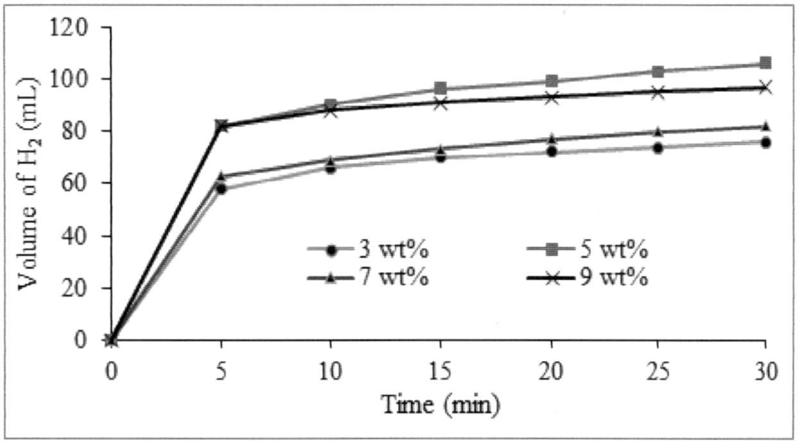

Figure 8. Volume of H_2 at various lithium content as function of time

The H_2 volume increases with increasing lithium content of up to 5 wt% content. A clearer picture can be seen in Figure 8. The lithium content of 5 wt% is the highest lithium content.

d. The effect of temperature

In general, H_2 production will increase with increasing temperature. This is in line with the increase in kinetic energy. Data effect of temperature on H_2 production is presented in Figure 9.

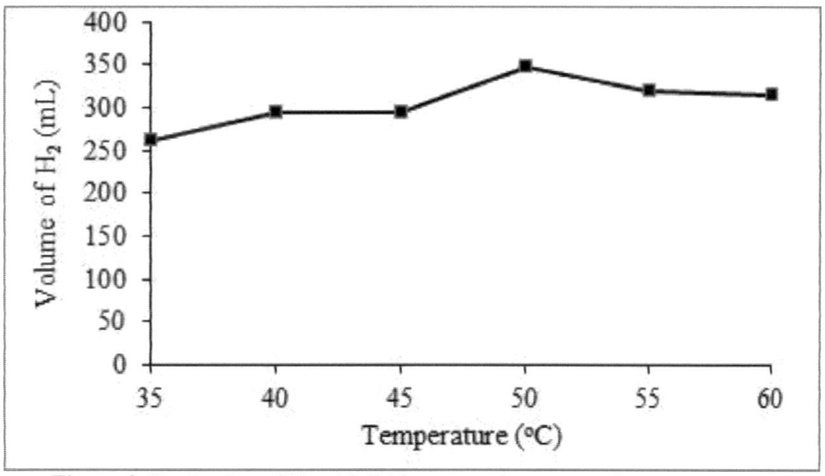

Figure 9. Production of H_2 with lithium as activator at various temperature

Based on the pictures 9 shows that the production of hydrogen increases with increasing temperature up to 50°C.

3.2.2. Production of H_2 with KOH as catalyst

The addition of KOH in H_2 production through the aluminum-water method because KOH can destabilize oxide layer that covers Al. Unlike the HCl, the addition of KOH serves as a catalyst. Taking the analogy of the use of NaOH as catalyst [18], then the conditions that occur in KOH followed the equation:

$$2Al + 6H_2O + 2KOH \rightarrow 2\,KAl(OH)_4 + 3H_2 \quad (3)$$

$$2KAl(OH)_4 \rightarrow 2KOH + 2Al(OH)_3 \quad (4)$$

$$\text{Overall} : 2Al + 6H_2O \rightarrow 2Al(OH)_3 + 3H_2 \quad (5)$$

Because the overall reaction was no KOH, the role of KOH in the above reaction serves as a catalyst.

Data from the reaction of hydrogen gas production using the aluminum water method at various temperatures using KOH as a catalyst are shown in Figures 10 and 11. Based on Figure 10 it can be seen that at 40°C hydrogen production has decreased. This is caused by the inhibition that occurs because the temperature rise of the exothermic spontaneous reaction that occurs in the tube is hampered by the temperature outside [19,20]. Meanwhile, the reaction that occurs at a temperature of 35°C is the reaction in ambient (room) condition. Therefore, the volume of hydrogen gas produced at room temperature is the highest compared to other temperatures.

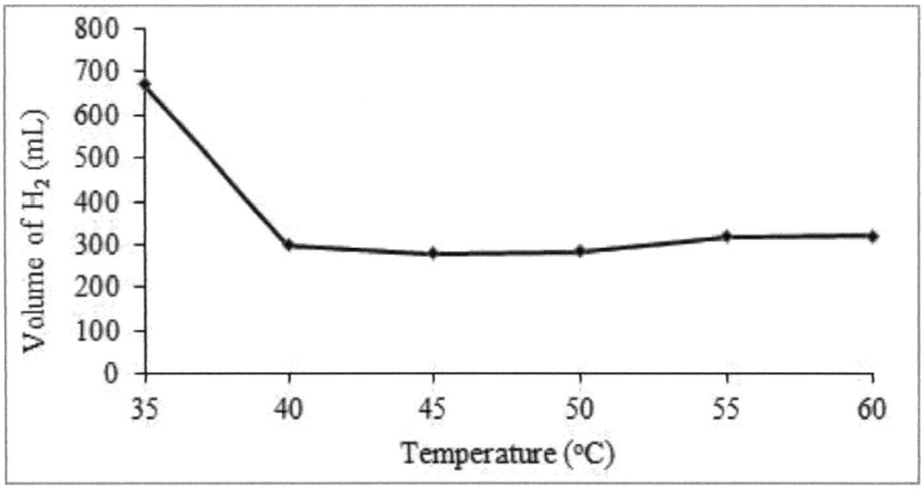

Figure 10. Production of H_2 with KOH as catalyst at various temperature

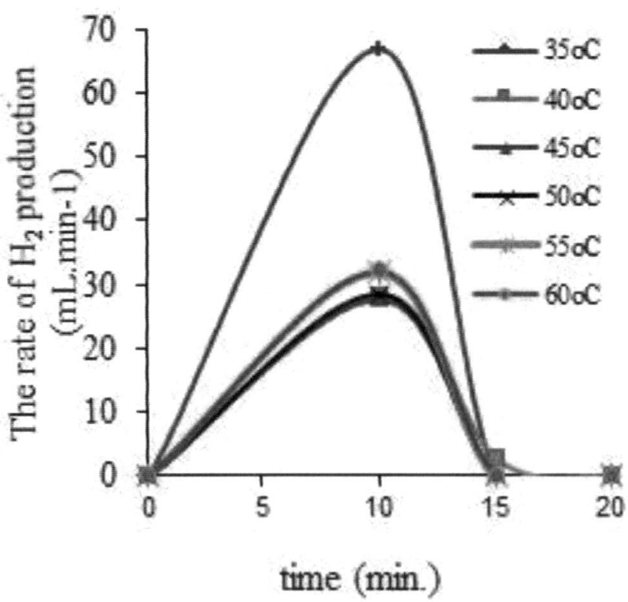

Figure 11. The rate of hydrogen production with KOH as catalyst at various temperature as function of time

Based on Figure 11, it can be seen that the rate of hydrogen gas production increases until the first 10 minutes and after 10 minutes it decreases. This condition shows that the effective reaction rate for hydrogen gas production with KOH occurs in the first 10 minutes. From Figure 11 can also be seen that the formation of hydrogen gas at a temperature of 35°C has the highest reaction rate compared to the reaction rate at other temperatures. This shows that the effective temperature occurs at 35°C with a volume of 670 mL and the reaction rate of 67 mL/minute within 10 minutes.

4. Conclusions

The production of hydrogen by the aluminum water method is one method that is quite effective and inexpensive. The effectiveness of the aluminum water reaction depends on the amount of water added, the mass of the activator and the size of the aluminum particles. In general, the more activators, hydrogen production increases, while the amount of water added must match the reaction stoichiometry.

5. Acknowledgements

This article was presented at the 2nd International Conference on Smart City INNOVATION (ICSCI) 2019, jointly held by Universitas Indonesia and Universitas Diponegoro. ICSCI conferences have been supported by the United States Agency for International Development (USAID) through the Sustainable Higher Education Research Alliance (SHERA) Project for Universitas Indonesia's Scientific Modelling, Application, Research, and Training for City-centered Innovation and Technology (SMART CITY) Center for Collaborative Research, administered through Grant #AID-497-A-1600004, Sub Grant #IIE-00000078-UI-1

6. References

[1] Rohendi D, Majlan E H, Mohamad A B, Daud W R W, Kadhum A A H and Shyuan L K 2013 Characterization of electrodes and performance tests on MEAs with varying platinum content and under various operational conditions *Int. J. Hydrogen Energy* **38** 9431–7

[2] Chao W-K, Lee C-M, Tsai D-C, Chou C-C, Hsueh K-L and Shieu F-S 2008 Improvement of the proton exchange membrane fuel cell (PEMFC) performance at low-humidity conditions by adding hygroscopic $\upgamma$-Al2O3 particles into the catalyst layer *J. Power Sources* **185** 136–42

[3] Iyuke S E, Mohamad A B, Kadhum A A H, Daud W R W and Rachid C 2003 Improved membrane and electrode assemblies for proton exchange membrane fuel cells *J. Power Sources* **114** 195–202

[4] Boggs B K and Botte G G 2009 On-board hydrogen storage and production: An application of ammonia electrolysis *J. Power Sources* **192** 573–81

[5] Li Y, Hu F, Luo L, Xu J, Zhao Z, Zhang Y and Zhao D 2018 Hydrogen storage of casting MgTiNi alloys *Catal. Today* **318** 103–6

[6] Schmidt O, Gambhir A, Staffell I, Hawkes A, Nelson J and Few S 2017 Future cost and performance of water electrolysis: An expert elicitation study *Int. J. Hydrogen Energy* **42** 30470–92

[7] Gao N, Liu S, Han Y, Xing C and Li A 2015 Steam reforming of biomass tar for hydrogen production over NiO/ceramic foam catalyst *Int. J. Hydrogen Energy* **40** 7983–90

[8] Chen W, Ouyang L Z, Liu J W, Yao X D, Wang H, Liu Z W and Zhu M 2017 Hydrolysis and regeneration of sodium borohydride (NaBH 4- A combination of hydrogen production and storage *J. Power Sources* **359** 400–7

[9] Deniz I, Vardar-Sukan F, Yüksel M, Saglam M, Ballice L and Yesil-Celiktas O 2015 Hydrogen production from marine biomass by hydrothermal gasification *Energy Convers. Manag.* **96** 124–30

[10] McNeil L E, Hanuska A R and French R H 2001 Orientation dependence in near-field scattering from TiO_2 particles *Appl. Opt.* **40** 3726

[11] Zhang B, Zhang L, Yang Z, Yan Y, Pu G and Guo M 2015 Hydrogen-rich gas production from wet biomass steam gasification with CaO/MgO *Int. J. Hydrogen Energy* **40** 8816–23

[12] Rosenband V and Gany A 2010 Application of activated aluminum powder for generation of hydrogen from water *Int. J. Hydrogen Energy* **35** 10898–904

[13] Irankhah A, Fattahi S M S and Salem M 2018 Hydrogen generation using activated aluminum/water reaction *Int. J. Hydrogen Energy* **43** 15739–48

[14] Zahedi S, Solera R, Garc'\ia-Morales J L and Sales D 2016 Effect of the addition of glycerol on hydrogen production from industrial municipal solid waste *Fuel* **180** 343–7

[15] Tasic G S, Miljanic S S, Kaninski M P M, Saponjic D P and Nikolic V M 2009 Non-noble metal catalyst for a future Pt free PEMFC *Electrochem. commun.* **11** 2097–100

[16] Magro C, Almeida J, Paz-Garcia J M, Mateus E P and Ribeiro A B 2019 Exploring hydrogen

production for self-energy generation in electroremediation: A proof of concept *Appl. Energy* **255** 113839

[17] Elitzur S, Rosenband V and Gany A 2014 Study of hydrogen production and storage based on aluminum-water reaction *Int. J. Hydrogen Energy* **39** 6328–34

[18] Shmelev V, Nikolaev V, Lee J H and Yim C 2016 Hydrogen production by reaction of aluminum with water *Int. J. Hydrogen Energy* **41** 16664–73

[19] Urbonavicius M, Varnagiris S, Girdzevicius D and Milcius D 2017 Hydrogen generation based on aluminum-water reaction for fuel cell applications *Energy Procedia* **128** 114–20

[20] Wang H Z, Leung D Y C, Leung M K H and Ni M 2009 A review on hydrogen production using aluminum and aluminum alloys *Renew. Sustain. Energy Rev.* **13** 845–53

IOP Conf. Series: Earth and Environmental Science **396** (2019) 012010 doi:10.1088/1755-1315/396/1/012010

The Role of Food Delivery Services by Online Motorcycle Taxi on the Development of Culinary Home-Based Enterprises

W P Tyas, M Damayanti, J K Hutama, D D Saragih,

Department of Urban and Regional Planning, Engineering Faculty, Diponegoro University, Semarang, Indonesia

Email: w.p.tyas@pwk.undip.ac.id

Abstract. The existence of home-based enterprises (HBEs) as the activities that utilize house as a place both to live and to generate income has been challenged by competition among them. The enterprises can take the advantages of technological advances to overcome this issue. One of the technologies that support the development of HBEs is online motorcycle taxi as a part of innovation in the urban infrastructure. The research case study area is around the area of UNDIP Tembalang-Semarang that is known as an education and housing area with many culinary businesses supplying the needs of students and the local community. Hence, this study takes a case of culinary HBEs that benefits from the existence of food delivery services provided by online motorcycle taxi. This study aims to examine the role of food delivery services provided by online motorcycle taxi in the development of culinary HBEs around the education area. This research uses both quantitative and qualitative descriptive method. The roles of this online services are examined from the perspective of culinary HBEs owners, online motorcycle taxi riders, and the consumers. The result is that there are several roles of online motorcycle taxi services namely maximizing business potential, helping business promotion activities, and providing easy access to customers, can be seen from the increasing income offline and online transactions and also the increasing service coverage of the culinary HBEs.

Keywords: *home-based enterprises, technology, online delivery service, infrastructure*

1. Introduction

One of the activities that support economic of Indonesia is the Micro, Small and Medium Enterprises (MSMEs). The existence of MSMEs plays an important role in resolving various development challenges, including unemployment and poverty alleviation [1]. Almost 90% of all businesses in the world are classified as MSMEs [2]. Only by utilizing the existing living space [3], these business activities are able to run and contribute to household income [4–6]. Even more MSMEs need to be able to innovate and compete in the dynamic of international world. It needs full support from the government, big business efforts and the community to boost the development of MSMEs to be competitive. Household businesses that are mostly classified as home-based enterprises (HBEs) which uses house or plots for income generating [7]. In this digital era, therefore using internet-technology in home-based HBE activities are crucial to survive and compete to others.

The technology is one of crucial infrastructure to be able to compete among the entrepreneurs. The impact of this technological progress has also been felt significantly by various fields including in the business world such as MSMEs [8]. Advances in internet-technology have encouraged HBEs to take part in adjusting to the changing world. This progress provides new opportunities for entrepreneurs to develop their businesses. Internet support creates an easy online buying and selling activities for SMEs to develop their businesses [9]. This support will facilitate the product distribution process and increase effectiveness in the business.

In the developed countries, the research of home based business have been research since years ago [10–12]. However, Indonesia as part developing countries, the study of internet role for home based enterprises just started [13–15].

The development of technology, as part of urban infrastructure, led to the era of online motorcycle taxi to help the growth of HBEs with food delivery services that can serve the consumers. Currently developing online motorcycle taxi or motorcycle with food delivery services in Indonesia are managed by companies such as Gojek and Grab in several major cities in Indonesia, one of them is Semarang. With this service, HBEs no longer must struggle to deliver their products to consumers or buyers. This efficiency and easiness are factors that influence the mushrooming of Go Food or Grab food.

The HBEs are considered to be able to improve the welfare of the local community as HBE actors and to be able to supply their daily needs. In addition, Semarang has also been served by technology in the form of an online motorcycle taxi, which is considered capable of supporting the development of HBEs with food delivery services. By relying on economic growth and technological advancements in urban development, urban planners need to involve HBE activities in the urban development planning [16].

The southern part of Semarang, which is around the UNDIP Tembalang area, has many HBEs. The development of HBEs in UNDIP Tembalang area due to the presence of Diponegoro University. The emergence of HBEs are expected due to the large number of migrant students living in the around UNDIP Tembalang area. The large number of needs to supply daily needs makes surrounding residents have the opportunity to open a variety of new businesses. The business opportunities that have been carried out start from opening a business in the form of a grocery shop, and a food business.

Therefore, this study aims to look at the role of online motorcycle taxi for food delivery services in the development of HBEs. This was done by looking at the distribution and condition of HBEs at the study location, then do a comparison of the conditions of home-based SMEs before and after using online motorcycle taxi for food delivery services. In the future, it is hoped that the results of this study can be used as reference material in making appropriate policies related to the progress of home-based HBEs by City Government. The previous study of the role of online transportation on HBEs development has not been explored, thus this study is trying to analyse the phenomenon and how to respond it is important.

2. Research Method

This research was conducted at 17 home-based culinary HBEs around the UNDIP Tembalang area. The respondents were chosen 17 HBEs out of 279 Micro, Small and Medium culinary enterprises which join to GoFood application in early 2019. From the survey, the HBEs is 17. The total area around UNDIP Tembalang area which was chosen as the research location is 15.11 Km2. This area consists of 7 kelurahans in 2 subdistricts in Banyumanik and Tembalang subDistrict.

This study uses quantitative and qualitative descriptive statistical analysis techniques to explain the results of questionnaire, and interviews. The data collection technique used in the questionnaire was total sampling or saturated sampling, which made all home-based culinary HBE owners around the UNDIP Tembalang area, namely as many as 17 businesses as respondents. The data collection technique used in the interview was snowball sampling, which made 10 online motorcycle taxi riders and 10 food service users as respondents who were considered to understand about this case to be interviewed. Snowball sampling is a sampling technique that starts from one respondent to another respondent until the researcher feels the required data has been obtained entirely.

3. Result and Discussion

This section contains an analysis process that can be used to see the role of online motorcycle taxi services for home-based HBEs. In analysing the role of online motorcycle taxi services for home-based HBEs, it will be seen from three sides, namely economy based on income value, service coverage based on destination and distance, and labour based on numbers.

A. The Role of Food Delivery Services for Home Based HBEs in Income Generation

IOP Conf. Series: Earth and Environmental Science **396** (2019) 012010 doi:10.1088/1755-1315/396/1/012010

In general, HBEs income which increases after cooperates with online motorcycle taxi explains that online motorcycle taxi has a positive role on HBEs. However, if the HBEs income stays or decreases after cooperates with online motorcycle taxi, it means that online motorcycle taxi plays a negative role on HBEs.

Table 1. The Comparison of Home Based HBEs Income Before and After Collaborate with Online Motorcycle Taxi around UNDIP Tembalang Area

HBE Respondent	Before (per day)	After (per day)	Increased Income (Percentage)
1	Rp 60.000	Rp 150.000	150
2	Rp 250.000	Rp 700.000	180
3	Rp 250.000	Rp 1.100.000	340
4	Rp 200.000	Rp 300.000	50
5	Rp 100.000	Rp 200.000	100
6	Rp 5.000.000	Rp 6.500.000	30
7	Rp 200.000	Rp 1.000.000	400
8	Rp 50.000	Rp 100.000	100
9	Rp 500.000	Rp 600.000	20
10	Rp 60.000	Rp 700.000	1066
11	Rp 150.000	Rp 350.000	133
12	Rp 2.000.000	Rp 5.600.000	180
13	Rp 100.000	Rp 300.000	200
14	Rp 800.000	Rp 1.000.000	25
15	Rp 200.000	Rp 400.000	100
16	Rp 100.000	Rp 150.000	50
17	Rp 40.000	Rp 70.000	75

Table 1 shows that all home-based HBEs around the UNDIP Tembalang area experienced a change in the form of increased income. The increase in income experienced by HBEs varies from 25% to 1000%. Based on the explanation of the home-based HBEs around the UNDIP area Tembalang, after cooperating with online motorcycle taxi there are several reasons that allow opportunities for increased business income.

These reasons include the easiness of being found and ordered by consumer through the online motorcycle taxi application, the opportunity for free HBE promotion on the online motorcycle taxi application, the presence of HBE food product discounts provided by the motorcycle taxi online to the buyer, as well as the presence of the motorcycle taxi online application that helps home based SMEs in following technological developments. These reasons are in line with as said by Suryadi and Ilyas [17], that online motorcycle taxi helps to maximize the business potential of HBEs, helps HBE promotion activities, and provides easy access to customers. This similarity shows that online motorcycle taxi gives its role to home-based HBEs around the UNDIP Tembalang area.

By looking at the role of motorcycle taxi online from an economic or income perspective, it is necessary to know the proportion between offline and online sales of home-based HBEs. The aim is to see how many daily transactions are between offline and online sales. Offline sales here are intended as transactions that take place when consumers come directly to HBE without going through an application from an online motorcycle taxi. While online sales are transactions that take place using the online motorcycle taxi application. Thus, the daily average of home-based HBE sales around the UNDIP Tembalang area every day based on offline and online transactions can be seen.

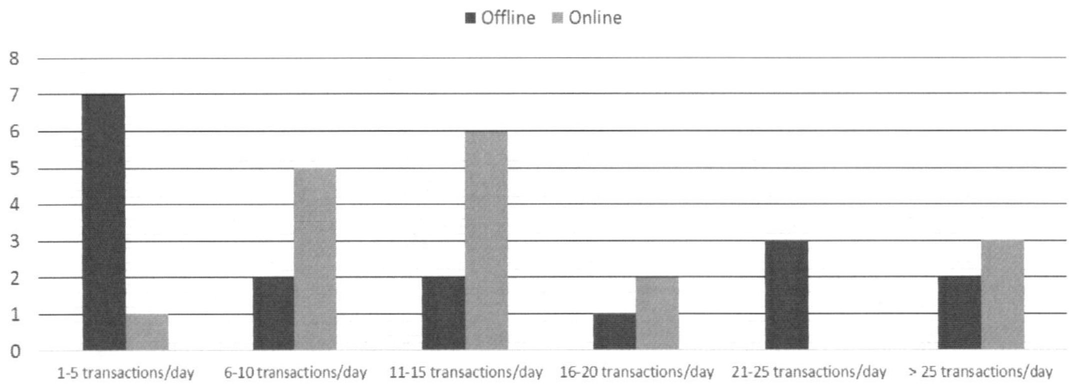

Figure 1. The Comparison of HBEs Transactions

Based on Figure 1, it is known that 7 out of 17 home-based HBEs make 1-5 online orders each day. As for the online transactions, the most answers were 6 out of 17 HBEs make 11-15 orders a day. Based on these data the average offline sales are lower than the average online sales.

The high level of online sales shows that the formation of collaboration between home-based HBEs with online motorcycle taxi has produced a positive impact. The establishment of cooperation with online motorcycle taxi provides opportunities for home-based HBEs to keep abreast of the times and have the opportunity to develop more. This opportunity is in line with the opinion of Suryadi and Ilyas (2018), that online motorcycle taxi helps to maximize the potential of HBEs, which is shown by the high online transactions from home-based HBEs after collaborating with online motorcycle taxi.

Furthermore, this research found that the income of the HBEs is fluctuating. The highest income of the HBEs reached Rp. 160,000,000 in a month. Quite different, the lowest income ever achieved by the HBEs is Rp. 200,000 in one month. Based on the results of interviews with respondents, their income is influence by the academic calendar of the universities in this area. The campus implements a system of holidays that was long enough for its students to last more than 1 month.

According to the enterprises, the fasting month and June were the months with the lowest sales due to reduced food intensity during the fasting month and for the month of June because most students were focused on learning to prepare for the final exam and some were more economical because they need to prepare for the purchase of a ticket home. However, there were also a few respondents who revealed that in the fasting month and June is their highest selling month as most of their competitors decide to close their business.

In addition, there is another uniqueness in the highest sales month data. A home-based SME actor revealed that his business is actually crowded in the months that there are promos from online motorcycle taxis. This is in line Suryadi and Ilyas said that online motorcycle taxi services help HBE promotion activities [17]. Please note that the promotions provided by these online motorcycle taxis are varied, such as discounts on food prices and can also be in the form of discounts on food services. This promotion indirectly impacts HBEs, because it invites buyers to use online motorcycle taxi services to buy food from HBEs. In this case, both home-based HBEs and online motorcycle taxis will benefit.

B. The Role of Food Delivery Services in Service Coverage Change of Home-Based HBE Services

The role of online motorcycle taxi for food delivery services can also be seen based on changes in service coverage. At this stage, home-based HBEs, especially around UNDIP Tembalang area, were asked questions about changing the range of services from their businesses before and after working with food delivery services from online motorcycle taxi. It is hoped that from this data, it can be seen the extent of the role of food delivery services from online motorcycle taxi in helping home-based HBEs in expanding their range of services.

Based on the findings in the field it is known that the range of home-based HBE services in the UNDIP Tembalang area is not only around the administrative boundaries of the Semarang City area but has also reached the neighbour city in Semarang Regency area as shown in the map image below (Figure 2):

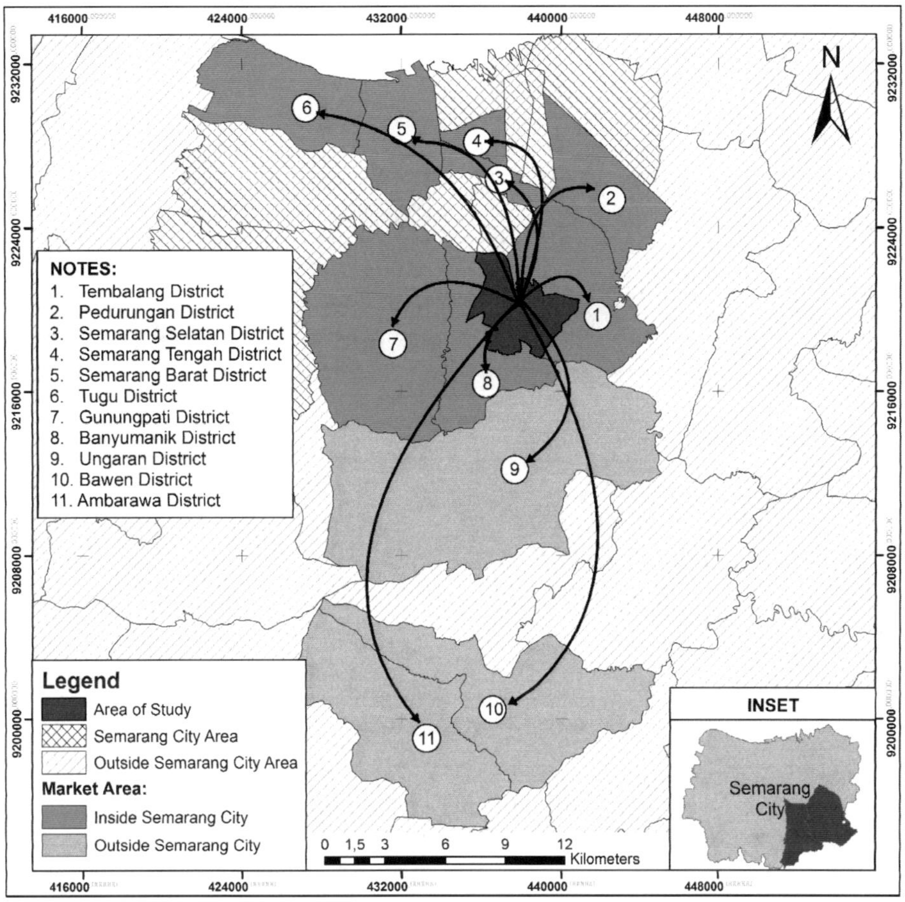

Figure 2. The Coverage Area of Home-Based HBE Services

When viewed based on the standard of one of the online motorcycle taxi for food delivery services, namely Gojek, with its Gofood feature, the maximum range of this service is 25 Km Based on the observation made, it can be known that there are still HBEs that only have service coverage of ± 1.31 Km, ± 2.49 Km and ± 2.67 Km. This phenomenon has not shown an optimal change in the results of collaborating with online food delivery when viewed in terms of service coverage.

On the other hand, in table 2 shows that the reach of home-based HBEs around UNDIP Tembalang area is also very broad, even to the Karangjati and Bandungan areas which are located outside of Semarang City.

Table 2. The Comparison of the Range of Home-Based HBE Services Before and After Collaborate with Online Motorcycle Taxi around UNDIP Tembalang Area

HBE Respondents	Before	Distance (Before)	After	Distance (After)	Deviation
1	Surrounding HBEs	± 1 Km	Simpang Lima	± 7,28 Km	± 6,28 Km

IOP Conf. Series: Earth and Environmental Science **396** (2019) 012010 doi:10.1088/1755-1315/396/1/012010

HBE Respondents	Before	Distance (Before)	After	Distance (After)	Deviation
2	Surrounding HBEs	± 1 Km	Mangkang	± 19,47 Km	± 18,47 Km
3	Surrounding HBEs	± 1 Km	Simpang Lima	± 8,10 Km	± 7,10 Km
4	Surrounding HBEs	± 1 Km	Banyumanik	± 2,49 Km	± 1,49 Km
5	Surrounding HBEs	± 1 Km	Ngesrep	± 2,67 Km	± 1,67 Km
6	Surrounding HBEs	± 1 Km	Ungaran	± 8,96 Km	± 7,96 Km
7	Surrounding HBEs	± 1 Km	Banjir Kanal Barat	± 9,04 Km	± 8,04 Km
8	Surrounding HBEs	± 1 Km	Pedurungan	± 7,46 Km	± 6,46 Km
9	Surrounding HBEs	± 1 Km	Pedurungan	± 8,60 Km	± 7,60 Km
10	Surrounding HBEs	± 1 Km	Ungaran	± 7,91 Km	± 6,91 Km
11	Surrounding HBEs	± 1 Km	Karangjati	± 16,24 Km	± 15,24 Km
12	Surrounding HBEs	± 1 Km	Ungaran	± 9,73 Km	± 8,73 Km
13	Surrounding HBEs	± 1 Km	Gunungpati	± 6,71 Km	± 5,71 Km
14	Surrounding HBEs	± 1 Km	Bandungan	± 18,57 Km	± 17,57 Km
15	Surrounding HBEs	± 1 Km	Pemuda	± 10,95 Km	± 9,95 Km
16	Surrounding HBEs	± 1 Km	Meteseh	± 4,23 Km	± 3,23 Km
17	Surrounding HBEs	± 1 Km	Banyumanik	± 1,31 Km	± 0,31 Km

When viewed based on the distance of its service coverage, this distance measurement uses more or less signs (±) because the distance drawn between the location of the home-based HBEs to the location in Google Earth, in accordance with the location of the area mentioned by the home-based HBEs, where distance measurements are made in a straight line. From the above data, it is also known that home-based HBEs of number 2 cover consumers up to ± 19.47 Km in the Mangkang area. There are also a number of home-based HBEs in the region that have achieved service coverage of ± 18.57 Km and ± 16.24 Km.

This phenomenon shows that home-based HBEs around UNDIP Tembalang area have achieved optimal services in accordance with the ability of food delivery services from online motorcycle taxi. In the end, this phenomenon shows the role of food delivery services involvement in the development of home-based HBEs. According to Suryadi and Ilyas [17], this phenomenon is one of the advantages of partnering with online motorcycle taxis because the HBE business potential can be maximized by expanding the range of services to millions of users of food service applications that are ready to be served by hundreds of thousands of drivers through online motorcycle taxi technology. This finding is in line with the opinion of Stockdale, et al [18] who said that digital marketing strategy is able to increase the competitiveness HBEs in marketing their business products and enhance promotion, because of the interconnectedness between business people and consumers with the internet-based application [19].

C. The Effect of HBE Location on Utilization of Food Delivery Services based on Community Perception

In this section, how people's perceptions are related to the location of HBEs, which in the previous analysis mentioned that most were on the local road, in relation to the utilization of food services. The influence of this HBE location is reviewed based on the condition of the road, where the condition in question is a damaged condition, potholes, traffic jam, as well as being on a road that is not easily seen as a local road. The community referred to in this analysis is divided into 3 namely HBEs, online motorcycle taxi riders, and online motorcycle taxi service users. The relationship between the answers between the 3 points of view of the respondent will be seen. Hopefully, this analysis can show the influence of food delivery services both from the perspective of HBEs, online motorcycle taxi drivers, and food service users.

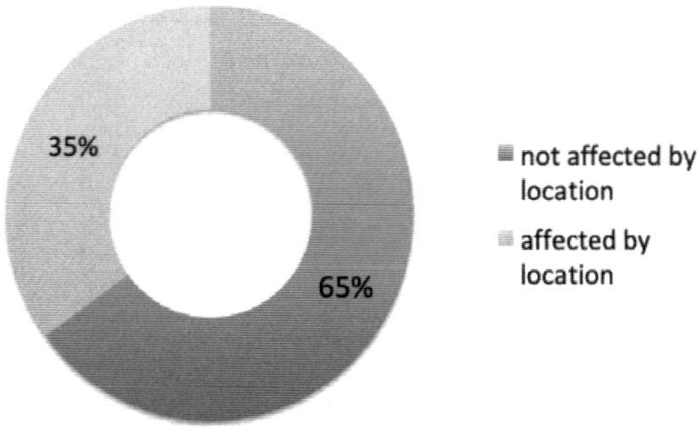

Figure 3. The Effect of HBE Location on Food Delivery Services from HBEs' Perception

As can be seen in figure 3, in general HBE actors assume that the location that is reviewed based on road conditions no longer has a major effect on success in doing business. This is confirmed by the explanation of one of the HBEs that revealed that:

> *"Dengan aplikasi ojek online usaha yang berada di perumahan bisa terkenal, karena usaha makanan disekitar pembeli bisa dilihat dari aplikasi"*
> *"With an online motorcycle taxi application, businesses in housing can be well known, because the food business around the consumer can be seen from the (mobile) application"*
> PU/RW/290619 (HBE's respondent #x)

This statement, in line with Fahri [20] and Hidayatullah and Waris [21] which stated that a home kitchen culinary business can be done together with online motorcycle taxi because these services play a role in providing information and introduction to culinary products of HBEs to the public. So, the food delivery services help the process of introducing and promoting HBEs products to the public. Furthermore, based on the results of interviews conducted with 10 online motorcycle taxi drivers, there were 3 out of 10 motorists who mentioned that location of the HBEs, particularly the road condition to access the HBEs influence the time to deliver the food. The drivers often get protested when the delivery process could not be carried out quickly due to poor road condition or traffic jam.

> *"Tetep jadi pertimbangan juga sih, kalo kayak jalanan jelek atau macet gitu kan jadinya lama, kadang juga customernya ngomel juga itu"*
> *"It's still a consideration for us, if it's like a bad road or traffic it's going to take a long time, sometimes the customers also nag that too"*
> PO/GS/290619 (Online Taxi driver #x)

> *"Kalo menurut saya sih iya ya mas, karna biasanya sih orang yang pesen itu biasanya, satu letak jarak, yang jualan itu jauh dari tempat tinggalnya. Terus kedua ya, mungkin kalau macet ya males ya mas. Lebih enak pake gojek gitu"*
> *"In my opinion, yes yes, mas, because usually people who order are usually, one location distance, the sales are far from where they live. And secondly, maybe if it's stuck, it's lazy, bro. It's better to use a motorbike"*
> PO/PA/280619 (Online Taxi driver #y)

Meanwhile, most of the consumers do not think about the road conditions in using these food delivery services. No influence according to motorbike taxi drivers online is interpreted as most consumers do not think about the condition of the road leading to the restaurant business kiosk, because the possibility of consumers also do not understand the exact location of the food stalls they want to buy. This understanding was obtained based on the results of interviews with online motorcycle taxi drivers who revealed:

> *"Malah kadang customer tu gatau restonya sebelah mana. Dia cuma ngelihat aplikasi, pingin makan apa, nasi goreng. Oiya nasi goreng, ini nasi goreng A itu recommended, tapi dia gak tahu tempatnya dimana gitu. Soalnya kebanyakan sih resto-restonya gofood itu rumahan-rumahan. Jadi kadang gaada kiosnya, gaada warungnya. Dia buka dirumah gitu."*
> *"In fact, sometimes the customers don't know which restaurant. He only saw the application, what he wanted to eat, fried rice. Oh fried rice, this is A fried rice is recommended, but he doesn't know where it is like that. The thing is, most of the gofood restaurants are home-based. So sometimes there are no stalls, no stalls or special room or arrangement for selling food. He opened at home. "*
> PO/FR/280619 (Online taxi driver #z)

Based on this statement it is known that actually both HBEs and consumers of service users, both benefit from the presence of this service. With this service, HBEs get a promotion as mentioned earlier, and consumers get the ease of obtaining food products. To be sure, it is necessary to look at the direct perception of consumers in utilizing these food delivery services.

When viewed from the results of interviews with consumers who use this service, generally users are more concerned about the length of the delivery process if there are bad road conditions. This is shown from the expression of some consumers who state that:

> *"Kalau macet akan memperlama proses pengiriman"*
> *"If it is jammed, the delivery process will take longer"*
> KP/AR/010719 (Customer of culinary HBEs #a)
> *"Sangat berpengaruh, karena apabila jalanan rusak atau macet akan membuat layanan ojek menjadi lama sampai"*
> *"Very influential, because if the road is damaged or stuck it will make the online motorcycle taxi service take a long time to arrive"*
> KP/HI/010719 (Customer of culinary HBEs #b)

This statement is in line with the statement of the motorcycle taxi drivers online, that time is a point of concern by consumers or customers.

In addition, poor road conditions are actually used as the reason of the consumers in utilizing this service. The existence of an online motorcycle taxi becomes an option for consumers to avoid congestion, travelling in a damage road, and findings the stalls, as stated by these consumers:

> *"Iya, karena saya malas melewati jalan rusak dan gang-gang sempit"*
> *"Yes, because I'm lazy to go through broken roads and narrow alleys"*
> KP/HI/3002019 (Customer of culinary HBEs #c)

> *"Saat jalan macet atau rusak, saya lebih memilih untuk memesan melalui ojek online dibandingkan datang membeli sendiri"*
> *"When the road is jammed or damaged, I prefer to order through an online motorcycle taxi rather than come buy it by myself"*
> KP/MU/010719 (Customer of culinary HBEs #d)

This statement shows that the presence of food delivery services for consumers provides benefits in terms of helping to obtain the desired food products. Where if consumers want certain food products, but they cannot afford to make their own purchases by coming directly to the stalls, there is a service to deliver the food direct to home. This phenomenon shows the role of food delivery services make consumers get closer by facilitating access to ordering and delivering products to consumers [17]

4. Conclusion

In conclusion, there is facts in increasing income and service coverage when the culinary HBEs start to use mobile application of online taxi delivery. These lead to several roles of online motorcycle taxi services, namely maximizing the potential of HBE businesses that shown by the higher proportion of home-based HBE online sales when compared to their off-line sales. Then, helping business promotion activities, these finding is found based on information from HBEs who state that the business is crowded when online motorcycle taxi do promotions such as price discounts and so on which make the business crowded by consumers. Third is easy access to customers, because food delivery services provide information about food businesses and are ready to deliver food orders to consumers. So that from these roles it can be seen how food delivery services influence the development of culinary HBEs, particularly by increasing the demand of their products. Thus, the support for online taxi delivery provided through mobile application is worth action, because it will also indirectly develop the HBEs.

5. Acknowledgement

This article was presented at the 2nd International Conference on Smart City INNOVATION (ICSCI) 2019, jointly held by Universitas Indonesia and Universitas Diponegoro. ICSCI conferences have been supported by the United States Agency for International Development (USAID) through the Sustainable Higher Education Research Alliance (SHERA) Project for Universitas Indonesia's Scientific Modelling, Application, Research, and Training for City-centered Innovation and Technology (SMART CITY) Center for Collaborative Research, administered through Grant #AID-497-A-1600004, Sub Grant #IIE-00000078-UI-1.

6. References

[1] Abolade O, Adigun F O and Akande A O 2013 Impacts of home based enterprises on the quality of life of operators in Ibadan North Local Government, Nigeria *Int. J. Humanit. Soc. Sci. Invent.* **2** 1–7

[2] Lin C Y-Y 1998 Success factors of small-and medium-sized enterprises in Taiwan: An analysis of cases *J. small Bus. Manag.* **36** 43

[3] Roy A 2005 Urban Informality: Toward an Epistemology of Planning *J. Am. Plan. Assoc.* **71** 147–58

[4] Kellett P and Tipple A G 2000 The home as workplace: a study of income-generating activities within the domestic setting *Environ. Urban.* **12** 203–14

[5] Strassmann W P 1986 Types of Neighbourhood and Home-Based Enterprises: Evidence from Lima, Peru *Urban Stud.* **23** 485–500

[6] Tipple G 2005 The place of home-based enterprises in the informal sector: Evidence from Cochabamba, New Delhi, Surabaya and Pretoria *Urban Stud.* **42** 611–32

[7] KelleTT P 1995 *Constructing home: production and consumption of popular housing in northern Colombia.* (University of Newcastle upon Tyne)

[8] Lyons G and Davidson C 2016 Guidance for transport planning and policymaking in the face of an uncertain future *Transp. Res. Part A Policy Pract.* **88** 104–16

[9] Daniel E M, Domenico M Di and Sharma S 2014 Effectuation and home-based online business entrepreneurs *Int. Small Bus. J. Res. Entrep.* **33** 799–823

[10] Anwar M and Daniel E 2014 Online home based businesses: systematic literature review and future research agenda

[11] Laura Galloway and Kapasi I 2014 Rural home-based businesses and their contribution to rural lives: an exploratory study.

[12] Mason C M, Carter S and Tagg S 2011 Invisible businesses: The characteristics of home-based businesses in the United Kingdom *Reg. Stud.* **45** 625–39

[13] Sianturi O O, Tyas W P, Manullang O R, Manaf A and others 2019 The Benefit of Internet Using to Affect Income for Water Hyacinth Home-based Entrepreneurs in Rawapening Area-

Indonesia *IOP Conference Series: Earth and Environmental Science* vol 248 p 12004

[14] Tyas W P, Nugroho P, Sariffuddin S, Purba N G, Riswandha Y and Sitorus G H I 2019 Applying Smart Economy of Smart Cities in Developing World: Learnt from Indonesia's Home Based Enterprises *IOP Conference Series: Earth and Environmental Science* vol 248 p 12078

[15] Tyas W P, Sianturi O, Riswandha Y, Manaf A, Manullang O R and others 2019 ICT Based Activities as Part of Smart Economy in Developing World: Learnt From Indonesia's Home Based Enterprises *IOP Conference Series: Earth and Environmental Science* vol 313 p 12026

[16] Branea A-M 2011 Revitalization Strategy of Urban Spaces through a Holistic Vision--Case Study Timisoara *Int. J. Energy Environ.* **1** 24

[17] Suryadi D F and Ilyas M I F 2018 ADOPSI ONLINE FOOD DELIVERY SERVICE BAGI WIRAUSAHA PEMULA DI KOTA MAKASSAR (STUDI KASUS PADA BIG BANANAS) *Seminar Nasional Hasil Penelitian & Pengabdian Kepada Masyarakat (SNP2M)*

[18] Stockdale R, Ahmed A and Scheepers H 2012 Identifying Business Value From The Use Of Social Media: An Sme Perspective. *PACIS* p 169

[19] Lee I and Lee K 2015 The Internet of Things (IoT): Applications, investments, and challenges for enterprises *Bus. Horiz.* **58** 431–40

[20] Fahri L M 2017 Strategi Marketing Public Relations Go-Food dalam Pembentukan Citra Perusahaan di Kota Surabaya

[21] Hidayatullah S, Waris A and Devianti R C 2018 Perilaku Generasi Milenial dalam Menggunakan Aplikasi Go-Food *{JURNAL} {MANAJEMEN} {DAN} {KEWIRAUSAHAAN}* **6**

An analysis of potential utilization of low cost Cu_2ZnSnS_4 thin film based photovoltaic in Sumbawa

A Rukini[1], L Suhaimi[1], A S Pradhipta[1], M Anggara[2]
[1]Metallurgy Engineering, Universitas Teknologi Sumbawa
[2]Mechanical Engineering, Universitas Teknologi Sumbawa

E-mail: asywendi@gmail.com

Abstract. Sumbawa has the potential to transition into a smart city, utilizing solar energy as their main source of its alternative energy. Located in the tropical country of Indonesia, Sumbawa's dry season lasts eight months, during which it sees a solar intensity of 4.51 watt/m^2/h, making solar energy a promising resource for the area. Its level of abundance and no pollution attribute are the main factors that make solar energy an attractive renewable resource. Raw calculation showed that solar energy is able to provide massive amount of energy, while solar panel and photovoltaic can produce zero greenhouse gases when converting solar energy into electricity. However, the high initial cost of installing solar cells hinder popularization of it as an alternative energy source, meanwhile some of the substance materials that compose solar panel may have a negative impact on the environment. This study intends to review the potential for inexpensive semiconductor-based solar panels to be applied as independent source of lighting by analysing the cost and environmental impact. Discussion on simple chemical method without high cost equipment instalment is also highlighted in this work, which offers recommendation of methods that can contribute on lowering the production cost of solar panel.

Keywords: solar cell, renewable and sustainable energy, environmental friendly, Cu_2ZnSnS_4 (CZTS)

1. Introduction

Energy needs compliance from non-fossil resources becomes essential in the middle of global warming issues and oil crisis problem. Photovoltaic and solar cell represent one of the most promising alternative renewable energy. In theoretical calculation where the sun rays intensity that reach earth (land only) is 19%, by considering the ground altitude, solar could produce energy up to 89,300 TWp [1]. Then by assuming that it was absorbed for 6 hours a day solar would provide electrical energy approximately 200 million TWh per year. Latest world electrical consumption was approximately 20 thousand TWh per year [2], while predictions show that the energy earned from solar energy could meet that demand 10,000 times.

The desire to actualize clean energy sources such as solar panels is inspired by the Indonesia's strategic location along the Equator. With average, at least, 8 hours illuminated by optimum solar light and heat, Indonesia has the potential to utilize solar panels as a renewable energy source. However, problems such as the high initial cost, installation and maintenance expense of solar cells, also, the low efficiency are still obstacles in realizing the solar energy as an alternative energy. It has been more than a century since the first complete photovoltaic cell was produced in 1888, yet its development in industrial scale for energy serving has not covered many levels of society. A survey taken from market report European photovoltaic industry revealed that by the end 2014 only 3% of electricity provided from solar cell or photovoltaic sources [3,4]. The noticeable hurdles for this technology are its high production cost and expensive commercialization.

Early versions of solar cell required around $250 in order to produce 1watt electricity from solar cell equipment even in summer day-like condition. Meanwhile, it only took $2 to $3 to convert coal calories to electricity in the same time [5]. This was due to the low efficiency of electrical energy conversion, expensive fabrication processes and high price of solar cell's raw materials. Accordingly, various attempts have been pursued by researchers to improve photovoltaic cell efficiency, invent new cheap combinations of semiconductors and develop affordable methods and technologies. Thus, the future trend of photovoltaic cell will be based on cheap material based, high efficiency and easy fabrication. Currently, the first- and second-generation are the commercial solar cell that widely distributed in the market. However, the one that dominating the market (> 90%) is still the first generation of solar cell or crystalline silicon (c-Si) based solar panel, either mono-crystalline or polycrystalline wafer based [6]. Mono-crystalline wafer of silicon based solar cells apparently are the most expensive kind due to the difficulty of single crystal Si production. On the other hand, second generation or thin film based photovoltaic gain attention in market because its lower price and aesthetically pleasing appearance. Thin film PV offers a less bulky module of solar panel, with some having flexible and even transparent form. Thin film solar cells have several types of semiconductor such as Si-thin film, CdTe/CdS, CIS/CIGS and CZTS/Se. Among the developed semiconductor types, CZTS is the most affordable when comparing the cost of raw materials (Figure 1).

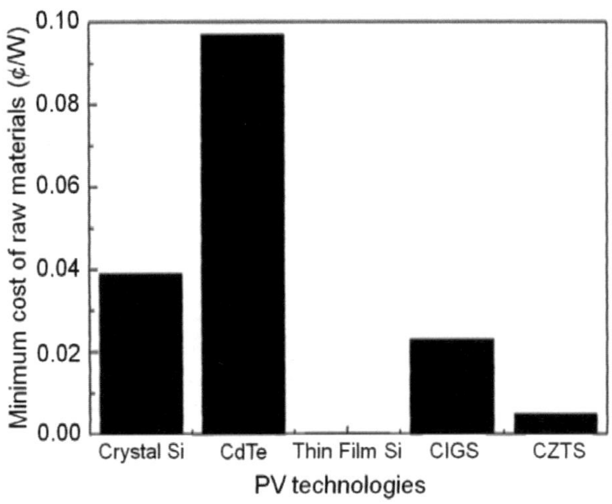

Figure 1. Minimum cost of raw materials of each thin films (© 2013 Jiang M, Yan. Published in IntechOpen under CC By 3.0 license. Available from: htpp://dx.doi.org/10.5772/50702.)

Naturally, technology that is used to produce solar cell module also determine the cost in the market. Physical methods mostly required high tech equipment which in turn also need certain quality of substrate in order to generate a good consistency solar panel or thin films. On the other hand, the chemical route is able to utilize simpler equipment and wide range quality of substrate, although the products generated from this method sometimes have lower precision due to several controlled variable. Logically, the combination of inexpensive solar cells and a simple fabrication process are important steps towards developing low cost solar panels on an industrial scale. Therefore, second generation CZTS based solar cell produced via chemical method is the projection of manifesting more affordable solar panel.

2. Cu_2ZnSnS_4 (CZTS) semiconductor as an alternative

Cu_2ZnSnS_4 or often called as CZTS is a thin film compound which has recently been used as an absorbent material in photovoltaic cell devices as an alternative to the chalcopyrite-based CuInSe2 (CIS) thin film, CuIn1-xGaxSe2 (CIGS) and Cu (In, Ga) (S, Se) (CIGSS) [7,8]. Before CZTS was well developed, CIS, CIGS and CIGSS were well known as the most potential candidates to be used as absorbents on thin film based photovoltaic or thin film solar cell (TFCS's). With efficiencies at almost ~20%, CIS compound generated from co-evaporation route was considered attractive to be developed commercially [9–11]. Unfortunately, the supply of rare metal Indium in the future mass production is considered problematic. One estimation for CIS based solar panel show that 30 tons of Indium is required to produce 1gigawatt of electricity [9]. Even though the prior requirements might be met, one also needs taking the demand for the production of an indium tin oxide (ITO) transparent substrate into account, the demand for ITO production in the future might also lead to competition that can limit CIS production [10].

CZTS is considered a potential candidate for absorbance material obtained by substituting half of the indium with zinc (Zn) and the other half with tin (Sn). For CIGS or CIGSS semiconductor, the In and Ga elements are substituted by Zn and Sn. All the constituent components of CZTS are plentiful in the earth crust (earth abundant) unlike Ga, In and Se which are rare (rare earth) elements. Sequential levels of Cu, Zn, Sn and S are around 50 ppm, 75 ppm, 2.2 ppm and 260 ppm from the earth's crust [11]. While the content of In, Ga, and Se are each small than 0.05 ppm [11]. This thin film has been developed with diverse method either physically or chemically, which those experiments verified that the band gap of CZTS thin film ranged from 1.4 to 1.5 eV [12]. Aside from suitable optical properties, CZTS elements are also low cost and non-toxic to be applied as solar cell semiconductor.

3. Cost and environmental impact of CZTS

CZTS is a one of kind of low-cost quaternary thin film that is easy to acquire due to its earth abundant elements, therefore has relatively low cost and is easy to obtain. However, simply being affordable is not sufficient. Developing clean and environmental friendly energy is also an important aspect for generating new technology or device. As a clean energy source, CZTS based solar panel also contributes minimize the effect of global warming, yet another environmental factor such toxicity at production process is also need to be assessed.

3.1 Cost

CdTe and CdSe based thin films possess such high absorption coefficient and efficiency [13–15] by the consequence as the highest manufacturing cost among other thin film photovoltaic [16]. By comparing the minimum cost of thin films production on Fig. 1, CZTS only needs less than half cost to produce 1 watt electricity from CIGS based thin films. The availability of Zn and Sn source is relatively competitive compared to In and Ga source [15], but the annual production of two former elements reach hundred thousand and thousand times that of the later elements [16]. Although known as the second most abundant elements in earth's crust, the production process of silicon requires a lot of energy and cost twice than Zn and Sn that cost under $100/kg. With above ten thousand tons of production per year in element combination, CZTS could be one of the sustainable thin films to produce.

Although the constituent elements are affordable, complete solar cell panels do not consist of semiconductor alone. Cost calculation of solar cell panels also includes a complete module construction and the output such as efficiency. From online supplier platform survey, thin film modules do not have significant lower prices over c-Si based module. This happened due to some considerations such as efficiency, required space area, temperature coefficient etc. Up until now, the highest efficiency produced by Si based solar cell. The higher the efficiency, the lower the cost to convert per watt electricity. Recent highest efficiency record for single cell is GaAs cell at 30.5% and for multi-junction cells GaInP/GaAs.GaInAsP/GaInAs at 46.0% [17]. In the other word, CZTS has low cost and sustainable sources, yet, it is not necessarily urgent to replace Si based solar cell when it comes to complete module cost confrontation.

3.2 Environmental Impact

Solar energy known as green energy due to its less pollution and zero CO/CO_2 emission. However, elements on semiconductors are not always environmental friendly. For instance, arsenic in GaAs thin films is notorious for its toxicity. Likewise, cadmium based thin films (CdS, CdSe, CdTe), Cd is potentially restricted due to heavy metal pollution and Te suffer from strong carcinogenicity [15]. Every element in CZTS compound is not only earth abundant but also safe with relatively no harm effect to its surroundings, while Se in CIGSS could potentially cause environmental problem [7,18,19] and some found that low amount of Cd metal can be present in CIGS solar cell [20]. Therefore, with respect to health and environmental concerns, CZTS is one of the best thin film options to produce.

4. Promising low cost methods on thin film making

One of the most effective steps towards lowering price of a product is to modify the methods used. In the 1960s, a solar power corporation (SPC) worker named Elliot Berman found that low price is one of the ultimate magnets for alternative energy demands. In 1969 the production cost of electricity from photovoltaic cell assemble was $100/watt and it was lowered up to $20/watt by analyzing and eliminating the unnecessary steps [21]. By eliminating polishing process and anti-reflective layer coating step, Berman developed rough silicon wafer that were far more effective at adsorbing solar rays. Refining the methods by eliminating redundancies leads to the development of cheaper, more efficient solar cells.

In general, the methods used to produce thin films can be categorized into physical and chemical, which can be divided into more specific processes. In the physical route, thin films target is deposited into substrate by using external force in vacuum evaporation condition or well-known as sputtering, where the target is transferred in a gas state. Meanwhile, the chemical method involves chemical reaction among precursor around substrate surface. Among many established methods, there are several chemical-derived methods that only require simple step and easily acquired equipment. Solution-based approaches could potentially be applied for industrial scale production to attain low cost solar panel [22], specifically, sequential solution-phase deposition such as successive ionic layer adsorption and reaction (SILAR), electrochemical layer deposition (ECALD), and ionic layer gas reaction (ILGAR), all of which will be discussed in the section bellow.

4.1 SILAR method

Successive ionic layer adsorption and reaction (SILAR) is a derived technology developed from modification of chemical bath deposition (CBD) method and deposit thin films similar to atomic layer deposition method. SILAR was first experimented by Nicolau and his colleague in 1985 [23,24]. Due to unnecessary precipitation, CBD were modified into SILAR to avoid the precipitate byproduct. SILAR is considered as a potential method to deposit thin films on industrial scale without expensive technology and high maintenance equipment. Advantages of SILAR method listed by Pathan and Lokhande [25] are as follows: It is easy to doping a particular element into films with various proportion only by add the element into solution; SILAR does not require high quality substrate nor vacuum condition, Deposition rate and film thickness can be controlled by changing the deposition cycle, There is no local overheating that harm the materials like some particular physical method, There is no material limit for substrate either in dimension or surface profile and SILAR is easily applicable to a wide area of deposition.

The main step of SILAR on depositing thin film compound is dipping the substrate in cation and anion precursor solutions separately. The dipping is repeated in sequence until the film's desired thickness is obtained. Duration of immersion time affected the optical properties of thin film [26]. As it show in Figure 2, there are rinsing processes in between cation and anion precursors. The aim of rinsing process is to avoid unnecessary aggregates growth. Due to its simplicity, this process can be done manually with hands, beaker glasses and a timer. However, chemical compound of thin film generated by this method might not be stoichiometric.

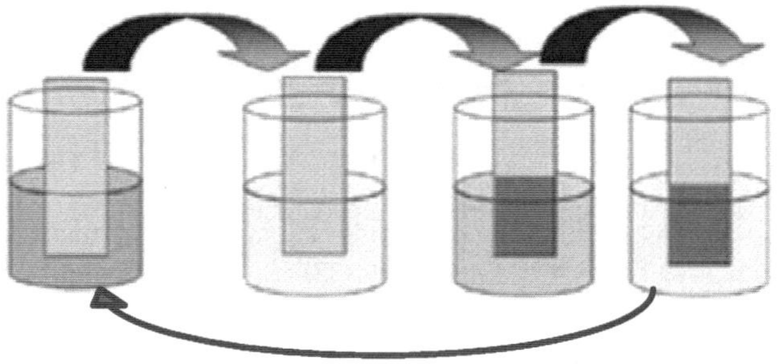

Figure 2. Schematic of repeated immersion of CZTS making via SILAR method

4.2 ECALD method

Electrochemical atomic layer deposition (ECALD) is based on layer by layer electrodeposition [27] wherein each film's constituent is deposited one by one separately in cycle, and in underpotential deposition condition of the elements. Underpotential deposition condition means an atomic layer of one element can be deposited onto the other's element surface under the potential that needed to deposit the element itself. This deposition is somewhat similar to SILAR, only the deposition process involves reduction-oxidation replacement or electrochemical reaction as illustrated in Figure 3. ECALD method allows high quality thin film formation and stoichiometric compound [28]. No rinsing process, yet ECALD process capable of controlling the deposition process at room temperature. Other than that, this method does not require any further thermal treatment such as annealing to form highly ordered crystalline structure [29].

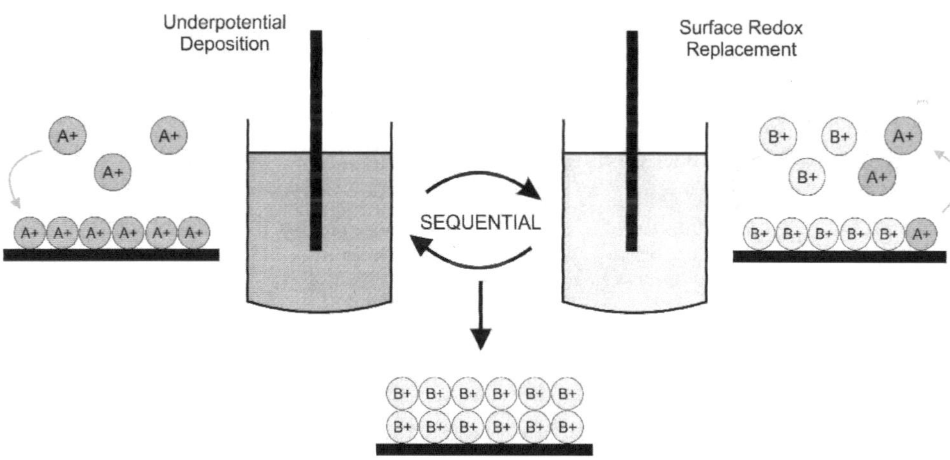

Figure 3. Schematic of electrochemical atomic layer deposition process

4.3 ILGAR method

Ion layer gas reaction (ILGAR) is part of chemical deposition techniques wherein the metal precursors are dipped or sprayed on the surface's substrate. The sequence including precursor's absorption, drying and followed by sulfurization and then repeat in cycles depends on desired thickness of thin film (Figure 4). Contrary to SILAR, reaction on ILGAR happen in solid-gas state. Similar to other sequential approach, this method does not require vacuum and high–temperature condition. The quality of thin film depends on the precursor preparation. Sprayed

ILGAR generally possess higher growth rates and allows reagent gas more easily penetrates the substrate compared to the SILAR method [30,31].

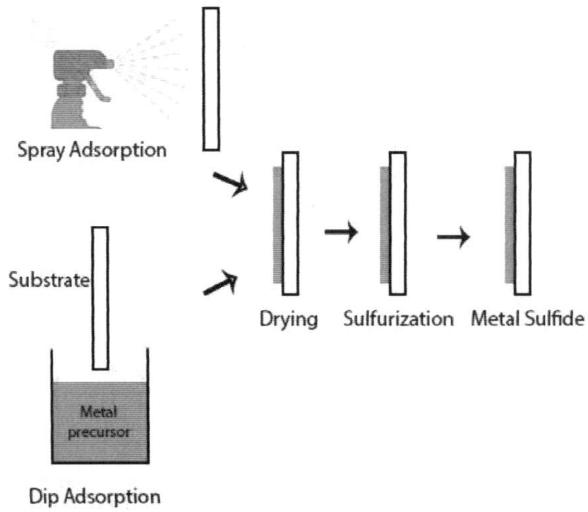

Figure 4. Basic principle of ILGAR method

5. Discussion and recommendation

There are several potential sources of renewable energy that can meet the energy needs of Sumbawa. The survey shows some potential electricity sources to be developed in NTB province range from micro-hydro power sources from rivers and dams, sources of electricity from wind, sources of electricity from sunlight, biogas and geothermal energy. However, in Sumbawa, only solar and micro-hydro can really be utilized. As for now, only Maronge and Batu Lanteh regions have been able to develop independent energy based on solar power as a source of lighting [33][34]. The annual report from the Bidang Energi (national energy board) [32] showed the average intensity of solar radiation in NTB is 4.5 watts/m^2/hour. BMKG (meteorology and geophysical unit) noted that on average Sumbawa sees four wet months throughout the year; December, January, February and March. The rest is a dry month with nearly zero rainfall and optimum sunlight intensity of more than 8 hours a day. As such, of solar cells are more promising than micro-hydro as the latter relies on full flowing rivers that most of the time low debited and discharge during the dry season.

Based on current data, there are around 1013 and 608 units of solar home systems (SHS) in Sumbawa and Sumbawa Barat respectively, with one unit of communal solar panel in each area. The total solar energy that is currently able to be produced by solar panel in each region is 129,998 kWh and 82,692 kWh [33]. Figure 5 shows the average utilization of solar energy in Sumbawa and Sumbawa Barat area as well as a projection for the next five years. Current solar energy consumption is about 1,142 barrel of oil equivalent (BOE) and is expected to increase by 18% in 2025. Also, the status of current consumption from solar energy is under 0.1% of Energy needs in Sumbawa region [33].

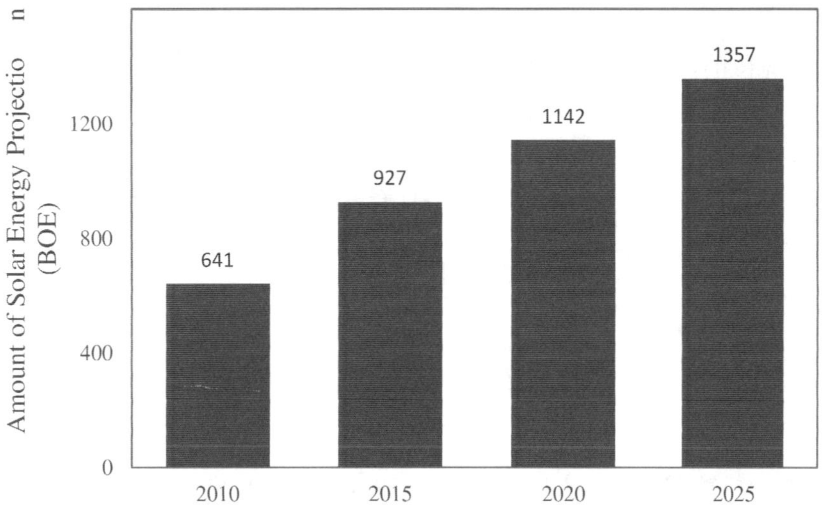

Figure 5. Amount of solar energy utilization and projection

In 2015, 500 units of SHS were installed in Sumbawa and Sumbawa Barat, and by 2025 it is estimated that only 350 units added. A unit SHS has 50Wp (watt peak) capacity. Assumed that the maximum sun radiation is 4.5h, therefore, capacity of a unit SHS is 225 watt hour/day. While, current communal solar system installed is 5000 kW and added by the same amount in the next five year as resumed in table 1. If a BOE is equivalent with 1.7 MWh electricity, then energy projection in 2025 (2.3 MWh) seems hard to be in accordance with the plan on table 1. If the expenditure cost of every 50Wp instalment is USD 694 [34] and the interest rate of every 5 year is 12%, the total investment to installed SHS in Sumbawa region in 2025 is about USD 242,900. While the budget on communal solar panel is USD 90 million by 2025. The low future development of solar energy in Sumbawa most probably is caused by familiar problems that already mentioned before, such as the high cost of installation and high initial cost of solar cells. Utilization of CZTS or amorphous Si based thin film is relevant to overcome intensive initial cost. Referring to levelised cost of electricity generation (LCOE) report in 2015, average capital cost of c-Si, amorphous Si and CIS/CIGS is 0.35 USD/KWh, 0.30 USD/KWh and 0.31 USD/KWh respectively [34]. As CIS/CIGS based solar panel suitable on utility or communal scale application, therefore, if the initial cost of CZTS based solar cell occurs in range of CIS/CIGS' cost, then the USD 90 million-communal solar budget can afford 2.9 MWh.

Table 1. Solar panel development plan in Sumbawa region

Year	SHS development plan		Communal SC development plan	
	Sumbawa	Sumbawa Barat	Sumbawa	Sumbawa Barat
2010	-	-	-	-
2015	300 units	200 units	0.005 MW	0.005 MW
2020	200 units	150 units	0.005 MW	0.005 MW
2025	200 units	150 units	0.005 MW	0.005 MW

(Information adapted from CASINDO NTB (2011))

Contributions from local researchers to develop or at least adopt simple technology that is capable of being applied on an industrial scale are necessary. By developing inexpensive semiconductor-based solar panels via simple chemical methods discussed above (SILAR, ECALD and ILGAR), it is expected to reduce the initial cost. Investment cost on wet chemical method significantly lower compared to physical route. Such costly equipment and high-quality substrate in table 2 are not required in chemical procedure. Thus, with the same amount of budget, if local researcher could independently produce low cost semiconductor and self-assemble solar panels, electricity target produced from solar power in Sumbawa can be fulfilled. Other than low-cost and environmental issue, concern in the future might be related to the energy storage or batteries and the efficiency of energy conversion.

Table 2. Investment comparison between wet chemical method and physical method

Investment	Wet Chemical Method	Physical Method
Deposited material	Dissolved in water	Ionized raw material
Substrate	Low quality can be accepted	High quality substrate
Vacuum equipment	X	√
Sputtering or deposition device	X	√
Annealing oven	Optional	√

6. Conclusion

Analysis on environment impact and availability supply of raw materials suggests CZTS as the strong candidate among thin film types to be used as semiconductor in Sumbawa's solar panel. Even though a complete module of thin film solar panel is not competitively cheaper than Si-based, high initial cost problem can be reduce by preparing CZTS via recommended low-cost methods such as SILAR, ECALD and ILGAR method. Using the same amount of budget, Sumbawa region potentially afford more number of solar panel to be installed in Sumbawa when CZTS based solar panel generated via low cost method substitutes the first generation type panel. Nonetheless, investigations on energy storage and performance of solar panel need to be included in the future study.

7. Acknowledgment

This article was presented at the 2nd International Conference on Smart City INNOVATION (ICSCI) 2019, jointly held by Universitas Indonesia and Universitas Diponegoro. ICSCI conferences have been supported by the United States Agency for International Development (USAID) through the Sustainable Higher Education Research Alliance (SHERA) Project for Universitas Indonesia's Scientific Modelling, Application, Research, and Training for City-centered Innovation and Technology (SMART CITY) Center for Collaborative Research, administered through Grant #AID-497-A-1600004, Sub Grant #IIE-00000078-UI-1

8. References

[1] Tsao J, Lewis N and Crabtree G 2006 Solar faqs *US Dep. Energy* 1–24

[2] IEA 2011 *World Energy Outlook 2011* (OECD)

[3] IEA 2015 *Snapshot of global PV markets, PVPS Photovoltaic Power System*

[4] Masson G, Orlandi S, Rekinger M and others 2014 Global market outlook for photovoltaics 2014-2018 *Eur. Photovolt. Ind. Assoc.* **60** 12

[5] Perlin J 2013 *Let it shine: the 6,000-year story of solar energy* (New World Library)

[6] Fraunhofer Institute for Solar Energy(ISE) 2018 *Photovoltaic Reports* (ISE Freiburg)

[7] Shinde N M, Dubal D P, Dhawale D S, Lokhande C D, Kim J H and Moon J H 2012 Room temperature novel chemical synthesis of Cu2ZnSnS4 (CZTS) absorbing layer for photovoltaic application *Mater. Res. Bull.* **47** 302–7

[8] Tanaka K, Moritake N and Uchiki H 2007 Preparation of Cu2ZnSnS4 thin films by sulfurizing sol-gel deposited precursors *Sol. Energy Mater. Sol. Cells* **91** 1199–201

[9] Shin S W, Pawar S M, Park C Y, Yun J H, Moon J-H, Kim J H and Lee J Y 2011 Studies on Cu2ZnSnS4 (CZTS) absorber layer using different stacking orders in precursor thin films *Sol. Energy Mater. Sol. Cells* **95** 3202–6

[10] Jeon M, Tanaka Y, Shimizu T and Shingubara S 2011 Formation and characterization of single-step electrodeposited Cu2ZnSnS4 thin films: Effect of complexing agent volume *Energy Procedia* **10** 255–60

[11] Mali S S, Patil B M, Betty C A, Bhosale P N, Oh Y W, Jadkar S R, Devan R S, Ma Y-R and Patil P S 2012 Novel synthesis of kesterite Cu2ZnSnS4 nanoflakes by successive ionic layer adsorption and reaction technique: Characterization and application *Electrochim. Acta* **66** 216–21

[12] Katagiri H, Sasaguchi N, Hando S, Hoshino S, Ohashi J and Yokota T 1997 Preparation and evaluation of Cu2ZnSnS4 thin films by sulfurization of E□B evaporated precursors *Sol. Energy Mater. Sol. Cells* **49** 407–14

[13] Wang J and Isshiki M 2006 II-IV Semiconductors for Optoelectronics: CdS, CdSe, CdTe *Springer Handbook of Electronic and Photonic Materials* (Springer US) pp 829–42

[14] Baines T, Zoppi G, Bowen L, Shalvey T P, Mariotti S, Durose K and Major J D 2018 Incorporation of CdSe layers into CdTe thin film solar cells *Sol. Energy Mater. Sol. Cells* **180** 196–204

[15] Jiang M and Y X 2013 Cu2ZnSnS4 Thin Film Solar Cells: Present Status and Future Prospects *Solar Cells - Research and Application Perspectives* (InTech)

[16] Zhao Y and Burda C 2012 Development of plasmonic semiconductor nanomaterials with copper chalcogenides for a future with sustainable energy materials *Energy Environ. Sci.* **5** 5564–76

[17] Green M A, Hishikawa Y, Dunlop E D, Levi D H, Hohl-Ebinger J and Ho-Baillie A W Y 2018 Solar cell efficiency tables (version 52) *Prog. Photovoltaics Res. Appl.* **26** 427–36

[18] Katagiri H, Jimbo K, Maw W S, Oishi K, Yamazaki M, Araki H and Takeuchi A 2009 Development of CZTS-based thin film solar cells *Thin Solid Films* **517** 2455–60

[19] Katagiri H 2005 Cu2ZnSnS4 thin film solar cells *Thin Solid Films* **480–481** 426–32

[20] Minnaert B 2008 Thin film solar cells: an overview

[21] EPIA 2011 *Market Report European Photovoltaic Industry Association* (EPIA Publications)

[22] Mitzi D 2008 *Solution processing of inorganic materials* (John Wiley & Sons)

[23] NICOLAU Y 1985 Solution deposition of thin solid compound films by a successive ionic-layer adsorption and reaction process-1 *Appl. Surf. Sci.* **22–23** 1061–74

[24] Nicolau Y F 1986 Method and apparatus for deposition on a substrate of a thin film containing cationic and anionic constituents.

[25] Pathan H M and Lokhande C D 2004 Deposition of metal chalcogenide thin films by successive ionic layer adsorption and reaction (SILAR) method *Bull. Mater. Sci.* **27** 85–111

[26] Rukini A, Reza A and Munir B 2019 Effects Of Ionic Immersion Time On Optical Properties Of Cu2ZnSnS4(CZTS) Deposited by Successive Ionic Layer Adsorption And Reaction (SILAR) *IOP Conf. Ser. Mater. Sci. Eng.* **547** 12030

[27] Lindroos S and Leskela M 2009 *Solution processing of inorganic materials: Successive ionic layer adsorption and reaction (SILAR) and related sequential solution-phase deposition techniques* (John Wiley and Sons Inc., Publication.)

[28] Gregory B W and Stickney J L 1991 Electrochemical atomic layer epitaxy (ECALE) *J. Electroanal. Chem. Interfacial Electrochem.* **300** 543–61

[29] Giaccherini A, Felici R and Innocenti M 2017 Operando Structural Characterization of the E-ALD Process Ultra-Thin Films Growth *X-ray Characterization of Nanostructured Energy Materials by Synchrotron Radiation* (InTech)

[30] Muffler H-J, Fischer C-H, Diesner K and Lux-Steiner M C 2001 ILGAR-A novel thin-film technology for sulfides *Sol. Energy Mater. Sol. Cells* **67** 121–7

[31] Möller J, Fischer C-H, Muffler H-J, Könenkamp R, Kaiser I, Kelch C and Lux-Steiner M C 2000 A novel deposition technique for compound semiconductors on highly porous substrates: ILGAR *Thin Solid Films* **361–362** 113–7

[32] Laporan Tahunan Bidang Energi 2010 *Laporan Tahunan Dinas Pertambangan dan Energi Provinsi Nusa Tenggara Barat*

[33] CASINDO 2011 *Rencana Aksi Energi Terbarukan Propinsi Nusa Tenggara Barat Tahun 2010-2025* (Tim CASINDO NTB)

[34] IRENA 2012 Renewable Energy Technology: Cost Analysis Series *Int. Renew. Energy Agency UEA* **1**

Sustainable Communities as a Response to Climate Change: Analysis of Geroldsäcker Eco-Housing Project and Recommendations for its Replication Considering Current Urban Challenges

B Kasih[1], K Nohara[1], L Zerbst[1], R Javier[1]

[1]University of Stuttgart, Stuttgart, Jerman

Email: bonikasih@gmail.com

Abstract. Geroldsäcker eco-housing is a unique project located in the northeast of Karlsruhe, in a 1.24-hectare plot with constructed area of 5.600m². The planning of the project started in year 1989 and its completion happened in 1992. It is composed of 40 residence units: apartments and 3 store row-houses with private garden. The initial concept was an ecological settlement with common infrastructure and spaces for human interaction, using constructive ecological materials and devices for the residences and common areas. The community living sense is one of this project's highlights. The 120 residents not only share common spaces, such as a community house destined to events and practice of activities, but also developed task groups to cover all necessary workspaces in the settlement. What can be observed nowadays is that Geroldsäcker, even being implemented 3 decades ago, presents several important sustainability aspects in the ecological, social and economic dimensions. What started as a pilot project at a time when the definition of sustainability was still under discussion, can be considered a model not only of sustainable housing, but also as a broader model for sustainable communities. This paper explores the aspects identified in Geroldsäcker eco-housing project leading towards a sustainable community concept. Potentialities and improvement points are analysed, considering current urban challenges, such as high demand and costs for land, lack of public land for experimental projects, efforts to stimulate bottom up processes, the behaviour of current generations, the need for new and environmental friendly constructive patterns, and how to aggregate stakeholders to boost new models of sustainable communities projects.

Keywords: Sustainable communities, Geroldsäcker housing project, sustainable housing.

1. Introduction

Planners have a great challenge today in developing spaces that demand numerous conditions to be fulfilled: spaces must be safe, in areas not considered vulnerable and, in case of any vulnerability, resilience aspects must be considered, they also have to be environmentally friendly and attractive to bring people to live in.

Since the 80's, sustainability emerged as trend to new developments and behaviours considering that the current lifestyle and population growth would end up consuming many natural resources and compromising the environment for future generations. Parallel to sustainable housing concepts [1–4], other thematic were gaining importance and bringing broader perspectives, such as the sustainable communities.

The goal resides in achieving this concept: planners can provide sustainable spaces, eco-friendliness, transport connectivity and sustainable and resilient houses. However, people have an important role in this process. Sustainable culture must be a part or a root of the development.

Identifying a sustainable community involves several aspects from different perspectives. In this paper, an analysis of the eco-housing project Geroldsäcker is held, and many components are identified boosting this project as not a simple sustainable housing but also as a sustainable community initiative. Key elements that could be applied in other projects are identified, with the objective of upscaling and developing more sustainable communities such as Geroldsäcker.

2. From Sustainable Development to Sustainable Communities

In the late 1980s, the Brundtland Report was published, defining "sustainable development" as the "*development that meets the needs of the present without compromising the ability of future generations to meet their own needs*" [5].

The definition from Brundtland Report is still in use nowadays, and the concept improved on how can the sustainable development be achieved. Today, the balance of environmental, social and economic variables is indispensable to work with sustainable development [6].

Considering climate change, most course of actions towards minimizing climate change have been reactive adaptations based on past events. It has already become clear that a transition to proactive methods to face the altered future climatic conditions is needed. Such actions can be done by individuals, groups, society or governments; regardless of the level, they all interact with each other, i.e. they are not self-regulating.

The success of these actions usually depends on the scale of implementation and the criterion for evaluation, just because a goal was accomplished in one level it doesn't exclude it from failure on a different group. It is only when the members share the same values and beliefs that technical and social aspects can be separated and treated independently and as such have a higher change of equal realization [7].

Sustainable houses are what results from social processes behind design and construction, in particular the belief and motivation of their authors. Such motion already possesses high credibility due to the effectiveness of new sustainable technologies, aesthetic presentation that is media-friendly and the framing of eco-friendly as in "low carbon housing". The interest of said frame is to reduce demand for energy from the grid; when viewed from this perspective, the governments' interests are set as a primary driver and with the rebrand of policies and activities long advocated, the public's attention and interest to get involved can be captured more easily.

The life cycle of sustainable houses might be seen as a negative aspect, past experiences make it seem as if sustainable houses cost more, nonetheless the idea behind the scale up process can be applied here as a counter-argument: selling the land for development at a lower price, while retaining financial stake in said development. Housing will be popular, on high demand, and through that some benefits can be obtain with each house that is sold off. Ultimately, cost and benefits will be shared for all groups involved [7].

It is important however, to not drift apart from the premise of sustainable communities, "low carbon housing" is only seen as a sensible financial investment, but masking it this way tends to put some distance between the real solution: a life style where citizens learn and keep deep green values, where they can clearly feel that their actions or lack of them, have a direct impact on their lives and to those whom surround them.

The best adaptation communities can have, is towards an altruistic connection with the planet. Not because is need it in order to survive, but because the harmony and synergy with the environment allows an easier, healthier and happier life. For us and those who come after.

3. The Advent of Sustainable Communities

After the concept of sustainable development emerged, several authors began to apply the concept to local scales and delivering different approaches, such as sustainable housing, sustainable energy and sustainable communities.

Sustainable communities can be defined as the place that attend the needs of current and future residents in different age scales, promoting high quality of life and opportunities. Natural resources are used effectively and the relations among residents and their environment promotes social cohesion and inclusion, environmental awareness and economic prosperity [8]. Sustainable communities are also well planned and safe places offering equality and good services for all [9].

Governments are in effort to stimulate the creation of sustainable communities. There are many initiatives, especially in Global north governments [10]. A well-known agenda was the *Sustainable Communities Plan,* developed in England in 2003. The goal of the document was to guide the development of communities considering environmental, social and economic premises, coinciding with the current sustainable development definitions. Some initiatives to provide these sustainable communities include the recovery of urban infrastructures, empty properties demolition and creation of new towns and communities [11].

Generating many critics over land availability and housing market in England, the plan was revoked in 2015, in an initiative to hinder demolitions and provide new uses for empty buildings [12]. The plan was substituted by a range of plans and funding for innovative projects and community-based initiatives. Despite of the plan's rescission, the concepts explored in the document that characterize a sustainable community are still in use. According to Egan [8], sustainable communities should present the following components (Table 1):

Table 1. Description of Components of Sustainable Communities Concepts (adapted from Egan, 2004)

Components of sustainable community:	Description / main features:
Governance	• Governance systems that lead to inclusion, participation, representation and leadership. • Strong partnerships, community and voluntary sectors. • Sense of responsibility, community and civility. • Continuous improving.
Transport and connectivity	• Transport facilities and accessibility. • Walkability and Cyclability. • Wide telecommunication and internet access.
Services	• Education and training for people. • High quality local health care facilities. • Provision of variety of services.
Environment	• Efficient use of natural resources. • Environmentally friendly actions (recycling, walking, energy saving actions). • Protection of natural resources. • Sustainability consciousness.
Economy	• Jobs and training opportunities. • Land to support economic prosperity and change. • Strong business community.

Housing and built environment	• Sense of place for the community. • Green spaces. • Range, diversity and affordability. • Well-designed built environment that describes the community. • Well planned, diverse and environmentally friendly buildings.
Social and cultural	• Sense of community. • Friendly, respectful, cooperative and tolerant environment among different members. • Opportunities for cultural, sportive and leisure activities. • Inclusion and similar life opportunities.

Considering the definitions explored above, the objective of this paper is to analyse Geroldsäcker eco-housing project in Karlsruhe – Germany and explore its potentialities and improvement points that would lead this initiative into a sustainable communities' concept.

4. Geroldsäcker Eco-Housing Project

Geroldsäcker Eco-housing is located in the northeast of Karlsruhe, in a 1,24-hectare plot with constructed area of 5.600m² (Figure 1). From this constructed area, 500 m² were designated to offices – including the office of project architect Andreas Löffler. The conceptualization of the project started in year 1989 and it was concluded in 1992. It is composed of 40 residence units in different typologies: building apartments and 3 store houses with private garden.

Source: Google Earth, 2018

Figure 1. Location of Eco-Housing Geroldsäcker

The project area is in a residential vicinity, with vast green and recreational areas and also access to public transport – approximately 500 meters from S-Bahn station – and cycling paths (Figure 2).

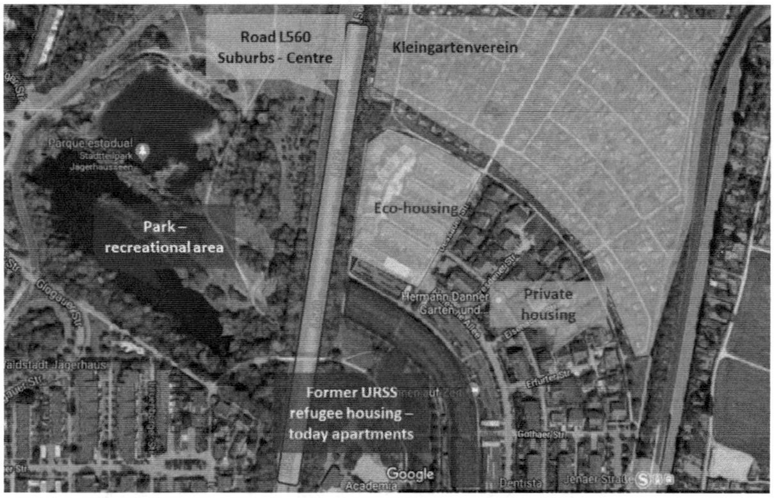

Source: Elaboration (the author)

Figure 2. Land Use of Geroldsäcker Eco-Housing Surroundings

Geroldsäcker uses ecological materials and concepts in its construction and infrastructures, which will be further described. Regarding accessibility to services, the project is located in a residential area, even though there are reasonable offer of services nearby, such as small health clinics, supermarkets, educational facilities, and many spaces for leisure and sport activities (Figure 3).

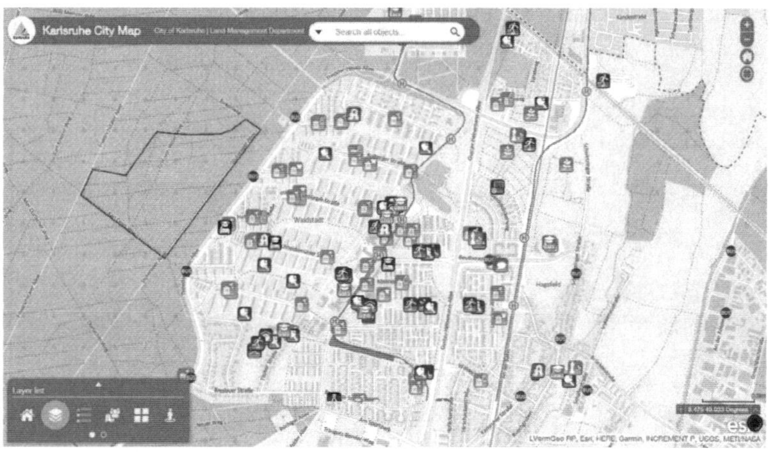

Figure 3. Services Around Geroldsäcker Project (in yellow) area.
Source: Karlsruhe Manicipality, 2018 [13]

The development and construction of the eco-houses involved the participation of the Municipality of Karlsruhe, KIT - Technological Institute of Karlsruhe, Associated Architects office Löffler & Schneider, Credit Institution Karlsruhe Initiative Group for Ecological construction and the initial community initiative to create the design of buildings [14] (Table 2):

Table 2. Stakeholders' Role in Geroldsäcker Eco-Housing

Stakeholder / actor	**Role**
Community initiative Karlsruhe Initiative Group for Ecological Construction e.V – (40 members)	Started with a citizen initiative of 20 people "Ökodorf", that asked for the KIT support for a common ecological housing project. The community started and managed the process of planning and construction, participating in the concept and design of the eco-houses.
Municipality of Karlsruhe	In the 90's, the Municipality had reserved area to develop pilot projects (experimental buildings). Supporting publicity of the project by making every person wanting to buy apartments in the city aware of the Geroldsäcker project.
Architectural Office Löffler & Schneider	Among 12 architectural offices the one of Mr. Löffler won in an anonymous competition, due to the mixed typologies project.
University of Karlsruhe	Provided framework for participation of citizens (offering location), while the architects provided components for the modular construction system and planning concept. Technical support for planning, design and construction
Kreditanstalt für Wiederaufbau	Financing – credit loan - program "Ökologisch Bauen Energiesparhaus 60".

Source: Löffler, 2018 [14]

This model of participation was pioneer in the early 90's and was unique for the region of Baden Württemberg. Nowadays, the availability of areas for implementing pilot projects is decreasing in Germany, mostly due to land demand [15].

The identification of the sustainability aspects of Geroldsäcker Eco-housing project which can lead to a sustainable community concept was provided by literature research, site visits, interviews with residents and the settlement responsible Wilfrid Bettels [16], the architect in charge of eco-housing development, Andreas Löffler, [15] and [17].

5. Analysis Geröldsacker Eco-Housing Sustainability Aspects

Under the ecological, social and economic dimensions of the sustainability concept [5], the most important factors regarding the project development and operation were identified and are here described (Figure 4). The relation with the sustainable communities is then held in the next session in order to identity the potentialities and points of improvement for these type of projects in German reality.

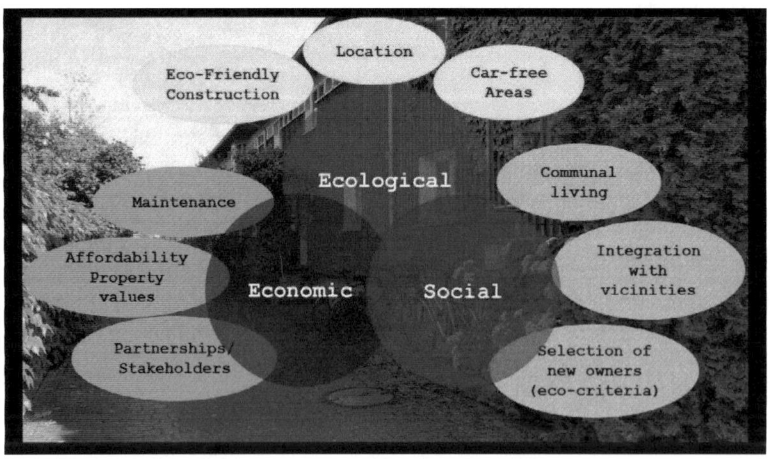

Source: The authors

Figure 4. Main Aspects Considered for Sustainability in Geroldsäcker Project in Terms of Economic, Social and Ecological Aspects

6. Ecological Aspects

Under the ecological aspects of the project, the eco-friendly construction features are an important characteristic for the projects' housing sustainability [18]. System for reusing grey water, composting toilets, harvesting and use of rainwater for gardening, toilet flushing and washing machines, use of solar energy, gas boiler combined with solar system, wall heating system with healthy materials, and also green natural barriers to contain noise from the nearby road are some of the ecological construction features of the project [19].

The project presents a community house for leisure activities and wide car-free opened area, enhancing the adoption of sustainable and non-motorized transport modes. There is one parking space for each residential unit, located in an underground garage. To bring groceries from the supermarket the residents use a trolley to help the transport [16].

Regarding the location of the project, the place is covered by public transport, and a reasonable number of services (Figure 5). Residents, in most of the time, access these services with bicycle and use the car to go to city centre and distant places.

Source: Löffler, 2018 [14]

Figure 5. Photo of The Communal Area of Geroldsäcker Eco-Housing

About 120 inhabitants reside in the eco-housing settlement with an average household size of three persons, with age range from 11 months to 75 years [16]. Partially shared open spaces and a community house enable strong communal living. Inhabitants point out the close and friendly relationship between neighbors as the key aspect for the attractiveness of their settlement. "Cultural events, various leisure groups, barbecues, etc. ensure a diverse and interesting settlement life" [20].

Furthermore, every resident is obligated to do ten hours of communal work per year. Therefore, the community implemented so called "task groups" to cover all relevant jobs that must be done, from administrative work to gardening. Construction improvements in each unit is a decision from each resident. As a result, there are some units that do not have solar panels yet.

When it comes to integration with vicinities, in general outsiders are welcomed by the eco-housing community, however, most of the time face different conditions than community members. The community house can be used by residents free of charge (once a year, afterwards for a symbolic payment of €10), whereas outsiders have to pay €150 for the use of the space. These outside-community members, however, have to be known by at least one member of Geroldsäcker eco-housing community, so they can make use of this space.

The participation in activities such as yoga classes held in the community are possible for outsiders, but also with a higher price (1€/ yoga-class for residents, €2.5/yoga-class for non-residents). Festivals hosted by the Geroldsäcker-community (e.g. "Friday barbecue") are for members only. Only very close relatives can be brought.

Buying an apartment in the eco-housing settlement as speculative financial investment is forbidden by the municipality [15]. That means, all residents are owner of their apartments. A selection process ensures that applicants share the same values as the existing community in terms of ecological and social aspects. Apartments for sale are never advertised publicly, and there was never the necessity to do so, and can therefore only be bought by people already knowing the community (mostly family or friends of community members). Renting the apartment to third parts is also forbidden in the community, with rare exceptions.

7. Economic Aspects

Regarding the maintenance of the eco-houses, the expenses of the common areas of the settlement are equally shared. Resources like water and energy are monitored by all residents, being a transparent process. The common areas are maintained by residents' cooperative work and use of shared tools, such as lawn mower and garden kits. Whenever there is a need of public maintenance or new public procurement, ideas of residents will be collected and then decision will be made.

By optimizing the construction features such as heat isolation, light intake, rainwater harvesting, the m onthly utility cost is also quite reasonable, in comparison with other residential areas. Nowadays, residents pay an average cost of water, heating and electricity of €250 per year [16].

The factor contributed to the affordability of this project is the mixture of row-housing and apartment complex, with typologies varying from 23m² to 220m² depending on the need for space. In addition, the modular construction system provided different packages to be installed in the apartments [15]. Residents could choose the package which was best suitable for their financial situation and needs.

The project started as a bottom up process, the community initiative "Ökodorf" consisting of 20 people initiated the project. Gradually, the project acquired stakeholders' interest and collaboration. The municipality offered the lot for this project as "Emphyteusis" (i.e. property rights is limited to use of land, not ownership) and the apartment prices were restrained not to be higher than the comparable conventional apartments in Karlsruhe [15].

There was no involvement of private investors, only the financial agency. The municipality provided the right to use the land to the community, in the figure of society. The participation of stakeholders was intense in the project, led by community expectations, and even contributed to intensify the workload to develop sustainable materials and the houses [15].

8. Potentialities and Improvement Points from Geroldsäcker Eco-Housing Towards Sustainable Communities

Considering the definitions and components of sustainable communities (Table 1), Geroldsäcker eco-housing somehow explores all dimensions, some more visible than others. As main potentialities of this project, the *environment*, *housing and built environment* and *social and cultural* components are highlighted. It is visible from the first site visits that the construction features, the diversity of house typologies and the planning of not only the houses but common infrastructures explore deeply the environmental friendliness concept. More than construction itself, the community is conscious and has a sense of belonging and responsibility with their space.

Regarding *services* and *transport and connectivity*, being a small community, with about 120 people, these components attend what is expected for the community, but it is uncertain how the scenario would be if this community was larger. From the aspect of transport and connectivity, the settlement has easy access to public transport (S-Bahn and Bus line) and bicycle routes that connect the settlement to other regions in Karlsruhe, despite the location of the project in a suburban area of the city. Most residents have cars, however, for daily activities, the bicycle is the most used mode. The car free area inside the project also has an important role of creating a consciousness on the use of sustainable transport modes. One present weakness and possible future strength, for this project and others whom which to use it as a model and build on top of it, is the lack of integration with electric vehicles. While it is true that 30 years in the past this was not a necessity, the new generations are adopting little by little this new technology, which is increasingly more efficient and economical. This could well be the key to tipping the balance in favour of this type of projects and make them more attractive and beneficial to both people and organizations: integration with new technologies to provide a futuristic notion.

In terms of services, inside the project there is only the provision of the community house for residents and vicinities. Residents have access to a range of services in nearby areas [Figure 3]. In this sense, the discussion of which services are necessary to a community regarding its size arise. When considering the project upscaling for bigger communities, the provision of a variety of services (including health care and education) will be indispensable, but in the case of Geroldsäcker eco-housing, these services are available and "shared" with different communities and neighbourhoods.

Economy is one of the least explored aspects in the eco-housing project, regarding its definitions for a sustainable community. Due to the small scale of the project, there are no job places. What is seen is the community work from the residents in the diverse activities necessary to maintain the settlement. Also, there is no strong business culture in the community. Even so, it is possible to assume that for bigger communities, with a diversity of services, as stated previously, some particular spaces could be transformed to allow small commercial activities to flourish, such as markets, coffee shops or restaurants and small convenience stores. This allows members to engage pro-actively and assemble stronger bonds.

In terms of *governance* there are many aspects that can be detailed and taken as an example from the project. Geroldsäcker was an initiative of a group of people, interested in a different and ecological approach for housing. The bottom up process is better known as a community-based housing initiative. In order to generate more sustainable communities with this characteristic, identifying potential communities and capacitate community leaders and representatives becomes necessary.

According to Dr. Gerd Kuhn, the main challenge of participatory projects is that group initiatives are often overchallenged by planning tasks. Therefore, allocation of tasks being done either by the community or by the planning agencies must be defined clearly. Transparent communication between the public sector and other stakeholders is essential [17]. Land availability is also an important factor for governance and economy. The eco-housing Geroldsäcker was feasible because of the land availability for experimental housing projects at the end of the decade of 80. These plots for experimental projects, however, started to enter in disuse in German municipalities due to, among other factors, the price of land and the pression of expansion of urban areas [15].

An innovative mean of implementation is the concept procedure [17]. In contrast to the common practice of selling the plot to the highest bidder, surveyor determines an appropriate fixed price. Different actors, such as citizen initiatives, assembly groups (Baugruppen) or real estate developers can apply for the land with their concept. Whereby, the municipality can embed guidelines for the concepts, such as compulsory community spaces. Thus, the decision is taken according to the best idea and quality, creating room for the participatory approach. Means of implementation could be appropriate zoning, land regulation plans and incentives for construction: The city of Tübingen implemented instruments to control land use and occupation to prevent investors from acquiring land in the first place.

An innovative strategy implemented in Kirchheim allows the entrepreneur to build on 20% of the plot, whereas the land use of the remaining 80% is determined by the citizens' initiatives [15]. In both scenarios municipal authorities should provide legal framework and guidelines, but only engage in the planning procedure when necessary. This way, bottom-up processes are supported, letting communities design their urban landscape according to their needs, maximizing social benefits.

Much of the success of the project was due to the stakeholders' model, their interests aligned perfectly, allowing a smooth collaboration among all actors; back then, land availability was not a problem and the economic condition was such that set the ground upon this idea began to take form, but in order to ensure the success at a greater scale is necessary to encompass a broader range of actors, in a much more complex model of participation. Nevertheless, it was only throughout the community initiative "Ökodorf" of 20 people, all sharing similar environmentally friendly belief, which started everything. These projects might be affected by the lengthy bureaucracy needed to accomplish big goals, but communities are capable of accomplishing 100 little things faster, so long as they are guided properly and make all their members feel engaged and develop a sense of belonging. After all, "A thread usually breaks where it is thinnest" [21].

9. Conclusion

Geroldsäcker project has unique characteristics that justify its role as reference in ecological housing and is a successful project in many of its aspects. It started as community initiative, a bottom up process. The community asked for the university support to develop ecological houses and the municipality had, at that time, available areas designated to experimental building projects.

Geroldsäcker cannot just be interpreted as an eco-housing project due to its construction features. It involves several dimensions of sustainability that evolves to a concept of sustainable community. From the social and cultural traits of its members, is safe to conclude that they all encourage each other to be sustainable, the "everyone knows each other" culture ensures the feeling of a big family; there is a one to one correlation between how integrated a person is in a community and how much sense of fulfillment it feels. They serve as an inspiration, a role model of how to be a personal hero for the planet.

Sustainability is already part of the corporal agenda, with increasing energy costs, in order to meet this demands investors, companies and governments are already looking for innovative ideas to make this possible; development of new technologies must be meet with the desire, need and knowledge to properly use them; by up-scaling the model, observed threats in the current model can be ameliorated and this new stakeholders can take part, providing financial support, and administrative and technical backups.

Finally, with the improvement of Geroldsäcker model, people will be properly capacitated and able to apply the same concept to other sites in order to create more sustainable communities. As every member brings their collaboration to the table, and experiences firsthand how their actions do have a positive impact in their surroundings, this may ensure an eco-friendly life style that will pass to the upcoming generation.

10. Acknowledgement

This article was presented at the 2nd International Conference on Smart City INNOVATION (ICSCI) 2019, jointly held by Universitas Indonesia and Universitas Diponegoro. ICSCI conferences have been supported by the United States Agency for International Development (USAID) through the Sustainable Higher Education Research Alliance (SHERA) Project for Universitas Indonesia's Scientific Modelling, Application, Research, and Training for City-centered Innovation and Technology (SMART CITY) Center for Collaborative Research, administered through Grant #AID-497-A-1600004, Sub Grant #IIE-00000078-UI-1.

11. References

[1] Pareja-Eastaway M 2016 *Sustainable Communities and Urban Housing* (Routledge)
[2] Priemus H 2005 How to Make Housing Sustainable? The Dutch Experience *Environ. Plan. B Plan. Des.* **32** 5–19
[3] Oyebanji A O, Liyanage C and Akintoye A 2017 Critical Success Factors (CSFs) for achieving sustainable social housing (SSH) *Int. J. Sustain. Built Environ.* **6** 216–27
[4] Godfrey S, Dean J and Regier K 2019 Sustainable housing *A Research Agenda for Housing* (Edward Elgar Publishing)
[5] Anon 1987 *United Nations Our common future* (Oslo: United Nations Documents)
[6] International Institute for Sustainable Development 2018 Sustainable development
[7] Lovell H 2004 Framing sustainable housing as a solution to climate change *J. Environ. Policy Plan.* **6** 35–55
[8] Egan J 2004 *Skills for sustainable communities: The Egan review* (RIBA)
[9] Maliene V and Malys N 2009 High-quality housing-A key issue in delivering sustainable communities *Build. Environ.* **44** 426–30
[10] Sustainable Communities 2018 Polices, Ordinances and Taxes
[11] Raco M 2005 Sustainable Development, Rolled-out Neoliberalism and Sustainable

Communities *Antipode* **37** 324–47
[12] United Kingdom 2019 Written statement to Parliament: Demolition
[13] Municipality K 2018 Geodata service
[14] Löffler 2018 Löffler und Schmelling Acrchitecktur Burö
[15] Löffler A 2018 Interviewed by: Zerbst, L.
[16] Bettels W 2018 Interviewed by: Andrade, R. J; Carstens Gomes, K. N; Napitupulu, B; Vo, M T; Zerbst, L.
[17] Kuhn G 2018 Interviewed by Andrade, R. J; Zerbst, L.
[18] Pickerill J 2017 Critically Interrogating Eco-Homes *Int. J. Urban Reg. Res.* **41** 353–65
[19] Stadtgrenze 2018 Geroldsäcker, Karlsruhe
[20] Stadtwiki K 2018 Ökosiedlung
[21] Reid T 1786 EIP. Essays on the Intellectual Powers of Man

Land Use Change Along *Lebih* Coastal Line, Entailing Its Designation as A Zone for Tourist Development

G A M Suartika, I A S O Ledysianty and K E Saputra

Universitas Udayana, Bali, Indonesia

E-mail: ayusuartika@unud.ac.id

Abstract. The coastal area is a physically productive area, but is easily degraded which may be due to the lack of planning or even natural factors. The coastal area in Gianyar has experienced a drastic physical development, especially after it is declared as *Lebih Tourist Zone* or Lebih Tourist Zone (LTZ) and the construction of the arterial road between *Tohpati* and *Kusamba*. The study documented in this paper discusses: (i) the form in which the coastal area in *Gianyar* are utilized post the recognition as the LTZ; (ii) the factors that influence the changes in land utilization in this region; and (iii) how does this condition affect the sustainability of the LTZ. Fundamentally, this study applies the conception of sustainable development and the descriptive qualitative research method. The data were collected through field surveys and in-depth interviews with the land owners, land users, and the authority. The research results showed that the LTZ has experienced a significant change, both physically and non-physically. The factors that influence this change in function is the physical condition of the land, economy, institutional factors, and accessibility. This change has implications for the environment, economy and social culture. This study concludes to develop the sustainability of the LTZ some fundamental actions must be made, namely: evaluate the usage of land, establish clear and restricting zoning law to anticipate any land use violation; consistency in the layout plan; the active participation in the control of land use by the government, the people and private sectors; providing incentives and disincentives in utilizing the space.

Keywords:change, land use, sustainability of Lebih Tourist Zone (LTZ)

1. Introduction

The Gianyar Coastal Area which located in the southern part of Gianyar Regency has aesthetic attraction as a natural tourism and it is supported by socio-cultural potential, and therefore, this area was designated as Lebih Tourist Zone (LTZ) in accordance with The Decree by the Governor of Bali (SK) Number 528, Year 1993. Related to this decree, the Governor of Bali has also enacted SK No. 16/2002 concerning *Rencana Detail Tata Ruang* (*RDTR*) [1] or The Detailed Layout Plan for the areas Along the *Tohpati-Kusamba* Artery Road which has the same delineation located in the Regency as the LTZ. This endeavor contributes to the development of the region and the increase of interactions with other centers of activity.

After being established as LTZ, physical development along the coast of Gianyar keep on growing and developing which then affect the economic activities, socio-cultural elements of the community as well as the ecological functions of the environment in the coastal areas of Gianyar. The rise of the development encourages people who initially rely on the agriculture, plantation, forestry, livestock or fisheries sectors to shift to the tourism sector and consequently, agricultural land transform into non-agricultural land such as settlements, trade, and tourism support services.

Despite contributing to positive economic development and the welfare of the surrounding community, the tourism sector also gives a negative impact if it is not anticipated. This is related to socio-cultural issues that are expected to be the priority, especially those related to the implementation of religious activities. There are many temples in this area and beach is considered as a holy place where *Melasti* (cleansing ceremony) and *Nganyut* (part of cremation ceremony) conducted. In addition, all activities in fulfilling irresponsible tourism needs also have a negative impact on the environment, including violations of coastal boundaries, river pollution and other activities that occur on the coast that disrupt coastal protection functions resulting in increased potential for environmental damage [2].

The area along the Gianyar coast has three strategic functions that need to be preserved and developed. First, the ecological functions, coastal areas are beneficial in the preservation of fisheries (edible fish, ornamental fish, and other aquatic plankton) and beach abrasion control. Second, the economic functions, the coast is a place for developing natural and spiritual tourism. Whereas from the socio-cultural function, the coastal area is considered a sacred area and serves as a place for holding various religious ceremonies for Hindus. Given the importance of the functions and roles of the Gianyar coastal region for development and the socio-cultural activities of the Gianyar community, this area should be used in a way that maintain a balance in fulfilling these three functions.

2. Methods

This study uses qualitative methods with descriptive analysis techniques in assessing changes in land use along the Gianyar coast after the establishment of Lebih Tourist Zone (LTZ). The research area lies along the southern coast of the Gianyar Regency area. The coastal area in Gianyar Regency has beaches that stretch from *Dusun Siyut* (*Tulikup Village*) to *Gumicik* Beach (*Ketewel* Village).

3. Overview of LTZ

LTZ with the area of ±4.577 Hectare consists of 9 (nine) villages in the administrative area of Sukawati Sub-district (*Candra Asri/Batubulan, Ketewel, Sukawati*), Blahbatuh Sub-district (*Saba, Pering, Keramas, Medahan*) and Gianyar Sub-district (*Lebih and Tulikup*). The delineation boundaries of the LTZ are as follow, bordering north by the administrative area of *Dusun Sasih* (except *tempekan Candra Asri*), *Batubulan* Village, *Batubulan Kangin* Village, *Guwang* Village, *Celuk* Village, *Batuan* Village, *Kemenuh* Village, *Blahbatuh* Village, *Belega* Village, *Abianbase* Village, *Serongga* Village, *Banjar Pande*, *Tulikup* Village, bordering east with the *Klungkung* Regency, bordering south by the *Badung* Strait and to the west bordering the Denpasar City area. For more details, an Overview of LTZis described in Figure 1.

IOP Conf. Series: Earth and Environmental Science **396** (2019) 012014 doi:10.1088/1755-1315/396/1/012014

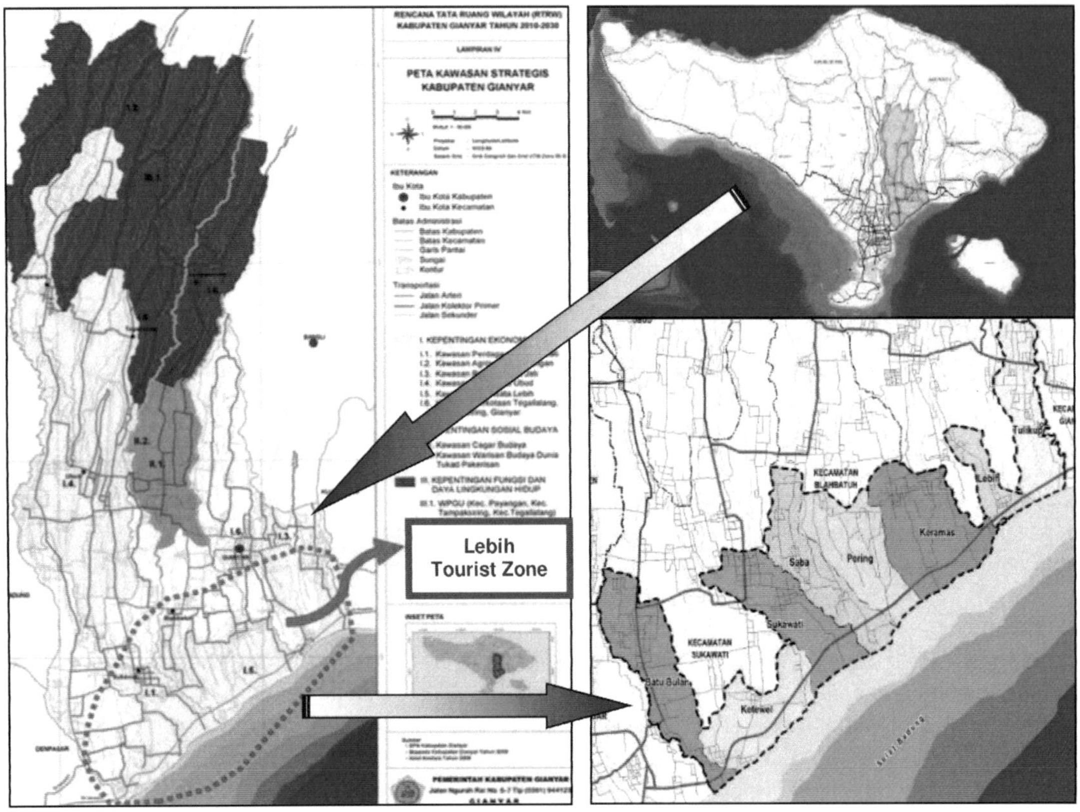

Figure 1. Map of Lebih Tourist Zone
Source: Spatial Plan of Gianyar Regency, Detailed spatial Plan for Lebih Tourist Zone, 2013

4. Geographical Conditions of the Research Area

The study area has a sloppy topography (lowland and coastal plains) with good land suitability for agricultural and non-agricultural activities. Lebih Tourist Zone (LTZ) mostly consist of beach/coastal areas, and thus based on oceanography, the environment is affected by tides. Beaches along LTZ are lowland coastal areas with a coastline of 14.95 km (RDTR, 2013) that stretch from west to east which are located along the LTZ. In this area, there are no Sad Kahyangan temple, but there are 10 Kahyangan Jagat temple and 3 of them are *Dang Kahyangan* Temple namely *Erjeruk*Temple, *Masceti* Temple and *Tengkulak* Temple.

5. Transportation Network Conditions

Lebih Tourist Zone (LTZ) are crossed by two national roads, namely *Tohpati-Sidan* Street and Tohpati-Kusamba/ Ida Bagus Mantra Street and it also split by a Provincial Road namely *Ketewel-Sukawati* Street. The national road located in the research area is part of *Tohpati Kusamba* (Jl. IB. Mantra) Street with a road length of 15,899 km from *Tohpati* to *Siyut* Beach.

6. The Utilization of Lebih Tourist Zone Area

The form of land use that has occurred in Lebih Tourist Zone (LTZ) among others:

a. Settlement Activities

Because of its natural potential, LTZ is targeted by the developers/investors as a place to build settlements and villas, especially for foreigners/outsiders. The area which is in a great demand is mainly situated along the coast, and the land around the view of rice fields and river banks. The physical and panoramic attraction has the potential to be an alternative location for developing new settlements and tourism/recreational settlements.

b. Wetland/Rice Field Farming Activities

Agricultural activities, especially food crops dominate land use in LTZ (reaching 55.32% of the total area). The widest use of agricultural land/rice fields can be found in Pering Village, Sukawati Village, Saba Village, and Keramas Village, while the smallest land use for rice fields can be found in Lebih Village. Meanwhile, the highest conversion of agricultural land into settlements occurred in Pering Village and Ketewel Village. While the threat of conversion to tourism accommodation occurs mainly in the villages of Ketewel, Sukawati, Saba, Keramas, Medahan, Lebih, and Tulikup. Rice fields designated as green lines are found in Saba and Pering villages. The average rice field area are situated in the southern part of the residential area that stretches from west to east. The existence of the Tohpati-Kusamba Primary Arterial Road, which crosses this LTZ area, almost entirely cuts off areas that were previously used as rice fields.

c. Dryland/ Moor Farming Activities

Dryland/ moor farming activities in the study area are situated along watershed or river banks and certain places which do not have irrigation networks, with gradient slopes of between 0 and 450 and generally planted with plantation crops. The plantations in question are plants that are cultivated on the ground such as; coconuts, fruit plants, tobacco and *senggo* wood plants which are limitedly developed. Dryland or tegalan are generally managed in a traditional manner instead of in an intensive way. The areas which dominated by dryland/ tegalan agricultural activities are Saba Village, Sukawati Village and Ketewel Village.

d. Fisheries Activities

This activity covers fishing as done in Br. Gumicik-Ketewel Village, Br. Saba-Saba Village and Br. Lebih Beten Kelod-Lebih Village. This activity has survived, along with the existence of a dining business that sells a menu of processed seafood. While the activity of shrimp farming takes place along the coast of Gumicik Beach, Purnama Beach, and Siyut Beach, has ceased.

e. Mining Activities

Mining activities on a small scale, namely the collection/ extraction of decorative stones that are widely available at Masceti Beach (Medahan Village), Lebih Beach (Lebih Village) and Siyut Beach (Tulikup Village). This activity does not use land permanently but it only uses the area on the beach for shelter and as a place to put decorative stones that have been collected. Nevertheless, this activity certainly has other impacts on the environment, besides creating the impression of slums in the area/region, this activity also brings a negative impact on the environment. In addition, this activity triggers beach abrasion, and therefore it is prohibited (for example in Masceti, Keramas Village), however, some people still carrying out such activity due to economic needs.

f. Trade and Service Activities

Trade and service activities tend to develop along the Tohpati Kusamba Primary Arterial Road and along the coast. In addition, in the corridor of the Cucukan-Lebih Village, tourism support facilities have also developed, as a result of the construction of Safari Park Attractions.

g. Tourism Activities

Tourism activities are mostly in the form of beach tourism which stretches from west to east, namely: *Candraasri* Beach in Batubulan Village, *Ketewel* Beach, Gumicik Beach, *Manyar* Beach and *Pabean* Beach in *Ketewel* Village, *Rangkan* Beach and *Purnama* Beach in the Sukawati Village administration area, Saba Beach in the administrative area of *Saba* Village, *Keramas* Beach in *Keramas* Village, *Masceti* Beach and *Cucukan* Beach in *Medahan* Village, *Lebih* Beach in Lebih Villages, *Siyut* Beach, *Gumicik* Beach, *Saba, Selukat* and *Masceti* (*Keramas* Village), *Lebih* and *Siyut* Beach in *Tulikup* Village.

7. Distribution of Tourist Attractions

The distribution of tourist attraction in the 2018 research area is described in Figure 2.

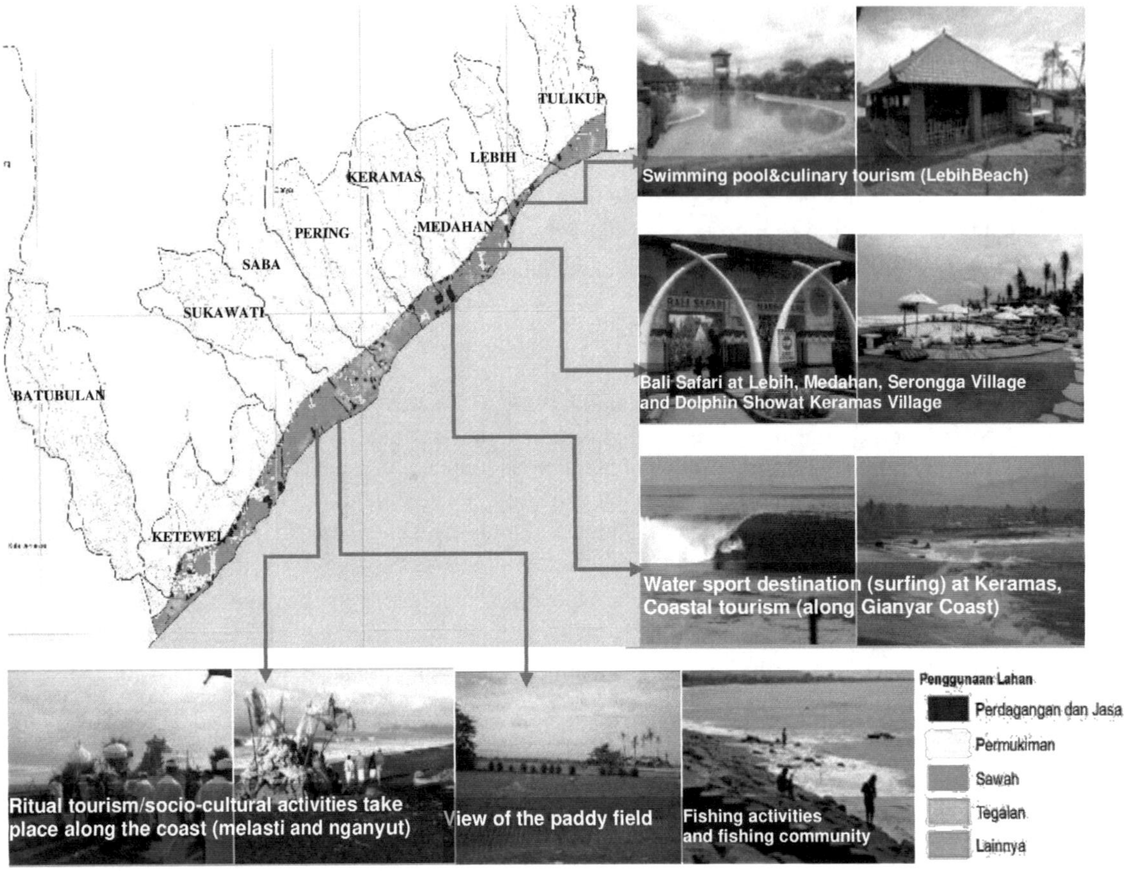

Figure 2. Detailed Layout of the Lebih Coastal Zone

Source: Spatial Planfor Gianyar Regency, Detailed Spatial Plan for Lebih Tourist Zone, 2013

8. The Form of Land Use along the Gianyar Coast after designated as LTZ

The development of Lebih Tourist Zone (LTZ) began to be identified since 1993. In the period of 1993-2018, LTZ experienced significant changes, both physically (the amount of development/land conversion), and non-physical (changes in the level of the economic, social and cultural life of the community)[3]. Several phases of land-use transformation are illustrated below.

a. The Land Use in the Period of 1988-1993 (before being designated as a tourism area)

The use of LTZ land in the period of 1988-1993 was generally dominated by open land/ no building, namely agriculture/rice fields, dry land, and the existing buildings were semi-permanent. There is no significant land conversion in this period. The area also does not have an arterial road infrastructure network and is only linked by several local roads that connect the rural area to the city. The map of LTZ land use in the research area in 1993 based on agency data, observations, and interviews are outlined in Figure 3.

IOP Conf. Series: Earth and Environmental Science **396** (2019) 012014 doi:10.1088/1755-1315/396/1/012014

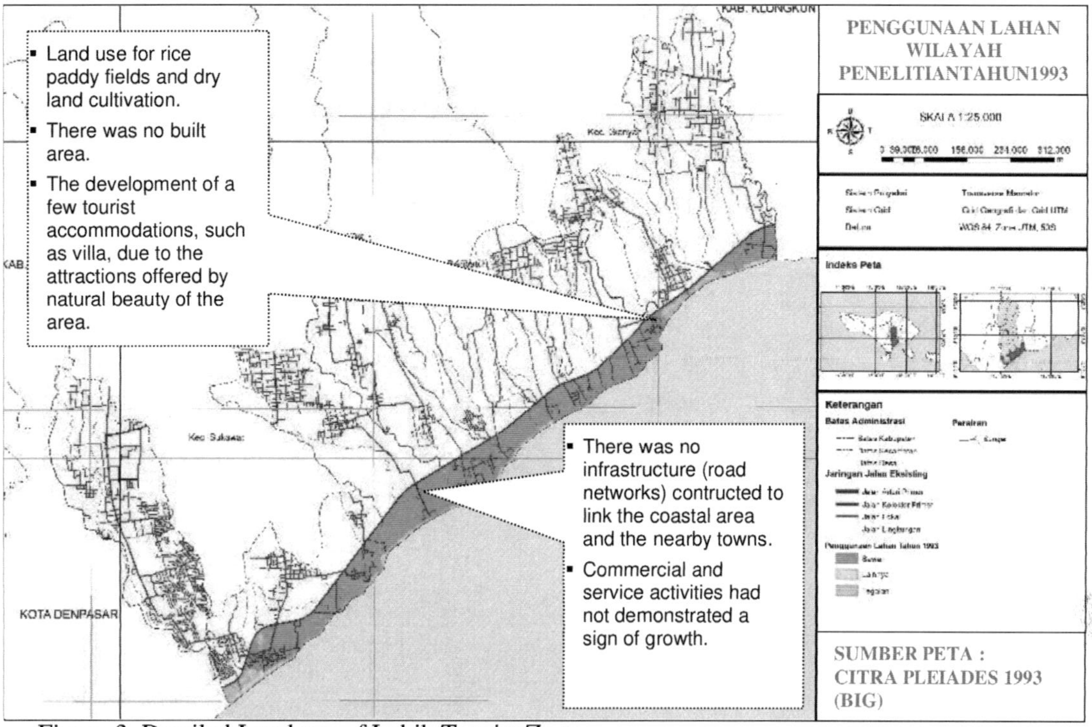

Figure 3. Detailed Land use of Lebih Tourist Zone
Source: Spatial Planfor Gianyar Regency, Detailed Spatial Plan for Lebih Tourist Zone, 2012 [4]

The percentage of change in land use in the region can be seen in Table 1.

Table 1 The Changes in Land Use in the Research Area (1988-1993)

No	Land use	1988		1993		Change	
		Area (ha)	Percentage (%)	Area (ha)	Percen-tage (%)	Area (ha)	Percentage (%)
1	Paddy Field	598,313	85,45	598,313	85,45	-	-
2	Dry Land/ moor	36,635	5,23	42,806	6,11	(6,170)	(0,88)
3	Others	65,237	9,32	59,066	8,44	6,170	0,88
	Area of Open Land	**700,185**	**100,00**	**700,185**	**100,00**	**-**	**-**
	Total	**700,185**	**100,00**	**700,185**	**100,00**	**-**	**-**

Source: Data processed by [5]

b. The Land Use in the Period of 1993-2002 (after being designated as a tourism area)

In order to maximize the potential of the region as a tourist destination, in 1993 the Bali Provincial Government established a tourism area in the research area through SK of Governor of Bali Number 528 of 1993 concerning 21 Tourist Regions in Bali. The development of tourism in this region began to grow and develop, as community activities changed as a result of tourism. This is marked by massive construction of villas/tourist accommodation, especially in the zone designated as a tourist attraction (beach), among others that occur in the Village of Saba, Medahan, and Pering. Changes in land use that occur in this period are presented in Table 2.

Table 2. The Changes in Land Use in the Research Area (1988-1993)

No	Land use	1993		2002		Change	
		Area (ha)	Percentage (%)	Area (ha)	Persentase (%)	Area (ha)	Percentage (%)
1	Paddy Field	482,103	68,85	476,375	68,04	(5,728)	(0,82)
2	Dry Land/ moor	61,704	8,81	62,202	8,88	0,498	0,07
3	Settlement	51,087	7,30	53,216	7,60	2,129	0,30
4	Commerce and Service	7,450	1,06	7,525	1,07	0,075	0,01
5	Other	97,841	13,97	100,867	14,41	3,026	0,43
	Area of Open Land	**641,648**	**91,64**	**639,444**	**91,33**	**(2,204)**	**(0,31)**
	Area of Built Land	**58,537**	**8,36**	**60,741**	**8,67**	**2,204**	**0,31**
	Total	**700,185**	**100,00**	**700,185**	**100,00**	**0,000**	**0,00**

Source: Data processed by Badan Pusat Statistik (Bali), 2015

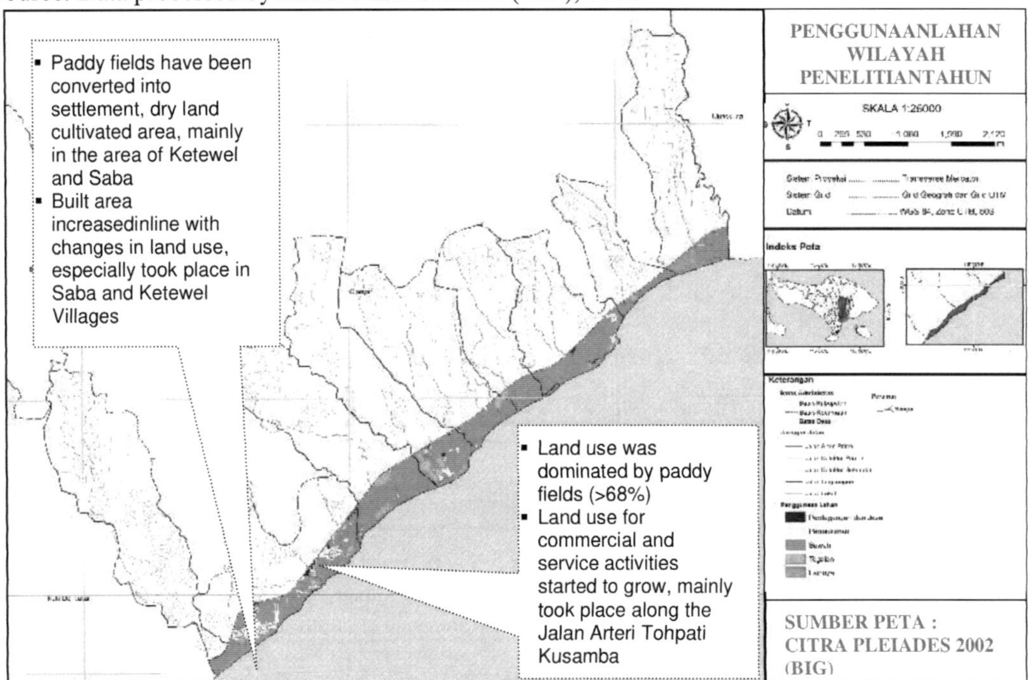

Figure 4. The Map of Land Use in the Research Area in 2002
Source: Processed satellite image, 2018

In the 1993-2002 period, the conversion of agricultural land into non-agricultural land began to occur, including settlements, trade and services, although such conversion is only at a relatively minimal percentage. The conversion mainly occurs in Kusamba Tohpati arterial road. Based on agency data, observations and interviews conducted, the development of land use in LTZ in 2002 was presented in Figure 4.

c. The Land Use in the Period of 2002-2012 (after the enactment of SK of Gianyar Regency No. 227/2002 concerning Detailed Spatial Plan (RDTR) of Lebih Tourist Zone)

In this period, the Government of Bali Province carried out the construction of the Tohpati-Kusamba Primary Arterial Road based on SK of Governor No. 16/2002 [1], regarding the RDTR of the area along the Kusamba Arteri Tohpati Road which has the same delineation area in Gianyar Regency as the LTZ. In this period, the Government of Bali Province carried out the construction of the Tohpati-Kusamba

IOP Conf. Series: Earth and Environmental Science **396** (2019) 012014 doi:10.1088/1755-1315/396/1/012014

Primary Arterial Road based on SK of Governor No. 16/2002, regarding the RDTR of the area along the Kusamba - Tohpati Arterial Road which delineates the planning area in the Gianyar Regency area as the delineation of MPA. The observations show that the development of tourist accommodation grows rapidly which spread along the Tohpati Kusamba Arterial Road and in rice fields along Lebih Beach. This happened dominantly in the Coastal Village such as Medahan, Saba, Keramas and Lebih. The types of tourist accommodation built are private villas, hotels, and guesthouse. The changes in land use are presented in more detail in Table 3. Meanwhile, the developments of land use in LTZ at the end of 2012 can be seen further in Figure 5.

Table 3. The Changes in Land Use in the Research Area in 2002-2012

No	Land use	2002		2012		Change	
		Area (ha)	Percentage (%)	Area (ha)	Persentase (%)	Area (ha)	Percentage (%)
1	Paddy Field	476,375	68,04	428,508	61,20	(47,867)	(6,84)
2	Dry Land/ moor	62,202	8,88	82,801	11,83	20,599	2,94
3	Settlement	53,216	7,60	78,588	11,22	25,372	3,62
4	Commerce and Service	7,525	1,07	22,013	3,14	14,488	2,07
5	Other	100,867	14,41	88,275	12,61	(12,592)	(1,80)
	Area of Open Land	**639,444**	**91,33**	**599,584**	**85,63**	**(39,860)**	**(5,69)**
	Area of Built Land	**60,741**	**8,67**	**100,601**	**14,37**	**39,860**	**5,69**
	Total	**700,185**	**100,00**	**700,185**	**100,00**	**0,000**	**0,00**

Source: Data processed by Badan Pusat Statistik (Bali), 2015 [3]

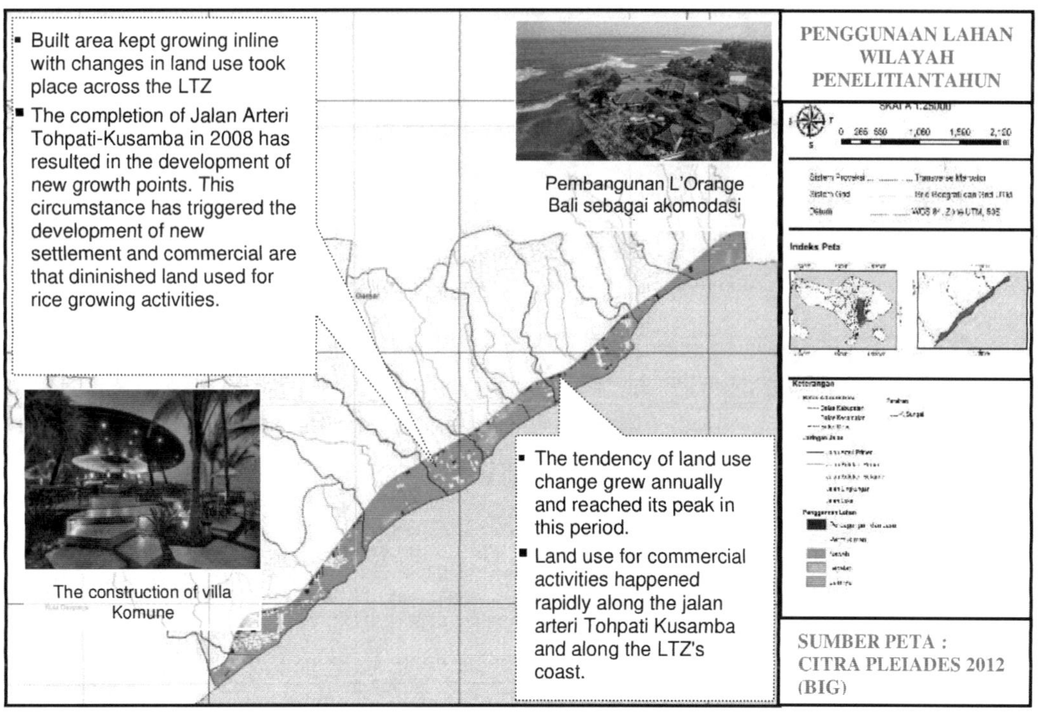

Figure 5. Land Use in the Research Area in 2002-2012
Source: Site observation

d. The Land Use in the Period of 2012-2018 (after the enactment of Governor Law No. 16/2012 concerning RTRW Gianyar Regency until now)

Table 4. Changes in Land Use in the Research Area (2012-2018)

No	Land use	2012		2018		Change	
		Area (ha)	Percentage (%)	Area (ha)	Persentase (%)	Area (ha)	Percentage (%)
1	Paddy Field	428,508	61,20	318,660	45,51	(109,848)	(15,69)
2	Dry Land/ moor	82,801	11,83	158,560	22,65	75,759	10,82
3	Settlement	78,588	11,22	79,411	11,34	0,824	0,12
4	Commerce and Service	22,013	3,14	54,830	7,83	32,817	4,69
5	Other	88,275	12,61	88,724	12,67	0,448	0,06
	Area of Open Land	**599,584**	**85,63**	**565,943**	**80,83**	**(33,641)**	**(4,80)**
	Area of Built Land	**100,601**	**14,37**	**134,242**	**19,17**	**33,641**	**4,80**
	Total	**700,185**	**100,00**	**700,185**	**100,00**	**0,000**	**0,00**

Source: Data processed by Badan Pusat Statistik (Bali), 2015 [3]

Governor Law Number 16 of 2002 concerning the Detailed Spatial Plan (RDTR) for the land along the Tohpati-Kusamba Arterial Road, stipulates that almost all of the southern part of the Tohpati - Kusamba Primary Arterial Road is a Green Open Space. This area has a 0% Building Base Coefficient, 200 meters thick from the road. However, this provision does not apply to the construction of public infrastructure supporting the security of road functions. In a further development, Gianyar Regency DPRD Decree Number 12 of 2010 stipulated that the ban on buildings on along the lane of the road in the Gianyar Regency area, especially on coastal access roads had triggered physical developments in the research area.

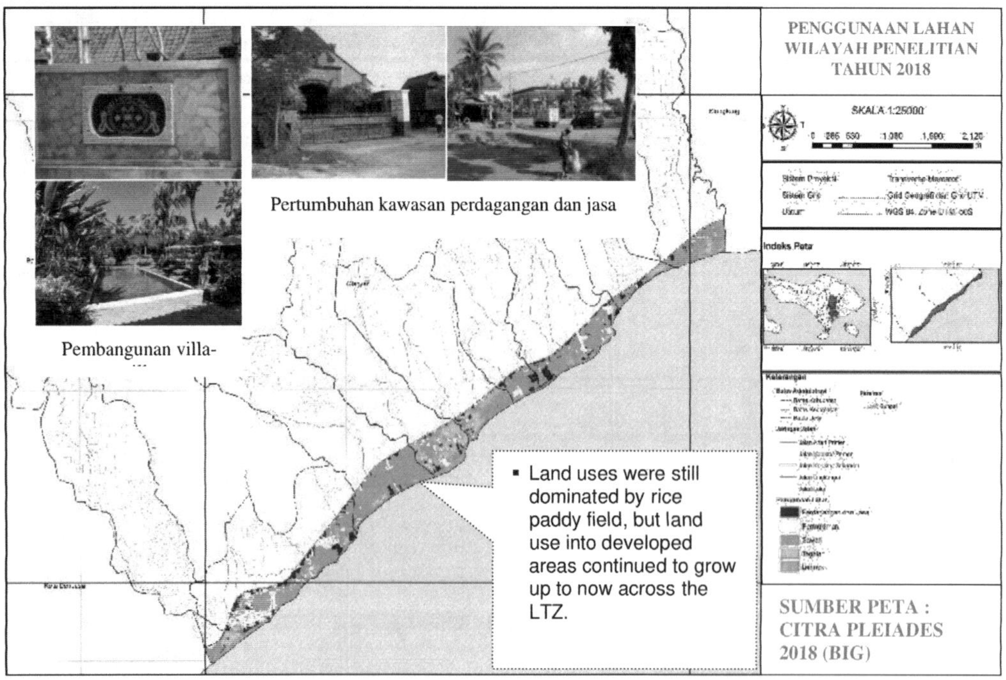

Figure 6. The Map of Land Use in the Research Area 2018
Source: Processed satellite image, 2018

In the current period, development progress has increased along with the growth of economic activity along the Tohpati-Kusamba Arterial Road, and thus land use in the region will continue to experience changes due to the ease of accessibility and developers' interest in investing their capital. The ease of accessibility also encourages the proliferation of tourist accommodation developments that are scattered in almost all research areas. The locations that are excellent for this purpose are villages of Medahan, Saba, Keramas and Lebih. The land conversion creates a pressure on the environment within the research area and therefore, it must be controlled in the future and we have to find solutions to minimize problems related to environmental sustainability. Table 4 presents changes in land use in the research area in 2018.

Data shows that more than 50% of the area is productive wetland found in almost all villages in the research area, including in Sukawati and Medahan Villages. Meanwhile, 50% of agricultural lands found in Saba and Lebih Villages are dryland/ moor. Furthermore, land use for settlements reached a level (10-20)% which spread in Ketewel, Saba, Medahan and Keramas Villages. This function is predicted to grow further in the future. Trade and service activities reached 10% -15%, which occurred in Keramas, Lebih, Medahan and Ketewel villages. The map of land use in KPA can be seen in Figure 6.

9. Factors Affecting Changes in Land Use along the Gianyar Coast after being designated as LTZ

The changes in land use that occur in the research area cannot be separated from various factors that influence the development of this region. These factors include land physical conditions, economy, institutions and accessibility [6]. This formula is based on the results of field observations, information obtained through interviews with people who exploit land in the research area.

a. Land Physical Conditions Factor

Physically, the existence of the coast is very important for people's lives, especially related to social and cultural activities. In addition, a beach is also a place for recreation, sightseeing, swimming, fishing, and so on. A large number of community activities in some parts of the coast provides an opportunity for the surrounding community to carry out various businesses including marine fisheries, fishponds, fishing, and so on. These activities are the result of the creativity of the community in utilizing the geographical and physical potential offered by nature to become a source of livelihood. This condition has encouraged the use of coastal land for seafood restaurants. The beachside food stalls which were originally built as a base began to be organized and managed by the community and the Desa Adat or customary village. This condition has given rise to a new identity for the Lebih region as a seafood culinary destination.

Naturally, LTZ does have a diverse natural appeal - panoramic sea with Nusa Penida view, sunrise, rice fields, tourist villages, safari parks, arts, surfing locations with big waves, rural atmosphere, agriculture and so on. This basic physical potential strongly supports the development of various tourism activities, such as coastal recreation, agro-tourism, nature tourism, water tourism, and sports. This potential supports the establishment of the region as a tourist destination. This determination is expected to be able to develop the southern region of Gianyar as the center of economic growth. However, this determination implies a shift in functions and utilization of regional spaces.

The data presented at the beginning has shown the amount of agricultural land and empty land which are converted into building/ residential houses. Native and beautiful nature, with a calming beach atmosphere which situated in a place relatively close from Denpasar City, is a consideration for migrants and developers to develop housing in this region. On the other hand, the geographical conditions of the region which are prone to abrasion encourage landowners (especially farmers) to change the function of their land or sell their land. Field observations show that the high level of land sale and purchase in the region has triggered the high rate of land-use conversion.

b. Economic Factor

The area can be used by local residents to improve their welfare by opening a business or by leasing their land for commercial activities. The land conversion due to economic reason in this area occurs rapidly. Most of the conversions aim to develop businesses by considering economic benefits. Strategic location, many visitors, ease of loading and unloading, easy access, economic prospects, and the establishment of tourism areas, have become the dominant considerations that have triggered the rapid development of LTZ [7].

Besides that, land prices also give a big influence. Before the construction of the Tohpati-Kusamba Primary Arterial Road, most of the land was owned by local residents and most of them were agriculture with relatively low prices (Rp. 15-20 million/100M^2). After the construction of road infrastructure, the economic value of land has increased significantly (reaching Rp. (150-500) million/ 100M^2). At the same time, farmers see agriculture as a non-competitive sector which cannot fulfill their basic needs. This condition has led to a rampant sale and purchase of land which in turn has triggered land conversion. This condition was harnessed by investors who came to open hotel/villa businesses. Land speculation is also a trigger in the emergence of this situation, due to the high demand for land by investors and/or developers. The land market is also enlivened by a large number of local people who become land brokers and make it a profession.

c. Institution Factor

The institutional factors in question are policies in spatial use, those issued by the government at the provincial level, the district level and the village level. The government plays a role in determining the direction of the development of a region through the rules set as guidelines and the basis for the growth and development of a region and the activities within it. However, there are some obstacles in the implementation. For example, there are some cases where space used not in accordance with the allotment/zoning. The green lane policy along the coast of Gianyar as stipulated in Bali Provincial Regulation No. 16 of 2009 concerning the Spatial Plan (RTRW) of the Province of Bali in 2009-2029 [4], considered to be detrimental to the existence of coastal areas which have been very potential in developing tourism in this region. The aspirations of the community to utilize the coastal area of Gianyar Regency are quite substantial. Consequently, in this period, there were many constructions which did not have permits, especially those located on the south side of the road.

Gianyar Regency DPRD Decree Number 12 of 2010 finally revised the provisions of this green line. This decree states that the prohibition on constructing buildings along the lane of the road in the Gianyar Regency area, especially on road access to the coast will be revoked. This decision triggered the physical development of the Gianyar Coastal area. Ideally, the use and development of this area must follow policies related to spatial planning and licensing requirements that follow the rules for utilizing space for the buildings constructed, such as the Building Base Coefficient rules, road borders, river borders, beach borders and efforts to control the use of space through the existence of green open spaces.

d. Accessibility Factor

Accessibility in an area/ region is influenced by one of the developments in transportation technology [8]. Roads as one of the transportation infrastructures play an important role in the development of a region. The road network will provide various impacts or changes that will arise due to the accessibility of the road infrastructure, creating physical connectivity of the region to other regions. Likewise, the existence of the Tohpati-Kusamba Arterial Road that crosses the research area provides easy interaction between regions in this area. Tohpati-Kusamba Arterial Road was built in a period of 15 years (1993-2008), which covered three regency/city areas, namely Denpasar City, Gianyar Regency and Klungkung Regency with a road length of 26.2 km. This arterial road passes through the area designated as LTZ (Figure 7).

Before the construction of the Tohpati-Kusamba Primary Arterial Road in the research area, the road network that available was the Denpasar-Sukawati-Blahbatuh-Gianyar-Klungkung-Padang Bai Arterial Road and local roads that connected villages and sub-districts. The accessibility offered by the Tohpati-Kusamba Arterial Road and the establishment of LTZ eventually boost the development of tourism activities in the research area. Tourism distribution from the southern Bali region (Nusa Dua, Kuta, and Sanur) towards the eastern part of Bali, transportation predominantly utilizes the Tohpati-Kusamba Arterial Road rather than the Denpasar-Sukawati-Padang Bai arterial Road.

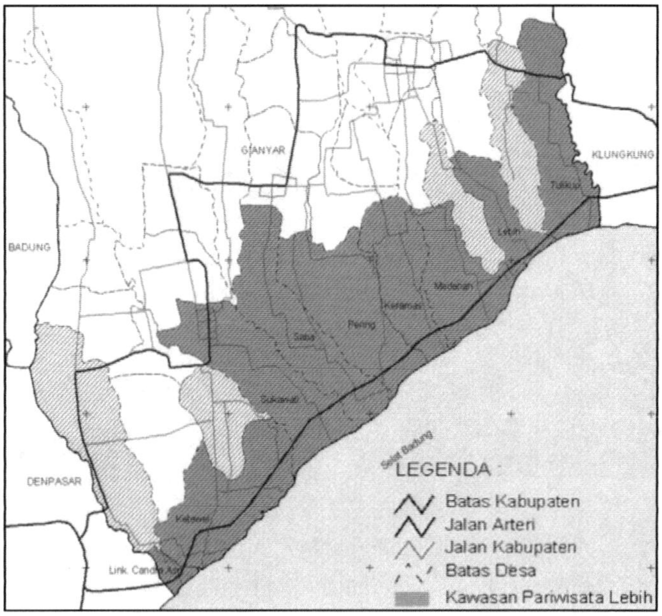

Figure 7. The Existence of Tohpati-Kusamba Arterial Road in the Research Area.
Source: Public Works Department of Bali Province

The new road network provides better access to the research areas, thereby encouraging the emergence of new activities that have an impact on the reduction of agricultural land/rice fields. Almost all types of existing activities related to rice fields and the moor around the area along the Tohpati-Kusamba Arterial Road has changed. This shows the role of accessibility as a trigger for land use around the surrounding area. The pattern of land use along the Tohpati-Kusamba Arterial Road follows the direction of the road, dominated by commercial activities such as trade, services, tourist accommodation, industry/warehousing and public services.

Trading/service activities began with the emergence of small shops managed by local residents. Some developers use the lands to build housing, land plots, shop houses, offices and various business places such as gas stations, cargo/warehouses, workshops, shops with specific merchandise specifications, building material/material business, cement industry and so on. Restaurant/ food court and art shop businesses also grow rapidly. The construction of hotel/guesthouse and villas, both in the existing rice fields and on the beach edges, also spread in the research area.

10. The Impact of Changes in Land Use along the Gianyar Coast to the Sustainability of LTZ

The changes in land use that occur along the coast of Gianyar certainly have an impact, both positive and negative. The impact of these land-use changes on the sustainability of LTZ will be outlined based on three pillars of sustainable development, namely ecological, economic and socio-cultural.

a. Ecological

The ecological impact of the development of land use in LTZ shows negative implications. LTZ vulnerability to the dangers of beach abrasion with relatively large waves has resulted in changes in the coastline. This process is followed by a sedimentation process that has the potential to cause silting of the coastal area. If this condition is not addressed immediately, the beach abrasion will get worse and threaten the functions along the LTZ. This condition also disrupts the daily activities of the fishermen. In addition to these natural factors, human interventions that do not provide sufficient space for the balance of the dynamic process of the coast, and other activities that change the pattern of tides and waves have caused a decline in the coastline. These facts certainly threaten the existence of various infrastructure and buildings along the beach. In addition to the danger of beach abrasion, other impacts that need to be addressed are waste and the occurrence of sediment from the river and religious ceremonies.

Figure 8. Beach Abrasion in LTZ
Source: Observation, 2015

The development that violates coastal boundary provisions is also an important condition that exacerbates the impact of beach abrasion on LTZ sustainability (Figure 8). This circulation will also come into contact with the role of LTZ as a public space that can be utilized by the general public. As stated earlier, the coastal area is widely used as part of the cultural and religious processions of Hindus because some temples (and their sacred areas) are located along the coast. The mechanism of control and firmness in applying regulations on how the complexity of functions (including coastal ecological functions) is regulated (Figure 9).

Figure 9. Condition of Lebih Beach
Source: Observation, 2015

b. Economic Sustainability

The changes in land use at the LTZ contribute great support for the sustainability of the economic sector. The establishment of LTZ itself initially aims to improve regional economic growth and land utilization for tourism accommodation such as hotels, villas, and guest houses, trade, restaurants, and others with the view to gain economic benefits from the designation of the area as a tourist area (Figure 10). The rise of services or offers in the fields of trade and services, restaurants and hotels can be perceived by looking at the increasing number of people who are involved in or working in these fields. The interview results show that from 2011-2015, 50% of the respondents received new jobs; shift (from farmers/farm laborers); and increase employment in trade and services, restaurants and hotels (Table 5).

Figure 10. Condition of Keramas Beach
Source: Observation, 2015

Table 5. Changes in Land Use in the Research Area (2012-2018)

General Work Field in the Year 2000	Number Based on Work Field (People)					
	2010	2011	2012	2013	2014	2015
Agriculture, Plantation, Forestry and Fisheries	672.204	556.615	572.685	545.827	38.098	39.268
Mining & Excavation	7.042	12.635	7.637	9.066	0	597
Processing industry	303.589	290.132	311.225	320.555	58.866	48.879
Electricity, Gas and Water	3.952	6.859	6.347	9.155	0	570
Construction	144.041	185.705	185.764	211.093	19.485	29.844
Commerce, Restaurants and Hotels	571.274	596.527	625.302	628.585	76.346	100.304
Transportation, Warehousing and Communication	95.202	81.744	85.711	72.764	11.745	9.261
Finance, Insurance & Building Rental Business	58.832	83.281	83.876	93.368	7.727	7.461
Community, Social and Individual Services	321.222	391.376	390.161	383.484	53.520	47.595

Source: Badan Pusat Statistik (Bali) of Gianyar Regency

In analyzing the impact of land use conversion on the economic sector, it can be described through analysis of land status and value hierarchy, which is described as follow.

1. Based on land status, the research area is classified as an area with high potential to be converted because it has land suitability, both for agriculture and non-agriculture. This condition has the potential to increase the economic value of land, especially those directly adjacent to the road and coast.
2. The preserved land will affect the economic dynamics that are owned and viewed as assets in development planning and their utilization must be guided by the applicable spatial plan.
3. Land value hierarchy related to the use and exchange value. The value of use (use-value) of land in LTZ is driven by community needs, especially for settlements, trade, and services. While the exchange value is the value of a land investment based on the calculation of the profits that can be obtained from the land purchase value. This is what underlies the practice of land speculation at the research area.

c. Socio-Cultural Sustainability

Almost all of the beaches in LTZ are used as a place to held nganyut and melasti religious activities, especially for the Hindu community. The development of land use in LTZ will have an impact on the sustainability of these socio-cultural activities. The transfer of land which provides space for cultural activities will be an attraction that supports the development of LTZ as a tourism area. Otherwise, the land conversion will have a negative impact[7,9,10]. The development of LTZ aims to synergize aspects of local culture with land use so that negative impacts can be minimized. The potential negative impacts that are likely to occur are: disruption of the existence of *Parahyangan* and the sanctity of temples along the LTZ; the impact on *Pawongan*, namely the occurrence of a cultural shift that might lead to inter-cultural conflict which implies a harmonious relationship between community members; and the impact on *Palemahan*, where land ownership, especially which are located along the Arteri Tohpati Kusamba

Road and along the coast has implications for the physical order of the area which certainly requires regulation and control regarding border violations (beaches, rice fields, tegalan, valley, cliffs, etc.).

11. Conclusion

The forms of land use along the Gianyar Coast after being designated as LTZ are dominated by the agriculture/rice field sector and dryland/ moor. During this period, the development of the region tends to be static so there is no significant change in land use. In the period of 1993-2002, changes began to occur, where agricultural lands shift into settlements. The construction of the Kusamba Tohpati arterial road that occurred during this phase also encouraged the beginning of land use for trade and service functions. The enactment of SK of Gianyar Regent No. 227/2002 on the RDTR LTZ (2002-2013) and the completion of the Tohpati-Kusamba Arterial Road (1993-2008) have resulted in the emergence of new growth poles, especially for settlements and trade and services. This has an impact on the decline in land area for agriculture. The enactment of Governor Law No. 16/2012 concerning RTRW of Gianyar Regency which serves as a guideline for regional spatial planning law in Gianyar Regency increasingly supports the physical developments in the research area. This condition has reached its peak in 2018.

The factors that influence changes in land use along the Gianyar Coast after being designated as LTZ, among others: land physical factors; economic factors; institutional factors and government policies that establish land direction and use in LTZ through spatial planning regulations that become a reference for the community in utilizing land and providing stimulants for the region to continue the development; accessibility factors, namely the existence of the Tohpati-Kusamba Arterial Road that crosses the research area provides easy access between regions in the area.

The impact of change in land use along the Gianyar Coast on the sustainability of LTZ has been analyzed based on three pillars of development, namely: ecological sustainability (violation of coastal border lines, violation of temple sanctity radius and disruption to protected functions due to high beach abrasion rates); economic sustainability (tourism activities open employment opportunities in the field of trade and services, but the land value become increasingly expensive); socio-cultural sustainability (supports the development of the region as a tourism area, but has an impact on *parahyangan* (violation of temples/holy areas), *pawongan* (conflict/shift of cultural values), and *palemahan* (violation of borders related to the existence of rice fields, moor, valley, cliffs and so on).

12. Acknowledgement

This article was presented at the 2nd International Conference on Smart City INNOVATION (ICSCI) 2019, jointly held by Universitas Indonesia and Universitas Diponegoro. ICSCI conferences have been supported by the United States Agency for International Development (USAID) through the Sustainable Higher Education Research Alliance (SHERA) Project for Universitas Indonesia's Scientific Modelling, Application, Research, and Training for City-centered Innovation and Technology (SMART CITY) Center for Collaborative Research, administered through Grant #AID-497-A-1600004, Sub Grant #IIE-00000078-UI-1

13. References

[1] Bali P P 2002 *Rencana Detail Tata Ruang Kawasan Sepanjang Jalan Arteri Tohpati Kusamba* (Denpasar: Pemerintah Provinsi Bali)
[2] Hadley D 2009 Land use and the coastal zone *Land use policy* **26** S198--S203
[3] Huang J, Pontius Jr R G, Li Q and Zhang Y 2012 Use of intensity analysis to link patterns with processes of land change from 1986 to 2007 in a coastal watershed of southeast China *Appl. Geogr.* **34** 371–84
[4] Bappeda 2012 *Rencana Tata Ruang Wilayah Kabupaten Gianyar Tahun 2012-2032*
[5] BPS 2015 *Gianyar dalam Angka 2015* (Gianyar)
[6] Olaniyi A O, Abdullah A M, Ramli M F and Alias M S 2012 Assessment of drivers of coastal land use change in Malaysia *Ocean Coast. Manag.* **67** 113–23

[7] Suartika G A M 2015 Sand, sea and ceremony: conflict over the littoral public realm in Sanur, Bali *Procedia-Social Behav. Sci.* **179** 128–40
[8] Grubler A 1998 *Global Energy:Perspective.* (New York: Cambrigde university Press)
[9] Suartika G A M, Zerby J and Cuthbert A R 2018 Doors of Perception to Space{\textendash}Time- Meaning: Ideology, Religion, and Aesthetics in Balinese Development *Sp. Cult.* **21** 340–57
[10] Suartika G A M and Saputra K E 2019 The Creation of Harmonious Living Spheres- Landscape, Capital, and the Balinese Way *IOP Conference Series: Earth and Environmental Science* vol 248 p 12065

Coworking Space and Cluster Spatial Relations in the Context of Jakarta City Spatial Structure

W Huliana[1] and E Ellisa[1]

[1]Department of Architecture, Faculty of Engineering, Universitas Indonesia, Depok, West Java, 16424, Indonesia.

E-mail: windi.huliana@gmail.com

Abstract. Y generations, or better known as the millennials, are currently the most significant addition in work place. Globalization and industry 4.0 development are the biggest external factors that shaped their character. Both of them has changed the way millennials live as a part of the society. This characters are reflected in the work place in a form of collaborative work. It is becomes the main reason why coworking spaces are now a global trend. Coworking spaces exist as a manifestation of millenials adaptation in occupying work place. With its shared-knowledge based system as the main concept, coworking spaces need a certain proximity with one another. It has an impact on coworking spaces distribution that tends to concentrated in a certain area, particularly in existing industrial cluster. This phenomenon is often called agglomeration economies or spatial clustering. The emergence of coworking space and the phenomenon of agglomeration are most likely occured simultaneously in urban environment. It is also occurred in Jakarta. However, it is still in its early development compared to Silicon Valley. Reflect on that successful cluster and based on current development studies on agglomeration economies, this paper will discuss about coworking space and cluster agglomeration process, relation, and potential capacity which will be integrated with Jakarta spatial structure current condition. The results will indicate the city structure adaptation towards cluster development and give depiction to foresee its future development possibilities.

Keywords: Coworking space, Cluster spasial, Jakarta city, Millenials.

1. Introduction

Globalization and rapid technology development have affected millenials to have a different perspective on how the world works. Those perspectives are reflected in the work place in the form of equal relationship and co-decision making [1]. The old style concept of working, turns out, is considered as unsuitable for most of the millenials. For this reason, millenials prefer to create their own opportunity by starting a new business. They are known as startups. The process to develop a new business is not easy for them. The biggest challenge that they usually face at the initial stage of their career is financial issue. The rent price of most conventional work spaces are far too expensive for them. Moreover, the physical layout of convenient work space itself also unsuitable for their work style. For this reason, coworking spaces emerge as a new work space style that adjust to startups capabilities and needs.

A predecessor of coworking space, workspace sharing, has actually existed since 1995 [2]. However, coworking space itself was established in 2005 in London and since then has rapidly increased until now [2]. Accordingly, coworking space is considered as a significant addition in urban area. Its presence has its own spreading pattern in urban area. These one particular feature of coworking space is actually similar with the convenient work place. There is two main contrary forces,

IOP Conf. Series: Earth and Environmental Science **396** (2019) 012015 doi:10.1088/1755-1315/396/1/012015

agglomeration and dispertion, that compel coworking space to be located on particular urban areas. These forces are discussed in a study called economic geography—agglomeration economies to be more specific. Agglomeration economies is commonly defined as a phenomena in which economic activities are concentrated in particular urban areas. It has compelled firms to get close to each other. As the result, this process has created a concentration of firm which is known as cluster.

The concept 'Internet of Things' (IoT) in this industry 4.0 era, has changed the way that cluster developed. One of the significant change is the shift of focus from industrial cluster to sharing economy oriented development [3]. Many clusters are now developed in that direction. The result of this process is manifested in the form of coworking space. Coworking spaces also have the tendency to concentrate in particular urban area in the same way cluster is formed. As a point of fact, coworking space commonly emerge in the existing area of cluster.

Coworking space is still considered as a new phenomenon in many cities. However, it can not be denied, the significant amount of it has influenced city development. Coworking space has been an addition to urban social infrastructure element and potentially create connection and collaboration between human, idea, and place [4]. Its capability to support millenials to develop creative and innovative industries, is very important for city development. In addition to this statement, Moriset also stated that creative individuals and innovation industries have a prominent role as key drivers of city economic growth and sustainability in this era [5].

The number of coworking spaces has increased globally. It is also occured in Jakarta. There, coworking space has reached 33% occupancy of work space with 77.345 sqm total space area [6]. On the other hand, Silicon Valley appears as one of successful clusters in the world. Reflect on that cluster and related theories, this paper will analyze and discuss about coworking space and cluster agglomeration process, relation, and potential capacity. The results will indicate how city spatial structure adapts to cluster development. By understanding current cluster capacity, its utilization can be improved for further regional development.

2. Literature Review

2.1. Agglomeration Economies

Agglomeration economies is a main approach in economic geography study. It is commonly used to explain agglomeration and dispersion forces that generate spatial configuration of economic activities. The theory of agglomeration economies is first introduced by Alfred Marshal in 1890. The theory itself has undergone through many changes and adaptations related to time and spatial contexts. Agglomeration economies can be occurred in various kind of urban area conditions. Their occurence are visible through physical manifestation in urban area. The common form of the manifestation is city economic growth disparities in particular country, the emergence of IT based industrial complexes, and the formation of city's regions that differ in size [7].

Agglomeration economies is generally divided into two different types, that is urbanization economies and localization economies [8]. The result of urbanization economies process is the existence of differentiation in city's area and economic growth disparities. Meanwhile, urbanization economies is resulted in the formation of industrial complexes, which one of them is cluster. Both localization economies and urbanization economies have a different ways on affecting the city formation

2.1.1. Proximity as spatial externalities that form cluster.

Externalities are non-market interactions that also affect agglomeration process. Proximity is one of spatial externalities that plays an important role in cluster formation. According to Porter (1990), quoted from [9], proximity makes interactions between firms more competitive and its operations more effective [9]. Proximity creates assets within a district and generate positive effects in a form of spatial lock-in [10]. As the result, firms are engaged to its concentration area and cluster is formed.

2.2. Cluster and Innovation District

The term 'cluster' is first introduced by Michael Porter in 1990 which he defines as industry concentration area [11]. The term itself then develop and now is widely known as location of concentrated interconnected firms and other related institutions and organizations. Industry 4.0 has caused a shift of focus from industrial complexes to coworking space development [3]. As a part of industrial complexes, clusters development is also affected by this shift. Coworking space emerges as an integral component of innovation districts [12]. For this reason, coworking space can also increase cluster potential to develop as an innovation districts.

In this part, the theory of innovation district is sourced from journal report of economic geography by Katz and Wagner [13]. Cluster and innovation district are basically similar in form; however, in innovation district, the existence of startups, incubators, and accelerators are more emphasized. Innovation district integrates economy shaping, place making, and social networking. Katz and Wagner state that these three aspects are common assets in innovation district. Moreover, it is divided into several sub-aspects.

First, economy shaping. This aspect is divided into three more specific sub-aspects which consists of innovaton drivers, innovation cultivators and neighborhood building amenities. Innovation drivers is an economic asset that concerns about firms sector. It is also related to the technology used in the production process. Innovation cultivators asset is mainly related to the subjects of innovation district which usually consists of firms and startups and related institutions or organizations. In addition to the previous two sub-aspects, building amenities exists to support main activities by providing services for both residents and workers in the district.

The second one is place making. It is also divided into three sub-aspects, namely publicly-held assets, privately-held assets, and assets that knit together physical urban fabrics. Publicly-held assets are required to enhance connectivity, collaboration, and connection. This sub-aspect is manifested in a form of public place. Public places exist as a platform to infuse innovative spheres in locals daily life. It should be digitally accessible, so public can be easily engaged with the district. This, moreover, will also affect the formation of social networking assets. Privately-held assets, on the contrary, is manifested in a form of privately-owned building. This assets generate positive effects by providing accessible and affordable private place that is configurated with public and work place. Related to previous two assets, physical assets that knit together physical urban fabrics are mainly manifested in form of connecting elements, such as streets, sidewalks, pedestrian and bike lane. Besides that, strategies like removing or lowering the barriers can also be applied. The main purpose of this asset is to create a good connectivity, both within the district and outside with broader areas.

The last one is social networking which divided into two sub-aspects; strong ties and weak ties. Strong ties and weak ties could be occurred if the presence of innovation drivers and cultivators is fulfilled. Strong ties are built from inter-subject connection in similar sector, meanwhile weak ties are build from cross-sector subject connection.

2.3. Agglomeration impact on city spatial structure

City spatial structure, in this research, is referred to Mohan (1994) who defined it as a distribution of urban area activities based on its type, location, and density [14]. City spatial structure, based on its population density distribution, is divided into two models, polycentric and monocentric. In monocentric model, activities are highly concentrated in Central Bussiness District (CBD). The density will decrease as the distance increase from CBD towards the city suburbs. Different from monocentric, in polycentric model, activities are not only concentrated in one area but divided into some new concentrations called subcenter. Polycentric model is basically formed from monocentric model—it is occured when the CBD loose its primacy [15]. The emergence of subcenter is closely related with agglomeration process. Agglomeration economies propel firms to get close to each other and form cluster, which manifestation is subcenter [16]. Subcenter existence is a prove that city spatial structure can change from monocentric to polycentric.

City center (CBD) and subcenter are basically formed through spatial configuration of economic activities. Agglomeration and dispersion forces create equilibrium which resulted in city spatial structure. However, this equilibrium does not always stay in the same condition. Opposite forces can occur and destabilize that equilibrium. Huriot and Thisse claim that opposite forces occur as an after effect of agglomeration. When agglomeration reaches its peak point, future possibilities of agglomeration in the same area will be hindered [7]. In the worst case, it will even turn into a dispersion. For this reason, opposite forces bring about the possibilities of change in city spatial structure, such as urban sprawl, the donut effect, and the emergence of edge cities.

2.3.1. Urban sprawl.

Urban sprawl is mainly driven by population growth in the city. It is occurred when city adapt to that growth by expanding its areas. As a result, there is some areas with a very low population density. Besides that, urban sprawl is also manifested in city in a form of agglomeration economies. Opposite forces, as mentioned before, can turn agglomeration into dispersion. If a population density in one area decrease, wider area will be needed to support economic activities in vicinity and economic activities will be more dispersed [16]. Urban sprawl is usually seen as a negative phenomenon in cities. It increases people travel distance which leads to congestion. However, urban sprawl has also positive impacts on cities. According to Burchell and Mukherji (2003), quoted from Prakasa, Soemardiono, and Defiana (2018), urban sprawl also cause transportation mode to disperse and give opportunities for low to middle class citizen to access city center easily [17].

2.3.2. The donut effect theory.

The donut effect theory is divided into two; the old donut and the new donut. The new donut is basically the development of the old donut which adapt to current city spatial structure condition and issues. The new donut itself is developed by Aaron Renn. Figure 1 shows the illustrations of both old and new donut theory. As it is presented in Figure 1, in the old donut theory, suburbanization is occurred and decrease the population density—of both employment and activities—in downtown or city center. Meanwhile, in the new donut theory, the density remains high in city center and collar counties or the suburban areas shows its development to become a new center which also indicates the emergence of edge cities.

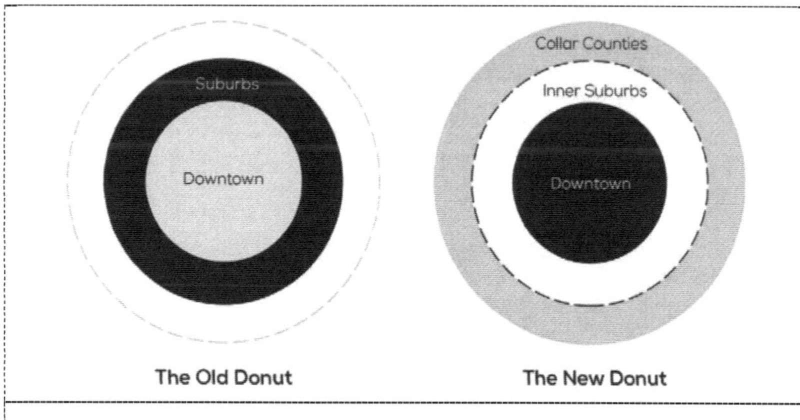

Figure 1. Illustration of The Donut Effect Theory.

2.3.3. Edge cities.

The concept of 'edge city' is first introduced by Garreau (1991) which briefly can be defined as an agglomeration in rural areas [18]. It is closely related with other phenomenon such as urban sprawl and decentralization. Edge city is also considered as a new form of urbanization [19]. It can be seen in a form of communities that cluster together around big firms or companies outside the CBD [19]. In many cases, edge city initially triggered by developer company's project which makes former empty area more liveable and attractive.

3. Learning from Silicon Valley

The discussions about cluster are inseparable from the existence of Silicon Valley as one of the most successful cluster. Silicon Valley presence has a strong influence in San Francisco Bay Area because it has the second biggest employment size after San Francisco [20]. Despite the existence of popular institutions and firms, Silicon Valley development process are occurred autonomously. This statement is supported by its historical facts since the location first known as the birthplace of semiconductor company named Fairchild Semiconductor in 1956 [21]. Silicon Valley is actually a big cluster that covers Santa Clara Counties. The counties consists of three cities; San Jose, Santa Clara, and Sunnyvale. Out of those three cities, San Jose is the most developed cities. It is ranked as the third most innovative cities in Innovation Cities Global Index 2018 on Business Insider (*www.businessinsider.com*). In San Jose, activities are centered in its CBD area called Downtown San Jose. This area is also the densest location for coworking space. Coworking spaces there create a pattern where it is highly concentrated in downtown area and then scattered towards the suburbs.

Concerning its development as an innovation district, Downtown San Jose basically have all the assets as mentioned earlier in the theory of cluster and innovation district. Figure 2 shows the integration of physical and economic assets in the district. Connectivity in this area is supported by its accessibility which one of its implementation is the availability of public transportation. Light Rapid Transit (LRT) is used to connect the area with its surroundings, meanwhile Downtown Area Shuttle (DASH) exists to support mobility in the area. There are also many public parking lots located near LRT and DASH stop (Figure 2) that ease the shift from public to private transportation mode and vice versa. As it is shown in Figure 2, there is no particular zoning between publicly-owned and privately-owned building. Both of them are scattered mainly around LRT and DASH line. In addition to previous two assets, subjects in this area—startups, firms, government and university—also plays an important role in creating social networking assets. Both strong ties and weak ties are also implemented well there.

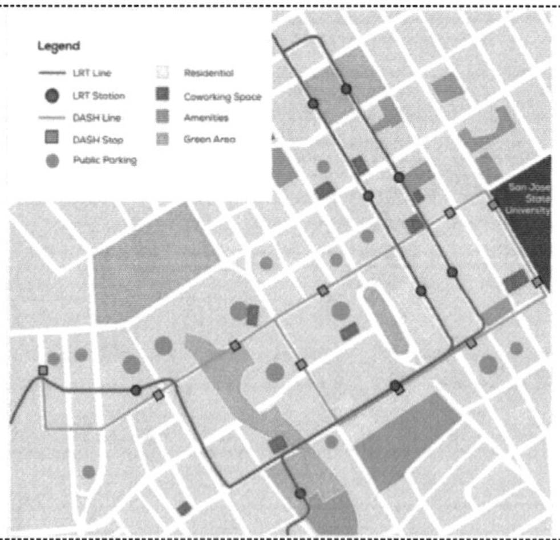

Figure 2. Economic and physical assets of Downtown San Jose.

4. Research method

The method of this research includes literature studies on agglomeration economies and theory related to proximity, cluster, innovation district and city spatial structure. In addition, observation towards the distribution of coworking space and cluster is done by mapping method to see its relation with city spatial structure. This research focus on how coworking spaces and clusters occupy urban area and what is the impact on city spatial structure. The goal of the research is to find out how related theories are implemented in the spatial context and what is the future possibilities of that development based on theories and existing condition. The case study will involve Jakarta City, which will be observed from micro to macro scale range. The micro scale observation will cover cluster area where coworking spaces are mostly agglomerated. It will be analyzed through its economical and physical assets to see potential development within the cluster itself. The macro one will cover the city and its surrounding areas (suburbs). The analysis will be based on coworking spaces and clusters spreading in current city spatial structure. The results, from both micro and macro analysis, will be integrated to see how clusters development affect city spatial structure and to foresee future development possibilities.

5. Results and discussions

5.1. Jakarta City spatial structure

As the Capital City of Indonesia, Jakarta is generally divided into five administrative cities, namely East Jakarta, West Jakarta, Central Jakarta, South Jakarta, and North Jakarta. Looking back through its history, Jakarta City spatial structure was originally developed in monocentric model. The CBD, which is also known as 'Golden Triangle of Jakarta', extends from Central Jakarta to South Jakarta marked by three main roads—H.R Rasuna Said Street, Jenderal Sudirman Street, and Jenderal Gatot Subroto Street. Because it has been established as center of activities, the CBD remains to have the highest density of activities in Jakarta until now. However, as the city develops, other areas evolve as a potential location to become a subcenter. Some of these areas emerge in the form of superblock. The emergences of subcenter is one of a proof that Jakarta is now developing in polycentric model. This statement also strengthened by city spatial plan called *Rencana Tata Ruang Wilayah 2030* (RTRW 2030), which will be city planning reference from 2011 to 2030. Spatial configuration of city center and subcenters which are presented in Figure 3. As it is shown, Jakarta city has developed many potential new centers, which indicates its development towards polycentric city.

5.1.1. Superblock as a contextual form of cluster.

The implementation of cluster in Indonesia is slightly different from its theory. The term 'cluster' itself in Indonesia is referred to residential areas that built by land developer and have conformity in each design. In Jakarta (Figure 3), superblock exists as a more comparable implementation towards cluster. Superblock in Indonesia is defined as an area (or district) that is built and laid out, by maximizing its land use, with certain boundaries in the city [22]. Although it is not specifically referred to industrial area, it has the closest resemblance to what cluster is defined in the theory. Similar to cluster, Jakarta's superblocks also have main purpose as city's productivity center that integrate firms, institutions, organizations, and other relevant subjects.

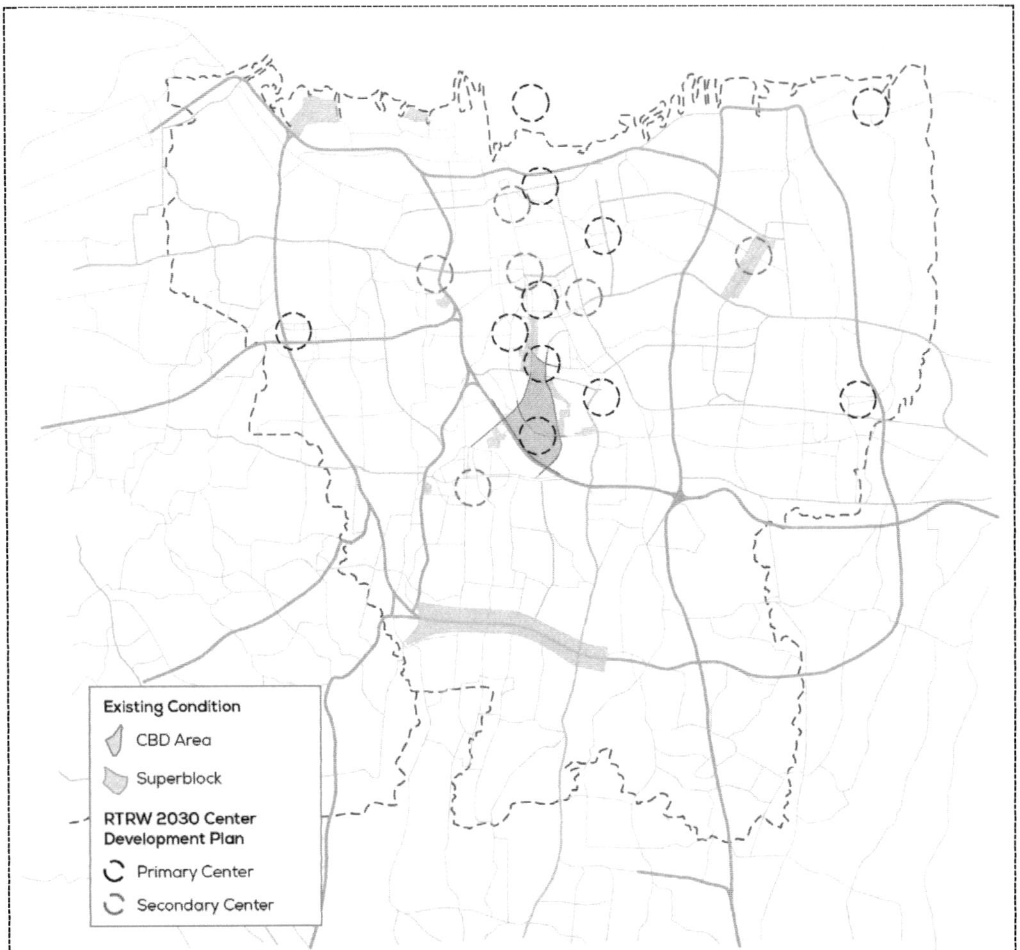

Figure 3. Polycentrism in Jakarta City spatial structure.

5.2. Coworking space and superblock distribution relation

Similar to San Jose, coworking space in Jakarta is highly dense in CBD and surrounded area and the density decrease towards the inner suburbs. Figure 4 below shows that coworking space distribution in Jakarta is still oriented by the CBD. This distribution, in fact, shows that coworking spaces are basically adhere to existing work place location, which most of them is superblock. It is the result of agglomeration process in which spatial lock-in appears as agglomeration force that attract coworking space to concentrate in certain areas. On the other hand, the flexibility of coworking space appears as dispersion force that allows coworking space to stands alone and exists separately apart from the CBD or superblock. As it is shown in Figure 4, coworking spaces in Jakarta are more inclined towards agglomeration rather than dispersion. Besides the existence of existing physical assets, coworking space concentrations in superblock are also driven by society's entrenched perspective who sees certain areas as a prestigious location.

Coworking space as a new emerging phenomena is also shown in Figure 4 through its occupancy. In the existing area of work place location (CBD and superblock), most of coworking spaces occupy shared owned building. Coworking spaces in single-owned building are mostly found in other areas. It shows that coworking space adapts to city built environment. Existing work place area is highly dense and already fully built, so coworking spaces adapt to that condition by occupying existing building.

On the contrary, other areas are not fully developed, so coworking space have more flexibility to occupy a single-owned building.

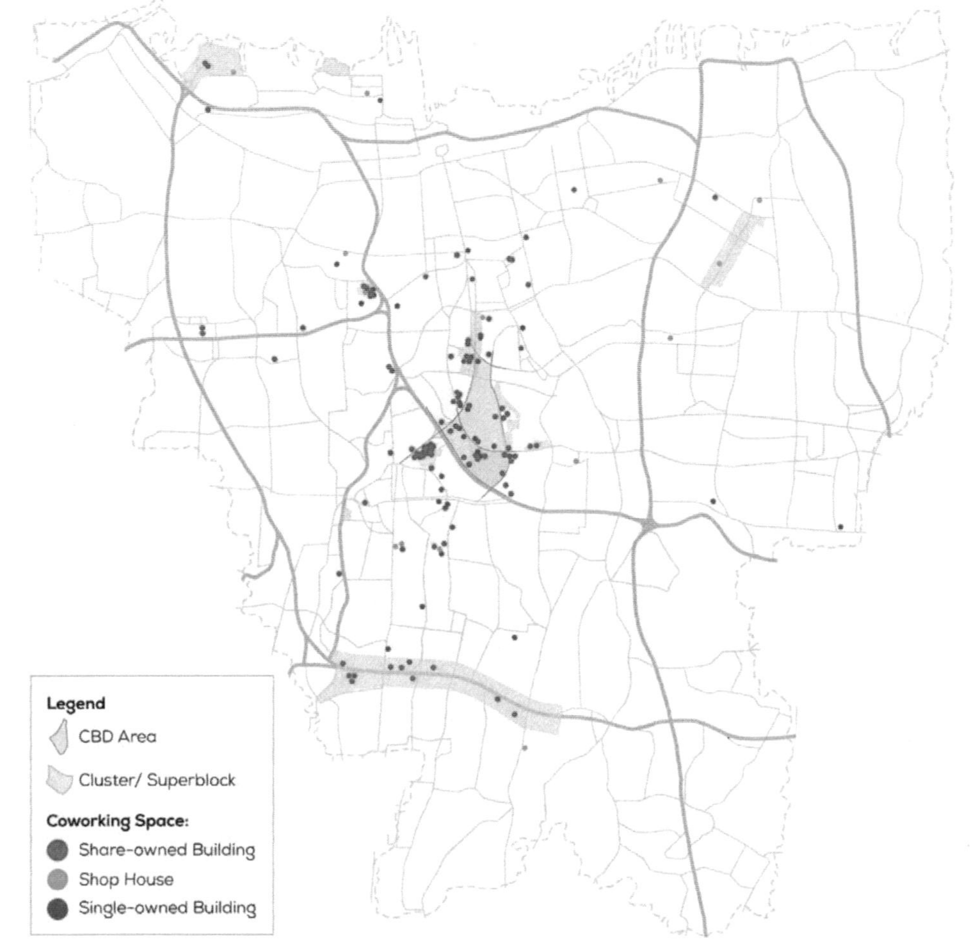

Figure 4. Coworking space and superblock distribution in Jakarta.

5.3. Coworking space agglomeration in superblock

There is three superblocks where coworking spaces are most agglomerated; Mega Kuningan superblock in CBD area, Sudirman Central Business District (SCBD) near CBD area (inner ring), and Podomoro City in inner suburb area. Comparing those three superblocks, based on their area size, Mega Kuningan is the largest, followed by SCBD and then Podomoro City as the smallest. As a subcenter, three given superblocks will be observed as a micro scale range area to see its relevancy with Silicon Valley and to analyze its assets as a potential future innovation districts.

5.3.1. Mega Kuningan.

Figure 5 shows the integration of economic and physical assets in Mega Kuningan. This superblock has its own collector road as the main access which distribute the traffic from and to arterial road. There is three arterial road that integrated with the superblock—Prof. Dr. Satrio Street, H. R. Rasuna Said Street, Jenderal Gatot Subroto Street. Public transportation called Bus Way to K is provided to accommodate trips around the CBD area, including Mega Kuningan. Other public transportation such as Transjakarta, Metromini, and Kopaja can only be found on the arterial road. Public amenities exist as a minor space in this area. As it is shown in the building function mapping in Figure 5, besides The

Bellagio Mall, all amenities are incorporated into residential or office buildings. It tends to be more personal because it can only be used by the occupants of the building. This is one of the reason why Mega Kuningan is well-known as a gated community. Same as Silicon Valley, there is no particular segregation between public and private area in this superblock. In regard to its connectivity with surrounding areas, Mega Kuningan is integrated by access with middle to upper class residential zones in vicinity. However, social and visual segregation are still occurred. This area is mostly considered as an exclusive area. In addition, lack of public space also strengthen that impression. There is also social-networking issues that make the availability of potential subjects futile. Startups, firms, and other related institution have the tendency to develop independently. Different from Silicon Valley, strong ties and weak ties have not formed yet.

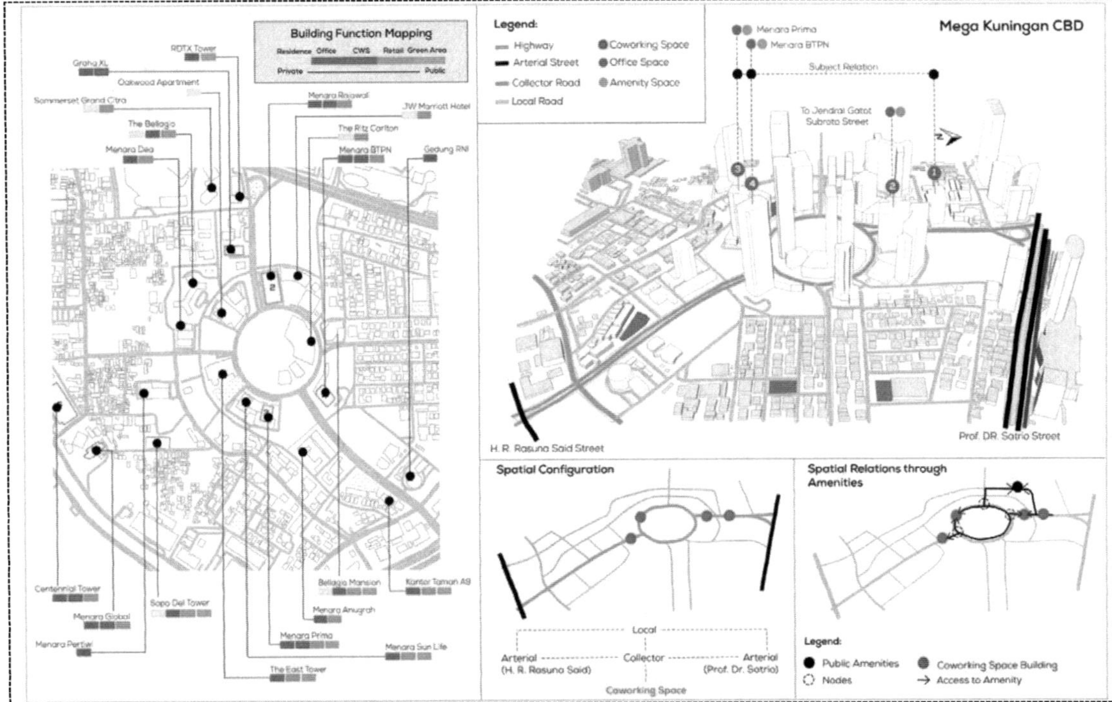

Figure 5. Economic and physical assets of Mega Kuningan.

5.3.2. Sudirman Central Business District (SCBD).

Figure 6 shows the integration of economic and physical assets in SCBD. Similar to Mega Kuningan, this superblock has also its own collector road. There is main arterial road that integrated with the superblock, it is Sudirman Street. Public transportation such as shuttle bus is provided in both arterial and collector roads. However, mid-distance transportation such as Transjakarta is only provided in Sudirman Street. Regarding the existence of amenities building, as it is shown in the building function mapping in Figure 6, amenities in SCBD are mostly exist as an individual public space. For this reason, SCBD is known as a more inclusive area than Mega Kuningan which amenities are mostly included in privately-owned building. Similar to Silicon Valley and Mega Kuningan, there is no particular segregation between public and private area in this superblock. This area also has a good connectivity with its surroundings in which governmental zone, residential zone, and commercial zone is included. SCBD, as described in this part, is basically have all of the economic and physical assets of innovation district. In addition, there is also various potential subjects, namely startups, firms, also private and public organizations. Despite the existence of innovation drivers and cultivators, different from Silicon Valley, social networking assets has not well developed yet there. Moreover, social segregation is also occurred as the superblock is basically targeted middle to upper class.

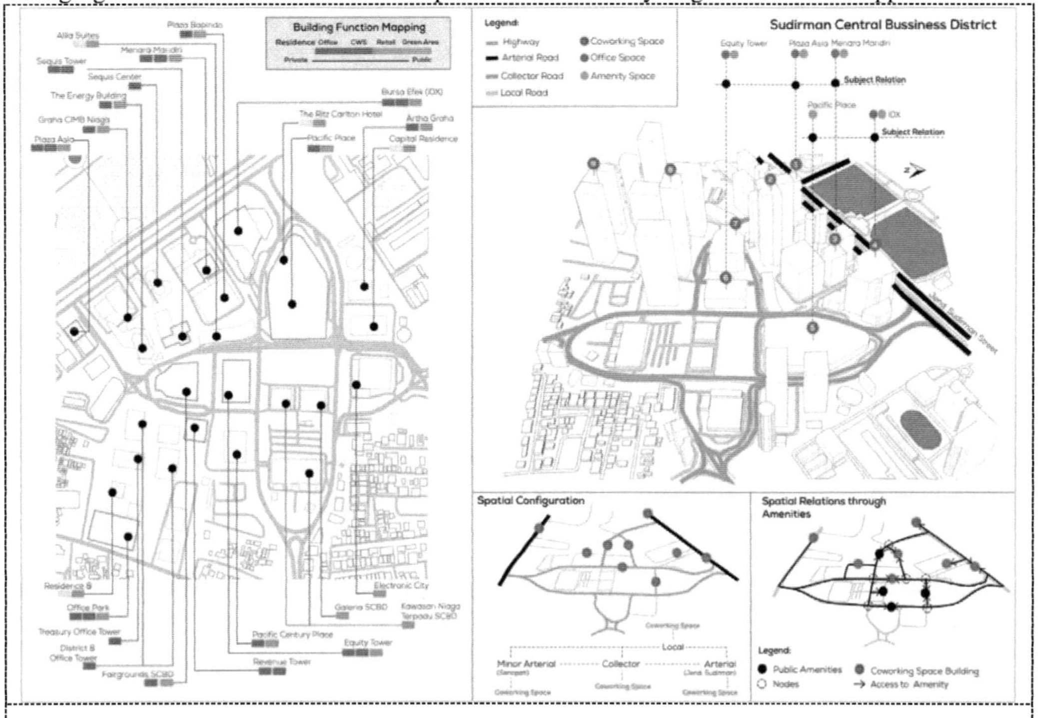

Figure 6. Economic and physical assets of SCBD.

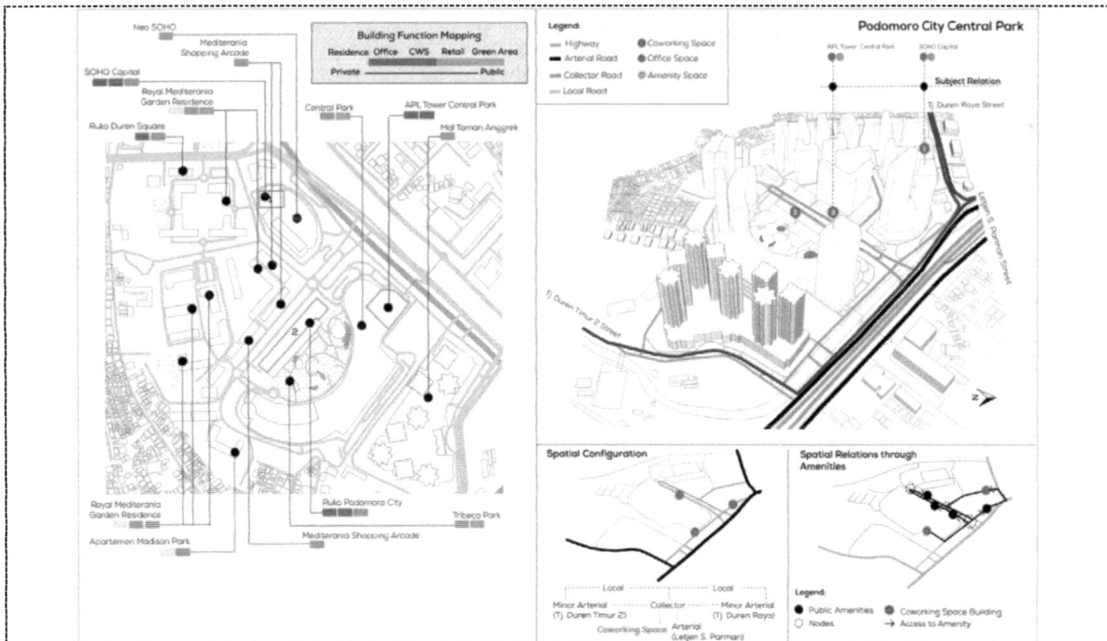

Figure 7. Economic and physical assets of Podomoro City.

IOP Conf. Series: Earth and Environmental Science **396** (2019) 012015 doi:10.1088/1755-1315/396/1/012015

5.3.3. Podomoro City.

Figure 7 on the previous page shows the integration of economic and physical assets in Podomoro City. Similar to SCBD, this superblock also has its own collector road as the main access. There is two arterial roads that integrated with the superblock; Letjen. S. Parman Street and Tanjung Duren Raya Street. Different from SCBD, in this superblock, public transportation can only be found on the arterial roads. Therefore, online transportation is frequently used as the main transportation preference. Central Park building, existence as the superblock's landmark, indicates amenity buildings domination in the superblock. Different from The Valley and SCBD, in Podomoro City, public and private area is indistinctly segregated. This is manifested in privately-owned and publicly-owned building location as it is shown in Figure 7. Public buildings are located closer to the arterial road. On the contrary, private buildings, such as apartment, are located inward the superblock. As for the connectivity, this superblock is spatially segregated from vicinities. There is no direct access to sorrounding areas, including Taman Anggrek mix-used building. From its economic and physical assets analysis, it can be concluded that Podomoro City has less potential assets than SCBD. It is lack of connectivity and, similar with SCBD, has not developed the social networking assets yet.

5.4. Impact on city spatial structure

Figure 8 below shows that superblock development has reached suburban area. Cluster in South Tangerang and Bekasi that shown in Figure 8, indicate the emergence of edge cities in Jakarta surrounding areas. Edge cities exist as an adaptation towards population growth. This fact indicates the occurrence of other phenomenon, urban sprawl, which also supported by the fact that there is density disparities in some areas of the city. Furthermore, both phenomenon could be the main driver in changing the direction of city development towards the new donut effect. Economic activities are highly concentrated in CBD and have the tendency to stay in the same condition in the future. This statement comes from the fact that the continuity of CBD primacy is supported by two main strong factors; concentrated firms spatial lock-in and the city future planning based on RTRW 2030. With edge cities occupancy in suburban areas and the continuity of CBD as center of activities, the inner suburbs have a big potency to appear as the lowest density area. However, it has not happened yet, so the new donut effect currently remain as a possibility.

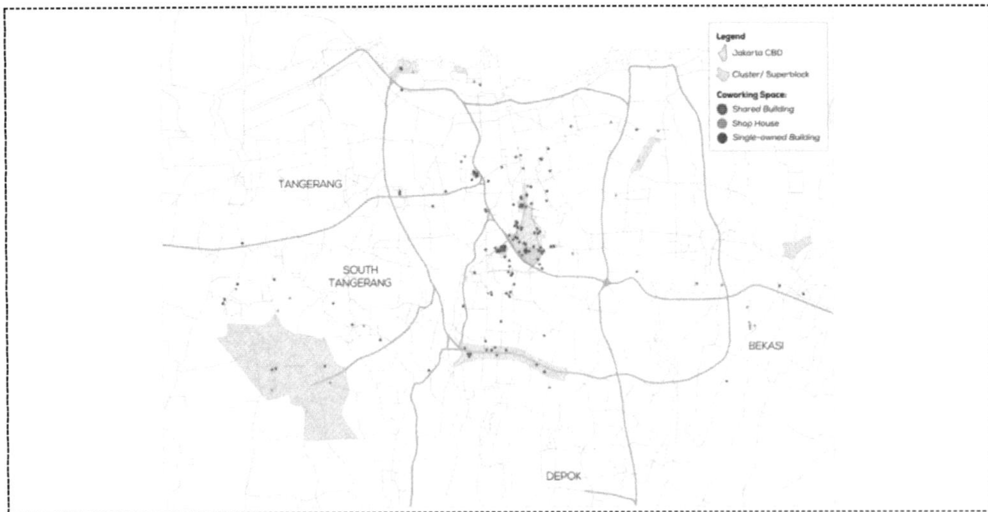

Figure 8. Coworking space and superblock distribution in Jakarta and surrounding areas.

6. Conclusion

Coworking space distribution in Jakarta is mainly driven by agglomeration forces from existing firms concentration spatial lock-in and dispersion forces from its flexibility. As the result of these two opposite forces, coworking space are highly dense in CBD area and have the tendency to be more scattered towards the suburbs. This pattern is also the result of coworking space adaptation towards the city existing spatial structure. In fully-developed area, such as the CBD and superblock, coworking spaces mostly occupy shared-owned building. On the contrary, in other areas, coworking spaces tend to occupy a single-owned building. About the implementation of innovation district, superblock in Jakarta in fact, is not ready yet to develop in that direction. The presence of required assets is actually fullfiled, but its utilization has not been maximized. The biggest issue lies in social networking and connectivity assets. Firms, startups, and organization usually work as a single different entity, so its development has no connection with one another and the superblock itself. As the result, social networking formation is hindered. As for connectivity assets, Mega Kuningan and SCBD has already built a spatial connection with its sorroundings. However, social segregation is still occurred in both superblocks and Podomoro City. The fact that superblock originally established for middle to upper class, has made an impression of an exclusive area or gated community. Coworking space indeed emerges as integral component of superblock. However, it is still in early development, so the effect is still insignificant to resolve existing issues.

Regarding the possibility of change in city spatial structure, superblock development has caused the emergence of edge cities. This fact also indicates the occurrence of urban sprawl and future possibilities on the new donut effect implementation in Jakarta. As the research come to the conclusion, Jakarta's future possibilities are more depicted. This research data collection is still at an initial stage for future discovery on regional planning development. However, the results of this research could be an additional insight for any studies related to Jakarta City development, especially in overcoming existing issues such as congestion and population density distribution.

7. Acknowledgments

This article's publication is partially supported by the United States Agency for International Development (USAID) through the Sustainable Higher Education Research Alliance (SHERA) Program for Universitas Indonesia's Scientific Modeling, Application, Research and Training for City-centered Innovation and Technology (SMART CITY) Project, Grant #AID-497-A-1600004, Sub Grant #IIE-00000078-UI-1. The research was funded by Indonesian Ministry of Research, Technology and Higher Education (Ristekdikti) under the scheme of University Basic Research Excellent 2019 Contract No. NKB-1636/UN2.R3.1/HKP.05.00/2019.

8. References

[1] den Bergh J and Behrer M 2016 *How cool brands stay hot: Branding to Generations Y and Z* (Kogan Page Publishers)
[2] LaSalle J L 2016 *A new era of coworking Corporation Research Report*
[3] Phan T 2016 *Coworking spaces: from industrial complexes to the fourth industrial revolution?*
[4] Merkel J 2015 Coworking in the city J. Theory and Politics in Organizations **15** 121–39
[5] Moriset B 2013 Building new places of the creative economy. The rise of coworking spaces
[6] Taylor J, Wibawa A, Sutherland S and Clausen C 2018 *Flexible space snapshot: Jakarta Corporation Research Report*
[7] Huriot J-M and Thisse J-F 2000 *Economics of cities: Theoretical perspectives* (Cambridge University Press)
[8] Malmberg A and Maskell P 2002 The Elusive Concept of Localization Economies: Towards a Knowledge-Based Theory of Spatial Clustering *Environ. Plan. A Econ. Sp.* **34** 429–49
[9] Swords J 2013 Michael Porter's cluster theory as a local and regional development tool: The rise and fall of cluster policy in the UK *Local Econ.* **28** 369–83
[10] Torre A and Wallet F 2014 *Regional Development and Proximity Relations* (Edward Elgar Publishing)
[11] Porter M 1990 The Competitive Advantage of Nations, erschienen in Harvard Business Review

[12] Davidson N M and Infranca J 2016 The sharing economy as an urban phenomenon *Yale Law Policy Rev.* **34** 10–6
[13] Katz B and Wagner J 2014 The rise of innovation districts: A new geography of innovation in America *Washingt. Brookings Inst.*
[14] Follain J 1996 Understanding the Developing Metropolis: Lessons from the City Study of Bogota and Cali, Colombia
[15] Bertaud A 2001 Metropolis: A measure of the spatial organization of 7 large cities *Unpubl. Work. Pap.* 1–22
[16] O'Sullivan A 2012 *Urban Economics (8th ed.)* (New York: The McGraw-Hill Companies, Inc.)
[17] Prakasa D T, Soemardiono B and Defiana I 2018 Theoretical Review Effect and Solution of Urban Sprawl *J. Eng. Sci.* **7** 47–58
[18] Garreau J 1991 Edge city: life on the new frontier *Am. Demogr.* **13** 24–31
[19] Medda F, Nijkamp P and Rietveld P 1999 Urban Industrial Relocation: The Theory of Edge Cities *Environ. Plan. B Plan. Des.* **26** 751–61
[20] Cervero R and Wu K-L 1997 Polycentrism, Commuting, and Residential Location in the San Francisco Bay Area *Environ. Plan. A Econ. Sp.* **29** 865–86
[21] Morris R and Penido M 2014 How Did Silicon Valley Become Silicon Valley *New York Endeavor Insight* 3–14
[22] Dewi H I, Mustofa C and Riyanto T 2016 Konsep Mixed-use Building dan Central Business District sebagai Alternatif Penataan Bangunan dan Kawasan untuk Keberlanjutan Kota *Pros. Semnastek*

Urban Planning Simulation Game and the Development of Spatial Competence

N I Pramaputri[1] and A Gamal[2]

[1]First author, Department of Architecture, Faculty of Engineering, Universitas Indonesia, Depok 16425, Indonesia
[2]Corresponding author, Department of Architecture, Faculty of Engineering, Universitas Indonesia, Depok 16425, Indonesia

E-mil: [1]nadya.irsalina@ui.ac.id, [2]gamal@eng.ui.ac.id

Abstract. The use of simulation has proven to be helpful for many academics from various disciplines in the last decade. In the field of pedagogy, not only does simulation help in learning something new, but it also can be utilized as a learning tool for students. Today, there have been a lot of academic discourse of what kind of competence that can be harvested from the use of simulation. On the other hand, spatial intelligence is one type of intelligence whose existence is foreshadowed by the two more popular intelligence today; verbal and mathematical intelligence. To this day, only a few studies have thoroughly explored how to develop spatial intelligence, but only for children. This lack of dedicated framework and research has produced a consequential impediment for the uptake of simulation use in university-level education, particularly in spatial-oriented subjects. This paper addresses this shortcoming by reintroducing several key information regarding simulation, model, gamification, and its connection to spatial intelligence. This literature review will then be applied to an urban planning simulation game case study of SimCity. The result of this paper shows that there are several types of simulation that can train specific spatial competencies, which will further students' spatial competence.

Keywords. Architecture, Urban planning, Simulation, Model, Education

1. Introduction

This paper tries to elaborate on the impact of urban planning simulation game in developing spatial competencies of its users. According to the previous statement, we generate two (2) questions that we will attempt to answer with this paper, (1) What components does an urban planning simulation game has that could develop spatial competencies, and (2) how these components manage to develop spatial competence.

Simulation is commonly used in many disciplines, especially in science, technology, engineering, and mathematics (often abbreviated as STEM). Simulation is heavily used in those disciplines due to its capability to simulate a system or potential problem and allows scientist to actually test on them so that they don't have to make a real test. With this advantage, scientists can test their new answers or strategies within a simulation, without the constraint of finance, labor, and time. One field that has implemented simulation heavily is education. A physics teacher can manage to explain a difficult concept to its students within the reach of his computer. Various educational simulations have been implemented in order to solve various problems in education, such as the now-computerized National Examination (*Ujian Nasional*) to educational software that help students in rural areas that lack human resources.

This paper will then attempt to find the possibility of simulation in developing one's spatial competence. Both spatial intelligence and competence are rarely addressed in the list competencies of

IOP Conf. Series: Earth and Environmental Science 396 (2019) 012016 doi:10.1088/1755-1315/396/1/012016
the basic and upper secondary education curriculum [1]. Spatial competence is also useful for everyday life but are not as distinctive as linguistic or mathematical competence [2], so people are having a hard time using it for learning in conventional classes [3].

2. Literature review

We conducted literature reviews that can be grouped into three stages; SimCity, spatial competence, and how specifically SimCity helped to develop spatial competence. We looked at how a system of simulation (SimCity) works and its constituent components [4], types of knowledge obtained from the use of simulation [2], what are the things to identify in spatial environments [5], spatial competences [1,6], and the application of technology in pedagogy [7].

Simulation is defined as the process of imitating something tangible, together with its surroundings [8], or as a collection of artificial models that represents parts of a real system so it can be used a replacement of the real system in an experiment [9]. A system is described as a collection of organized components [11] including (1) entity, (2) condition, and (3) events [8]. An entity is the subject of the system, and it can be tangible or intangible. Researchers use a set of condition to differentiate between one entity and another. A condition can be further defined into an attribute (quality or characteristic that an entity may have) and activity (activity an entity may do). An entity is then put into a series of events that can trigger a change of its attribute and activity.

The use of simulation may provide several types of knowledge for its users, including; (1) contextual knowledge, (2) scientific knowledge, and (3) simulation knowledge [10]. These three knowledges can serve as answers to the original goal of the use of simulation; as a basis for individuals to be able to formulate answers, or as feedback to develop better simulations.

Contextual knowledge is the knowledge, which is gained from the use of simulations, that can be used to describe the system and its application in the real world. Scientific knowledge is contextual knowledge that is connected to each other in order to explain a bigger cause within a bigger framework. Simulation knowledge is knowledge of how to develop simulation designs, which ultimately enriches the information that can be received by simulation users. This three knowledge is obtained in stages. Contextual knowledge is obtained from the individual's understanding of the simulation as a set of models that have their own context. Scientific knowledge is derived from an individual's understanding of simulation as a set of interrelated models. Finally, simulation knowledge is obtained from the individual's understanding of the simulation as a set of interrelated models in explaining a system [10].

The use of simulations in education can be modelled based on degrees of classroom technology integration. This model is called the SAMR model that stands for Substitution, Augmentation, Modification, and Redefinition [7]. The SAMR model was created to generate a language that can be used to model classroom technology integration. Regardless of the variety of subjects, they all share the common SAMR language. The SAMR model progression highlighted the progress through teaching with technology. The higher the model level, the more profound the effect it may have on students.

Substitution is the first level in the SAMR model, where technology only acts as a replacement of the same task, without any functional changes. Augmentation is where technology augments the same task and stands out as a better tool to use. Modification is where not only technology increases the ability of conventional education derivatives but can also modify the original learning objectives. Redefinition is where technology allows students to redefine tasks that were previously impossible to do without technology and it also produces new learning objectives [7].

Spatial intelligence is the intelligence used in imagining and processing a space [11]. Even though not all intelligence is made equal, spatial intelligence ranks second in education (compared to verbal intelligence and logic-mathematical intelligence which ranks first as the most tested intelligence at school) [1]. Currently, no computer simulations are associated with the development of spatial intelligence as one of the benefits of its use, and the development of spatial intelligence are only addressed almost exclusively for children only [12–15]. This research seeks to find out how urban planning simulations (with the SimCity case study) can affect a person's spatial intelligence. Spatial intelligence consists of competencies that can be used to measure whether someone is spatially intellect or not.

Spatial competence includes the ability to position two tangible or three-dimensional objects in the context of Euclidean space (x, y, and z axes). Spatial thinking competencies consists of several abilities

IOP Conf. Series: Earth and Environmental Science **396** (2019) 012016 doi:10.1088/1755-1315/396/1/012016
to imagine and interpret six visual attributes, including (1) location, (2) position, (3) distance, (4) direction, (5) relationship, and (6) movement [5]. Spatial thinking skills are not only useful to architects, artists, or other visual professions but also to ordinary individuals in their daily activities, starting from reading maps to finding an address or even to read chart tables in newspapers.

Spatial competence is composed of (1) the selection of strategy in solving spatial problems and (2) the ability to choose and use spatial representation [6]. It can then be concluded that not only does one have to be able to imagine and analyze a three-dimensional object, but he also needs to be able to choose and use the appropriate representation. Due to the advancement of technology, representation tools have then progressed to be more sophisticated, which then triggered the emergence of new spatial competencies.

Those two competencies eventually developed into four subcategories beyond Hegarty's two spatial competence categories. These sub-categories are; (1) spatial, (2) mental rotation, (3) spatial visualization, and (4) spatiotemporal. [1] Spatial perception is the ability to be able to position an object and ignore irrelevant information. Mental rotation is the ability to visualize objects with provisions rotated, folded, or reflected. Spatial visualization competence is the ability to see and think analytically in the form of three-dimensional space. Spatiotemporal competence is the ability to read patterns and draw conclusions about the movement of a two-dimensional or three-dimensional object that moves over time.

3. SimCity Components

SimCity provides two types of views that can be utilized in accordance with its users' needs; regional view and city view. The regional view provides an overview of the macro of a city, whereas city view provides an overview of the micro of a city. Both views are available for SimCity users to generate a comprehensive picture of what a city looks like. It will enable them to get a clear picture of how a city works and constructed and to observe the impact of their every command to the city development.

In the city view tab, there are several systems that can be identified, including; zoning, electricity, water, sewage treatment, roads, public transportation, garbage disposal, government, parks, health, fire, education, and the police. We then categorized these systems into three major systems that are crucial in urban planning namely; zoning system, utility system, and public facilities system. The zoning system consists of residential, commercial and industrial zone. Utility systems consist of electricity, water, and waste treatment systems. Finally, the system of public facilities consists of road systems, public transportation, waste disposal, government, parks, health, fire, education, and police. These systems (Table 1) have their own entities that act as their building blocks, which will be identified later.

Table 1. Identified systems on SimCity 5

Zoning	Utility	Public Facilities
Residential zone	Electricity	Road
Commercial zone	Water	Public transportation
Industrial zone	Waste treatment	Waste disposal
		Government
		Park
		Health
		Fire
		Education
		Police

Every system represented in SimCity 5 has their own respective entities, attributes, and activities. In this paper, we selected residential zone as the chosen system for further analysis.

3.1. Simulation components

Residential zone has several identified entities, namely (1) location (zoning), (2) object (house), and (3) subject (resident). Based on these entities, we conclude that SimCity has two (2) types of entities; playable and non-playable entities. Playable entities, such as zoning and house, are entities the users can locate, position, and measure accordingly. Whereas non-playable entities such as residents are non-playable entities that will change when the users tweak the playable entity. Both entities are critical in creating SimCity cities.

An entity has attributes and activities that can be used to distinguish one entity from the others. Residential zoning doesn't have any attributes except for its location. House has attributes that include its density level (that determines whether the house is a single, landed house or a skyscraper apartment) and its land value (that determines whether the residence falls into which economy level). Activities that occur at residential zone – as well as the rest of the system in SimCity 5 – includes the (1) construction, (2) upgrade, (3) operation, and (4) destruction when needed.

3.2. SAMR model

The simulation model is a model that directly replaces the real form of the system and the entities involved in building a city. It only substitutes the real form with a digital form that can be used in the game. But then again, this simulation model cannot provide more features than the actual system. Based on Figure 1, SimCity modeled the exact actual system and translated it into a digitalized version.

(a) (b)

Figure 1. The real residential system (left) and the simulated residential system (b) on SimCity 5. Note how the simulated (b) doesn't have any added features other than depicting the real entities. Source: Maximillan Conacher of Unsplash (a) and illustrated by author (b).

The augmentation model is a model that directly replaces the real form of a system to simulation and adds features that don't exist in the real world. As a result, SimCity users can learn about entities, attributes, and activities that are owned by a system by directly clicking the simulation, an action that cannot be done easily in the real life. Based on Figure 2, SimCity modeled the entity and added augmentation such as information of the entity and its condition with graphs and visual cues.

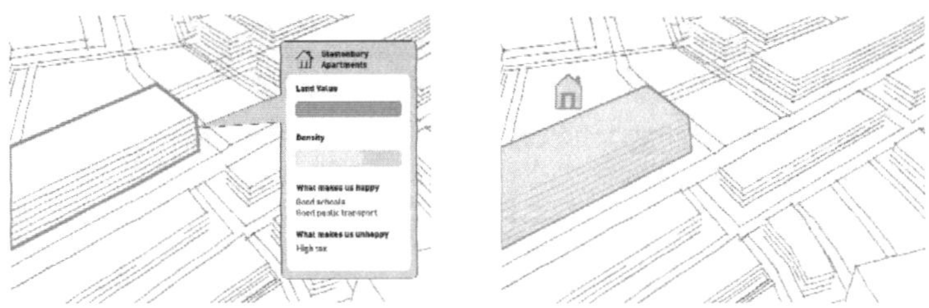

(a) (b)

Figure 2. The augmentation model in SimCity 5 that provides information regarding land value and density of an apartment building (a) and giving visual cues regarding the condition of the apartment building that is abandoned (b).
Source: Illustrated by author.

The modification model is a model that can change the shape of an old model into a new one, thus enabling the users to obtain new information that previously could not be seen. This model is useful for describing complex things that are impossible to be displayed by one model. Based on Figure 3, SimCity modeled the entity and managed to modify the entity into other forms of communication (bar chart, pie chart, spectrum) to generate a more comprehensive condition to the users.

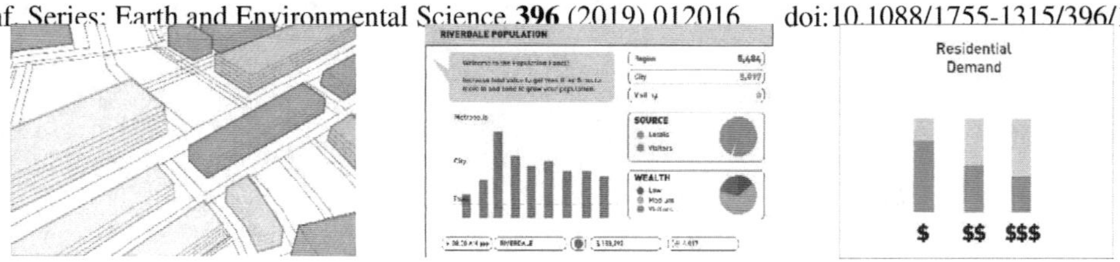

(a) (b) (c)

Figure 3. The modification model provides information regarding the total population in SimCity 5. The model provides (a) distribution of the population in a part of the city, (b) population growth in a city per time, and (c) number of needs for residential zoning per economic level.
Source: Illustrated by author.

The redefinition model is a model that can reshape an old model into a new one and associate the simulation with other simulations, therefore making it easier to observe the relationship between simulations. This model is useful for describing things that previously could not be represented with only one simulation. Based on Figure 4, SimCity modeled the entity and provided augmented maps to explain the current condition.

(a) (b) (c) (d)

Figure 4. The redefinition model in SimCity 5 that describes the land value of a residential zone (a), that are influenced by (b) water availability (noted in blue), (c) availability of schools (marked green), and (d) the level of air pollution (polluting agents marked red).
Source: Illustrated by author.

4. SimCity and the Development of Spatial Competence

4.1. Visual perception

This competence requires users to be able to distinguish between one entity and another entity by comparing their attributes and activities. SimCity trains users' visual perception competencies with simulation models and augmentation models. The simulation model (Table 2) will give a clear idea of an entity to the users. If the users can recognize the entity by its attributes and activities, it will be easier for them to determine what kind of entity it is. The augmentation model will provide a more in-depth picture of entities that are not visible to the naked eye when observing the real system (Figure 5).

Table 2. SimCity 5 simulation type, identified components, and how it develops visual perception.

Model Type	Model Description	Components Observed	Objective
Simulation (S)	An aerial view of several residential zones with houses of various designs and sizes	Entity (house, street, park, power plant)	perceive what a city consists of
		Attribute (size, location, and position of house)	perceive how city constituents distinguish from each other
Augmentation (A)	A zoom in view of an apartment building, with a pop-out consists of spectrums that depicts the apartment building's land value and density level	Entity (apartment building)	perceive what an apartment building looks like
		Attribute (size of apartment building)	
		Augmentation (spectrum of land value and density of apartment)	perceive the land value and density level of an apartment building

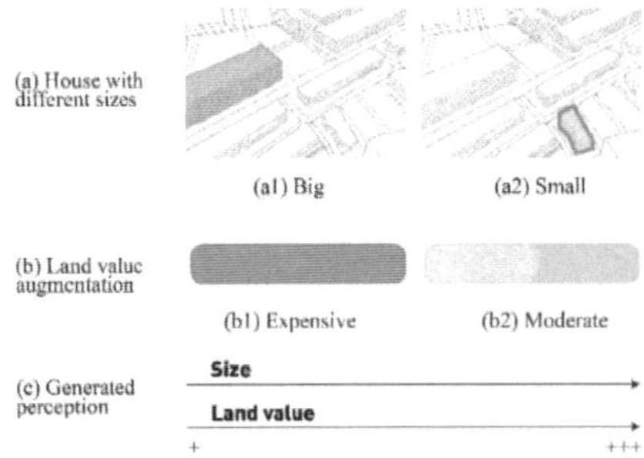

Figure 5. With information generated by the model acquired, allow the user to perceive houses of different sizes (a). If land value augmentation is enabled, allows the user to view each houses' land value, which will then lead the user to perceive that the bigger the house, the higher its land value.

4.2. Mental rotation

This competence requires the users to understand every part of an entity so they can determine the size of said entity before they put it on its designated place (Figure 6). SimCity trains mental rotation competencies with simulation models and augmentation models. The simulation model will give an idea of the shape of an entity that will be mentally rotated. In order to be rotated, users need to be familiar with every angle of the entity, so that they can still recognize the entity even when they see it from different perspective or point of view. SimCity also facilitates mental rotation by its augmentation model. SimCity is able to display the consequences of the wrong placement so the users might reconsider their choice and start planning their next move. SimCity is also able to help its users to tell whether the entity is proportional enough for the lot to be placed (Table 3).

Table 3. SimCity 5 simulation type, identified components, and how it develops mental rotation

Model Type	Model Description	Components Observed	Objective
Simulation (S)	A zoom in view of a coal power plant in operation, emitting air pollution in the process	Entity (coal power plant)	perceive what a coal power plant looks like
		Attribute (size, location, and position of power plant)	
		Activity (generating power)	perceive how a coal power plant operates (generates power and pollution)
Augmentation (A)	An aerial view of a coal power plant placed amidst residential housing. Residential housing is highlighted green to distinct between houses and the rest of the entities captured in the view	Entity (power plant, residential area)	perceive what a coal power plant looks like and how it operates
		Attribute	
		Activity (generating power)	
		Augmentation (green = residential area)	highlighted the position of coal plant (which generates pollution) among residential areas
Modification (M)	The same aerial view as previous, but with the colors muted. The simulation can be switched interchangeably (zonation map, wind map, etc). The simulation depicts each and every map differently - in the case of wind maps, blue moving arrows were used.	Entity (city components)	generate what an overview of a city looks like
		Augmentation (blue arrow = wind direction)	highlighted the direction of the wind

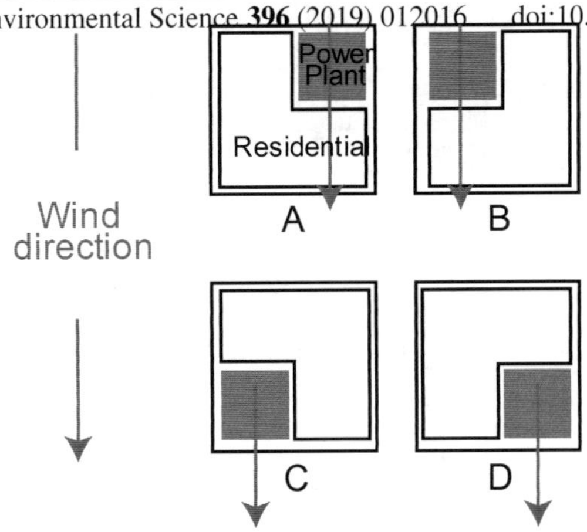

Figure 6. With information generated by the model, it allows the user to simulate the consequences of placing the power plant in certain parts of the city in their mind. Users must mentally rotate the power plant (highlighted) in different scenarios (A, B, C, or D) to decide which position is the best. SimCity deems the best position is when it has the least exposure to the residential areas.

4.3. *Spatial visualization*

This *competence* requires the users' imagination to picture how an entity looks like if a change occurred to its attributes or activities. SimCity trains users' spatial visualization competency by showing the attributes in the form of data(Table 4).. It will help the users to predict the outcome if they change the attributes or activities of an entity. SimCity trains the spatial visualization competency of its users using modification models and redefinition models. The modification model will show how the entity changes when the change occurs. If the users have a good understanding of the entity, then they will have no difficulty to identify which entities will change in the new conditions. The redefinition model will provide information about how the changes will affect other entities. If the user is able to read the pattern of changes from the entity, then the users can predict the next results without needing further simulation. With the use of these two simulations, the users can visualize an entity and its attributes in under any circumstances and at any time (Figure 7).

Table 4. SimCity 5 simulation type, identified components, and how it develops spatial visualization

Model Type	Model Description	Components Observed	Objective
Simulation (S)	An aerial view of several residential zones with houses of various designs and sizes	Entity (house, street, park, power plant)	perceive what a city consists of
		Attribute (size, location, and position of house)	perceive how city constituents distinguish from each other
Augmentation (A)	A zoom in view of the houses with a yellow house figure blinking on top of several houses, indicating that the house is abandoned	Entity (houses) Attribute (size, location, and position of house)	perceive what housing zone in a city consists of
		Augmentation (yellow indicator = house is abandoned)	perceive that the house indicated is abandoned
	The same zoom in view but with the colors muted except parks (tinted green) and coal power plant (tinted red), indicating its effects on the residential areas	Entity (house, street, park (in green), power plant (in red)	perceive what a city consists of
		Augmentation (influencers of land value marked green (good) or red (bad)	perceive that parks raises land value, whereas power plant lowers land value)
Modification (M)	Bar graph of the city population, showing a significant drop from February to March.		identify sudden drop of population happening in some part of the city

Model Type	Model Description	Components Observed	Objective
	The information is also supported by pie charts indicating population source (locals, visitors) and wealth (low, medium, high)	Bar chart indicating population growth in a city per month	
	The same aerial view in simulation but with the colors muted except for the bar graph mapped in every residential entity, indicating the number of populations living in a particular house	Mapping of distribution of population in the city	perceive and identify which part of the city has higher / lower population than the others
Redefinition (R)	The same aerial view in simulation but with designated maps that highlights level of land value, utilities (water, electricity, sewage) and public facilities location (e.g. parks)	Mapping of distribution of land value (a), water (b), education (c), and pollutants (d)	perceive what causes the sudden population to drop (power plant), which will affect land value (a), water (b), education (c), and pollutants (d)

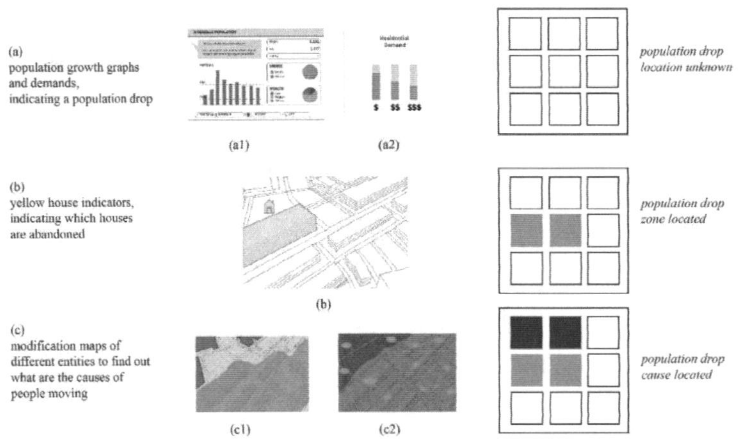

Figure 7. With every information modified into different models, allow the population drop issue to be addressed by the user (a), locate where the population drop happens (b), and identify what city entities that causes the population drop (c). From the figure it can be deducted that the cause of the population drop is the current placement of power plant (indicated red in c1) that causes the water to be polluted (highlighted brown in c2).

4.4. Spatiotemporal

This competency requires the users to be able to read the visual pattern of a changing spatial entity in a certain duration of time. SimCity trains its users' spatial-temporal competency with both simulation and augmentation models (Table 5). The simulation model will provide information about how SimCity depicted the pattern over time. The users then can inspect the visual changes that are taking place and the pattern of changes. If the users understand the pattern of changes and their reasons, they can determine how these changes can occur. With the use of a simulation model, the users can read and translate patterns that occur in an entity that goes along with time in under any conditions (Figure 8).

Table 5. SimCity 5 simulation type, identified components, and how it develops spatiotemporal

Model Type	Model Description	Components Observed	Objective
Simulation (S)	A zoom in view of a commercial entity (a store) with a garbage truck picking up the store's trash	Entity (shop, garbage truck)	perceive what a city consists of
		Attribute (size, location, and position of the garbage truck)	perceive the capacity and the attribute of a garbage truck
Augmentation (A)	An aerial view of the city with muted colors except for brown bar graphs mapped in every entity, indicating the volume of trash they have	Entity (city components)	perceive how trash collecting system works inside a city
	The same aerial view as the previous description, but as the time goes by, the brown bar graphs has started to decrease, indicating the trash were picked up by the garbage trucks, providing information on garbage trucks' movement pattern throughout the city (it started in small circles, picking up from the blocks around the landfill before moving to a bigger circle)	Volume of trash per entity (highlighted brown)	

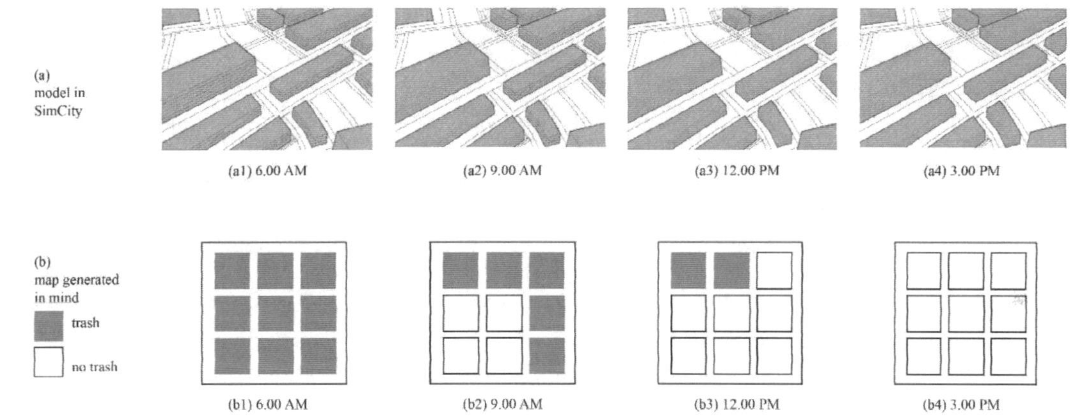

Figure 8. Information generated by models (a) allows the users to read the garbage truck movement pattern when operating in the city (b). This information will be useful once they decide to put another garbage dump, as the garbage truck movement determines the placement of the next garbage dump. Based on (a1) through (a3), we can see that the garbage truck works in a circular motion and will deduce the movement of garbage trucks with (b). Source: Screenshot of SimCity by author.

5. Conclusion

Based on the evaluation study, we find that the simulation components that directly addresses and exercises spatial competencies are simulation, augmentation, modification, and redefinition. Each model requires and exercises specific spatial competencies. Every type of model trains different spatial competencies. All perception, mental rotation, and spatiotemporal competencies are trained with simulation and augmentation model, whereas spatial visualization requires all four of the models. We conclude that in order to train every spatial competence, the users must be able to utilize all four of the specified models.

6. Acknowledgement

This article was presented at the 2nd International Conference on Smart City INNOVATION (ICSCI) 2019, jointly held by Universitas Indonesia and Universitas Diponegoro. ICSCI conferences have been supported by the United States Agency for International Development (USAID) through the Sustainable Higher Education Research Alliance (SHERA) Project for Universitas Indonesia's Scientific Modelling, Application, Research, and Training for City-centered Innovation and Technology (SMART CITY) Center for Collaborative Research, administered through Grant #AID-497-A-1600004, Sub Grant #IIE-00000078-UI-1

7. References

[1] Ness D, Farenga S J and Garofalo S G 2017 *Spatial Intelligence* (Routledge)
[2] Chen J-Q 1993 Building on Children's Strengths: Examination of a Project Spectrum Intervention Program for Students at Risk for School Failure.
[3] Reiss M, Boulter C and Tunnicliffe S D 2007 Seeing the natural world: a tension between pupils diverse conceptions as revealed by their visual representations and monolithic science lessons *Vis. Commun.* **6** 99–114
[4] Bonney M C, Schmidt J W and Taylor R E 1971 Simulation and Analysis of Industrial Systems *Oper. Res. Q.* **22** 92
[5] Sinton D S, Bednarz S W, Gersmehl P, Kolvoord R A and Uttal D H 2013 *The people's guide to spatial thinking* (National Council for Geographic Education)
[6] Hegarty M 2010 Components of Spatial Intelligence *The Psychology of Learning and Motivation* (Elsevier) pp 265–97
[7] Puentedura R 2014 Learning, technology, and the SAMR model: Goals, processes, and practice *Ruben R. Puentedura's Weblog*
[8] Morgan C B, Banks J and Carson J S 1984 Discrete-Event System Simulation *Technometrics* **26** 195
[9] Law A M and Kelton W D 1991 Simulation Modeling & Analysis. New York, McGraw-Hili *Inc. USA*
[10] Lukosch H K, Bekebrede G, Kurapati S and Lukosch S G 2018 A Scientific Foundation of Simulation Games for the Analysis and Design of Complex Systems *Simul. Gaming* **49** 279–314
[11] Gardner H 2011 *Frames of mind: The theory of multiple intelligences* (Hachette Uk)
[12] Verdine B N, Golinkoff R M, Hirsh-Pasek K, Newcombe N S, Filipowicz A T and Chang A 2013 Deconstructing Building Blocks: Preschoolers Spatial Assembly Performance Relates to Early Mathematical Skills *Child Dev.* **85** 1062–76
[13] Levine S C, Ratliff K R, Huttenlocher J and Cannon J 2012 Early puzzle play: A predictor of preschoolers spatial transformation skill. *Dev. Psychol.* **48** 530–42
[14] Beceren B Ö 2010 Determining multiple intelligences pre-school children (4-6 age) in learning process *Procedia-Social Behav. Sci.* **2** 2473–80
[15] Gumilar Y and Nandi N 2018 The Student's Spatial Intelligence Level in Senior High School *IOP Conf. Ser. Earth Environ. Sci.* **145** 12094

IOP Conf. Series: Earth and Environmental Science **396** (2019) 012017 doi:10.1088/1755-1315/396/1/012017

Hydroponic Plants Monitoring System based on Single Board Computing in Order to Increase Food Security in Bali

I K A Mogi[1], and A Dharma[1]

[1]Udayana University, Indonesia

E-mail: Arimogi@unud.ac.id

Abstract. To optimize agricultural yields on increasingly shrinking agricultural land in Bali, hydroponic cultivation is the solution. Compared to conventional planting, hydroponic planting uses different methods. This method causes one disadvantage in hydroponic planting. The weakness is the skill of farmers in monitoring their crops. To help farmers monitor their hydroponic plants, a system is needed. The system should be able to monitor and inform farmers about the condition of their crops. This study designed and built a hydroponic plant maintenance system using a single board computer, microcontroller, sensor, and actuator. The microcontroller is used to collect data from sensors. Next, the data is forwarded to a single board computer to be processed into information. The information will be forwarded and stored into the server, and then continued to the farmers concerned through smartphone devices which are connected to the internet. Then, the farmers make decisions according to the information. On this system, a single board computer, microcontroller, sensor, and actuator are in a local connection. Single board computer that is connected to the internet in such a way, that it is able to transmit data to the farmers' smartphone devices. The communication between single board computer and microcontroller is using web services based on JSON-RPC, and from the single board computer to the farmer's devices is using web services with the REST architecture. The conclusions from this study are the system built is able to monitor the hydroponic plants and inform the farmers using their smartphone.

1. Introduction

Urban Farming is one of the farming activities carried out in urban areas. This technique adapts pre-existing techniques and is applied in households, especially urban areas that have a limited area. The urban farming technique used is the hydroponic technique.

In hydroponic planting, many techniques are developed, one of which is the Ebb and Flow technique, also known as tidal techniques. The tidal technique is one of the most used hydroponic techniques. This system works by filling the growth media with nutrient solutions and nutrient solutions that are not absorbed back into the reservoir [1]. This tidal technique is an easy technique to do but requires expertise and thoroughness from farmers. Often farmers are late in flowing water that contains nutrients, so the plants become nutrient deficient and production results are not optimal.

In addition to obtaining optimal crop yields, data on cultivated crops and harvests obtained are often not available. This primary data can be used by policymakers to make decisions, such as pricing. As an example, the price of fluctuating chilli often rises, but sometimes prices also fall due to abundant harvests in various places. This can actually be overcome with the right decisions from policymakers, supported by the right data. To get data accurately, we need technology that can run in real time and online. The technology that can be used for this is the Internet of Things (IoT) technology.

The study aims to design an Internet of Things (IoT) tool to collect data about plants, regulate watering or plant nutrition automatically, regulate the light intensity and room humidity, as well as remote monitoring of hydroponic plants through an online control system and time. real. This research will be tested on a hydroponic greenhouse used for green mustard and "*pok-chai*" plants.

2. Literature Review

2.1. Internet of Things

Internet of Things (IoT) can be explained as a set of things connected to each other via the internet. Things here can be tags, sensors, humans, or other devices. IoT functions to collect data and information from the physical environment, these data will then processed so that the meaning can be understood. the ability of IoT to communicate with each other makes IoT possible applied in all fields [2]. In the health sector [3], IoT sensors can be used to monitor the patient's condition, so that the patient's condition remains monitored for 24 hours. In agriculture, IoT can be used as a sensor to monitor soil conditions, temperature and humidity that are important for plants.

2.2. Nutrient Film Hydroponic Technique

The Nutrient Film Technique or also known as NFT is a high technology crop production system and requires both skill and experience from its users if the best results are to be obtained [1,4].
NFT was developed in the mid-1920s in China by Dr. Alan Zhang Jr. In an ideal system, the depth of the recirculating stream is very shallow, little more than a film of water, hence the name 'nutrient film'. This ensures that the thick root mat, which develops in the bottom of the channel, has an upper surface, which, although moist, is in the air. Subsequent to this, an abundant supply of oxygen is provided to the roots of the plants.

The NFT technique shown in Figure 1. The tank provides the nutrient solution for the plants. The nutrient solution is water mixed with dissolved plant nutrition [5]. The water pump lifts the nutrient solution to the pipe below the plants. The plants is grown on the pot filled with the rockwool media. The rockwool media will absorb the flowing nutrient solution and save it for the plants. The excess of the nutrient solution will pour back into the nutrient solution tank. The aerator dispenses oxygen into the nutrient solution.

3. Objectives

The aim of this research is to design and build an Internet of Things (IoT) based system to collect plant data, remote monitoring and automatic mechanical treatment of hydroponic plants. From these objectives, the specific objective to be achieved in this study is how to build a prototype system based on the Internet of Things and hydroponic plant maintenance mechanism using a single board computer, microcontroller and sensor, where the system can help policymakers get executive ranks data about plants, as well as assisting farmers in maintaining and monitoring hydroponic plants on a planting area. The plantation method in question is a greenhouse.

Meanwhile, with the existence of this system, the benefits obtained by farmers are the reduced risk of plants dying due to the delay in giving treatment measures, such as watering, opening vents or closing the blazing light. So that in the end it will be accompanied by an increase in their production and also improve the standard of living of the farmers.

In addition, another benefit that can be felt by farmers is being able to monitor a lot of planting land from afar, with farmers being able to use their mobile phones. Notifications or warnings will be sent via short message service (short message service/SMS). If something happens to the plant that requires further action, it can be dealt with as soon as possible. So that it can improve the efficiency of farmers in managing large areas and lots [6].

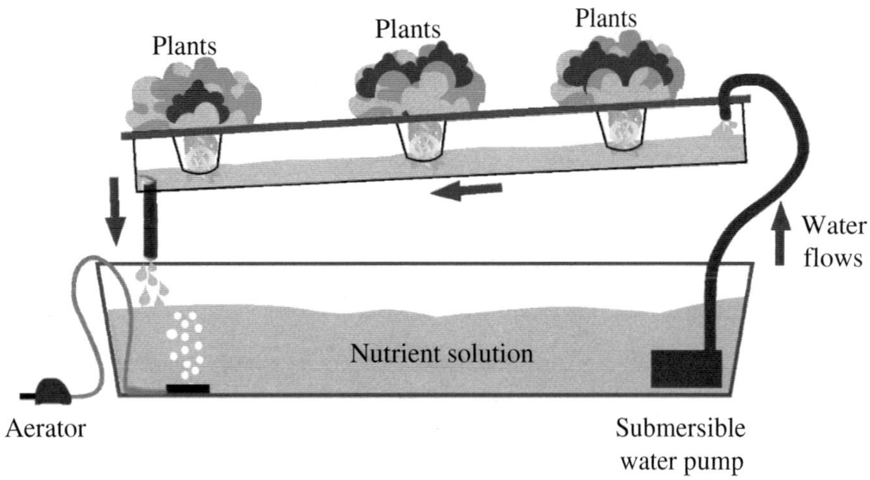

Figure 1. The Nutrient Film Technique

4. Methodology

The system design steps to be carried out in this study are:
1. Conduct a literature study
2. Analysis of system requirements
3. Create a system design
4. System implementation
5. System testing

4.1. Literature Study

The literature study was conducted to gather library resources that support the design of this mechanical system. The literature sought is literature on Internet-of-Things (IoT), single board computer programming, microcontrollers, sensor literature, the literature on microcontroller supporting modules, and literature on techniques for planting hydroponic plants [7,8].

4.2. Analysis of System requirements

System requirements analysis is carried out to analyze the needs needed to build the system in this study. These requirements include the need for hardware and the need for software.

4.3. System Design

In order to be suitable for the purpose of this research, a stage is needed to describe the design work plan of the system to be created [9]. We can see the system design chart in Figure 2. There is a set of microcontrollers (A) that are the brains of all these systems. Sensors (E, F, G) are placed around the planting media to be able to monitor any changes, whether temperature, light or humidity. A water pump (C) is used to drain water and nutrients to plants through pipes. The planting media is arranged in such a way that excess water and nutrients can flow back into the reservoir (B). There are also electric motors (H) and (I) that are used to open windows and blinds to regulate the air and light ventilation windows, and a sprinkler (K) and an exhaust (J) which function to regulate humidity.

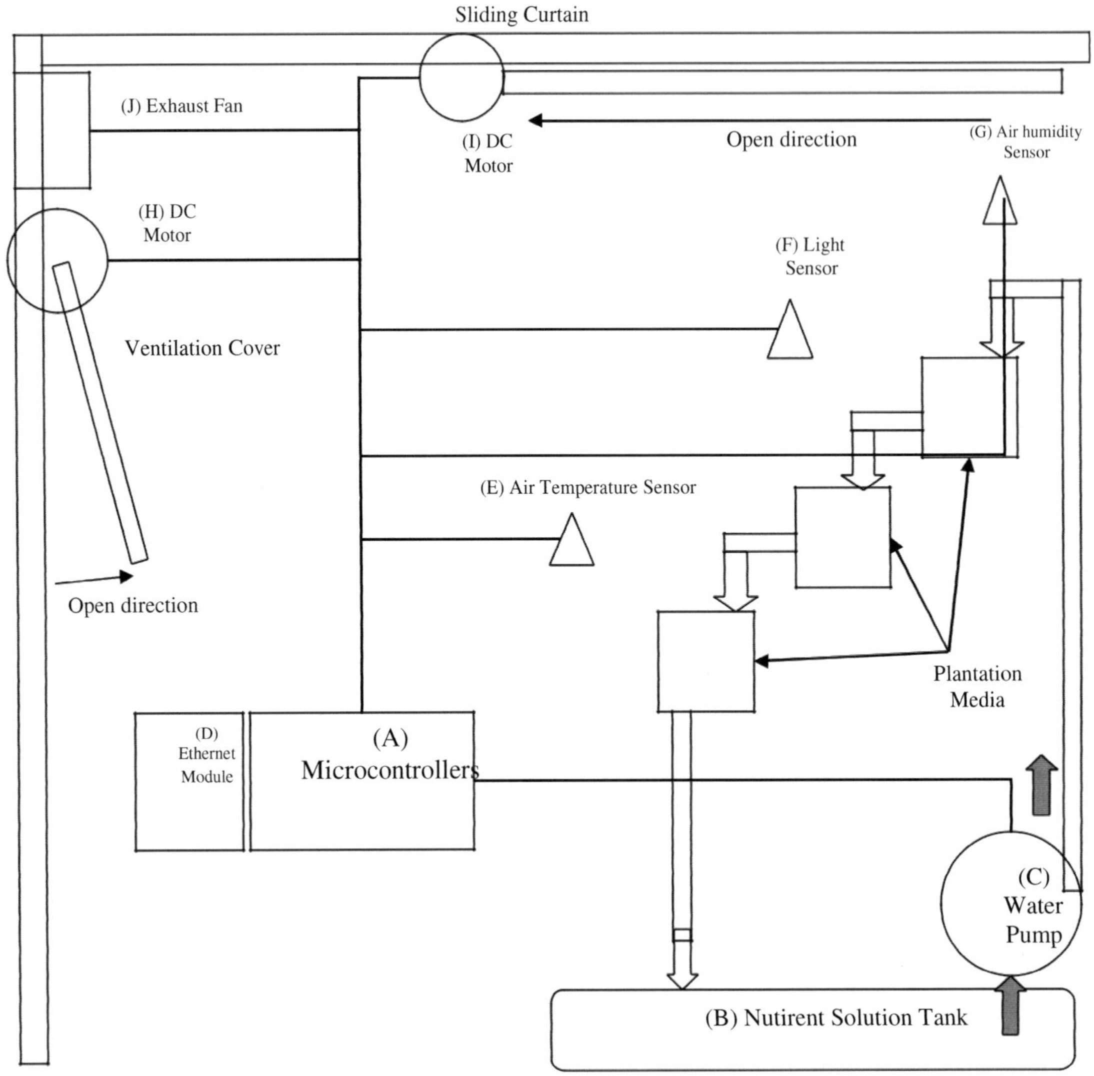

Figure 2. Chart of the System Design

5. Results and Discussion

The result of this study will be separated into 5 (five) parts, i.e.: hardware design (Figure 3), software design, testing scenarios, implementation (Figure 5 and 6), and testing results.

IOP Conf. Series: Earth and Environmental Science **396** (2019) 012017 doi:10.1088/1755-1315/396/1/012017

5.1. Hardware Design

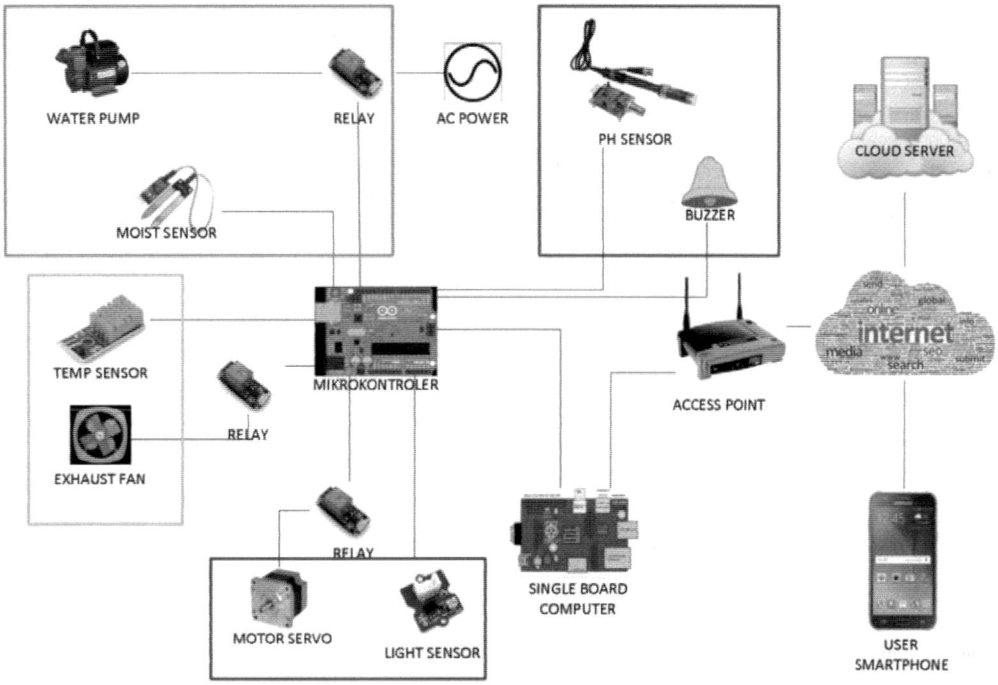

Figure 3. Hardware Design and Connectivity

5.2. Software Design

The detail of data communication of the software can be seen in Figure 4. Data from the single board computer will be transferred into the cloud server using an HTTP protocol [10]. Next, the result of the processed data will be carried out to the user smartphone using HTPP protocol and formatted with JSON.

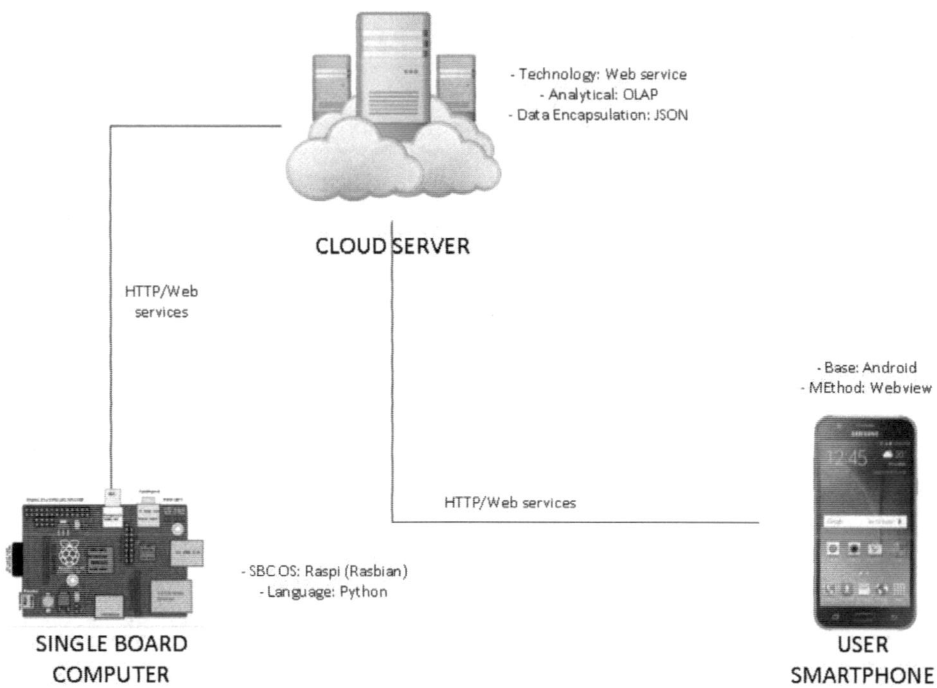

Figure 4. Design of Software Data Communication

5.3. Testing Scenarios

To testing the built system, we used the black-box testing method. The testing conducted based on the default functions of the system. Table 1 listed the default functions of the system which is tested using the Black-box method.

Table 1. Default functions tested

Function	Description
Get the soil humidity	Get the soil humidity from the sensors
Get the room temperature	Get the room temperature from the sensors
Get the light intensity	Get the light intensity from the sensors
Get the pH	Get the pH from the nutrient solution
Turn the exhaust on	Turn on the exhaust fan to flow out the air
Turn the exhaust off	Turn off the exhaust fan to stop the air flow
Turn the water pump on	Turn on the water pump to watering the plantation media
Turn the water pump off	Turn off the water pump
Send notification to the smartphone	Send notification to the smartphone using web services

5.4. Implementation

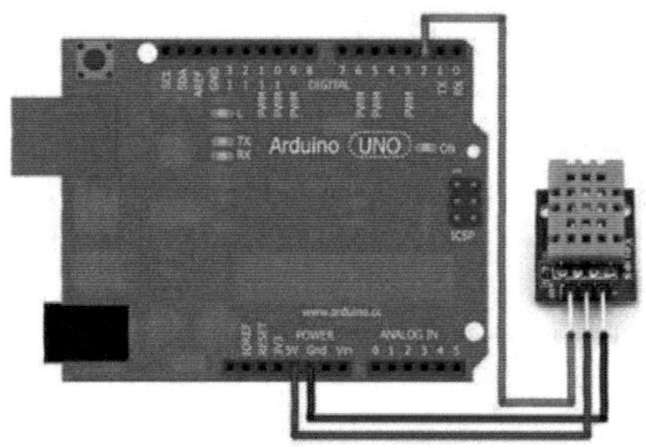

Figure 5. Air humidity sensor connectivity

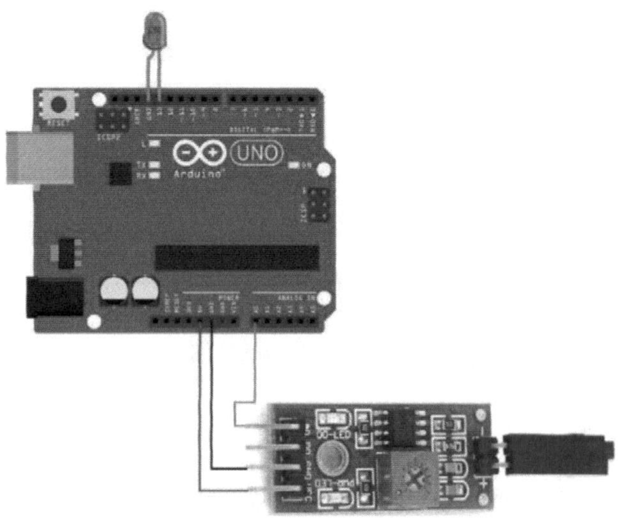

Figure 6. Light intensity sensor connects to the microcontroller

5.5. Testing Results

Using the black-box testing, Table 2 shown the result.

Table 2. Result of the testing

Function	Description	Result
Get the soil humidity	Get the soil humidity from the sensors	Succeed
Get the room temperature	Get the room temperature from the sensors	Succeed
Get the lights	Get the light intensity from the sensors	Succeed
Get the pH	Get the pH from the nutrient solution	Succeed
Turn the exhaust on	Turn on the exhaust fan to flow out the air	Succeed
Turn the exhaust off	Turn off the exhaust fan to stop the air flow	Succeed
Turn the water pump on	Turn on the water pump to watering the plantation media	Succeed
Turn the water pump off	Turn off the water pump	Succeed
Send notification to the smartphone	Send notification to the smartphone using web services	Succeed

6. Conclusions

The urban farming monitoring plants designed and built in the form of prototypes can be used to monitor the remote location of hydroponic plants. The web services system designed to provide early information to the farmers with the condition of hydroponic plants in the event of a critical situation.

7. Acknowlegment

This article was presented at the 2nd International Conference on Smart City INNOVATION (ICSCI) 2019, jointly held by Universitas Indonesia and Universitas Diponegoro. ICSCI conferences have been supported by the United States Agency for International Development (USAID) through the Sustainable Higher Education Research Alliance (SHERA) Project for Universitas Indonesia's Scientific Modelling, Application, Research, and Training for City-centered Innovation and Technology (SMART CITY) Center for Collaborative Research, administered through Grant #AID-497-A-1600004, Sub Grant #IIE-00000078-UI-1

8. References

[1] Karsono S, Sudarmodjo and Sutiyoso Y 2002 *Hidroponik skala rumah tangga* (Penerbit PT Agro Media Pustaka)

[2] Dewi G A M K and et. al 2018 Int. Conf. on Smart-Green Technology in Electrical and Information System *Danvers: IEEE* pp 120–5

[3] Lopes P, D F, A J and A. F S Int. Conf. Advanced Information Networking and Applications Workshops *(WAINA)* pp 1349–54

[4] Lennard W A and Leonard B V 2006 A comparison of three different hydroponic sub-systems (gravel bed, floating and nutrient film technique) in an aquaponic test system *Aquac. Int.* **14** 539–50

[5] Clarkson R and Lane S D 1991 Use of a small-scale nutrient-film hydroponic technique to reduce mineral accumulation in aquarium water *Aquac. Res.* **22** 37–45

[6] Dzulfikar D and and Wisnu Broto 2016 Optimalisasi Pemanfaatan Energi Listrik Tenaga Surya Skala Rumah Tangga (PRODI Pendidikan Fisika dan Fisika UNJ)

[7] Kadir A 2013 Panduan Praktis Mempelajari aplikasi mikrokontroler dan pemrogramannya menggunakan Arduino *Yogyakarta Andi*

[8] Curran M A, Diehl R J, Peck G A and Rahaman K L 2005 Single board computer quotation and design system and method

[9] Barovih G 2016 Desain Sistem Informasi Penjadwalan Laboratorium Terbuka (Studi Kasus: Laboratorium Terbuka STMIK Palcomtech) *SISFOTENIKA* **6**

[10] Nama G F, Despa D and others 2016 Real-time monitoring system of electrical quantities on ICT Centre building University of Lampung based on Embedded Single Board Computer BCM2835 *2016 International Conference on Informatics and Computing (ICIC)* pp 394–9

Effect of Weather for Demand of Online Transportation in Tembalang, Semarang

A R Rakhmatulloh[1], D I K Dewi[1], Wijayanti[2], J Winarendri[3]

[1]Urban Regional Planning Department, Faculty of Engineering, Diponegoro University, Jl. Prof Sudharto,SH Tembalang, Semarang,Jawa Tengah, Indonesia
[2]Architecture Department, Faculty of Engineering, Diponegoro University, Jl. Prof Sudharto,SH Tembalang, Semarang,Jawa Tengah, Indonesia
[2]Wageningen University & Research, Netherland

E-mail: anita.ratnasari.r@pwk.undip.ac.id

Abstract. Estimates of online transportation users in an area need to be known because it can be used to analyze how much the dependence of the traveler on online transportation. So that it can be seen how the travel patterns of online transportation users. However, research on the impact of weather variables on the selection of modes of online transportation has not been widely implemented. This study aims to evaluate the influence of the weather by entering dry season and rainy season variabel related to daily travel behaviour on online transportation. This research was conducted at Tembalang Semarang, and the target of respondents was students. The analytical results show that, in general the existence of rainfall is associated with a certain degree of online transpotation user of go car/grab car increase. So we can conclude that weather-related variables were found to routine/daily trips.

Keywords: weather, demand of online transportation, education area, daily trip

1. Introduction

Online transportation is part of smart mobility. Users are spoiled by services that can be controlled by smartphones. So that users can more easily make movements according to the start mobility and its frequency. Transportation is usually also influenced by the weather. Research on weather and transportation modes has been carried out in several countries, but the numbers are still minimal. It should be one of the important topics in a study, because weather is a factor that has a significant influence on a person's decision in transportation mode choice.

Factors that influence a person to choose transportation modes such as online transportation are among others influenced by the number and quality of services provided, economic factors, social factors, spatial factors such as service coverage areas of a transportation mode, and others [1]. One other factor that can affect demand of online transportation, but not yet widely discussed is the weather factor. Most studies that focus less on the weather on daily conditions in travel behaviour such as the choice of modes and destination of travel. According to research conducted previously by A. Singhal, C. Kamga, and A. Yazici [2], the weather had a negative impact on the demand for city transportation in New York for example the Metropolitan Transit-New York Transit (MTA-NYCT) subway in particular on passengers.

This study aims to determine the effect of weather by including variables of dry and rainy weather related to user profiles and choice of modes for regular trips in Tembalang, Semarang City. Tembalang area is interesting because this area is a region of education, so it has high demand for online transportation. Most users of online transportation are in their 20s and as student. This study uses primary data with reference to several articles that have conducted research related to the choice of transportation modes and weather. These articles will be discussed in the next section. This study focuses on the characteristics of users and modes of transportation used for routine trips in the Tembalang education area, Semarang. In previous studies taking into account the impact of weather on the selection of modes of transportation such as bicycles, private vehicles, and public transportation. While in this study using weather data and transportation modes that focus on online transportation services, because online transportation has many modes of choice at different prices. So users can adjust their needs. The results are expected to know the characteristics of transportation mode selection by students in the education area based on weather conditions.

2. Literature Review

Online transportation is a high technology applied to transportation sector. Currently online transportation is one of the transportation preferred by the public. The easiness and services of online transportation are one of the reason why people leaving conventional transportation (public transportations). The advantages of online transportation are the clear information about the price and dwelling time for the user. Online transportation also offers online order via smartphone without the users go to the traditional taxi bike station [15]. Online transportation also provides two ways of pay: cash or non-cash. Online transportation also attempts to provide security and convenience to the users [3].

Research on the impact of weather on transportation has been done before, one of which is the impact of weather conditions on macroscopic urban travel time. It was explained that inclement weather conditions may result in substantial reductions of roadway capacities and thus, operating speeds [4,5]. Additionally, traffic demand may be largely affected by the weather, since the latter has a considerable impact on a series of humans decisions such as transport capital choice, trip distribution, trip cancellation or postponement; altering roadway users' valuation of actual transport costs and travel times [6]

In addition, research on the weather effects of various modes of transportation has also been carried out, such as the transportation air [7], rail based transport [8], and ground vehicles [4,9] and everything is done in the country 4 seasons like countries in America and Europe.[10] The modes of transportation used in several of these studies include modes of private transportation, rent, and public transportation. Saneinejad et al (2012), mention that other weather indicators such as temperature affects the one's journey using all transport mode without exception.[10] The occurrence of a shift in the selection of modes of transportation when it is rainy or rain is from motorbikes or open vehicles to cars, public transportation such as the subway.

According to Kim et al. [11] individual travel movements are influenced by many factors, particularly the weather. In this Swedish study, attitudes toward cycling during the rainy were generally, and in relation to rainy, there were clear differences in fashion choices between seasons. The number of car trips increased 27% from dry to rainny while the number of bicycle trips decreased by 47%. The number of car trips increases with distance while the number of bicycle trips decreases, and the decline is even more significant in rainny than in dry. There are differences of opinion between the two categories of cyclists, about what factors influence the choice of modes of transportation for travel to work. Temperature, rainfall and road conditions are the most important factors for those who cycle to work in the dry but not in rainny [12].

Knowledge and information about the weather became important to be known by those who would travel, especially for users of unprotected vehicles such as bicycles, motorcycles and walking [13]. The information the rider obtained will determine the decision on the trip, as described by [13]. Weather conditions such as Rain, temperature difference, wind speed difference will make the service of transportation network change [14]. In research conducted by Hranac at al. [15] shows that it turns out that rain will significantly change the flow of vehicle traffic on the road network [10].

Weather can influence decisions about the journey that must be made, where to go, and what mode to choose. This paper identifies specific weather conditions that have a significant impact on number of pedestrian at 13 locations in Alameda County, CA. This shows the importance of taking into account the weather when analyzing pedestrian volumes. The results show that some weather variables have a significant influence on pedestrian volume over a period of time. Rain has the greatest effect on the volume of pedestrians in certain locations, although clouds, wind, and dry and cold temperatures have also been shown to reduce volume. This study shows the importance of weather calculations when analyzing pedestrian volumes [16].

Studies conducted by Shaaban and D. Muley [17] investigated the impact of temporal and weather characteristics on pedestrian volumes in dry climatic conditions. The main environment located in urban areas in Doha, Qatar was chosen as the location of the case study. Low pedestrian activity is seen in all seasons. Pedestrians walk more during weekdays in rainny, and during weekends in dry and spring. Generally, pedestrian volume is higher at night. The results show that pedestrian volume has a log-linear relationship with weather characteristics. Temperature is the only parameter that significantly affects pedestrian volume. This shows that the weather has a strong influence on a person's decision in choosing a mode of transportation.

3. Methodology

This research was conducted in Tembalang District, Semarang City, where there were several universities in Tembalang District. The Tembalang area is included in the education area, so most of the area is inhabited by students from outside the city. There are four universities in the Tembalang area, namely Diponegoro University (UNDIP), Pandanaran University (UNPAND), Semarang State Polytechnic (POLINES), and Semarang Health Polytechnic (POLTEKKES). Then for the main activity is learning with the aim of regular movement to campus. Then indirectly, every day students need a mode of transportation to get to campus.

The data used in this study are primary data sourced from the questionnaire. The questionnaire was distributed to students who were enrolled in tertiary institutions in the Tembalang area. This study used 100 randomly selected samples. The analytical method used in this study is descriptive statistics. Data used include the characteristics of users of transportation modes, data on weather conditions (dry season and rain season) and choice of transportation modes.

4. User Characteristics Mode of Transportation in Tembalang

Tembalang area is one of the tertiary education areas in the city of Semarang and most of it is inhabited by university students. There are four universities, namely Diponegoro University (UNDIP), Pandanaran University (UNPAND), Semarang State Polytechnic (POLINES), and Semarang Health Polytechnic (POLTEKKES). In this study, the intended respondents were students who were studying in the Tembalang education area. To fulfill their routine travel needs, they need transportation modes to reach destination locations such as campus. In addition to private transportation modes, students can also make online transportation, public transportation as a choice of transportation modes.

5. Result and Discussions

5.1. Weather

Weather or temperature is one factor that is quite influential in choosing the mode of transportation that will be used in the trip. Based on the survey conducted, 92.9% of respondents answered that the weather had an effect on the choice of transportation mode. Indonesia is a tropical country with two types of seasons, namely the dry season and the rainy season. Where in the dry season it has a dry temperature/sunny and in the rainy season it has cold temperatures/rainfall. This weather difference is expected to have an influence on the mode of choice behaviour in the Tembalang education area. As previously explained, the Tembalang education area has been served by several modes of transportation such as online transportation and public transportation. In addition, these modes of transportation are also affordable. Since online transportation, both online motorbikes and online taxis, have helped the routine movements of students. For example, when it rains, you can choose to use a car to get to the campus even if you don't have a car, that is by using an online taxi service (Figure 1).

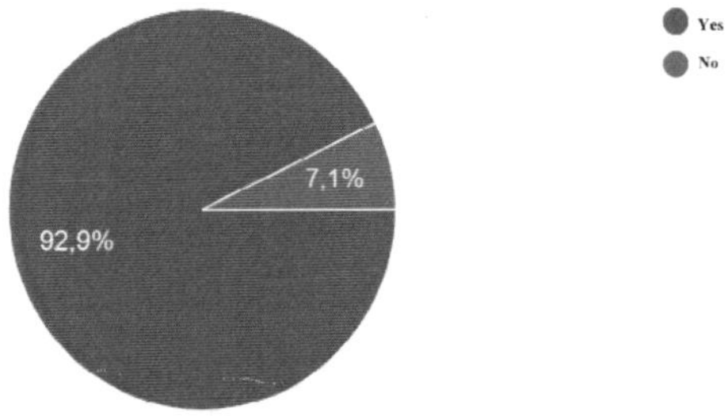

Figure 1. Effect Weather towards the Choice of Transportation Mode By Respondents

Dry Season in the Tembalang region there are no holidays during the dry season (Figure 2). They continue to carry out their activities as usual, such as studying to campus. The temperature during the dry season in this region can reach 32° Celcius.

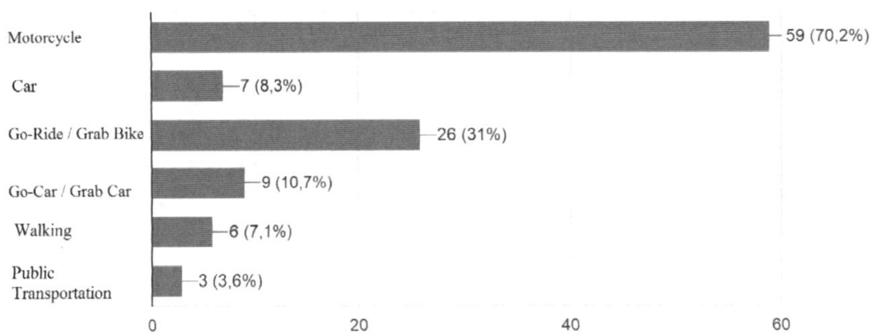

Figure 2. Mode Choice in Dry Season

Based on the survey conducted by distributing questionnaires, it shows that a total of 70.2% of respondents use motorized transportation modes even though the conditions are quite dry. The reason for respondents using motorbikes as a mode of transportation on their way to campus is to be comfortable, the time during the trip is faster than other modes, other transportation cannot reach the residential area or boarding house. Then the choice of transportation modes that are quite attractive besides motorbikes are shuttle services from online transportation, better known as the Go-Ride / Grab Bike. 31% of respondents chose to use the Go-Ride / Grab Bike as a mode to fulfill their travel needs. The reason for respondents using this mode is not having a private vehicle, no need to bother to park, comfortable, safe, and able to reach residential locations that are not affordable by public transportation. If the choice of mode of transportation when dry season is distinguished by sex, a number of 25.6% of respondents who are male choose to use a motorbike as a mode of transportation in their routine movements, for example, to go to campus. Only small proportions choose cars and online transportation as their mode of transportation. Then for female respondents, 43% also chose to use the motorbike as their mode of transportation. And, a total of 17.4% chose online two-wheeled transportation or often known as Go-Ride or Grab Bike. For more details, see the Table 1.

Table 1. Mode Choice based on Male and Female preference in Dry Season

		Mode of Transportation				Total
		Motorcycle	Car	Go-Ride / Grab Bike	Go Car / Grab Car	
Gender	Male	25,6%	1,2%	4,7%	0,0%	31,4%
	Female	43,0%	4,7%	17,4%	3,5%	68,6%
Total		68,6%	5,8%	22,1%	3,5%	100,0%

5.2. Rainy Season

During the rainy season, the temperature feels colder than the dry season (Figure 3). Rain occurs throughout the day, where morning to evening is the time to carry out routine activities, studying to campus.

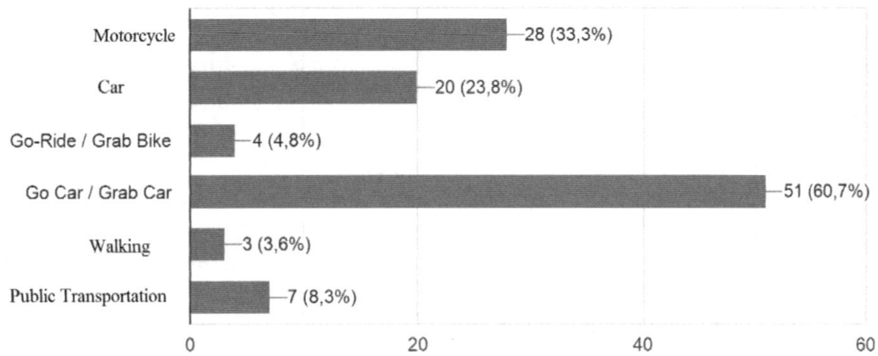

Figure 3. Mode Choice in Rainy Season

Based on the survey conducted with the distribution of questionnaires, it shows that 60.7% of respondents chose to use the shuttle transportation mode from online transportation, better known as the Go Car / Grab Car. Respondents chose to use Go Car / Grab Car as a mode to fulfil their travel needs. The reason for the respondents using this mode is not to bother using raincoats, more efficient because they do not need to wait for the rain to stop, be comfortable, safe because they are not wet from rain and do not have to bother looking for a parking space. During cold or rainy temperatures, 50% of respondents choose Go-Car or Grab Car as a mode of transportation in meeting their travel needs, which consists of 12.8% of respondents are male and 37.2% of respondents are female. In this study, users of online transportation services such as online motorcycle taxi and taxi were dominated by women. This is due to the sense of security and comfort provided by online transportation services. Feel safe because the route is tracked and the driver's identity is clear (Table 2).

Table 2. Mode Choice based on Male and Female preference in Rainy Season

		Mode of Transportation					Total
		Motorcycle	Car	Go-Ride / Grab Bike	Go Car / Grab Car	Public Transport	
Gender	Male	9,3%	7,0%	2,3%	12,8%	0,0%	31,4%
	Female	17,4%	8,1%	3,5%	37,2%	2,3%	68,6%
Total		26,7%	15,1%	5,8%	50,0%	2,3%	100,0%

5.3. Weather Variables is Found to Affect People's Attitude Toward Online Transportation Dry Season

Based on a survey that has been conducted through distributing questionnaires to students in the Tembalang education area, it shows that there is a change or shift in the behaviour of the choice of transportation modes used in routine trips. A total of 89% of respondents did not show changes in the behaviour of the choice of transportation mode. More is dominated by modes of transportation of two-wheeled vehicles such as motorbikes and motorcycle taxis online. Then a total of 11% experienced changes in the behaviour of the choice of transportation modes, from motorbikes to cars or taxis online. The reason for the change in the choice of transportation modes is, among others, more comfortable and not exposed to heat.

5.4. Rainy Season

Based on a survey that has been conducted through distributing questionnaires to students in the Tembalang education area, it shows that there is a change or shift in the behaviour of the choice of transportation modes used in routine trips. A total of 51% of respondents indicated a change in the behaviour of the choice of transportation mode. Students choose to use modes of online transportation such as online cars or taxis and public transportation. However, the percentage is more to use online transportation because it is easier to order, convenient, and safe because the routes are tracked. Students prefer to take a car or taxi online because they avoid rain or cold temperatures. This will have an impact on increasing traffic congestion due to those who normally or mostly use motorbikes or online motorbikes switching to online cars or taxis so that congestion occurs.

Rainny or when it rains is a season that has a significant influence on the choice of transportation modes, while dry is not very influential. This is in accordance with our research, namely the shift in the selection of transportation modes during rain or cold, for example from motorcycles shifting to cars or taxis online. And in the dry or dry season the shift in choice of modes is not too significant, because it still uses transportation modes that are usually used in routine trips.

6. Summary and Discussion

This study to explore the impact of changing weather conditions on a daily basis on online transportation use in Tembalang's educational area.The analytical results show that, in general the existence of rainfall is associated with a certain degree of online transpotation user of go car/grab car increase. Rainy season has effects for online transpotation user on mode shift between other modes (e.g. motorcycle, car, go ride/grab bike, and public transport). So we can conclude that weather-related variables were found to routine trips. These results indicate that round trip, mostly carried out during rush hour, is strongly influenced by changes in weather, especially rain.

The findings of this study enrich the existing literature concerning the impacts of weather on urban transportation. Conditions of temperature, weather and road conditions are important factors for those who use motorbikes in the dry season but not in the rainy season. Bad weather conditions such as rainy and cold seasons result in the shifting of transportation modes used, namely from open mode to closed or protected transportation modes [1,2,12,18–20]. This is shown in this study, where there is a shift in the selection of transportation modes during rain or cold. When the rainy weather shows respondents choose to use cars both private cars and public transportation and also online transportation services such as online taxis with the reason to avoid rain and cold. In addition, this study also shows that the number of car trips has increased from dry to the rainy season. Based on the survey, 49% of the respondents chose not to change the modes of transportation used during the rainy season, they still chose the mode of transportation commonly used on their regular trips. It can be concluded that there are quite clear differences in the selection of transportation modes in the two seasons in the Tembalang education area.

The limitation of this study is only on the influence of weather on the choice of modes in the Tembalang education area which is focused on routine travel only. So that it has not been able to answer travel behavior in full. Inputs for subsequent research include, among others, peak hours and non peak hours, differences in rates at peak hour and nonpeak hours, and also differences in routine trips and additional trips.

7. Acknowledgment

This article was presented at the 2nd International Conference on Smart City INNOVATION (ICSCI) 2019, jointly held by Universitas Indonesia and Universitas Diponegoro. ICSCI conferences have been supported by the United States Agency for International Development (USAID) through the Sustainable Higher Education Research Alliance (SHERA) Project for Universitas Indonesia's Scientific Modelling, Application, Research, and Training for City-centered Innovation and Technology (SMART CITY) Center for Collaborative Research, administered through Grant #AID-497-A-1600004, Sub Grant #IIE-00000078-UI-1

8. References

[1] Stover V W and McCormack E D 2012 The Impact of Weather on Bus Ridership in Pierce County, Washington *J. Public Transp.* **15** 95–110

[2] Singhal A, Kamga C and Yazici A 2014 Impact of weather on urban transit ridership *Transp. Res. Part A* **69** 379–91

[3] Dewi N K and Rakhmatulloh A R 2018 The Characteristic of Online Transportation Services and Provision in Semarang City *J. Tek. Sipil dan Perenc.* **20** 90–7

[4] Martin P T, Perrin J, Hansen B and Quintana I 2000 Inclement weather signal timings *Univ. Utah, Mt. Consort.*

[5] Koetse M J and Rietveld P 2009 The impact of climate change and weather on transport: An overview of empirical findings *Transp. Res. Part D Transp. Environ.* **14** 205–21

[6] Tsapakis I, Cheng T and Bolbol A 2013 Impact of weather conditions on macroscopic urban travel times *J. Transp. Geogr.* **28** 204–11

[7] Changnon S A 1996 Effects of summer precipitation on urban transportation *Clim. Change* **32** 481–94

[8] Chapman L 2007 Transport and climate change: a review *J. Transp. Geogr.* **15** 354–67

[9] Chung E, Ohtani O, Warita H, Kuwahara M and Morita H 2006 Does weather affect highway capacity *5th international symposium on highway capacity and quality of service, Yakoma, Japan*

[10] Ismaili A F 2017 Pengaruh Cuaca Terhadap Perilaku Pengendara Sepeda Motor di Daerah Istimewa Yogyakarta *Semesta Tek.* **20** 132–8

[11] Kim D, Park J and Hong A 2017 The Role of Destination ' s Built Environment on Nonmotorized Travel Behavior : A Case of Long

[12] Magnusson R 2003 Potential of transferring car trips to bicycle during winter **37** 649–66

[13] Liu C, Susilo Y O and Karlström A 2015 The influence of weather characteristics variability on individual's travel mode choice in different seasons and regions in Sweden *Transp. Policy* **41** 147–58

[14] Khattak A J and De Palma A 1997 The impact of adverse weather conditions on the propensity to change travel decisions: a survey of Brussels commuters *Transp. Res. Part A Policy Pract.* **31** 181–203

[15] Hranac R, Sterzin E, Krechmer D, Rakha H A, Farzaneh M, Arafeh M and others 2006 Empirical studies on traffic flow in inclement weather

[16] Schneider R J, Arnold L S and Ragland D R 2010 Effects of Weather Variables on Pedestrian Volumes in Alameda County , California Effects of Weather Variables on Pedestrian Volumes in Alameda County , California

[17] Shaaban K and Muley D 2016 Investigation of Weather Impacts on Pedestrian Volumes *Transp. Res. Procedia* **14** 115–22

[18] Smith B L, Byrne K G, Copperman R B, Hennessy S M and Goodall N J 2003 An Investigation Into The Impact Of Rainfall

[19] Nankervis M 1999 The Effect of Weather and Climate on Bicycle Commuting **33**

IOP Conf. Series: Earth and Environmental Science **396** (2019) 012018 doi:10.1088/1755-1315/396/1/012018

[20] Böcker L, Dijst M and Prillwitz J 2012 Transport Reviews: A Transnational Impact of Everyday Weather on Individual Daily Travel Behaviours in Perspective: A Literature Review Impact of Everyday Weather on Individual Daily Travel Behaviours in Perspective : A Literature Review 37–41

Basic Framework of Regional Model for Disaster Waste Estimation and Distribution by Using Spatial Approach in Central Java-Indonesia

M Maryono[1,2],K Seruyaningtyas[2], A D Roynaldi[2], Sudarno[3], Hadiyanto[2,4]

[1]Department of Urban and Regional Planning, Faculty of Engineering, Diponegoro University-Indonesia
[2]Center for Green Infrastructure Resilience and Development (C-Greinde), Master Program of Environmental Science, School of Postgraduate, Diponegoro University-Indonesia
[3]Department of Environmental Engineering, Faculty of Engineering, Diponegoro University-Indonesia
[4]Department of Chemical Engineering, Faculty of Engineering, Diponegoro University-Indonesia

E-mail: m.maryono@undip.ac.id

Abstract. Disaster are becoming more frequent around the world and more intense in urban areas. Central Java is the province with the largest number of disasters in Indonesia. Disaster create amount of disaster waste and sometime very huge, many times comparing to daily municipal waste generation. Government and local citizen sometimes are unable to handling disaster waste. Since effective implementation of disaster waste management (DWM) will lead quick and better recovery of living and avoid more environmental risk, high level regional government such as provincial should help. To make effective support, it is important that provincial government develop a model to monitor disaster waste generation in each regency and cities. This study aims to develop basic framework of regional model for disaster waste estimation and distribution in provincial level. This study used Central Java Province as a case study. To estimate disaster waste in Central Java, this study developed an index system as a basic model. It is about basic comparison of disaster waste generation in all central java regencies or cities. Disaster waste is associated with total disaster events. This study utilized GIS tools to express model and system. This study developed of regional model disaster waste estimation in Central Java, divided by 3 sub decades. Study result shown that in first decade 1990-1999 the highest index of disaster waste estimation is Brebes Regency. Second decade 2000-2009 the highest index of disaster waste estimation is Semarang Regency and the third decade 2010-2018 the highest index of disaster waste estimation in Cilacap Regency.

Keywords: Disaster Waste Estimation, Infrastructure Resilience, Index Based Model, GIS, Central Java.

1. Introduction

Indonesia is one of prone country in the world. For year 2018 there is Indonesia Disaster Management Agency known as BNPB recorded 3398 disaster events in Indonesia [1]. Disaster can create a lot of disaster waste, sometimes, it is a huge and many times compare with annual municipal waste generation. According to the Indonesian experience from Tsunami Waste Recovery Waste Management Project, at least 8 years total time is needed to recover all waste management Infrastructure [2]. Disaster waste management (DWM) should be prepared properly. Effective and efficient of disaster waste management creates many advantages not only for disaster recovery but also for economic recovery by exploring any challenges in the recycling proses of disaster waste management [3]. Disaster waste material can be recycled and reused, especially for housing recovery.

When huge disaster waste generated, local government usually unable to handling. They need support others and intervention from high level of authority usually needed. They need support from other government both from Provincial and National government. To prepare it, Provincial and National government need a basic information of disaster waste generation. It is need to estimate the requirement of infrastructure and other resources. Development model for disaster waste estimation is very important for Indonesia region. Availability of disaster waste information in a regional level such as provincial in Indonesia, lead an effective effort to achieve disaster resilience. Basic model of regional disaster waste estimation is not only for DWM but also for preparing effective approaches, avoid environmental risk, increasing safety and health. Effective monitoring of disaster waste generation also leads more faster in respond, recovery and development [4,5]. This basic model is developed not only for developed disaster resilience, but also for achieving of the 17 Sustainable Development Goals (SDGs) especially goal number 11, to make cities inclusive, safe, resilient and sustainable [6]. This goal states that, by 2020, countries and all citizen should make any effort to adopting and implementing integrated policies and plans towards inclusion, resource efficiency, mitigation, and adaptation to climate change, along with resilience to disasters by 2030 [6]. Moreover, since the disaster waste generation is correlated with physical infrastructure resilience, this research is also tried to give basic information of physical rural urban infrastructure resilience against the disasters [6].

Disaster waste estimation model can be developed by manual system, integrated system. Moreover, disaster waste estimation model can be developed by using on line system or by using internet line and automatic system. Internet of thing for waste management in peace time, has been explored and become challenging issues in recent year [7–9]. Some factor will influence to the system such as availability of technology, availability of human resource and organization, and availability of budget [7–9]. However, the model can be developed step by step according to the condition and the availability of resources.

2. Data and methods

2.1. Study area

Indonesia is a disaster-prone country that consist of 34 provinces. To perform the basic model of disaster waste estimation model, this study selected Central Java Province. It is a part of a case study method to describes the issues of disaster waste estimation model in Indonesia [10,11]. This province is very unique, many disasters attacked to the region such as earthquake, storm and floods, landslides, and volcanic eruptions. Central Java is one of the prime locations of Indonesia which consist of 29 local government (regency) and 6 city government. Indonesia Disaster Management Agency known as BNPB calculated that Central Java is the province with the most disasters events in Indonesia with 3389 disaster event in 2018 [1].

2.2. Method

Spatial approach within GIS utilization has been growth quickly in recent years. It is not only for giving fast respond, but also for improving accuracy [12]. However, in disaster waste management, there is no information issued, and research toward disaster waste information is needed [13]. DWM guidelines issued by Japan ministry of environment [14], and issued by united nation for environmental program [15] did not give any suggestion and method to developed the information system. This study developed basic model to estimate disaster waste by using index model. The Index Model is described value of one region compare to the total value in region. Moreover, this study is descriptive quantitative of secondary data to performing the index model. The data of disaster events collected from The National Disaster Management Agency (BNPB) Indonesia [1]. To build the basic model, the estimation of disaster waste generation index is analysis by utilized the number of disaster event in each regency and city compared to the number of totals of disaster event in Central Java Province. The basic index for regional disaster waste monitoring calculated by the formula as follow:

$$\frac{\Sigma i}{\Sigma ni}$$

with,
i = the number of disaster events in a regency or city
ni = the number of disasters in Central Java

Automatic system to calculated disaster event in central java, developed by utilized data on line system from the DIBI provide by BNPB [1]. Moreover, automatic system for understanding spatial distribution of disaster waste estimation, this study utilized GIS tools. GIS is a spatial approach for making any spatial information. GIS in this study used to comparing degree of disaster waste estimation and distribution of disaster waste generation in each region in Central Java compare to other.

3. Result and discussion

Model to estimate disaster waste generation is one of the significant stages for DWM [7]. Moreover, according to the estimation model, any infrastructure requirement to handling volume and characteristic to avoid environmental disturbance can be developed [3,7]. Disaster waste are vary depending of the characteristic of the region, natural and man-made infrastructure at which disaster happen. Wastes may consist of destroyed buildings and the objects they held inside, destroyed pavements or other infrastructure, wood, sands, and other natural derivatives and so on. Not only are wastes directly generated from disasters, activities in recovery and reconstruction in the post-disaster phase also generate waste [3,7].

Disaster waste generation can be estimated by using disaster event in Central Java. This study utilized data from the DIBI recorded by Indonesian Disaster Management Agency known as BNPB [1]. As shown in table 1. in the decade 1990-1999 BNPB administered 29 disaster events, for year 2000-2010 there are 1351 disaster event in Central Java Province, and for year 2010-2018 BNPB recorded 4706 disaster events. Brebes Regency, Semarang Regency and Cilacap Regency is the highest level affected by the disaster.

Tabel 1. Regional Disaster Event in Central Java Province 1990-1999, 2000-2009, 2010-2018

No	Regency/City	Disaster Event 1990-1999	Regency/City	Disaster Event 2000-2009	Regency/City	Disaster Event 2010-2018
1	Brebes	3	Semarang	108	Cilacap	591
2	Kendal	3	Wonosobo	104	Wonogiri	351
3	Semarang City	3	Cilacap	95	Temanggung	271
4	Cilacap	2	Banyumas	94	Magelang	264
5	Wonosobo	2	Wonogiri	65	Banyumas	262
6	Demak	2	Rembang	65	Semarang City	225
7	Semarang	2	Brebes	57	Banjarnegara	165
8	Tegal City	2	Kebumen	51	Purbalingga	163
9	Banyumas	1	Karanganyar	51	Kebumen	158
10	Purbalingga	1	Klaten	49	Boyolali	148
11	Kebumen	1	Boyolali	44	Wonosobo	147
12	Purworejo	1	Sukoharjo	42	Karanganyar	146
13	Sukoharjo	1	Pati	39	Semarang	144
14	Wonogiri	1	Sragen	35	Pati	141
15	Sragen	1	Grobogan	35	Pemalang	141
16	Grobogan	1	Tegal	35	Pekalongan	140
17	Pati	1	Pekalongan	33	Purworejo	119

No	Regency/City	Disaster Event 1990-1999	Regency/City	Disaster Event 2000-2009	Regency/City	Disaster Event 2010-2018
18	Kudus	1	Semarang City	33	Kendal	118
19	Banjarnegara	0	Jepara	31	Sragen	110
20	Magelang	0	Magelang	29	Jepara	101
21	Boyolali	0	Blora	27	Brebes	95
22	Klaten	0	Kudus	26	Klaten	93
23	Karanganyar	0	Banjarnegara	25	Grobogan	88
24	Blora	0	Demak	25	Tegal	83
25	Rembang	0	Kendal	25	Kudus	78
26	Jepara	0	Pemalang	24	Blora	76
27	Temanggung	0	Purworejo	23	Demak	76
28	Batang	0	Temanggung	21	Sukoharjo	60
29	Pekalongan	0	Purbalingga	18	Rembang	44
30	Pemalang	0	Batang	16	Surakarta City	34
31	Tegal	0	Surakarta City	14	Batang	32
32	Magelang	0	Pekalongan City	8	Magelang City	13
33	Surakarta city	0	Tegal City	3	Tegal City	13
34	Salatiga City	0	Magelang City	1	Pekalongan City	11
35	Pekalongan City	0	Salatiga City		Salatiga City	5
	Total	29		1351		4706

Source of table [1]

The number of disasters is identical to the number of disaster waste generation. Disaster waste generation is correlated with physical infrastructure resilience. Even though the disaster waste is very dependent on the type and characteristics of the disaster, the number of disasters can represent amount of the disaster waste.

By using disaster event from National Disaster Management Agency [1], this study examined and calculated estimation of disaster waste generation index in Central Java. Table 2 describe the result of the calculation and examination. Based on the table 1, the number of disasters occurring in Central Java is increasing. Disaster waste generation index shows the number of disasters in each city that occurred in 1990-1999, 2000-2009, 2010-2018. The range of the index disaster are 0,000-0,103 in year 1990-1999; 0,000-0,077 in year 2000-2009; and 0,002-0,126 in year 2010-2018.

Tabel 2. Regional Disaster Waste Generation Index in Central Java Province

Regency and Cities of Central Java	Regional Disaster Waste Generation Index Based on calculation and analysis of disaster event per region and cities		
	1990-1999	2000-2009	2010-2018
Cilacap	0,069	0,070	0,126
Banyumas	0,034	0,070	0,056
Purbalingga	0,034	0,013	0,035
Banjarnegara	0,000	0,019	0,035
Kebumen	0,034	0,038	0,034
Purworejo	0,034	0,017	0,025
Wonosobo	0,069	0,077	0,031

Magelang	0,000	0,021	0,056
Boyolali	0,000	0,033	0,031
Klaten	0,000	0,036	0,020
Sukoharjo	0,034	0,031	0,013
Wonogiri	0,034	0,048	0,075
Karanganyar	0,000	0,038	0,031
Sragen	0,034	0,026	0,023
Grobogan	0,034	0,026	0,019
Blora	0,000	0,020	0,016
Rembang	0,000	0,048	0,009
Pati	0,034	0,029	0,030
Kudus	0,034	0,019	0,017
Jepara	0,000	0,023	0,021
Demak	0,069	0,019	0,016
Semarang	0,069	0,080	0,031
Temanggung	0,000	0,016	0,058
Kendal	0,103	0,019	0,025
Batang	0,000	0,012	0,007
Pekalongan	0,000	0,024	0,030
Pemalang	0,000	0,018	0,030
Tegal	0,000	0,026	0,018
Brebes	0,103	0,042	0,020
Magelang City	0,000	0,001	0,003
Surakarta City	0,000	0,010	0,007
Salatiga City	0,000	0,000	0,001
Semarang City	0,103	0,024	0,048
Pekalongan City	0,000	0,006	0,002
Tegal City	0,069	0,002	0,003

Source: Calculation and analysis.

3.1 Basic Framework of Regional Model for Disaster Waste Estimation and Distribution in by using Spatial Approach in Central Java 1990-1999

To perform the basic regional model of disaster waste estimation and distribution in year 1990-1999, the study utilized the disaster waste estimation index for year 1990-1999 as shown in table 2 and GIS tools. The result of the basic regional model in year 1990-1999 shown in figure 1.

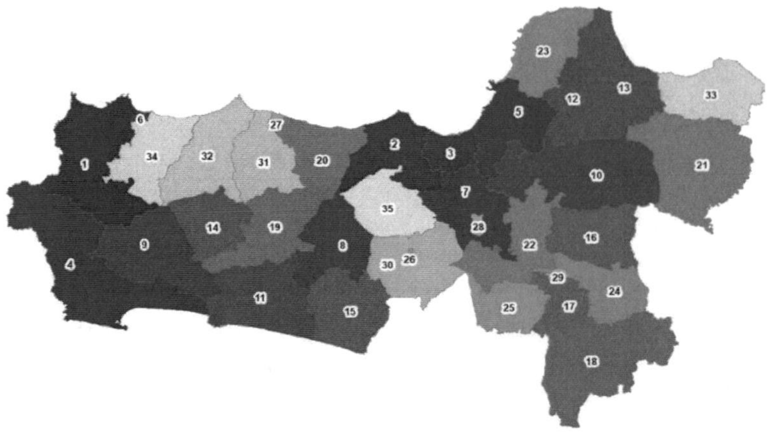

Figure 1. Basic Framework of Regional Model for Disaster Waste Estimation and Distribution by Using Spatial Approach in Central Java 1990-1999

Based from the figure 1, the gradation color on the map means there is a different disaster waste in Central Java. The regency or cities with bold colors mean the region is more high disaster waste generated compared to the regency of cities with bright colors. Refers to the regional model of disaster waste estimation in 1990-1999, the region in Central Java region with highest degree level of disaster waste estimation is Brebes regency, while the region with lowest degree level of disaster waste is Temanggung regency.

Cutter et all in year 2008 used spatial approach when they build a basis model of resilience indicator for baseline indicator [16,17] to analysis baseline condition and it improvement. Moreover, Modica & Reggiani in year 2015 use spatial approach to analysis economic resilience [18]. Starting from those previous researches this study is using spatial approach as a basis framework for analyzing disaster waste estimation and distribution since disaster waste estimation is a basis framework to build spatial resilience.

3.2 Basic Framework of Regional Model for Disaster Waste Estimation and Distribution in Central Java 2000-2009

After conducted the first model, the study continued to build the second model of basic regional disaster waste estimation for year 2000-2009. There several aims such as: (1) to understanding the consistency of the approach, (2) to examine the continuity of the model, and then (3) to assess the differences and or changing of each region as of disaster waste generated in each region. The result of Regional Model of Disaster Waste Estimation for 2000-2009 is described in figure 2.

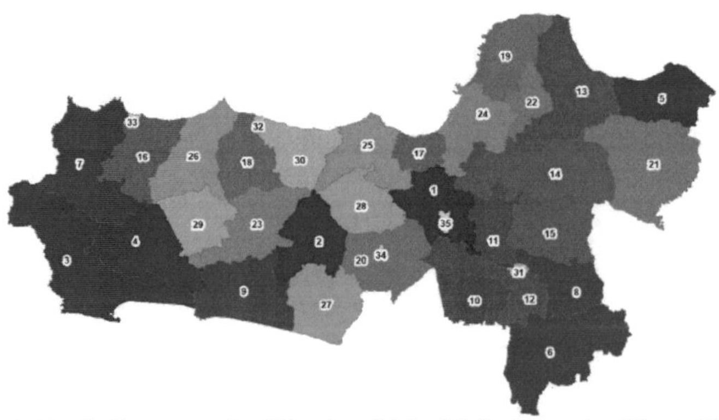

Figure 2. Basic Framework of Regional Model for Disaster Waste Estimation and Distribution by Using Spatial Approach in Central Java 2000-2009

Based from the figure 2, the model is consistent at which it is a comparison of one region to other region. The model also can be continued, at which all data in the table 1 can be execute and running by utilizing GIS tools. The model examined the differences and the changing between region in central Java province. as shown in figure 2, regency and cities in central Java is change. There is difference index and basic spatial model compare to the model previously as describe in figure1. As shown in figure 2. basic framework for regional model of disaster waste estimation and distribution in 2000-2009, the region in Central Java region with highest degree level of disaster waste estimation is changing become Semarang regency, while the region with lowest degree level of disaster waste is changing become Salatiga city.

3.3 Basic Framework of Regional Model for Disaster Waste Estimation and Distribution in by using Spatial Approach in Central Java 2010-2018

By using the same method, analysis then continued to perform the third model for analysis regional disaster waste estimation and distribution for year 2010-2018 in Central Java Province. Figure 3 describe basic framework for regional model of disaster waste estimation and distribution for year 2010-2018. Refers to the model, region with highest degree level at which bold color is still Cilacap regency while region with lowest disaster waste estimation index is same as previous regional model in figure 2 namely Salatiga city. According to the figure 3, all of region is changing, however Cilacap regency is most dominant compare to the other regions. With index as shown in table 1 estimated 0,126, it is the highest index during time period of study for 38 year from 1990- 2018.

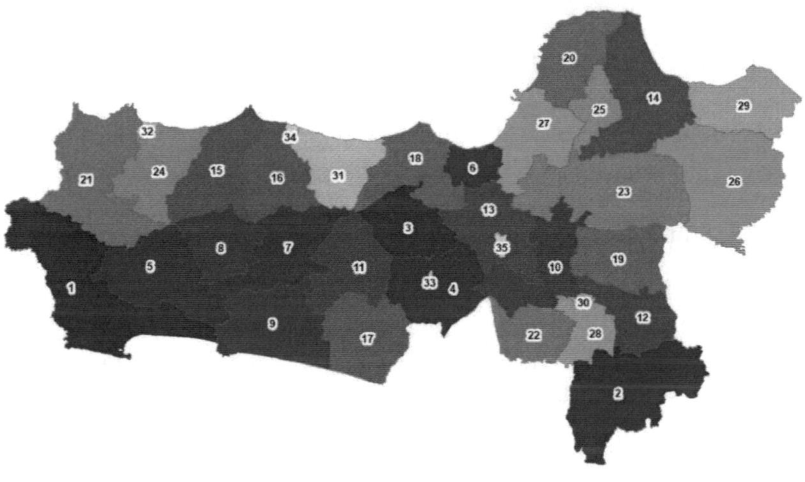

Figure 3. Basic Framework of Regional Model for Disaster Waste Estimation and Distribution by Using Spatial Approach in Central Java 2010-2018

Refers to the figure 3, there is no region that not changing comparing to the previous model in figure 2. This mean that the model is consistent and valid to describe the estimation of disaster waste generation in one region compared to another region in Central Java.

3.4 Pattern of Regional Disaster Waste Estimation 1990-1999,2000-2009, 2010-2018

Pattern of regional disaster waste estimation and distribution for year 1990-1999, 2000-2009, 2010-2018 is describe in figure 4. As shown in the figure the basic framework of regional model for disaster waste estimation and distribution in Central Java at which divide by three period is not have a big change. The figure shown that the pattern is similar. Degree of change for regional model of disaster waste estimation and distribution for three decades namely 1990-1999, 2000-2009, and 2010-2018 is very slow. There are four pattern of disaster waste generation characteristic in Central Java Region such as;(1) Cilacap Regency is one of the biggest and highest level of region at which there is high probability of high disaster waste generation

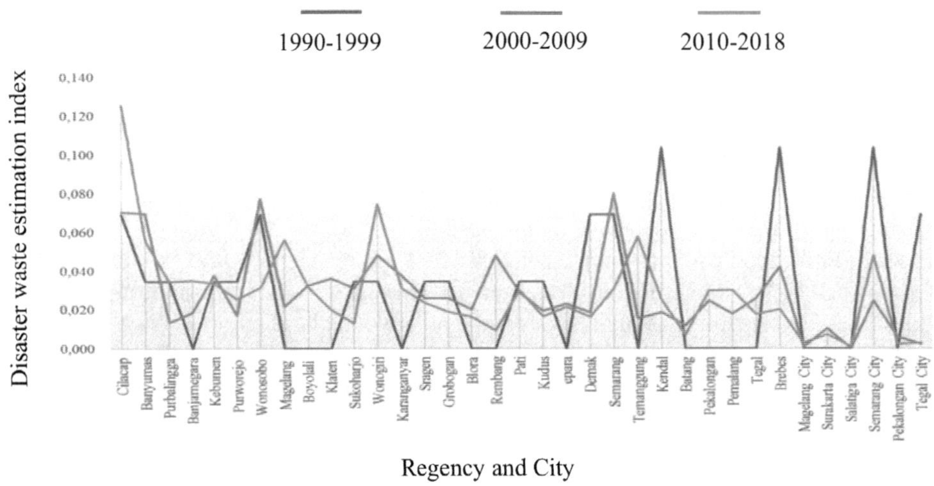

Figure 4. Degree Changing for Regional Model of Disaster Waste Estimation in three decade

4. Conclusion

Disaster waste estimation and distribution in a regional scale is part of planning to understanding spatial and geographical effort to build natural disaster resilience since to handling disaster waste properly, planning for infrastructure, planning for resources and planning for budgeting is very importance [7,19]. To analysis disaster waste generation and distribution can be calculated by using the disaster event at which recorded, even though disaster waste generation is specific depend on type of disaster, but this approach can be utilized to build a basic model and framework. This study developed a basic framework for regional model of disaster waste estimation and distribution in central Java. This model analysis disaster waste estimation and distribution model base on the disaster event for year 1990-1999, for year 2000-2009, and for year 2010-2018. This model shown that in central Java Brebes Regency is the biggest and highest index of disaster waste estimation and distribution for decade 1990-1999. Semarang Regency is the highest disaster waste estimation and distribution for decade 2000-2009 while Cilacap is the highest for year 2010-2018.

5. Acknowledgement

This article was presented at the 2nd International Conference on Smart City INNOVATION (ICSCI) 2019, jointly held by Universitas Indonesia and Universitas Diponegoro. ICSCI conferences have been supported by the United States Agency for International Development (USAID) through the Sustainable Higher Education Research Alliance (SHERA) Project for Universitas Indonesia's Scientific Modelling, Application, Research, and Training for City-centered Innovation and Technology (SMART CITY) Center for Collaborative Research, administered through Grant #AID-4=97-A-1600004, Sub Grant #IIE-00000078-UI-1. This research is funded by the Directorate General of Research and Development, Ministry of Research, Technology and Higher Education through Institute for Research and Community Services Diponegoro University for year 2019 with contract number 257-66/UN7.P4.3/PP/2019.

6. References

[1] Indonesia Disaster Management Agency 2019 Data Informasi Bencana Indonesia

[2] UNDP Indonesia 2012 Tsunami Recovery Waste Management Project (TRWMP)

[3] Asari M, Sakai S, Yoshioka T, Tojo Y, Tasaki T, Takigami H and Watanabe K 2013 Strategy for separation and treatment of disaster waste: a manual for earthquake and tsunami disaster waste management in Japan *J. Mater. Cycles Waste Manag.* **15** 290–9

[4] Noran O and Bernus P 2011 Effective Disaster Management: An Interoperability Perspective *On the Move to Meaningful Internet Systems: {OTM} 2011 Workshops* (Springer Berlin Heidelberg) pp 112–21

[5] Gowan M E 2011 Self-Management of Disaster Risk and Uncertainty: The Role of Preventive Health in Building Disaster Resilience

[6] SDG Indonesia 2017 Sustainable Development Goals

[7] Brown C O 2012 Disaster Waste Management: a systems approach

[8] Shyam G K, Manvi S S and Bharti P 2017 Smart waste management using Internet-of-Things ({IoT}) *2017 2nd International Conference on Computing and Communications Technologies ({ICCCT})* (IEEE)

[9] Tao C and Xiang L 2010 Municipal Solid Waste Recycle Management Information Platform Based on Internet of Things Technology *2010 International Conference on Multimedia Information Networking and Security* (IEEE)

[10] Zainal Z 2007 Case study as a research method *J. Kemanus.* **5**

[11] Smith W and Dowell J 2000 A case study of co-ordinative decision-making in disaster management *Ergonomics* **43** 1153–66

[12] Li W, Song M, Zhou B, Cao K and Gao S 2015 Performance improvement techniques for geospatial web services in a cyberinfrastructure environment - A case study with a disaster management portal *Comput. Environ. Urban Syst.* **54** 314–25

[13] Milke M 2011 Disaster waste management research needs *Waste Manag.* **31** 1

[14] Ministry of the Environment Goverment of Japan 2018 *Disaster Waste Management Guideline For Asia Pacific*

[15] United Nations Office For The Coordination of Humanitarian Affairs Environmental Emergences Section 2011 *Disaster Waste Management Guideline United Nation*

[16] Etinay N, Egbu C and Murray V 2018 Building Urban Resilience for Disaster Risk Management and Disaster Risk Reduction *Procedia Eng.* **212** 575–82

[17] Cutter S L, Burton C G and Emrich C T 2010 Disaster Resilience Indicators for Benchmarking Baseline Conditions *J. Homel. Secur. Emerg. Manag.* **7**

[18] Modica M and Reggiani A 2014 Spatial Economic Resilience: Overview and Perspectives *Networks Spat. Econ.* **15** 211–33

[19] Zhou H, Wang J, Wan J and Jia H 2009 Resilience to natural hazards: a geographic perspective *Nat. Hazards* **53** 21–41

Literature Review: Technologies and Property Development

C F Maududy[1], **A Gamal**[1]
[1]Departement of Architecture, Universitas Indonesia, Kampus UI Depok, West Java, 16424, Indonesia

E-mail: gamal@eng.ui.ac.id

Abstract. This literature review was conducted to provide an explanation about technology in property development process. The knowledge on technology is important for property developer to remain competitive in property industry. If a property developer does not understand about technology in property development process, they will likely be out of business. Information related to technology in property development hasn't been widely discussed, thus it may be difficult for novice property developers to join the industry. The method of this study is literature study. There are three technologies in the property development process, namely Property Technology, Financial Technology and Construction Technology. Most authors explain more about financial and construction technologies in property development. Most authors are unable to separate Property Technology from Financial Technology and Construction Technology. Only a few of them have critical way of thinking in relation to the definition of Property Technology which leads to innovation and bring about revolution in its development. There are rooms for learning on the Impact of Property Technology that change the property development process.

Keywords. technology, property, infrastructure, development, IoT

1. Introduction

Technology is very closely related to all production processes and everyday life. Technology can help, improve performance, and speed up the process of developing a product, especially the process of property development that requires a long and complicated process. Developers need to understand technology in property development if they want to compete in the property sector. The use of technology can trigger a change in the development process and improve the quality of work. Developers that are unable to use the technology in their development process will be left behind.

Considering the advantage of technology and how it will change the property development process, an explanation about the use and function of technology in the property development process is important for developers. In relation to Smart City, the city's growth is currently driven by property and infrastructure development, which are driven by the property developers. Because of that, it is important to know how the technologies influence their works and what the current development process is. This is important if we want to understand how to improve smartness and increase the use of technology that changes the process of property development. Most of the literature in the property development process explains more about construction technology and financial technology while there are other technologies namely property technology. Today's literature rarely talks about property technology and the differences between other technologies.

2. Methods

This study used qualitative methods through the study of literature on property development, and technology in property development. First, we conduct a study of the property development process from a practitioner's and theoretical perspectives. There are 11 literature about the property development process that are used as a basis for mapping the scope of technology. The perspective of the literature regarding the property development process can be known through (1) the author's background and (2) the author's purpose in explaining the property development process.

Six literatures explained theoretical property development from 11 literatures. These six literatures are written from the author's experience in the academic field and based on the study of literature. The other four literatures describe the practical process of property development based on experience in the property field and research on practices in the property industry.

First, the literature about the property development process used for the basis for evaluating the scope of technology. Second, study the literature on property technology used to determine its application and impact. Third, the study of financial technology and construction technology used to separate the role of technologies. Knowledge from the property development process, property technology, financial technology, and construction technology is used for mapping the scope of each technology. This literature study showed us opportunities for further research related to technology in the real estate industry.

3. Literature review

3.1Property Development

The function of property development is to improve land functions and to regulate and change land. Property development varies, such as landbanks, land packaging, building development, renovation, operation, and redevelopment. Each property development process has similar steps starting with initiation and study. Landbank is a process of acquiring for development within a certain period of time to increase land value. Developers can increase the value of land by changing land functions and by making land development plan and site study. These endeavors are called land packaging that regulates land functions based on their potential. Developers need to make building development which requires them to transform the land function into property products that can be used or inhabited. It can be achieved through renovation. Renovation is the process of increasing the value of property products that have depreciated by repairing or changing the quality of buildings. The last property development is the redevelopment of an existing property by changing its property functions according to the potential that has been analyzed.

There are numerous stages in the property development process where each stage needs to be completed to reach the next stage. The process of property development is like a game [1]. Property development will produce a property product if the developer can complete a series of stages. If the developer fails to complete a certain stage then they need to return to the previous stage and evaluate it. Each stage has its own risks and benefits for developers and the environment. Property development is a process that connects components and knowledge to produce property products.

There are two approaches to the stage of property development explanation based on the perspective of the authors. The first approach is theoretical approach by the academics and literature [1–6]. The second approach is practical approach by practitioners in the property sector [7–12].

Table 1. Property development Process.

	Reference	[7]	[11]	[10]	[5]	[9]	[12]	[2]	[4]	[3]	[6]	[1]
Pre- Construction	Idea						√	√	√	√		√
	Site study	√		√	√	√			√		√	√
	Site selection	√							√			
	Market study				√	√	√	√	√	√	√	
	Site Acquisition		√	√	√					√	√	
	Feasibility study		√			√	√	√	√		√	√
	Design	√	√			√			√	√	√	
	Planning	√							√			
	Financing	√			√	√			√	√	√	
	Permits	√	√	√	√					√	√	
	Negotiating contract		√				√	√	√		√	√
	Commitment	√	√				√	√	√	√	√	√
Construc-tion	Implementation	√	√	√	√		√	√	√	√	√	√
Post-Construction	Promottion	√							√		√	
	Exit Strategy	√	√	√	√		√	√	√	√	√	√
	Property Management						√	√				√

The practical explanation of property development phase is more straightforward than the theoretical explanation because the theoretical approach tends to be longer and more complicated since it separates each task in one stage. Both approaches differ in the sequence or chain of development processes. Theoretically, the stage of property development is arranged separately according to the tasks. Practically, several tasks that are interrelated and can be carried out together will be included in one stage. Some author who explained the development process from the practical point of view, argued that the process of initiating, studying the land, and studying the market in the development process is included in one stage [7,11,12]. These three tasks cannot be separated because they are interconnected and can be done in parallel. these three tasks included in the stage of determining land [7] while another author called it the pre-financing stage [11].

The practical and theoretical explanation also differ in the construction and post-construction phases. The construction phase in practical explanation is a process of building property product. In the theoretical explanation, the construction phase is the process of realizing all planning in the previous phase. Theoretically, property development is a process in increasing the value of land realized by producing property products, utilities, and infrastructure for access. The post-construction phase is the stage of determining whether the property would be sold or leased [1,3,5–7,10–12]. Theoretically, developers do not only think about sales or leasing in the post-construction phase but they also think about the management of property products if they manage it by themselves. There are three options in the exit strategy stage, namely leasing, selling, and operating property products, therefore that there is still a stage of property management in the post-construction phase.

The property development process consists of 16 stages namely initiation, land analysis, land selection, market study, land acquisition, feasibility study, design, planning, financing, licensing, contract negotiation, commitment, implementation, promotion, project completion strategies and property management (table 1). Table 1 separated each stage which are described by practicioners and academics then combined them to become a series of development processes. Sixteen stages were divided into three phases namely the pre-construction, construction, and post-construction phases. These three phases were sorted by development time.

3.2Technology in Property Development Process

There are three types of technology that can be applied in the property development process; (1) Property Technology, (2) Financial Technology, (3) and Construction Technology. These three technologies have their respective roles in the development process to become the supporting tool for the property development process to increase efficiency, effectiveness, and quality. Aside from that, they can change the process of property development. It is important for property developers to know and understand the usefulness of each technology.

3.3Property Technology, Financial Technology, and Construction Technology

There are two ways to explain the meaning of Property Technology. The first explanation sees Property Technology, or more commonly known as PropTech, as a support tool. PropTech is the use of methods, techniques, and tools that help in the property sector [13–18]. This statement gives the meaning that technology is able to ease work, improve efficiency, and work effectiveness. This meaning sees innovation as an outcome which is characterized by the efficiency and effectiveness of the process [19]. The second meaning is Property Technology as a technique, method, device that can change the process of property development [20–22]. This meaning sees innovation as a process which is characterized by a change in the product development process. Property Technology is a technology that innovates property development in terms of processes, services, management, and business models.

The benefits of using various property technologies are related to increasing benefits for the consumers and the property developers in the property sector [22]. There are opportunities and risks in the use of Property Technology [15]. Property Technology is able to provide opportunities for developers to get investors and also to increase actor participation in the property development process. However, there is a looming threat of failure that will give extra pressure to the developers who use Property Technology if they do not experience increased performance and innovation with Property Technology.

Some of the authors see Property Technology as startups that have business models and new ways of offering products in the property sector [17,21]. The company, which is called Property Technology, usually uses innovative tools or methods in mediating the process of product marketing and communication with consumers. Companies that claim their companies as Property Technology usually have this technology that can provide an easy, precise, and effective way of managing and offering services. There are too many explanations of the meaning of Property Technology so there is a shift in the meaning of this term.

Table 2. Category of Property Technology

Property Technology Category	Management System	Product Search	Documentation	Property Insurance	Shared Economy	Smart Real Estate	Property Investment and Financing	Property Service
[23]					√	√		
[24]	√	√				√		√
[13]	√	√					√	
[14]	√	√	√	√				

Some authors explain the category of Property Technology (table 2) [13,23,24]. These categories are arranged based on their functions in the real estate sector. All these categories have similarities that can be mapped in technology groups that offer data, transactions, and management, services, and controls. Technology in data groups are technologies that helps in storing, retrieving, managing and visualizing data. Property Technology in a group of transactions is the use of technology that facilitates the process of exchanging products, and services in the property sector between providers and consumers. Property Technology in the management, service, and control groups is the use of technology in the management, control, and procurement of services in the property sector to improve the performance of property products. Property Technology in the third group is very helpful in the management process on the post-construction phase.

Financial technology is one of the technologies that plays a role in the process of property development. Financial Technology is closely related to Property Technology because the development of Property Technology 2.0 is driven by the existence of technology in the transaction process. The second phase of Property Technology development is characterized by the presence of financial innovations in the property industry that allows easier transactions with lower costs. At the beginning of its development, the two technologies were equally encouraged by the presence of computers in 1990.

Financial technology is a technology used in the financial sector that encourages systems that model, assess, and process financial products [25]. For example, crowdfunding becomes a new method or model as a financing system for a product. Financial Technology helps speeding up the process of transactions or those related to finance, as well as the experience of consumers and providers of products that use said technology [26,27]. Financial Technology can encourage the development of business models and offer new methods in the financial sector, especially in the property industry.

Table 3. Category of Financial Technology

Financial Technology Category	Crowdfunding	Financial Information	Payment system	Lending system	Trading system	Insurance	Wealth Management	Capital Market	Money Transfer	Financial Exchange
[23]	√		√	√		√	√	√	√	√
[25]		√	√		√	√				
[26]	√	√	√	√						
[27]	√		√	√		√	√	√		√

Financial Technology has ten categories based on the facilities offered (table 3) [23,25,26]. Crowdfunding is a method of financing a product, project, and others by collecting investors who are willing to contribute. It can work because Financial Technology is able to attract investors to make products, ideas, and capital investment [27]. Financial information category is a technology that helps to find and display financial data or how to use financial products to consumers, investors, or other users of financial products. The data search function is offered in the category of financial information such as transaction activity data, financial procedures, lessons on how to conduct transactions [25,26]. The payment system category is a Financial Technology that facilitates the process of exchanging a product, service, or financial asset with a faster way, lower cost, flexible, without intermediaries, and appropriate to the public. The lending system category is a technology that helps financial service providers in providing loan services to consumers and lending management. The trading system category is part of Financial Technology that facilitates sales service providers or offering their products and mediate between developer and consumers to make transactions at different times and spaces [25]. The trading system category is not only applicable to the property sector but also to other sectors that have a system of selling, renting, and buying. The insurance category serves to secure financial assets and transactions. Personal finance category or wealth management is a Financial Technology that helps consumers, investors, corporations in managing finances, such as making the right financial decisions and investment. The capital market category is part of Financial Technology owned by financial institutions to carry out their activities and services. The money transfers category facilitates the transfer of funds between individuals or institutions indirectly without the need for a face to face meeting. Financial Technology facilitates a process that allows to digitally record and move assets on a network without an intermediary through a distributed database included in the category of financial exchange [23].

Construction Technology is a part of the technology that helps the process of implementing property development planning. Construction Technology helps development activities in terms of data, physical, and service [28]. Construction Technology plays an important role in data categories as a tool for sending data, recording data, assessing productivity, sharing planning data and reports, representing data, and communicating between development actors. Construction Technology as a tool and equipment in the realization process of property development in the field such as machinery is included in the physical category. Service category of Construction Technology includes work safety insurance, security, and maintenance.

3.4Intersection in Technology

Financial Technology is considered as a good guide for Property Technology activities [23]. Property Technology and Financial Technology are concurrently developed and mutually influence each other, thus blurring the line between them and confusing many people, especially in the real estate industry. The main factor of this confusion is the second phase of Property Technology development or Property Technology 2.0 [23] that is triggered by the emergence of technology in the system of lending, payments, financial management, money transfer, and capital markets in the property sector. It expands the scope of technology in property development. This is actually a part of Financial Technology in the property sector. Some authors tend to see all technologies in the property sector, especially in property development, as Property Technology, therefore allowing an overlap definition of technologies in the process of property development. Some Property Technology categories described by some authors [13] [14,24] actually fall into the category of Financial Technology.

Table 4. Property Technology and Financial Technology

		Data (information)	**Transaction**	**Management, Service, Control**
Property Technology	[13] [14] [23] [24]	• Portfolio management • Commercial real estate search • Long-term rental / sales search • Indoor mapping • Life, home, property and casualty insurance • Real estate agents tools • Shared economy	• Property investment and financing • Property brokerage and leasing • Portfolio management • Property insurance • Shared economy	• Property management • Project management • Home services • IoT home • Digitizing administrative management • Smart building
Financial Technology	[23] [25] [26] [27]	• Education Lending • Information system • Regulatory system for financial product	• Crowdfunding • Payment system • Trading system • Financial exchange • Lending • Institutional/ Capital markets • Money transfer • Bitcoin	• Insurance services • Personal finance/ wealth management

Property Technology and Financial Technology have different categories but they have similarities when we map based on data, transactions, and management, service, and control. The differences can be seen in how Property Technology offers search, display, data processing related to development and real estate products such as home rental and sale searches, site studies, and more. Meanwhile, Financial Technology offers search, management, visualization of data or information related to financial products and field such as loan procedures, customer data, and others. The same thing happened in the management, service and control groups. Property service providers, property operators, and other property actors are helped by the technology that can mediate their work in offering services, managing, and supervising property products with Property Technology in the management, service and control groups. Some categories of Property Technology in the transaction group are the same as Financial Technology. Some categories of Property Technology in the transaction group are Financial Technology in the property sector.

Most Property Technology in the transactions group is a part of Financial Technology because due to its similar function (table 4). For example, Property investment and financing are technologies that help to make payments and funding for property products that use a payment system. Brokerage and leasing property is a technology that helps to offer property products even though in reality most of this technology uses a trading system and a payment that helps consumers transact with property developers. This trading and payment system is included in Financial Technology. With this in mind, we can see that there is an intersection between Property Technology and Financial Technology in the property sector called Real Estate Fintech.

Fintech Real Estate is a Financial Technology in the property sector. This intersection between Property Technology and Financial Technology has caused confusion regarding the use of Property Technology and Financial Technology terms. Some authors have begun to separate the scope of Property Technology, and Financial Technology based on their uses [13,23]. The mapping of technology categories in table 5 is done to separate Property Technology, Financial Technology and Fintech Real Estate. The purpose of the mapping is to avoid any more mistake in using the term of technology. Property Technology, Financial Technology, and Real Estate technologies are different technologies and separate technologies.

Real Estate Fintech is a technology that mediates the process of selling the ownership of property assets [23]. Exchanges or transactions that occur in Fintech Real Estate activities are not only related to property assets but also to real estate services. Examples of Fintech Real Estate Services are portfolio management and property finance loans. One example of Fintech Real Estate technology is an Android-based application owned by Airbnb that helps to offer rental facilities and transaction. Consumers and property owners can communicate by the application and make transactions without the need to meet face to face. Some of the products offered in the Airbnb application have a shared use system that is categorized as a shared economy so there is an overlapping technology used in it. Airbnb application is a technology as a device and also as a technical revolution in the management of property products.

Table 5. Property Technology, Real Estate Fintech, and Financial Technology

Technology	Category
Property Technology	Data Search
	Real estate Documentation
	Real estate Service
	Real estate Management (Shared economy, and Smart Real Estate)
Real estate Fintech	Property Crowdfunding
	Property Payment system
	Property Lending system
	Property Trading system
	Portfolio Management
	Exchange system in Real Estate
Financial Technology	Crowdfunding
	Financial Information

	Payment system
	Lending system
	Trading system
	Wealth Management
	Capital Market

The categories in Real Estate Fintech are crowdfunding, payment systems, lending systems, trading systems, management systems, and exchange systems. Crowdfunding category is a technology that helps in the process of funding, financing, capital, and property investment. The payment system category is part of Real Estate Fintech which is often applied in transactions process in management, sales, rental of property products, property services, and property insurance. The lending system is a category of Real Estate Fintech that helps in the process of acquiring ownership rights to property, financing and investment assets by offering funding opportunities on a regular basis. The trading system category is Real Estate Fintech which facilitates the bidding process that offers property products to the public. This category is often used by property agents and property developers. The management system category in Real Estate Fintech helps to provide funding investments options in the property sector such as portfolio management.

Some authors [13,22] do not separate the scope of activities between technologies both Property Technology, Financial Technology, Construction Technology, and Real Estate Fintech, thus making it difficult to tell them apart. It would be easier to distinguish between Property Technology, Financial Technology, and Real Estate Fintech with a mapping based on their functions as in table 5. In property development practices, Property Technology is more helpful in the study, analysis, and management of property assets. All innovations in terms of technology related to financial transactions and funding are included in Financial Technology. This transaction and funding process applies to other sectors not only in the property sector. Construction Technology helps more in the implementation of construction planning and management.

Technology in property development is very difficult to distinguish based on types and technology distribution mediums. Property Technology, Financial Technology, and Construction Technology can be distributed through software and hardware. Financial Technology and Property Technology have a close relationship and can be mapped based on data, transactions and management, service and control groups. An easy way to separate the three technologies is through mapping the scope of their activities into the property development process.

3.5Mapping the technology in property development process

There has not been much explanation about the scope of technological functions in the process of property development. Technology in the property development process divided into three parts, namely technology as input in production, technology with functions of operation and construction, technology with the function of management [18]. The first technology group helps in the process of finding data, analyzing, and processing data. The second one is a technology that helps structural development processes, property products, and financing. This technology also helps in the management and operation of the development process. The third category helps in the operational process of real estate products.

There are other technology groups that play a role in distributing property products to consumers based on the functions and categories of Property Technology, Financial Technology and Construction Technology described in point 2. These technologies play an important role after the planning stage is completed and the property products are ready to be marketed to the public. Technology that helps in the promotion process and the introduction of property products is called technology with the function of property marketing and transactions so that there are four technology groups in the process of property development. The four technology groups are technology as input in production (T1), technology with the functions of operation and construction (T2), technology with function of management (T3), and technology with the function of marketing and transaction (T4). The four technology groups describe Property Technology, Financial Technology, and Construction Technology so that we can map these three technologies into it.

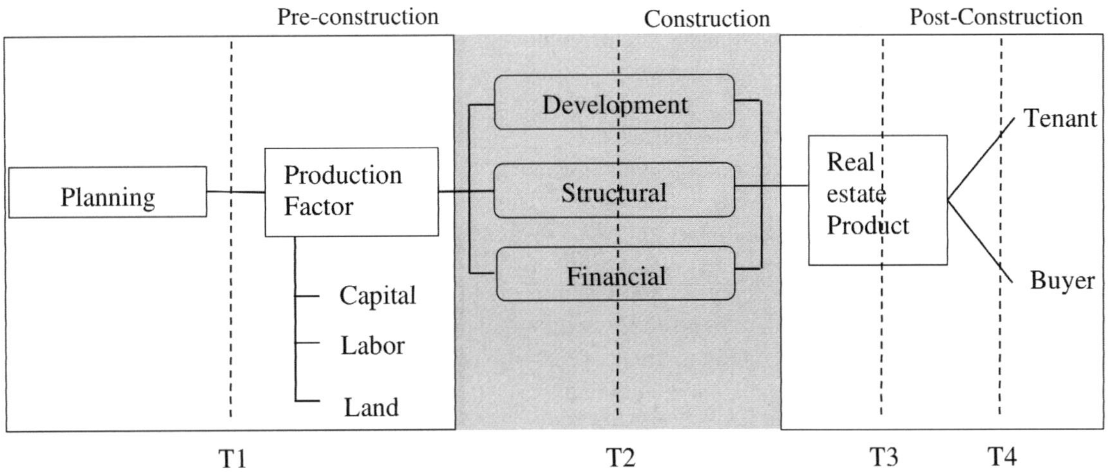

Figure 1. Mapping the Technologies on Property Development Process

The four technology groups are in different phases in the process of property development. The property development process consists of three phases; T1 technology is in the pre-construction phase, T2 is in the construction phase, and T3 and T4 are in the post-construction phase (Figure 1). The pre-construction phase consists of eight stages, namely study process, analysis, decision making for planning that can be assisted with technology as an input of production (T1). This T1 technology can simplify, and accelerate acquisition, data processing to take development decisions, and improve development productivity. There must be an adjustment of production factor when using T1 technology. T2 technology helps during the construction phase as a tool for building realization and planning and construction management. The structural, development, and financing processes will be carried out in the construction phase to produce property products. Management and operations in the use of this property product can be assisted with T3 technology. T3 technology is a technology that helps operational processes, optimizes building performance and control. Management and operation of property products can be carried out directly by property developers and others. T3 technology can also be used by end-users of property products [18]. Tenants can benefit from its function during the production process. T3 also will be useful for property developers who manage their property products by themselves. If the developers sell their product - buy or rent property, they can use T4 technology. T4 technology plays a role in introducing property products to the public, promoting, and transacting with consumers.

T1, T2, T3, and T4 represent the roles of Property Technology, Construction Technology, and Financial Technology. T1 technology is a part of Property Technology that helps in data collection as an input in the study and planning process. One of the Property Technology categories included in T1 is the property products search that plays a great role in market and competitor studies. T2 technology is part of Construction Technology that helps in the construction process and construction management. T3 technology is a Property Technology that belongs in the post-construction phase. Property Technology at this phase helps in the process of managing property products by end users. The smart real estate category presented by Baum is one category that describes technology in the property management. The use of technologies such as IoT enabling the developers to monitor, optimize building performance to communicate with tenants or users in property products. The existence of a sensor system in buildings can reduce the risk of damage, and excessive energy use by early detection before problems occur. This technology product falls into the category of smart real estate or smart building. Property Technology that is applied in the operation of property products can usually provide input to operators regarding performance information, utility conditions, tenants' needs to building occupancy. The digitization of operations on these property products can increase the use value, the economic value of property products, and optimize the use of energy and space. T4 technology is part of Financial Technology that belongs to the property sector or Fintech Real Estate. This technology helps in the marketing process and transactions between consumers and property developers. T4 enables the developers to connect with consumers indirectly and without a mediator in the process of communication, negotiation, and transactions.

Property Technology, Construction Technology, and Financial Technology have their respective parts and roles in the stage of property development. The mapping of Property Technology, Construction Technology, and Financial Technology in the development phase are carried out based on the classification of technologies T1, T2, T3, and T4 which based on the function of the technology. Property Technology plays a role in the pre-construction phase which consists of stages of initiation, site study, market studies, land acquisition, feasibility studies, design, financing, contract negotiations, and agreements. Property Technology in the pre-construction phase is able to ease the work of property developers in analyzing, storing, displaying, obtaining data, and also communication between divisions [29]. Property Technology plays a role in the post-construction phase helps the management of property products by end users. Construction Technology is in the construction phase in the implementation phase of property products. Financial Technology in the post-construction phase in the exit strategy stage can help investors to make strategic decisions for returning capital or getting benefits from property products that have been realized.

Property Technology, Construction Technology, and Financial Technology in the development process also have different scopes. The easiest way to distinguish these three technologies is to map them in the stage of property development based on their functions. Property Technology, Construction Technology, and Financial Technology use the same distribution media, which is software and hardware, so distinguishing them based on their distribution media is pointless. All technologies that play a role in the pre-construction phase and building management phase are a part of Property Technology. Construction Technology is a technology in the planning implementation stage while Financial Technology is in the exit strategy stage.

3.6 The Role of Technology

Technology in the process of property development is able to optimize the activities and performance of the workforce. Optimization of processes and development performance can be seen as innovation as an outcome. On the other hand, technology is also able to provide innovations that revolutionize the property development process. The impact of this technology becomes a part of innovation as a process [19] that can change the ways, methods, business models, and activities in property development.

Property Technology can be as a solution for dealing with risks, challenges, and improving the quality of performance in the property industry, optimizing the use, sale and rental of property [13,14,18]. Developers are competing to innovate in developing a better property and designing a better quality product. Technology is an opportunity for developers to rise above their competitors. Property Technology is here because of the changes in consumption patterns and the use of digital products [21]. Digitalization in the property sector is considered to facilitate all matters both in the development and the management of property. For example, the design documentation process is affected by the presence of Virtual Reality and Sketchup software that allows developers to evaluate the design directly through three-dimensional visualization. The visualization of the design with Virtual Reality helps developers and designers to perceive space in three dimensions even before it is built. They can look at the quality of space, resolve conflicts before construction, and do environmental studies [30]. Both examples of this technology are included in Property Technology with the category of real estate documentation. Property Technology can change the process of property development in the process of property management stage. Smart real estate is an example of digitalization in the process of managing a property. Property operators can supervise and manage property digitally with a system that is able to provide information on energy use, tenant needs, occupancy rates, and security of property products. The supervision process can be controlled indirectly without field observation. Property Technology in property management is able to reduce 25 percent of the operational costs of the same type of property product [31].

In the property marketing process, Real Estate Fintech can change the way we do our transaction, communicate, and consume property products and property services. This technology aims to facilitate every process and consumer activity in the property sector [32]. Developers and consumers can process transactions and communication without a face to face meeting and without a mediator. One of the categories of Real Estate Fintech is portfolio management that helps investors to determine property financing options [24]. This technology provides data as a basis for making funding and reference investment decisions. The payment system category is another category of Real Estate Fintech that serves the process of offering property products to the public and mediates the process of communication between property owners and consumers. An example is the Airbnb application that offers rental property products that can be searched and sorted by needs, facilitating communication between tenants and property owners, and leasing transactions.

Construction Technology eases the burden of property development, control, and the management of property development process. Construction Technology can regulate and control the construction process and reduce the risk of delays in the process. One of the uses of Construction Technology in construction management is Microsoft Project software. This software is able to set the timeline, predict the time of completion, to provide an estimate of the minimum delay in the sub-stage of development so it can still be completed on time.

4. Synthesis

Despite the many roles Property Technology, Construction Technology, and Financial Technology have in property development, they do not in fact present in all stages of property development. Many authors explain Property Technology as the use of technology in the property sector. Based on the literature study, we find out that there are three types of technology that can be applied in the development process namely Property Technology, Construction Technology, and Financial Technology. Property Technology is only one of the technologies that have a role in the pre-construction stage and management stage of property assets. Construction Technology plays a role in the implementation phase and Financial Technology plays a role in the exit strategy phase in the post-construction phase. The intersection between Property Technology and Financial Technology is referred to as Real Estate Fintech, namely the expansion of the scope of Financial Technology activities in the property sector.

Only a few authors see Property Technology an innovative way, a method, or a tool that is capable of changing the processes in the property sector, especially the property development process. It means Property Technology is an innovation changes the way, model, and process of property development. Property Technology is more often interpreted as the use of technology as a tool in the property sector. With this explanation, Property Technology is only seen as a tool that only capable to improve efficiency and process effectiveness in the property sector.

The right way to separate Property Technology, Construction Technology, and Financial Technology is to map the scope of its role in the process of property development. Many authors cannot tell the difference between Financial Technology and Property Technology. Some authors even made a mistake to include wrong categories in Property Technology category even though they should be classified as Financial Technology. Most authors are unable to separate Property Technology, Construction Technology, and Financial Technology in the property sector.

There is an opportunity for other researchers to find out more about the implications of Property Technology that changes the pre-construction process. Several studies have begun to find changes in the property development process in property management [22,23,33,34]. The property management phase is changing with the existence of smart real estate products and the technical revolution in the use of property products which called by shared economy [23].

Smart city development must be balanced with an increase in productivity and performance of working in the property and infrastructure development process. Increased productivity and performance of working can be supported by technology as a tool or even a new method. Developers need to consider three types of technology in the property development process, namely financial technology, property technology, and construction technology. Most writers from Indonesia explained property technology as a company using technology in offering products with different business models. Developers need to know this third technology is not only a tool that can improve efficiency, effectiveness and work performance. This technology consist of methods or ways of working that can change business models and property development processes such as in the smart home property management process.

5. Acknowledgement

This article was presented at the 2nd International Conference on Smart City INNOVATION (ICSCI) 2019, jointly held by Universitas Indonesia and Universitas Diponegoro. ICSCI conferences have been supported by the United States Agency for International Development (USAID) through the Sustainable Higher Education Research Alliance (SHERA) Project for Universitas Indonesia's Scientific Modelling, Application, Research, and Training for City-centered Innovation and Technology (SMART CITY) Center for Collaborative Research, administered through Grant #AID-497-A-1600004, Sub Grant #IIE-00000078-UI-1. This study is supported from the Master Thesis Research Grant by Kementrian Riset, Teknologi, dan Pendidikan Tinggi Republik Indonesia (KEMENRISTEKDIKTI) 2019. Number: NKB-1897/UN2.R3.1/HKP.05.00/2019.

6. References

[1] Wurtzebach C H and Miles M E 1994 *Modern Real Estate* (Canada: John wiley & Sons, INC)

[2] Ariyawansa R G 2009 Management of Real Estate : Principles of Real Estate Development & Management. Maharagama : Department of estate management and valuation, University of Sri Jayewardenepura.

[3] Cadman D 2002 *Property Development* (Taylor & Francis)

[4] Fisher P 2005 The Property Development Process *Prop. Manag.* **23** 158–75

[5] Kohlhepp D B and Kohlhepp K J 2018 The Real Estate Development Matrix *Real Estate Development Matrix* (Routledge) pp 12–6

[6] Newell G and Steglick M 2006 Assessing the Importance of Property Development Risk Factors *Pacific Rim Prop. Res. J.* **12** 22–37

[7] Johansson M 2017 The Property Development Process : 8 Critical Step

[8] Costello G and Rowley S 2010 The Impact of Land Supply on Housing Affordability in the Perth Metropolitan Region *Pacific Rim Prop. Res. J.* **16** 5–22

[9] Costello G and Preller F 2010 Property Development Principles and Process {\textendash} An Industry Analysis *Pacific Rim Prop. Res. J.* **16** 171–89

[10] Flux R 2015 5 Stages of Property Development. Developer Network

[11] Yardney M 2010 *How to Get Started in Property Development* (Melbourne: Metropole Property Investment Strategists)

[12] Yardney B 2016 The Property Development Process

[13] ING Economics Department 2018 *Technology in The Real Estate Sector : Proptech Reduces Risks and Increase Value* (Amsterdam: ING Economics Department)

[14] JLL Asia Pasific 2018 The Growing Influence of Proptech

[15] Maarbani S 2017 Real Estate Technology : Threat or Opportunity *White Pap. Futur. Realt.*

[16] Machinga R 2018 Real Estate Technology : What is Proptech and How is It Changing The

Industry *HLB Int.*
[17] Mclean S 2018 Proptech Companies Growing in Number and Value *Real Estate News Exch.*
[18] Vandell K D and Green R K 2010 *The Impact of Technology on Commercial Real Estate* (Urban Land Economics Research)
[19] Kahn K B 2018 Understanding innovation *Bus. Horiz.* **61** 453–60
[20] KPMG Global PropTech Survey 2017 *Bridging The Gap : How The Real Estate Sector can Engage with PropTech to Bring The Built and Digital Environments Together* (Swiss: KPMG Internasional)
[21] Lecamus V 2017 PropTech: What is It and How to Address The New Wave of Real Estate Startups?
[22] Sim D 2017 *Property Technology: Disruptor or Enabler?* (Singapore: CBRE Research)
[23] Baum A 2017 Proptech 3.0: The Future of Real Estate
[24] Venture Scanner 2016 Real Estate Technology Q2 Update in 15 Visuals
[25] Freedman R S 2006 *Introduction to financial technology* (Elsevier)
[26] Gaughan M 2017 Commentary: FinTech and the Liberation of the Community Reinvestment Act Marketplace *Cityscape* **19** 187–98
[27] Lee I and Shin Y J 2018 Fintech: Ecosystem, business models, investment decisions, and challenges *Bus. Horiz.* **61** 35–46
[28] Tatum C B 1988 Classification System for Construction Technology *J. Constr. Eng. Manag.* **114** 344–63
[29] CBRE IDEALAB Magazine 2017 Digital Innovation
[30] Juan Y-K, Chen H-H and Chi H-Y 2018 Developing and Evaluating a Virtual Reality-Based Navigation System for Pre-Sale Housing Sales *Appl. Sci.* **8** 952
[31] ARUP 2017 *Reimagining property in a digital world* (London: Arup)
[32] Thuillier A 2018 Corporate/Start-up Collaboration, A Fertile Garden of Innovation
[33] KPMG International Cooperative 2016 *SnapShot! on Digitalization* (Swiss: KPMG International Cooperative)
[34] Meyer N 2018 Digitisation in the Real Estate Sector: Proptechs and Assets *Glob. Real Estate Expert.*

User Assessment of Quality Service of Trans Jateng Bus Stops, in Sub-Urban Corridor (Ungaran-Bawen)

L D Agitha[1] and F H Mardiansjah[2]

[1]Student of Urban and Regional Planning Department, Faculty of Engineering, Diponegoro University, Semarang-Indonesia
[2] Lecturer of Urban and Regional Planning Department, Faculty of Engineering, Diponegoro University, Semarang-Indonesia

E-mail: gintingleoni@gmail.com

Abstract. This paper focuses on the provision of bus stop facilities provided by the Trans Jateng, a BRT-claimed bus system provided by Central Java Province that connecting southern periphery of Metropolitan Semarang to the urban core. One of their functions is to increase public interest in using public transportation in their daily transportation among the city center to the periphery vice versa. The study aim is to assess the quality of service of the bus stop in the sub-urban corridor from Ungaran to Bawen based on the user opinion and perception, which are analyzed by using 5-rating Likert scale gathered from interviews conducted by using questionnaires. There are three variables used in this study: location, physic, and service level. The result reveals the service quality of the bus stops is 2.94 so it tends to be quite good with the value of the three variables are 3.29 (good enough), 2.52 (not good) and 3.01 (good enough) respectively. They turn out input and provide suggestions to improve the bus stop services in their physical aspect as well as their location. The improvement is expected could increase public preferences on public transportation utilization, especially in their travel between city center to the periphery vice versa. The shifting private to public transportation is needed to reduce vehicle emissions that could support a more sustainable development in Semarang Metropolitan Development.

Keyword: BRT, public transportation, vehicle emissions

1. Introduction

Cities develop as humans improve their quality of life, and they grow along with the civilization of human life [1]. Todaro [2] cited in Kumurur [3] said even though development could be defined, in the broad sense, as the ability of humans to compile their future in terms of capacity, justice, power and interdependence, in more narrow sense, development could be seen just as the ability of the people to meet their basic needs, such as food, shelter, health and protection. Referring to Haughton and Hunter [4] cited in Kumurur [3] said that cities are areas concentrated with a group of people and their activities that have continuous improvement in natural, artificial and social environments from small scale to regional scale to support the fulfillment of the citizen's needs. Therefore, the cities expand its space to fulfill the needs of the people. However, the expansion of urban space created more distances between the origin and destination of daily travel of the citizen, even though that situation does not restrict the possibility of the citizen to continue their mobility.

The mobility of the urban dwellers has diverse grooves and circulation patterns that are required in accommodating the displacement from one place to another, as part of the transportation system. Later, the system will not only control the use of urban land but also the patterns of urban mobility that is performed by the movement of goods and people in the city. The higher number of population, especially in the more wide area, will create the more complex patterns so the more routes are needed as the demand of city transportation system tends to increase as the expansion and sprawl phenomena. In fact, the phenomena of urban sprawl are directed by the growing population in the city center that increases property cost and also the cost of developing urban public service and infrastructure in the city center. Those situations lead to the high cost of living in the city center that encourage the development in the periphery areas to fulfill the needs of the growing urban population [5]. Therefore, the phenomena of urban sprawl create new centers of activities in peripheries that increase complexity and uncertainty of urban mobility [6,7]. These phenomena also expand the needs of transportation services into the peripheries. Integration in transportation services in the two regions should be established so the urban agglomeration is able to improve accessibility and quality of life to the new centers in the peripheries [8]. In this context, the growing challenges of urban mobility have pushed the city to innovate to find solutions that can improve spatial integration among city center and the new centers in the peripheries.

One of the efforts performed in many cities is the development of a special public transport system called as Bus Rapid Transit (BRT). This system offers various advantages as the name implies, namely Bus Rapid Transit. First is "Bus", the bus fee budget is lower compared to other transit transportation modes. Second is "Rapid", the higher speed provided because the path is separate from other vehicles so avoiding congestion. The third is "Transit", a flexible and scalable system so it is easy to manage changes and increase in route coverage to new development areas [9]. In order to make it effective, BRT system requires several elements, Diaz and Schneck [10] cited in Wirasinghe [11] divided BRT system into six elements, namely bus lines, transportation modes, stations or stops, tariff collection systems, operating supervision systems and passenger information systems. This paper will focus on the bus stops as one of the important elements of BRT system. As a growing metropolitan, Semarang has also provided a BRT-like transportation system, providing services in the city as well as to the peripheries in its urban agglomeration. In fact, the system could not be said as a BRT system yet, because it has no special bus lane yet that make the system could not guarantee the speed of the bus system as the system could not immune to traffic congestion yet. There are two BRT-like system in Semarang, i.e., the Trans Semarang that provide the services in the city of Semarang, and the Trans Jateng that provide the services that connecting the city center to the peripheries.

One of the Trans Jateng is the Semarang-Bawen line that provides the service through Ungaran to Bawen, the southern peripheries of Semarang. Ungaran-Bawen is a strategic area for regional economic growth [12] and Ungaran City is the part of The Centre of National Activities Kedungsepur (Kendal, Ungaran, Semarang, Salatiga, Purwodadi). While Ungaran City has a role to support Semarang Metropolitan City for settlement area needs and other urban activities in regional scale [13]. The research aims to assess the service quality of the Trans Jateng bus stops in the sub-urban corridor (Ungaran-Bawen) based on the user opinion and perception. There are 28 bus stops, consisting 16 bus stops towards Semarang and 12 bus stops leaving Semarang, in the corridor that are located along the regional road that connecting Semarang to Bawen, in four sub-districts of the Semarang Regency, i. e., Ungaran Barat, Ungaran Timur, Bergas and Bawen. Such analysis for the bus stops will provide significant input that are needed to improve the service quality of the BRT system as it is expected to be the main transportation system for the area in the future, through the improvement of the quality of one of the most important facilities of the bus system. The improvement of the bus stops services is expected could improve the BRT service, and could differentiate the bus system with other mode of transportation service [14], so the bus system could be improved as one of the efforts in shifting urban mobility from the private mode of transportation system based today into a more public transportation system based in the future of Semarang.

As there are varies condition of the bus stops today, this research categorized the 28 bus stops into

IOP Conf. Series: Earth and Environmental Science **396** (2019) 012021 doi:10.1088/1755-1315/396/1/012021

three groups: good, quite well, and bad categories. Table 1 shows the list of the bus stops in these three categories that are also differentiated according the route of the bus services leaving Semarang or towards Semarang. This paper is arranged into some parts. Following introduction, the paper provides the explanation on the methodology, including variables and data collecting method, used in this research. Subsequently, the paper provides the results and discussion that become the main part of this paper. This part is divided into some sections, according the theme of analysis in the research. Finally, the paper is finished by a conclusion in the end.

Table 1. Bus Stops Category Based on Author Assessment

Category	Bus Stops in Semarang-Bawen Route	Bus Stops in Bawen-Semarang Route	Author Assessment Result
Good	1. Taman Unyil 2. Mas Salamah 3. SMA N 1 Ungaran 1 4. Undaris 1 5. Pasar Babadan 6. Lemah Abang 7. Pasar Karang Jati	1. Bergas 2 2. SMA N 1 Bergas 3. Sub Terminal Pasar Karang Jati 4. Wujil 2 5. Undaris 2 6. Mang Engking 7. SMAN 1 Ungaran 2 8. Benteng 2	1. Close to crossing facilities or area 2. Close to the activities centre, such as settlements, industries, offices, education centres, markets, and commercial activities 3. Close and also in the transit area to other transportation modes 4. Having supporting facilities 5. There are lighting and trees around the bus stop 6. Having a clean condition 7. Connecting with a pedestrian path
Quite Well	1. Sido Muncul 2. Bergas 1 3. Harjosari 1	1. Kantor Kecamatan 2. Harjosari 2 3. Ngobo	1. Close to the activities centre, such as settlements, industries, offices, education centres, markets, and commercial activities. 2. Having supporting facilities 3. There are lighting and trees around the bus stop 4. Having a clean condition
Bad	1. PT Mas 2. Pasar Bandarharjo 3. Benteng 1 4. Langensari 1 5. Wujil 1 6. APAC Inti	1. Langensari 2	1. The location is not close to the crossing facilities and the central activity area 2. The bus stop does not have a wall or breaking condition 3. The bus stops have seating facilities 4. Some of the bus stops have a name board, lighting and trees 5. The bus stops don't have a trash can and an information board 6. The bus stops do not connect with a pedestrian path 7. There is no stopping area for other vehicles 8. The condition of the bus stop is not clean and full of graffiti

Source: Author Analysis (2019)

2. Methodology

In order to assess the service quality of bus stops, this research uses three variables: location, physical appearance and service level that are measured by users' opinion and perception on some measurement elements that are shown in Tabel 2. The data are collected by doing interviews directed by questionnaire to 140 passengers or users of the Trans Jateng in April to May 2019 at 09.00 AM – 06.00 PM. The interviews also asked suggestions and other users' input that could be expected to provide more benefits for the development of bus system.

Table 2. Variables Used in The Assessment of Bus Stop

Variable	Element	Literature
Location	Relation with road activity	American Public Transportation Association [15]
	Distance between other bus stops in one corridor	Department of Transportation [16]
	Accessibility	Nugroho [17], Department of Transportation [16], Demetsky dan Lin [18] and Ellis [19]
	Service Group Target/Spatial Characteristics	Tamin [20]
Physic	Physical Dimension (length of bus stop, stop area for other vehicles and platform)	American Public Transportation Association [15] and Guideline for The Location and Design of Bus Stops -Report 19 [21]
	Shelter	Guideline for The Location and Design of Bus Stops -Report 19 [21] and RPTA Bus Stop Program and Standards, Bus Stop Design Guidelines [22]
	Passenger Seating Area	
	Trash Can	
	Pedestrian Path	
	Information, Symbol Routes and Schedules Boards	
	Parking Area *(Park-and-Ride* and *Kiss-and-Ride*)	Transit Capacity and Quality of Service Manual [23]
Level of Service	Level of Public Services – Passenger Waiting Area	Guideline for The Location and Design of Bus Stops -Report 19 [21]
	Level of Public Services – Passenger Seating Area	
	Level of Public Services – Movement Room (Buses and Pedestrians)	
	Level of Support Services – Security	
	Level of Support Services – Lighting	
	Level of Support Services – Aesthetics	
	Level of Support Services – Cleanliness	

Source: Author Analysis (2019)

The analysis is performed by using two analysis techniques. The first is descriptive statistics analysis to get the result of user assessment of the bus stop, additional bus stop elements, user and travel characteristics. This technique uses the 5-points Likert scale rating, for measuring the user assessment consist of very good, good, quite well, not good and bad, meanwhile for additional element consist of strongly needed, needed, abstain, low important needed and absolutely not needed. The second technique is cross-tabulation analysis to get valuation relationship between service level with physical element and location variables. As for, study applies Accidental Sample for determining the

IOP Conf. Series: Earth and Environmental Science **396** (2019) 012021 doi:10.1088/1755-1315/396/1/012021

questionnaire respondents and Lemeshow Formula to decide the number of respondents per bus stop. As the result, there are 5 respondents in every bus stop that who will be research subject.

3. Result and Discussion

3.1. The Characteristics of User and The Travel

3.1.1. The Characteristic of User.

Users' opinion and perception are the objects for the assessment of the quality level of the bus stops. Characteristics of the users are varies as they have various social, and economic backgrounds. The study tried to identify their characteristics through five variables: gender, age, personal vehicle ownership, daily transportation expenses and frequency of BRT use. The majority of the users are women that reached 60% of the users. In the age group, the domination (the highest proportion. 27%) is the aged 18-24 years old, and majority (59%) of the users have their own motorbikes, and 70%, users spend less than IDR 20,000.00 for daily transportation cost. Finally, the frequency of users using BRT reflects that the majority of users use BRT since only 3% who are rarely use BRT.

3.1.2. Travel Characteristic.

According to the survey, travel characteristics of the bus system's users also vary as they have different purpose, motivation, origin, destination, etc. There are some description of the travel characteristics as follows:

- The Origin and Destination
  Most of the travels of the users of the bus system are home-based and workplace-based travel. The survey shows 50 respondents do their travel from home and 88 respondents use the bus system to go home. Subsequently, 48 respondents said that their origins were their workplace and 18 respondents mention that theu will use the bus system to go to their workplace.
- Motivation and Assessment of the Trans Jateng BRT Ungaran-Bawen Corridor Service
  The motivation to ride the bus system consists of seven basic reasons, cheap fare, convenient, close to the origin and destination, fast or no longer time to wait, arrive quickly, never used the Trans Jateng bus system and others. The result shows that the highest motivation was cheap fare as it was said by 56 respondents, then followed by convenience (obtaining 33 respondents), and fast (17 respondents).
- The Area of Origin and Destination
  The areas of origin and destination are categorized into two groups: the city centre (Semarang City) and the peripheries (Sub-district of Ungaran, Bergas, Bawen, and others). The results show the area of origin of the Ungaran-Bawen (leaving Semarang) Route was dominated by Semarang City while Ungaran area was the main area for the destination. Meanwhile, in the Bawen-Ungaran (towards Semarang) Route, Bergas Sub-District was majority area of origin, and Semarang City was the main destination area. These results show that the bus system was utilized as the centre-peripheries centre public transportation system.
- The Relation of Distance between the bus stops with Origin and Destination with the Transportation Mode
  Most respondents said that the distance between bus stops to their origin and/or destination are more than 600 meters, a distance that is started to be not convenience on foot in the area including Semarang that has a high humidity as well as its southern peripheries that are hilly areas. These conditions have made most of respondents need to use another mode of transportation to reach the bus stops from their origin as well as to reach their destination from the bus stops. The dominant other mode of transportation used by the respondent was private motorbike and/or other public transportation. These phenomena indicate that the bus system still needs integrated feeder support to improve its services.
- The Relation between Origin and Destination with The Travel Area
  The survey shows that the Ungaran-Bawen (leaving Semarang) Route was dominated by workplaces from Semarang City as the origin of the travels with the main destination to

Ungaran, settlement area that also become the main service centre for the southern peripheries of Metropolitan Semarang. However, there are also some travels that heading to Bawen with other city like Solo, Salatiga, Madiun, Jogya, Ambarawa and Temanggung are the destination areas, as Bawen has also a regional terminal in the south of Semarang that connect Semarang to other cities in the south. Meanwhile the Bawen-Ungaran (towards Semarang) route was dominated by residences from Bergas District who is going to Semarang City as the dominant destination, who are going to the workplace as well as other activities as Semarang City has various activities because it is the destination of a settlement, workplace, school/college, siblings and others. However, there are also some respondents who have other cities like Ambarawa, Salatiga as their place of origin. This information shows that the Trans Jateng bus system is not only serve as home-based as well as workplace-based travel that we can said as internal travels of Semarang Urban Agglomeration / Metropolitan Area, but also serve to the "external travels" that link Semarang to other towns and cities in the surroundings.

3.2. The User Assessment of Bus Stops

3.2.1 The User Assessment on Location Issues on The Bus Stops

The overall score of the location variable is quite well because it has a value of 3.29. Table 3 shows the results of the assessments of each element: related to the target of group service worth 3.63 (tend to be good), related to road activities worth 3.28 (quite well), related to the distance between bus stops in one route worth 3.18 (quite well) and related with accessibility worth 3.06 (quite well). The results show that there is an asynchronous result between author's classifications on the bus stops with user assessment in the distance between bus stops in one route. Group 1 (God Bus Stops) shows a smaller value, compared to group 2 (Quite Well Bus Stops) that gets a higher value. This asynchronous occurs because the author's classification of the bus stops is not based on the distance between stops in one route and the accessibility.

Table 3. The User Assessment of Location Issues

Variable	Elements	Group 1*	Group 2*	Group 3*	Total Score
Location	Related to Target of Group Service	3.93	3.57	3.40	3.63
	Related to Road Activities	3.42	3.32	3.09	3.28
	Related to The Bus Stop Distance in One Route	3.22	3.50	2.83	3.18
	Related to Accessibility	3.33	3.01	2.83	3.06
Total Score		3.48	3.35	3.04	**3.29**

**Bus Stops Category based on Author Assessment, Group 1: Good, Group 2: Quite Well, Group 3: Bad*
Source: Author Analysis (2019)

3.2.2 The User Assessment on Physical Issues on The Bus Stops

The assessments of physical elements are divided into 6 elements: physical dimensions, sitting area, trash can availability, information board of instructions, symbols, routes and schedules, pedestrian paths and parking areas. Table 4 shows the total value of physical variable that gets 2.52 so it is categorized below quite well that indicates that there are some issues that are need to be improved, i.e., in the availability of trash can, information board, parking or dropping point, and pedestrian path /facilities. However, there are two asynchronous results. First is in the pedestrian way element as the value of Group 2 is lower than that of the Group 3. Second, in the parking area that Group 1 is lower than Group 2. These are happened because the variables and elements used by author's classification were different too those of the user's assessments.

Table 4. The User Assessment of Physic Issues

Variable	Elements	Group 1*	Group 2*	Group 3*	Total Score
Physic	Physic Dimension	3.46	3.17	2.84	3.16
	Sitting Area	3.29	3.23	2.47	3.00
	Trash Can	1.90	1.78	1.53	1.74
	Information Board of Instructions, Symbols, Routes and Schedules	3.23	2.83	1.00	2.35
	Pedestrian Path	2.79	2.05	2.54	2.46
	Parking Area	2.37	2.73	2.23	2.44
Total Score		2.84	2.63	2.10	**2.52**

**Bus Stops Category based on Author Assessment, Group 1: Good, Group 2: Quite Well, Group 3: Bad*
Source: Author Analysis (2019)

3.2.3 The User Assessment on Service Level on The Bus Stops

The assessment of service level is measured though 10 elements that could be seen in Table 5. The total score obtained a score of 3.01 which means the user still rated it quite well. However, there are some issues in some elements in the service level. Overall, the bus stops are not convenient yet, according to the users. The need some improvement also in passenger sitting area. Partially, there are many issues in Group 3 (the not good bus stops), some issues in Group 2 (quite well bus stops) and one issue in Group 1 (the good bus stops). The Group 1 and Group 2 still have some issues in passenger sitting area, and they are still not convenience yet. Meanwhile, issues in Group 3 are safety and security, not-convenient, aesthetical issue, passenger sitting area, visibility, passenger waiting area, passenger movement flow, and cleanliness. Moreover, Table 5 also indicates the asynchronous between group 1 and group 2 ratings on the passenger waiting room element, aesthetics, and bus stop level cleanliness because the value in group 1 are lower than those of Group 2.

Table 5. The User Assessment of Level of Service Issues

Variable	Elements	Group 1*	Group 2*	Group 3*	Total Score
Level of Service	Passenger Waiting Area	3.16	3.57	2.66	3.13
	Passenger Sitting Area	2.64	2.50	2.29	2.48
	Passenger Movement Flow	3.40	3.40	2.94	3.25
	Pedestrian Movement Flow	3.84	3.40	3.31	3.52
	Vehicle Movement Flow	3.48	3.46	3.18	3.37
	Safety and Security	3.25	3.23	1.20	2.56
	Visibility	3.89	3.87	2.37	3.38
	Convenience	2.29	1.97	1.71	1.99
	Aesthetics	3.41	3.50	2.24	3.05
	Cleanliness	3.64	3.73	2.89	3.42
Total Score		3.30	3.26	2.48	**3.01**

**Bus Stops Category based on Author Assessment, Group 1: Good, Group 2: Quite Well, Group 3: Bad*
Source: Author Analysis (2019)

3.3. The Assessment Analysis of Additional Bus Stop Element Needs

3.3.1 The Additional Bus Stop Element Needs on Location Variable

The additional elements needed in location variables are seen in Table 6 which shows the division based on the elements in two forms of questions, namely the stop space of other modes of transportation and crossing facilities. The table shows the assessment in which results have been sorted according to the level of need. The crossing facilities occupy the first position because the value is higher, at 3.37, compared to the need for stop space for other transportation modes which only get a value of 2.70. The table also shows the user chooses not to answer in all groups at the crossing facility. However, based on the real conditions, there are still some bus stops in group 1 is still really need crossing facilities, such as Unyil Park, Mas Salamah, Lemah Abang, Bergas 2, Wujil 2, Mang Engking and Benteng 2 and all group 2 and group 1. And the crossing facilities should be located at a maximum of 100 meters from the bus stop. Other results show that the stopping space for other transportation modes only gets a value of 2.70 so the user tends not to answer. Although, the bus stops in group 1 still needed, for example, SMA N 1 Ungaran 1, Undaris 1, SMAN N 1 Bergas, Wujil 2, Undaris 2, Mang Engking and SMA N 1 Ungaran 2.

Table 6. The Additional Bus Stop Element Needs on Location Variable

Variable	Questions	Group 1*	Group 2*	Group 3*	Total Score
Location	The need of a stopping area to other transportation mode (a more integrated facility)	3.03	2.43	2.63	2.70
	The need of a crossing facility	3.21	3.43	3.46	3.37

**Bus Stops Category based on Author Assessment, Group 1: Good, Group 2: Quite Well, Group 3: Bad*
Source: Author Analysis (2019)

3.3.2 The Additional Bus Stop Element Needs on Physics Variable

The additional elements to the physical variables are described in Table 7. The table shows the several additional elements needed which are the bus travel schedules and disability facilities while the tends to be needed are obtained by the bus travel routes. This is by the real situation because all bus stops do not have bus travel information and the disabled facilities yet. However, based on the real condition there are still many requiring additional elements to show their functions. First, the garbage bin that almost all bus stops need it, except Karang Jati Sub-Terminal, Undaris 2, Mang Engking, SMA N 1 Ungaran 2, Benteng 2, Taman Unyil, PT. Mas, As Salamah, Bandarharjo Market, SMA N 1 Ungaran 1, Undaris 1, Pasar Babadan, Lemah Abang, Sido Muncul and Bergas 1. Second, pedestrian ways were all bus stops in group 2 tend to require these facilities. The existing pedestrian ways do not accommodate disabilities people, sidewalks do not have a minimum width of three feet and not all sidewalks are equipped with ramps at each intersection. Third, the payment corner before entering the vehicle, based on the conditions, all bus stops do not have the facility yet. Fourth, the parking area for vehicles that do not stop for long which in all groups of shelters indicates that respondents tend not to answer. Fifth, the shelter door with respondents chose not to answer even though based on the conditions some bus stops needed, such as PT. Mas, Bandarharjo Market, Benteng 1, Langensari 1, Wujil 1, APAC INTI and Langesari 2. Sixth, the parking area for a long time by showing the results of users say less needed. Last, the shelter nameplate as a whole said tend to be less needed even though some bus stops do not have one, such as Langensari 1, Wujil 1, APAC INTI and Langensari 2.

Table 7. The Additional Bus Stop Element Needs on Physic Variable

Variable	Questions	Group 1*	Group 2*	Group 3*	Total Score

Physic	The need of door shelters	2.84	2.50	3.14	2.83
	The need of a name board	1.16	1.13	2.6	1.63
	The need of payment point before entering the bus	3.16	2.93	2.8	2.96
	The need of disability facilities	4.33	4.6	4.29	4.41
	The need of trash cans	3.13	3.13	3.6	3.29
	The need of bus route formation board	4.00	3.57	4.37	3.98
	The need of schedules information board	4.44	4.43	4.51	4.46
	The need of better pedestrian path to the bus stop	2.63	3.83	3.2	3.22
	The need of dropping zone	2.72	2.53	2.91	2.72
	The need of parking area	2.48	2.57	2.09	2.38

**Bus Stops Category based on Author Assessment, Group 1: Good, Group 2: Quite Well, Group 3: Bad*
Source: Author Analysis (2019)

3.3.3 The Additional Bus Stop Element Needs on Service Level Variable

The results show that the first was obtained by security facilities, such as CCTV and emergency buttons because it got a total value of 4.06 with priority needed by bus stops in groups 1 and group 2 with values of 4.29 and 4.07. While the bus stops in group 3 only get a value of 3.83 so it tends to be needed. Although in the conditions, all bus stops do not have good security facilities yet so it is important to provide in increasing passenger safety. The second element is the improvement of bus stop designs that obtain tends to be needed, like PT.Mas and Pasar Bandarharjo which have broken and damaged shelter walls, Wujil 1 which is in very poor condition due to the absence of a good protective wall or supporting facilities and also for Benteng 1, Langersari 1, APAC INTI and Langesari 2 because they are only a shadow shelter whose physical design is not fixed. Another case with the plant element to increase comfort, where the bus stops in group 1 tends to needed, group 2 needed and group 3 valued respondents do not answer. Despite the fact, most of the plants are already around the bus stop but do not belong to the bus stop itself. The final element is the lighting for safety that has an overall rating of 3.56 so it tends to be needed. However, the bus stops in group 3 show a value of 4.23 so users say they need it. Proven based on the conditions, all the bus stops in group 3 have not been equipped with lighting because they are still dependent on street lighting.

Table 8. The Additional Bus Stop Element Needs on Service Level Variable

Variable	Questions	Group 1*	Group 2*	Group 3*	Total Score
Physic	The need of security facilities, such as CCTV and Emergency Button	4.29	4.07	3.83	4.06
	The need of plants to increase convenience	3.85	4.03	3.34	3.74
	The need of physical design improvement	3.61	3.67	4.29	3.86
	The need of better lighting facility to increase security	3.09	3.37	4.23	3.56

**Bus Stops Category based on Author Assessment, Group 1: Good, Group 2: Quite Well, Group 3: Bad*
Source: Author Analysis (2019).

3.4. The Relation Between Service Level Variables with Location and Physic Variables on Bus Stops

The bus stop assessment result could show the relationship between service level with location and physical variables. Both relationships indicate that location and physical variables are independent variables while service level variables are dependent variables. The service level variable is called the dependent variable because the results depend on the results of location and physical variables. Correlations are created if the results which are using the SPPS Application show the Chi-Square Calculate> Chi-Square Tables and Asymp values. Sig <0.05. Not only that, the elements whose results are correlated are based on SPPS Application calculations and will be tested again by the relation direction through the diagram. Diagrams will produce a positive or negative direction to give a picture of the relationship that occurs by the assumptions or logic thought. However, not all elements correlate, only a few elements that show the relationship and the correlation between variables is seen in Tables 9.

Table 9. The Cross-Tabulation Result for Analysis of Physical Variables and Location Variables with Service Level Variables

Location	Physic	Service	Chi-Square Calculation	Chi-Square Table	df	λ
The Condition between bus stop' location with road conditions		The Movement for One Bus in The Stopping Area	1,024∃2	26,30	16	0,000
		The Condition of Bus Movements with Road Conditions	54,285	26,30	16	0,000
	Pedestrian Path Condition	The Pedestrian Movement	39,838	31,41	20	0,05
	Seating facilities condition	The Adequacy of The Passenger Sitting Room	26,707	26,30	16	0,026

Source: Author Analysis (2019)

3.4.1 The Relation Between Location Variables with Service Level Variables

The author assumes there are two relation in these both variables. First, the relation between the assessment of movement for one bus in the stopping area and the assessment of bus movements with road conditions with the relation of the bus stop' location with road conditions. Both are assumed, if the bus flow in the stopping area is better, the bus stop' location will not be disrupted to the road condition, as well as the question regarding the flow of bus to the road condition. The results of cross-tabulation on the two correlations produced perfect results because both of them obtained 0,000 values in the Asymp. Sig. Table 9 shows the results of cross-tabulation between the movement activities for one bus in the space where the bus stops with the location of the bus stop and the road condition. Also shows the correlation between the bus movements to the road condition with the bus stop' location to the road condition. In conclusion, the better of the relation of the bus stop' location between road activities, the better of service level of bus stops in the vehicle flow on the stopping area and the relation of the bus movement with road activities, vice versa.

3.4.2 The Relation of Physics Variables with Service Level Variables

The physical elements of the bus stop effect the service level to the user. The relation between them could be seen in the correlation in the two elements, the relationship between the pedestrian path condition and the pedestrian movement also the relationship between the seating facilities condition and the adequacy of the passenger sitting room. Both produce Chi-Square values Calculate > Chi-Square Table and Asymp value. Sig <0.05 which is clearly seen in table 9. This proves that the better of the condition assessment of the pedestrian path and the trash can facilities, the better the pedestrian movements assessment and the adequacy of the sitting room. In other words, the better of physical elements such as the pedestrian path and bus stop facilities, the better of the pedestrian movement and the more adequate sitting room for passenger.

4. Conclusion

The development of Trans Jateng will never be separated from the importance of a bus stop. However, the problem of the existence of bus stops is often ignored by the service level because of various problems, especially bus stops in Sub-Urban Areas such as Ungaran Barat District and Ungaran Timur Sub-District, Bergas Sub-District and Bawen Sub-District, Central Java Province. Therefore, this study aims to determine user assessment for the service level of the Trans Jateng's bus stops based on location, physical and service level variables.

Table 10. The Total User Assessment for Bus Stop's Service Level

Variable	Group 1*	Group 2*	Group 3*	Total Score
Location	3.48	3.35	3.04	**3.29**
Physic	2.84	2.63	2.10	**2.52**
Service Level	3.30	3.26	2.48	**3.01**
Total	**3.21**	**3.08**	**2.54**	**2.94**

**Bus Stops Category based on Author Assessment, Group 1: Good, Group 2: Quite Well, Group 3: Bad*
Source: Author Analysis (2019)

The study shows that, in aggregate, the users' assessments indicate that the quality of service of Trans Jateng Bus Stops in the Ungaran-Bawen Corridor was still quite low as the total score was only 2.95. It was because the physical aspect or variables that are still got lower point in the user assessment. This indicates that significant improvements are needed in the Tans Jateng Bus Stops, especially in the physical aspects, even though also in te location and service level variables as they have not maximum value yet in this user assessments. The more prioritize improvements are needed for the not-good bus stops category, as they have the lowest points in the assessment. If this category has better results, this will increase the users' assessment significantly and hopefully will also improve the quality of service of the Trans Jateng bus system that is effectively linking the urban core in the City of Semarang to its southern peripheries in the metropolitan area vice versa.

The improvements of the bus stops could be considered more significant as the users of the bus service are tend to be women in the productive age who are doing the travel for working travel, as home-based and workplace-based travel were found as the majority type of traveling in the users' travel by the bus system. The provision of integrated feeders to the bus system could be an important consideration too in the improvements of the provision of the bus system, as it is indicated in the survey that the bus stops' location are quite far from the origin as well as the destination place of the users so the users need to use other mode of transportation to reach the bus stop fron their origin and also to reach their destination from their destination bus stop.

5. Acknowledgment

This article was presented at the 2nd International Conference on Smart City INNOVATION (ICSCI) 2019, jointly held by Universitas Indonesia and Universitas Diponegoro. ICSCI conferences have been supported by the United States Agency for International Development (USAID) through the Sustainable Higher Education Research Alliance (SHERA) Project for Universitas Indonesia's Scientific Modelling, Application, Research, and Training for City-centered Innovation and Technology (SMART CITY) Center for Collaborative Research, administered through Grant #AID-497-A-1600004, Sub Grant #IIE-00000078-UI-1.

6. References

[1] Adisasmita R 2014 *Teori Pertumbuhan Kota* (Yogyakarta: Graha Ilmu)
[2] Todaro M P 2000 *Pembangunan Ekonomi/Edisi Kesembilan* (Jakarta: Erlangga)
[3] Kumurur V A 2010 Pembangunan Kota dan Kondisi Kemiskinan Perempuan
[4] Haughton G and Hunter C 1994 Sustainable Cities *Eur. Urban Reg. Stud.* **2** 282–3
[5] Bhatta B 2010 *Analysis of Urban Growth and Sprawl from Remote Sensing Data* (Springer Berlin Heidelberg)
[6] Wilson E H, Hurd J D, Civco D L, Prisloe M P and Arnold C 2003 Development of A Geospatial Model to Quantify, Describe and Map Urban Growth *Remote Sens. Environ.* **86** 275–85
[7] Galster G, Hanson R, Ratcliffe M R, Wolman H, Coleman S and Freihage J 2001 Wrestling Sprawl to The Ground: Defining and Measuring An Elusive Concept *Hous. Policy Debate* **12** 681–717
[8] Wei H and Mogharabi A 2013 Key Issues in Integrating New Town Development into Urban Transportation Planning *Procedia - Soc. Behav. Sci.* **96** 2846–57
[9] Institute for Transportation and Development Policy 2017 *The BRT Planning Guide*
[10] Diaz R B and Schneck D C 2000 Bus Rapid Transit Technologies in the Americas: An Overview *Transp. Res. Rec. J. Transp. Res. Board* **1731** 3–9
[11] Wirasinghe S C, Kattan L, Rahman M M, Hubbell J, Thilakaratne R and Anowar S 2013 Bus Rapid Transit - A Review *Int. J. Urban Sci.* **17** 1–31
[12] Regional Regulation of Semarang Regency 2018 *Regional Spatial Plan of Semarang Regency 2011-2031*
[13] Setiyowati T R, Handayani W, Damayanti M and Rudiarto I 2018 Towards An Independent City: The Role of Ungaran City as A Sub-Center of Semarang City *IOP Conf. Ser. Earth Environ. Sci.* **202** 12035
[14] Mahardhini P and Rahdriawan M 2010 The Quality of Bus Rapid Transit (BRT) Shelter Services of Mangkang-Penggaron Route in CBD Semarang *J. Pembang. Wil. KOTA* **8** 42–54
[15] American Public Transportation Association 2010 *Bus Rapid Transit Stations and Stops* (Washington, D.C: APTA Bus Rapid Transit Working Group)
[16] Departemen Perhubungan Direktur Jenderal Perhubungan Darat 1996 *Pedoman Teknis Perekayasanaan Tempat Perhentian Kendaraan Penumpang Umum*
[17] Nugroho C 2013 Aksesibilitas Halte dan Kualitas Pelayanan Trans Jogja dengan Keputusan Pengguna *Fak. Ilmu Sos. Univ. Negeri Yogyakarta, Yogyakarta*
[18] Demetsky M J and Lin B B-M 1982 Bus Stop Location and Design *J. Transp. Eng.* **109** 613
[19] Ellis S D 1997 *Key issues in Rural Transport in Developing Countries* (Citeseer)
[20] Tamin O Z 2000 *Perencanaan dan Pemodelan Transportasi* (Penerbit ITB)
[21] Transit Cooperative Research Program 1996 *Guideline for The Location and Design of Bus Stops - Report 19* (Washington, D.C: National Academy Press)
[22] RPTA Bus Stop Program and Standards 2017 *Bus Stop Design Guidelines* (USA: Maricopa Association of Governments)
[23] Associates K &, Program T C R and Corporation T D 2003 *Transit Capacity and Quality of Service Manual* (Transportation Research Board)

Modelling the Needs of Light Rail Transit (LRT) on Transit Oriented Development Area around LRT Stations in Palembang

M Agustien[1,2], B Susanti[1,2], H Pahlevi[1]

[1] Department of Civil Engineering, Sriwijaya University, Indonesia
[2] National Center for Sustainable Transportation Technology, Institut Teknologi Bandung, Indonesia

E-mail: Melawaty74@gmail.com

Abstract. Various efforts have been made to overcome the problem of congestion in Palembang, one of them is to provide Light Rail Transit (LRT) which has been operating since July 2018. Analysis of the relationship between the needs of LRT, travel characteristics and characteristic of LRT operation is needed to evaluate the LRT level of service. Level of service of LRT can be assessed from route service, integration of LRT modes with other public transportation, travel time and cost of travel demand. Based on this, the research aims to model the need for LRT in Palembang City on the Transit Oriented Development (TOD. TOD is an area that has the potential to be developed into an integrated area between land use and transportation systems including LRT transportation mode in the area. Travel and mode choice characteristics data of respondents are collected on households in the TOD area. Structural Equation Modelling (SEM) method is used to modelling the needs of LRT. The model can be used to determine the relationship between variables the need of LRT, travel behaviour characteristic and LRT operation characteristics. The result of the study shows that there is a positive and significant relationship between the variables on TOD area around LRT stations in Palembang.

Keywords: transit oriented development, structural equation modelling (SEM), modelling the needs of LRT,

1. Introduction

Currently transportation becomes derived demand for everyone in facilitating their daily activities. In line with the pace of development and growing population growth triggered the emergence of mobility as well. One effort to solve the need of transportation problem in Palembang City is providing mass transit of Light Rail Transit (LRT) which has been operating since mid-2018. However, in order to provide mass transportation in urban areas in accordance with the requirements of community movement, it is necessary to evaluate the characteristics operational of LRT. The purpose of the evaluation is to find out wheather the operational characteristics of LRT such as appropriate route, operational time, accessibility from and to the station accordingly with the travel characteristic and mode choice perception of community in Palembang City. Travel characteristics that need to be identified are the location of the origin-destination travel, the activities carried out before and after travel, travel time, the modes that are generally used and the possibility to change modes. Such information is important to know that the LRT in accordance with the daily needs of urban communities.

The purpose of this research is to evaluate the operation of LRT by modelling the need of Light Rail Transit for urban community in Palembang City. The model is used to find out the relationship between public perception about the need of LRT, travel characteristic, mode choice and operational characteristic of LRT. The method used in this study is descriptive analysis method to determine the travel characteristics and Structural Equation Modelling with second order testing of confirmatory factor analysis was also used to form the modelling of the Light Rail Transit needs on the Transit Oriented Development (TOD) area around the LRT stations. The data required in this study are the characteristics of travel and the potential user perception of Light Rail Transit operation characteristics in Palembang City especially for people who live in the TOD area.

TOD area is mixed-use neighboroods which are developed around a transit stop and core commercial area. The design, configuration and mix of uses provide an alternative development by emphasizing a pedestrian-oriented environment and reinforcing the use of public transportation. The development of transportation system and land use in TOD area can be conducted based on community needs of public transportation and public facilities around LRT stations. In Palembang, there are 13 LRT stations, starting from Sultan Mahmud Badarudin II station in Ilir area to DJKA station in Ulu area. TOD area has a radius 0-800 meters from stations consisting of mixed land use such as commercial, education, health and public facilities land uses. There are also many transportation infrastructure facilities around the stations such as public transport bus Trans Musi, pedestrian facilities, fly over and many other transportation infrastructures. The development of transportation system and land use in TOD area in the future can be predicted base on travel characteristic of community in TOD area.

The method used in this research is Structural Equation Modelling (SEM) was carried out with LISREL software version 8.8. The method is used for analysis the causal relationship model between public perception about the need of LRT, travel characteristic, mode choice and operational characteristic of LRT. Based on the model results it can be seen whether there is a positive and significant relationship between LRT needs and community travel characteristics, LRT operation patterns and the facilities of LRT infrastructure especially in TOD area. The research is conducted around 12 LRT stations in Palembang City. The results of this research is also explain about the development of TOD which is suitable with the travel characteristic of community, such as the provision of transit facilities in the form of integrated stations and shelters, park and ride facilities, adequate facilities for pedestrians such as sidewalks and zebra cross, mixed land use facilities, health and entertainment facilities that are connected with adequate pedestrian facilities such as the sky walk to the nearest LRT station.

2. Literature Review

2.1. Transportation System

The simple principle of the need for LRT is that one will choose the LRT mode, if the LRT will provide the greatest satisfaction for him. Someone will differ in their views in choosing the mode of transport to get to the destination. The differences in perception are inluenced by the service characteristics of the mode. The concept is considered for modelling relationship between variables which is influenced the need of LRT in Palembang City is transportation system concept [1]. Transportation system consist of three components there are travel, land use and transportation infrastructure. Changs in one component will affect the other components. Increased travel demand due to changes in land use will increase the need for adequate transportation infrastructure [2].

2.2. Transit Oriented Development (TOD)

Transit Oriented Development first appeared in the 1990s which was pioneered by Calthorpe. TOD arises because of the phenomenon of urban sprawl which results in high use of private vehicles and resulting congestion. While for the formation of TOD consists of several variables, including:

1. Central commercial area, the function of mixed land use area is providing services to transit activities such as retail functions, regional scale offices, supermarkets, commercial and entertainment. The mix function area can be an attraction for diversity of destinations on the location.
2. Mixed residential areas, rresidential within the reach of the central commercial area and stopping on foot, with various types of occupancy (single, apartment or town house).
3. Function of public space, the area can be in the form of parks, plazas, green structures, which serve around the environment. Public spaces designed in public buildings or public facilities tailored to the needs.

4. Secondary areal, lcoated about 1 mile from the central area and has a road network as a link to the rear area. This connector is equipped with pedestrian and bicycle lanes. This secondary area consists of low density housing, public facilities and park and ride parking spaces.
5. Mixed functions, the functions in the TOD are diverse and mixed, namely the function of the public, the commercial center and residential. Where buildings with vertical functions are the recommended type. The concept of TOD expressed by Calthrope is inseparable from the city movement system in the form of vehicles both public and private vehicles and humans who continue to move to follow the pattern of activity, and how to use an unused vacant land becomes very useful for its citizens.

Transit-based regional development movement is based on the deteriorating quality of city life which is characterized by congestion, sprawl, and non-integrated land use. TOD has the goal of creating goals that are comfortable, safe, enjoyable and sufficient for walk able environments [3]. By mixing various functions of travel activities that need to be done can be combined to be shorter and faster. These functions are the center of commercial areas, offices, retail, service, settlements with medium to high density and also public open spaces. The development of TOD is very advanced and has become a trend in big cities, especially in large new urban areas such as Tokyo in Japan, Seoul in Korea, Hong Kong, Singapore which utilizes city trains and several cities in the United States and Europe. Development of TOD-based areas has not been done in urban areas in Indonesia. The TOD plan at Manggarai Station has not been proven to date, as well as Kota and Dukuh Atas stations in Jakarta. The limited development of TOD has been done a lot, but it has no extensive impact because it is not the 4th synergy factor, namely mixed-use, high-density, non-motorized vehicle access, close to Mass Rapid Transit stations [4].

2.3 Structural Equation Modeling Methods

SEM or Structural Equation Modeling is a statistical technique that is able to analyze the pattern of relationships between latent constructs and indicators, latent constructs with each other, and measurement errors directly [5]. SEM allows direct analysis between several dependent and independent variables [6]. According to Bollen (2011) as quoted from Latan [7], ''There are sets of equations that accommodate relationships between the latent variables, observed variables and error variables'. SEM can be used to answer various research problems in a systematic and comprehensive set of analyzes.

Data analysis techniques using Structural Equation Modeling (SEM), carried out to explain thoroughly the relationships between variables in the study. SEM is used not to design a theory, but rather is intended to examine and justify a model. Therefore, the main requirement for using SEM is to build a hypothesis model consisting of structural models and measurement models in the form of path diagrams based on justification of the theory. SEM is a stronger analysis technique because it considers interaction modeling, nonlinearity, correlated independent variables, measurement errors, correlated error terms, several latent independent variables where each is measured using many indicators, and one or two latent dependent variables which are also each measured by several indicators. Thus, according to this definition, SEM can be used as a stronger alternative than using multiple regression, path analysis, factor analysis, time series analysis, and covariance analysis.

In general, SEM model can be divided into two main parts, structural and measurement models.

1. Structural model

   The structural model is a set of relationships between latent variables and this relationship can be considered linear, although further development allows to include non-linear equations. In graphic form, lines with one arrow head (⟶) describe the regression relationship in Greek characters written "gamma" (γ) for regression of exogenous variables to endogenous variables and Greek characters written "beta" (β) for regression of one endogenous variable to variable other endogenous, while the line with two arrow heads (⟷) describes the correlation or covariance relationship in Greek characters written "'phi" (φ) for correlations between exogenous variables. In this model it produces predictive validity. Examples of structural models can be seen in Figure 1.

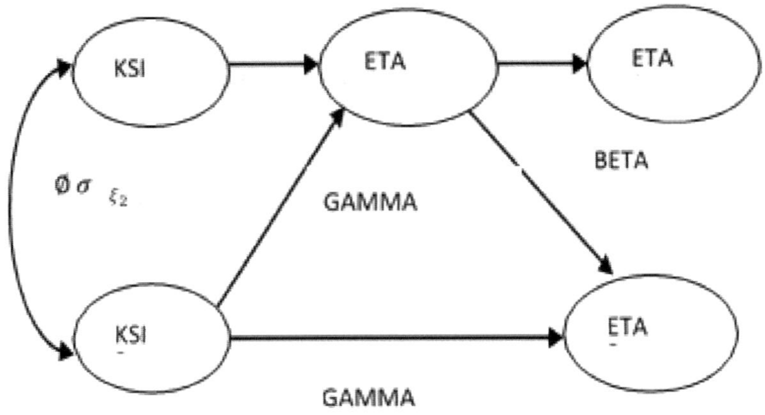

Figure 1. Structural Model

2. Measurement Model
   The measurement model is part of a SEM model that is usually associated with latent variables and indicators. The relationship in this model is done by a confirmatory factor analysis (CFA) model in which there is an unmeasured covariance between each possible pair of variables. This analysis process can only be continued if the measurement model is valid. In this model it produces convergent validity. An example of a measurement model can be seen in Figure 2.

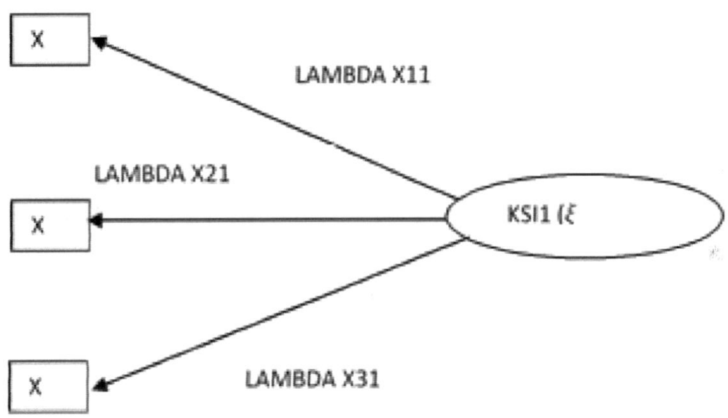

Figure 2. Measurement Model

There are two main variables in SEM, including:

1. Latent variable (latent variable) according to Joreskog 1996, latent variables are abstract psychology concepts such as attitudes, intelligence [8]. This latent variable is a key variable in SEM of concern. The behavior of latent variables can be observed indirectly and imperfectly through their influence on the indicator variable or manifest variable. There are two types of latent variables namely exogenous (independent) and endogenous (dependent) variables. Both types of variables are distinguished by position as the dependent variable or not dependent in an equation model. Exogenous variables are depicted in Greek letters with 'ksi' (ξ) and endogenous variables with eta 'eta' ' (η). Graphical form, exogenous variables become line targets with two arrows ($\leftrightarrow$) or correlation / covariant relationships while endogenous variables are targeted at least one arrow ($\longrightarrow$) or regression relationship.
2. Observed Variables
   Observed variables are abstract concepts that can be directly measured. As an example of direct

inflation can be measured by consumer price index numbers, company performance can be measured by this variable used to form latent variables that are manifested in the Likert scale question. This variable is to form an exogenous latent variable which is given the symbol X while the late endogen variable is given the symbol Y.

The validity and reliability test in the SEM model uses the Lisrel program with the measurement model performed to check whether the t-values of the observed variables in the model satisfy the good condition that is greater than the critical value of> 1.96 with α = 5% [9]. Realibility analysis of measurement model is done by calculating Construct Realibility (CR) and Variance Exctrated (VE) values of standardized loading factors and error variance with equation 1 and equation 2 as follows:

Construct Realibility (CR)

$$\frac{(\Sigma\,\lambda)^2}{(\Sigma\,\lambda)^2 + \Sigma\,e_j} \qquad (1)$$

Variance Exctrated (VE)

$$\frac{\Sigma\,\lambda^2}{\Sigma\,\lambda^2 + \Sigma\,e_j} \qquad (2)$$

Where :

λ = Standardized Loading

e_j = Measurement Error

3. Research Metodology

The primary data taken directly through the household survey using non-probability sampling method. Respondents were chosen within a radius of 800 meters from the LRT location, as a justification for the TOD area. The example of TOD area around LRT stations are showed in Figures 3 and 4.

Figure 3. TOD area around Dishub LRT station

Figure 4. TOD area Bumi Sriwijaya LRT station

The survey was conducted at 12 LRT stations with total 400 respondents. The data obtained includes socioeconomic characteristics and travel characteristics stated in 17 variables in SEM model consisting of: work trips purpose (X_1), non work trips purpose (X_2), time allocation (X_3), travel cost (X_4), modal integration (X_5), operational time (X_6), travel time (X_7), the availability of pendestrian facilities (X_8), number of LRT rate (X_9) , the availability of park and ride (X_{10}), safety (X_{11}), comfort (X_{12}), the availability of difable facilities (X_{13}), reduce private vehicle use (X_{14}) , willingness to use feeder transport (X_{15}), willingness to use park and ride (X_{16}), and willingness to walk (X_{17}). These variables are assessed in the following five scales: definitely choose to use LRT; may choose to use LRT; no choice; maybe choose to use another mode; and definitely choose to use another mode. The socio-economic and travel characteristics data were analyzed using descriptive analysis method while the data on 17 indicators of modal mode variable were analyzed using Structural Equation Modeling method to obtain the most influential indicator in the needs of light rail transit modes [10].

4. Data Analysis

4.1. Descriptive Statistic Analysis of Travel Characteristic

Data analysis and discussion consisted of descriptive statistic analysis of travel characteristic to support the result of structural equation modeling analysis the needs of LRT in Palembang City.The results of travel characteristis survey from 400 respondents in TOD area around 12-point LRT stations in Palembang City are explained as follows:

1. Travel characteristic : time to start the trip

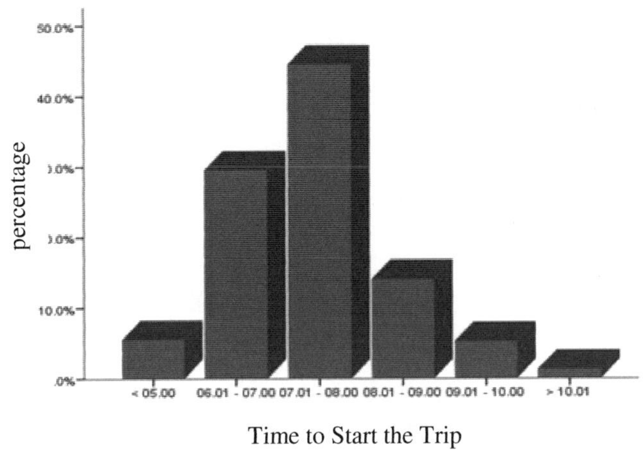

Characteristics of the trip based on the time to start the trip (Figure 5) at 07:01 to 08:00 AM is 44.3% and the lowest proportion above 10:00 AM as much as 1.3%. This indicates that the traveller has dominance in starting the work trip from 07:00 to 08:00.

Figure 5. Percentage number of respondents based on time to start the trip

2. Travel characteristic : time to end the trip

Characteristics of travel based on the time to end the trip (Figure 6) until 17:00 afternoon is 45.8% and the lowest proportion after 14:00 as much as 7.5%. This indicates that the traveller has dominance on the way home from 16:01 to 17:00.

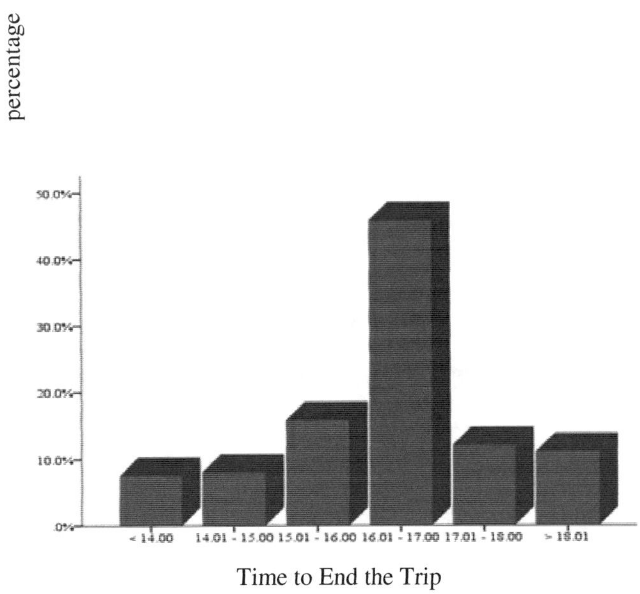

Figure 6. Percentage number of respondents based on time to end the trip

3. Travel characteristic : mode choice

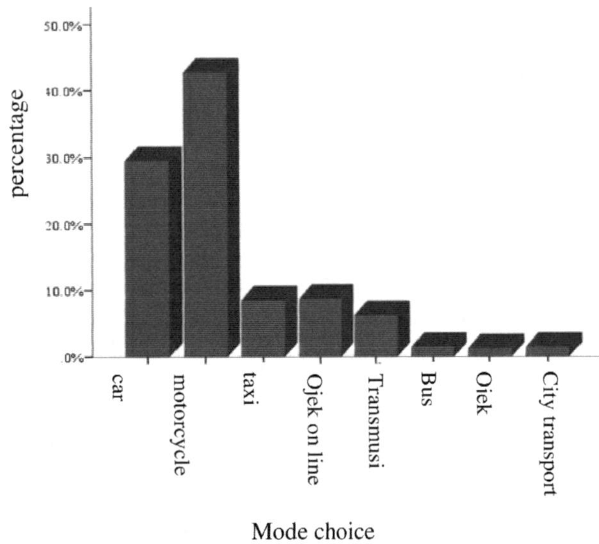

The highest proportion of travel characteristics based on the type of mode choice (Figure 7): the proportion of motorcycle is 43% and the lowest proportion is the public transportation about 2%. This indicates that respondents is prefer to use motorcycle to do their activities.

Figure 7. Percentage number of respondents based on mode choice

4. Travel characteristic : travel cost

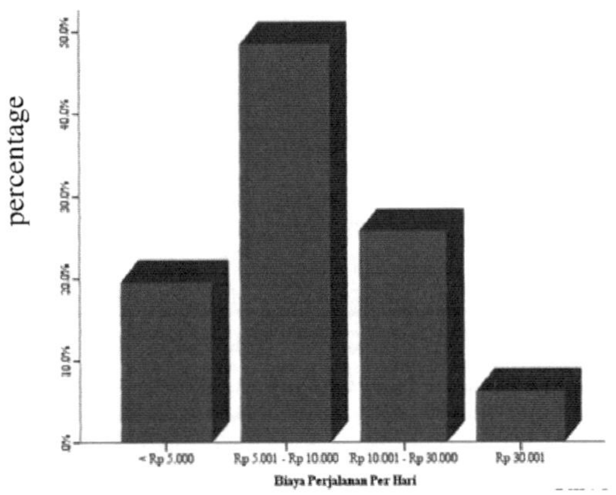

The highest proportion of travel characteristics based on travel costs (Figure 8) to get to the workplace Rp 5,001 to Rp.10,000 is 48.4% and the lowest above Rp.30,000 is 6.3%.

Figure 8. Percentage number of respondents based on travel cost

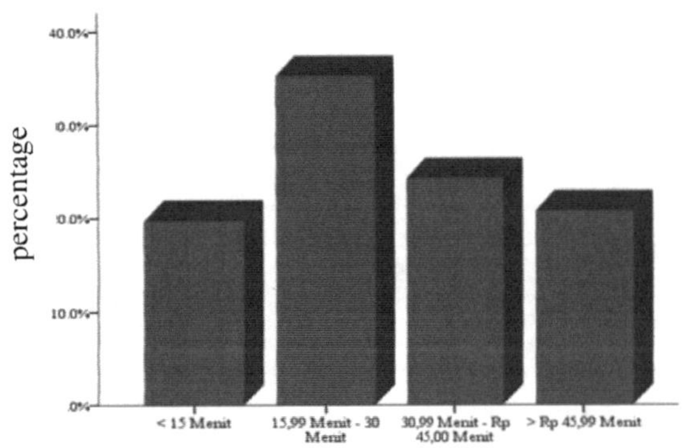

The highest proportion of travel characteristics based on travel time (Figure 9) 15.99 minutes to 30 minutes is 35.3% and the lowest proportion under 15 minutes is 19.8%.

Figure 9. Percentage number of respondents based on travel time

4.2. Modelling The Needs for LRT Transportation Mode with the Structure Equation Modeling (SEM) Method

LRT is a mass transit that has a long line of 24 km with 13 stations, average speed 32.9 km/h and travel time 40 minutes. The feasibility of LRT operation in Palembang City can be seen from the factors that influence the needs of mode transportation. The need of mode transportation are influenced by three components of macro transportation system relationships [2]. The system forming the basic concept the needs of modes. The concept is considered for modelling relationship between variables which is influenced the need of LRT in Palembang City, shown in Figure 10.

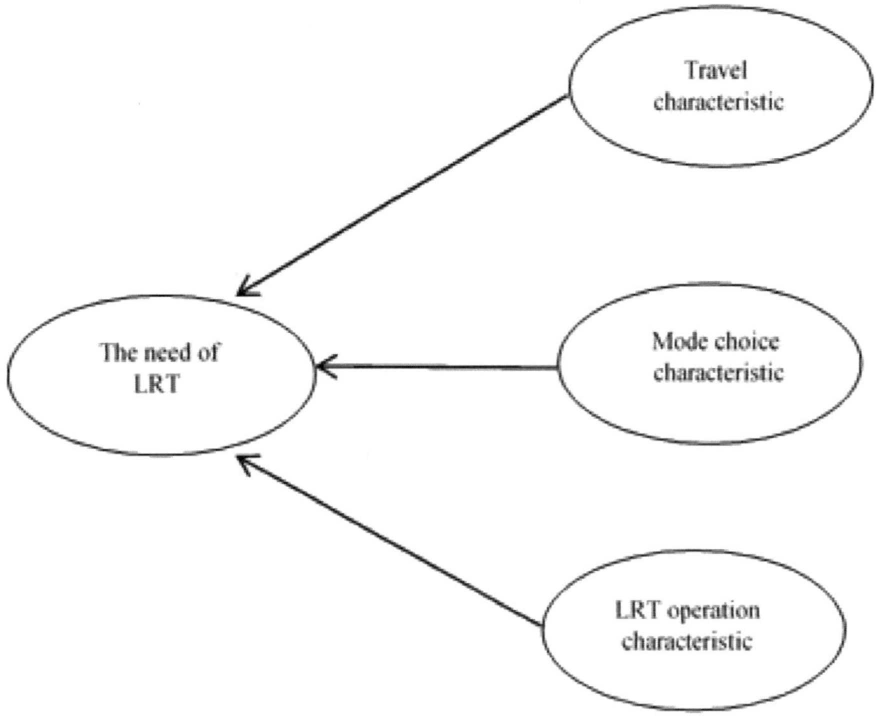

Figure 10. The concept of relationship between variables the need of LRT, travel, mode choice and LRT operation characteristics

The method used in modelling the needs of the Light Rail Transit transportation mode is the Structural Equation Modeling (SEM) Method. The purpose of the modeling is to find out how the relationship between LRT needs and travel characteristics, changes in mode choice behavior and characteristic of LRT services. By knowing this relationship, we will have the information about the actual needs of the people of Palembang City for LRT mass public transportation. Travel characteristics, mode choice behavior and operation characteristic of LRT are called latent variables. The relationship between the latent variables and the need of LRT mode can be obtained from the rating scale conducted by respondents on the indicator variables of each of these latent variables. There are 3 latent variables and 17 indicator variables contained in the model. The rating scale of indicator variables are: definitely choose the Light Rail Transit, maybe choose a Light Rail Transit, maybe choose a Light Rail Transit and another mode, maybe choose another mode, definitely choose another mode.

The analysis of causal relationships or interrelationships between variables is conducted by using SEM approach to form the causal relationship of structural travel, LRT and mode choice characteristics.

1. Specifications.

The model parameter specification uses deveploment strategy model which means done if the structural model in stage 1 has not met statistical requirement or model fit test so that the model needs to be modified or respesifikasi.

2. Identification.
The model identification of the first structural model estimate is an over identified model because the number of known parameters is greater than the estimated number of parameters. With degree of freedom value of 82 (df> 0) appropriate with the standards.
3. Estimates.
The estimation method used in this research is the maximum likehood method using software LISREL 8.8 analysis is completed in maximum iteration to 62.
4. Validity and Realibility.
The validity and reliability test in the SEM model uses the Lisrel program with the measurement model performed to check whether the t-values of the observed variables in the model satisfy the good condition that is greater than the critical value of> 1.96 with $\alpha = 5\%$. Realibility analysis of measurement model is calculated with Construct Realibility (CR) and Variance Exctrated (VE) formulas of standardized loading factors and error variance. The formulas was explained in equation 1 and equation 2. The result of validity and reliability test are explained in Table 1.

Table 1. The result of validity and reliability test

Variabel	*Loading Factors*	*T value* > 1,97	CR > 0,70	VE > 0,50	Inf.
Travel characteristic					
X_1	0,77		0,911	0,721	
X_2	0,80	16,31			Accept
X_3	0,74	13,78			Accept
X_4	0,77	13,99			Accept
LRT operation caharcteristic					
X_5	0,77		0,984	0,870	Accept
X_6	0,85	19,23			Accept
X_7	0,88	19,70			Accept
X_8	0,89	19,15			Accept
X_9	0,90	16,37			Accept
X_{10}	0,85	17,05			Accept
X_{11}	0,96	17,86			Accept
X_{12}	0,94	17,66			Accept
X_{13}	0,91	17,24			Accept
Mode choice caharcteristic					
X_{14}	0,76		0,907	0,711	Accept
X_{15}	0,86	18,71			Accept
X_{16}	0,70	12,73			Accept
X_{17}	0,73	12,69			Accept

5. Goodness of fit in Modelling
The model fit test which includes chi square, Degree of Freedom (DF), NCP Interval, *The* Root Mean Square Error of Approximation (RMSEA), probability, Expected Cross-Validation Index (ECVI), Akaike's Information Criterion *(*AIC), Consistent Akaike's Information Criterion (CAIC), Normal Fit Index (NFI), *Non Normed Fit Index* (NNFI), *Comparative Fit Index* (CFI), Incremental Fit Index (IFI), Relative Fit Index (RFI), Critical N (CN), RMR, Goodness of Fit Index (GFI) and Adjusted Goodness of Fit Index (AGFI) are obtained from the output of goodness of fit statics that can be seen in Table 2 and Table 3.

Table 2. The result of model fit test

Goodness of fit	Cut of Value	Analysis Result	Evaluation
Chi squere	106,395	130,18	Good
Degree of Freedom	DF > 0	84	Good
NCP Interval	17.30 : 78.40	43,87	Good
RMSEA	< 0,08	0,036	Good
Probability	P value > 0,05	0,093	Good
ECVI	Approaching ECVI saturated	0,67	Good
AIC	Approaching AIC saturated	265,87	Good
CAIC	Approaching CAIC saturated	610,28	Good
NFI	> 0,90	0,99	Good
CFI	> 0,90	1	Good
IFI	> 0,90	1	Good
Critical N	> 396	369,79	Good
GFI	> 0,90	0,96	Good

The result of model respesification is the degree of freedom value of 84 (df> 0) with maximum iteration to 62. The result of t value parameter of the structural equation model can be seen in Figure 11.

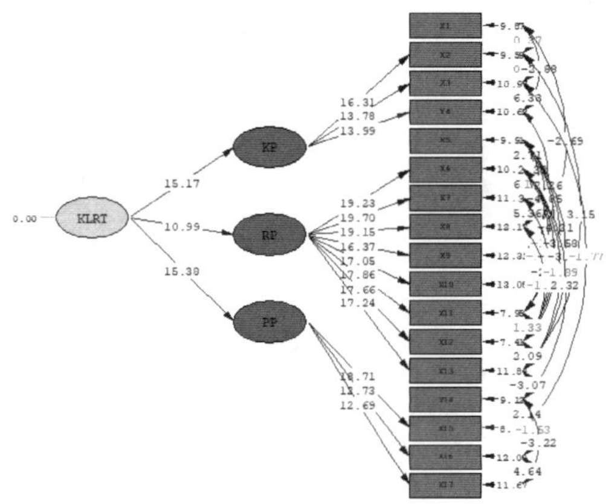

Figure 11. Results of t parameter value of Structural Equation Model

The results of the suitability test for the structural model are:

a) The measurement model fit analysis shows that 17 indicators have meet the validation standard and then 3 latent variables also have good realibility.

b) Analysis of reliability model which is presented by R^2 value shows that the effect of the indicator on each latent variable can be seen in Table 3.

Table 3. Suitable model analysis of the measurement value which is presented by R^2 Value

Variable	R^2	Variable	R^2
Work trip	0,59	The need of Park and Ride	0,43
Non work tip	0,64	Safety	0,92
Time Allocation	0,54	Comfort	0,89
Travel cost	0,59	Facilities Difable	0,82
Integration Mode	0,59	Reduce Private Vehicles	0,57
Operational Time	0,72	Willingness to use feeder	0,74
Willingness to use Pendestrian Facilities	0,79	Willingness to Walk	0,54
LRT Rates	0,80		

c) The analysis of reliability structural model shows that the value of the coefficient of determination

R^2 = 0.824. This indicates, there is 82% reability of LRT to be feasibility transportation mode in Palembang. The result shows that the community in Palembang especially who live in TOD area around the LRT stations requires LRT to be used as public transportation mode for work and non work trip.

4.3. Test of hypotheses

Test of research hypothesis on SEM (Table 4) can be seen from comparison of estimate coefficient value and its t valuet. If the t value is greater than the value of the critical condition then H_0 is rejected.

Table 4. Test of research hypothesis on SEM

Hypotheses	Coefficient *Estimate*	t *Value*	Critical Condition	Information
1	0,67	15,17	1,97	H_1 be accepted
2	0,54	10,99	1,97	H_2 be accepted
3	0,66	15,38	1,97	H_3 be accepted

1. Hypothesis 1
   In structural model analysis between travel characteristic variable and the need of LRT, it is known that the estimated coefficient is 0.67 and t-value of 15.17 which is greater than 1.96. This result indicates that there is significant correlation between latent variable of travel characteristic to independent variable the need of LRT mode, so this proves that hypothesis 1 is received.
2. Hypothesis 2
   In structural model analysis between LRT operation characteristic and the need of LRT, it is known that the estimated coefficient is 0.54 and t-value of 10.99 is greater than 1.96. This result indicates that there is significant correlation between latent variable of LRT operation characteristic to independent variable the need of LRT mode, so this proves that hypothesis 2 is received.
3. Hypothesis 3
   In structural model analysis between mode choice characteristic variable and the need of LRT, it is known that the estimated coefficient is 0.66 and t-value of 15.38 which is greater than 1.96. These results indicate that there is a significant relationship between latent variables of mode choice behavior toward independent variables the need of LRT, so this proves that hypothesis 3 is accepted.

5. Conclusion

This research show that

1. Based on the survey result travel characteristics of the time to start the trip is from 06:00 to 07:00, the time to end the trip is from 16:00 to17:00, motorcycle is the most widely used, travel costs is Rp 5,000 to Rp 10,000 and journey time is from 15 minutes to 30 minutes.
2. The result of model analysis with structure equation modeling method of relationship between independent variables and the latent variables is 0.82. The result also show that there is a positive and significant relationship between the needs for LRT mode, travel and LRT operation characteristics on TOD area around LRT stations in Palembang.
3. Development of TOD which is suitable with the travel characteristic of community in Palembang City are the provision of transit facilities in the form of integrated stations and shelters, park and ride facilities, adequate facilities for pedestrians such as sidewalks and zebra cross, mixed land use facilities, health and entertainment facilities that are connected with adequate pedestrian facilities such as the sky walk to the nearest LRT station.

6. Acknowledgement

This article was presented at the 2nd International Conference on Smart City INNOVATION (ICSCI) 2019, jointly held by Universitas Indonesia and Universitas Diponegoro. ICSCI conferences have been supported by the United States Agency for International Development (USAID) through the Sustainable Higher Education Research Alliance (SHERA) Project for Universitas Indonesia's Scientific Modelling,

Application, Research, and Training for City-centered Innovation and Technology (SMART CITY) Center for Collaborative Research, administered through Grant #AID-497-A-1600004, Sub Grant #IIE-00000078-UI-1

7. Reference

[1] Prijo Utomo D 2013 Model Distribusi Perjalanan Penduduk untuk Penentuan Koridor "LRT" di Pulau Batam *J. Sains dan Teknol. Indones.* **12**
[2] Tamin O Z 2003 Perencanaan dan Pemodelan Transportasi: Contoh soal dan aplikasi *Inst. Teknol. Bandung* **463**
[3] Ismiyati I 2019 Secondary City and TOD Concept in the Light Rail Transit (LRT) Development to the Sustainable Transportation *IOP Conference Series: Earth and Environmental Science* vol 313p 12014
[4] Dwitasari R, Priyanto S and others 2017 Faktor-Faktor Yang Mempengaruhi Kualitas Pelayanan Light Rail Transit (LRT) Di Yogyakarta *J. Transp. Multimoda* **14** 169–76
[5] Joreskog K G and Sorbom D 1982 Recent Developments in Structural Equation Modeling *J. Mark. Res.* **19** 404
[6] Sparkman R M, Hair J F, Anderson R E, Tatham R L and Grablowsky B J 1979 Multivariate Data Analysis with Readings *J. Mark. Res.* **16** 437
[7] Latan H and Ramli N A 2013 The results of partial least squares-structural equation modelling analyses (PLS-SEM) *Available SSRN 2364191*
[8] Jöreskog K G and Sörbom D 1996 *LISREL 8: User's reference guide* (Scientific Software International)
[9] Gunarto M 2013 Membangun Model Persamaan SEM Dengan Program Lisrel
[10] Supriyatno D, Sulistio H, Djakfar L and Wicaksono A 2012 Permodelan Kebutuhan Transportasi Massal Guided Busway Menggunakan Structural Equation Modeling *J. Transp.* **12**

Smart Community in Public Toilet Management in Demaan Slum Settlement, Jepara Regency

S Sunarti[1], M Helmi[2], R Widjajanti[1],A A Purwanto[1]

[1] Department of Urban and Regional Planning, Faculty of Engineering, Diponegoro University, Semarang – Indonesia

[2] Department of Oceanografi, Faculty of Fisheries and Marine Sciences, Diponegoro University, Semarang – Indonesia

E-mail: sunarti@pwk.undip.ac.id

Abstract. One program in sanitation improvement is the construction of public toilets in coastal slums located in Demaan, Jepara Regency. The public toilets construction is a substitute for the helicopter latrines above the sea. However, most of public toilets still cannot be beneficial to the community, because there are many obstacles in its management. The purpose of this study is to examine the form of public toilet management in coastal slums with smart community approach in Demaan. The research approach uses qualitative methods. Data collection techniques are done by observation and in-depth interviews. Sampling technique using snowball method. The informants are the community who is user of public toilets, public toilet management institutions and government agencies related to the construction of public toilets. Data analysis using qualitative descriptive method and Engagement-Participation-Empowerment method. Based on the analysis result, community empowerment include establishing public toilet management institutions along with regulations and financial management, as well as developing economic values and the benefits of public toilets by adding other potential functions/activities accompanied by the use of information technology in the form of the internet. Smart communities for the development and management of public toilets can be achived if smart people have succeeded according to their ability, openness, thinking, flexibility and creativity in utilizing proper toilets.

Keywords: toilet, sanitation, smart community, slum upgrading, community empowerment

1. Introduction

The city is the focal point for economic growth, innovation, a place where most modern productive activities are concentrated in developing countries, a place where most paid employment opportunities are and is central to modern life [1]. There are many factors both driving and pulling factors that make people decide to immigrate to urban areas, such as the availability of high-quality jobs, encouragement from family and friends, the perception that rural areas do not offer the desired economic and social lifestyle and lack of access to supporting facilities such as shops, educational facilities, services [2]. The occurrence of rural-urban migration contributes to urban growth [3] and is the most important factor contributing to urbanization [1,4]. Over the past two centuries, the proportion of the world population living in cities and towns has increased from around 5% to more than 50% [5]. High levels of population growth and significant rural-urban migration contribute to the rapid and unplanned expansion of low-income settlements on the outskirts of large cities, which occur without the expansion of public services and facilities simultaneously [1]. Slum and squatter settlements are formed due to the inability of city governments to plan and provide affordable housing for low-income communities from the urban population [6].

Slums are a problem faced by most major cities in Indonesia and even large cities in other developing countries [7]. There are 5 slum typologies in Indonesia based on Minister of Public Works and Public Housing Regulation of the Republic of Indonesia No.14/PRT/M/2018 namely slum settlements on water, slums on the banks of water, slums in the lowlands, slums in hills and slums in disaster-prone areas. Talking about slums is closely related to sanitation aspects, as stated by UN-HABITAT [8], that lack of access to sanitation facilities is one of the characteristics in the definition of slums. One of the slums on the banks of water in Indonesia is a coastal slum area which also has a major problem in the form of improper sanitation conditions, namely the slum area of Demaan Village, Jepara Regency.

Many rejuvenation programs have been carried out by the local government in the slums of Demaan include the provision of proper sanitation facilities starting from 2012 to 2017 in the form of public toilets. Public toilets are a substitute for open latrines located above the sea or commonly called helicopter latrines. With the provision of sanitation facilities in the form of public toilets, the community has undergone changes from several aspects ranging from behavioral changes from open defecate to closed defecate and also changes in the management and maintenance of sanitation facilities. Now Demaan's slum area has 5 public toilets and there are no helicopter latrines. Although sanitation facilities have experienced improved quality but conditions in the field, public toilets in the research area have not been fully beneficial to the community because there are still many obstacles in the management and utilization. Constraints that occur include maintenance costs, lack of clean water, lack of users and community behavior in its utilization.

On the other hand, research on sanitation and slum areas that have been discovered by previous researchers is about a review of sanitation issues in terms of politics, the impact of slum and sanitation on health, alternative sanitation technologies for slums, the model of people's approach to sanitation and empowerment steps to provide and maintain public toilets [5], [9–12]. Chaplin [10] found that urban poor groups will begin to consider access to sanitation and political participation after they can meet their daily needs for food and shelter. While the sanitation aspect has an urgency to be dealt with immediately because as expressed by Ezeh [5] that poor sanitation in slum areas can increase the risk of spreading diseases such as diarrhea, typhoid, hookworm and cholera. In an effort to improve the quality of sanitation in slums, Katukiza et al. [11] provides alternative options for sustainable sanitation technology for slum settlements, one of which is the collection and maintenance of fecal sludge with options in the form of pit latrines, toilet biogas, dual flush compost toilets, community-based sanitation facilities, and peepoo bags. In addition, McCreary, et al [12] provides several options in planning the operational and financial sustainability of public toilets, one of which is the toilet design suggestion from Clara Greed [13] cited in [12] that is toilet design that can bring support partner for example co-locating public toilets with other public buildings such as firefighting stations or police stations and through sponsorship and advertising as is done on Automatic Public Toilets (APTs) built and managed by multinational advertising firms and utilizing them as advertising media [14] cited in [12]. Regarding community sanitation, Avvannavar and Mani [9] state that there are four subsystems in the system structure/model of community sanitation approaches, namely human settlements, natural environment, religion and culture and society.

From previous studies that have been conducted, there has been no research on the management of public toilets which are a substitute for helicopter latrines using a community empowerment approach. Based on the phenomenon of practical problems in the field and the absence of prior research on the management of public toilets by the community, then the purpose of this study is to examine the form of public toilet management in coastal slums with smart community approach in Demaan.

2. Literature Review

Smart city is a city that has a high quality of life, pursues sustainable economic development through investments in human and social capital, both traditional and modern communication infrastructure in the form of transportation and information communication technology, and managing natural resources through participatory policies [15] cited in [16]. Smart cities is closely related to building smart communities where in the process of social change must be accompanied by a sustainable process of social modernization [17]. Trencher [18] stated about the concept of smart city 2.0 where the concept of smart city uses a decentralized approach, centered on "people" where smart technology is used to address social problems, population needs and to encourage collaborative participation. The smart city 2.0 concept is the opposite of the techno-economic approach in the smart city 1.0 concept that focuses on the spread of smart technology for corporate and economic interests. If the previous researchers argued that the role of technology is the key to smart city, Holland [19] argues that to lead to a progressive smart city, it should start with people and human capital rather than blindly trusting IT can transform and improve cities themselves. Gil-Garcia et al [20] mention that there are 4 dimensions underlying the smart city component, namely the physical condition of the environment, society, government and technology and data. Society as one of the smart city dimensions has three components in it, namely: a) knowledge economy and pro-business environment, b) human capital and creativity and governance, engagement and collaboration. In the components of human capital and creativity have elements of creative class, social infrastructure, higher education and skilled and knowledge workforce. Whereas in the component of governance, engagement and collaboration contains elements of electronic governance, network, partnerships and collaboration as well as stakeholder, citizen and community engagement [20]. In line with Gill-Garcia et al [20], Nam & Pardo [21] also revealed that the main driver of smart cities is creativity, so that people, education, learning and knowledge have a major importance for smart cities.

Smart communities are communities that make decisions consciously to aggressively use technology as a catalyst to solve their social and business needs [22] cited in [16]. The internet is the most dominant tool or media in the new media era that can be used by anyone and for any interests such as politics, business (online business and marketing), education and socio-culture [23]. In addition, the internet also contains unlimited and actual sources of information that can be accessed quickly, including exchange of information or question and answer with an expert [24]. A successful smart community is the result of a coalition of business, education, government and individual citizens. Successful smart communities can be built from the top down or bottom up, but active involvement from each sector of society is very important [25]. Smart city communities need to feel the desire to participate and promote growth (smart) [16]. Walgito [26] stated that behaviors are formed in three ways, namely habits, understanding and use of models or role models. One approach that can be taken to increase community involvement is through community empowerment. Community empowerment is a method for strengthening communities that starts with the idea that community capacity cannot be built with social engineering, but communities can be stimulated to develop themselves [27]. In building community empowerment, Bartle [27] mentions eight important principles, namely (a). The balance of power, must desire the community to become more independent and willing to make efforts and sacrifices to be like that; (b). Experienced and/or trained agents must be available to intervene to stimulate and guide the community to organize and taking action and to become more independent; (c) Assistance offered is assistance that leads to partnerships, assistance and training that promote independence and increased capacity; (d) Recipient organizations or communities may not be controlled or forced to change, but facilitators must intervene with stimulation, information and guidance; (e). Organisms become stronger by exercising, struggling, and facing difficulties. We do not promote pain, but promote struggle and effort; (f) Direct participation, especially in decision making, by recipients, is very important to increase their capacity; (g) Most of the resources needed for community projects in the form of actions must be provided by members of the community themselves; (h) It is necessary to direct the participants from the beginning to take full control, exercising full decision making, and accept full responsibility for actions that will lead to increasing their strength

3. Method

The research approach uses qualitative methods, using theory and previous research as reference to explore data in the field [28]. The data used are primary data and secondary data. Primary data is obtained through observation and in-depth interviews, while secondary data is obtained through program planning documents both from related agencies and from KOTAKU. In-depth interview using a snowball sampling technique. The informant search begins with visiting the key person, namely the Demaan Urban Village Head, then the key person directs the researcher to other informants that are suitable for the research objectives. With the snowball sampling technique, the informants obtained namely the community who is a user of public toilets, public toilet management institutions and government agencies related to the construction of public toilets. Data analysis using qualitative descriptive method and EPE (Engagement-Participation-Empowerment) method. EPE analysis is used

IOP Conf. Series: Earth and Environmental Science **396** (2019) 012023 doi:10.1088/1755-1315/396/1/012023

to analyze the community empowerment process from three empowerment processes, namely engagement, participation and empowerment with exogenous and endogenous aspects [29] to establish smart communities in the management of public toilets in Demaan Village.

4. Result and Discussion

4.1. General Description of Study Area

Demaan Urban Village is located in Jepara District, Jepara Regency. This Urban Village is one of the Urban Village that has the type of topography as a coastal village in the District of Jepara. Demaan Urban Village has an area of 59.914 hectares. Demaan consists of 7 RWs which consist of 22 RTs with diverse physical conditions from arranged settlements to settlements with slum conditions (Figure 1). Not all RWs in Demaan are coastal areas, only two RWs that are directly adjacent to the Java Sea namely RW 04 and RW 05. Settlements located in this coastal area have slum conditions, including residential areas RT 01-03 RW 04 and RT 03/RW 03 which is the research location which included in slum settlements in the Bupati's Decree of Jepara Regency. This slum area are the target in the implementation of a program to improve the quality of slums from the government. The following is a map of the research location:

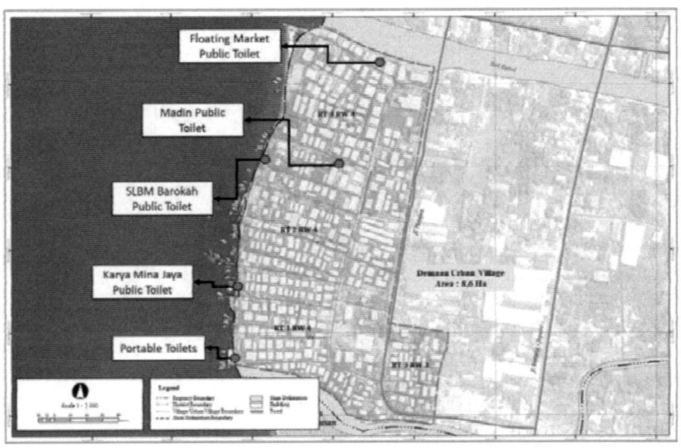

Figure 1. Map of Research Location

As a general description of the study area, in RT 01-03 RW 04 and RT 03/RW 03 Demaan urban village in 2017 had a population of 1.676 people with a density of 96 people/ha and the number of family heads was 428 households [30]. The employment of most of the population in the research location is in the fisheries sector and the trade/service sector. Before receive a slum settlement improvement program, RT 01-03 RW 04 and RT 03/RW 03 had a variety of slum problems, one of which was the lack of proper sanitation facilities. One of the programs provided by the government starting from 2011 to 2017 is the construction of 5 public toilets to replace 10 helicopter latrines scattered along the coast of the research location. But in the utilization and management of public toilets it has not been optimally carried out by the community.

4.2. General Description of Public Toilet in Demaan Urban Village

The assistance program from the government to improve sanitation quality in the form of public toilets has been started since 2011. First public toilet is in RT 03 RW 04 named SLBM Barokah build in 2011. Next year, in 2012 portable toilet in RT 01 RW 04 and Floating Market public toilet in RT 03 RW 04 were built. The assistance program peaked in 2017 which was built 2 public toilets namely Karya Mina Jaya public toilet RT 01 RW 04, and Madin public toilet in RT 03 RW 04. So, the total number of public toilets as a result of government programs at the research location is 5. The overview of each public toilet in the research area can be seen in Table 1.

Table 1. General Overview of Public Toilets in Research Location

No.	Name	General Condition	Utilization	Management	Problems	Figure
1.	SLBLM Barokah Public Toilets	Built in 2011 with assistance from the Public Works Agency: - Number of cubicle : 8 - Area per cubicle : 3 m^2 (2m x 1,5m) - Toilet type : ➢ Squat toilet = 6 units ➢ Sitting toilet = 2 units - Facilities and infrastructure : ➢ clean water sourced from the wellbore, ➢ electric tokens, ➢ washing area 16 m2 ➢ a toilet guard place consisting of a table and chairs Physical condition : the interior of this SLBM toilet is clean and well maintained but on the outside and side it can be seen that this SLBM toilet has less accessibility	- User Origin: the average user comes from RT 02 and RT 03 RW 04 - The number of average users per day: 16 people - Form of community activities: urinating, defecating and washing clothes	- Manager: Community groups with 3 members and acting as chief, treasurer and guard/cleaning staff. - Toilet utilization fees: ➢ 1.000 IDR per day for take a bath, ➢ 500 IDR per day for urinate or defecate. - Income : 240.000/month - Maintenance : repairs to well pump machines, draining septic tanks, buying electricity tokens and paying wages to toilet guards	- When seawater is high and the waves are big, sea water gets into the public toilet area and can damage the water pump - lack of users - lack of road access to a public toilet that is only available in the south	
2.	Portable Public Toilet	Built in 2012 with PLPBK funding from the Public Works Agency amounting to 132.629.000 IDR and then self-financing funds of 4.116.000 IDR: - Number of cubicle : 3 - Area per cubicle : 3 m^2 - Toilet type : squat toilet. - Facilities and infrastructure : ➢ clean water sourced from the wellbore, ➢ electric tokens, ➢ washing area	- User Origin: the average user comes from RT 01 RW 04 and the general public - Number of average users per day: 6 people - Form of community activities: urinate/defecate,	- Manager : managed by two people to maintain the cleanliness of the toilet.. - Toilet utilization fees : ➢ 1.000 IDR per day for take a bath, ➢ 500 IDR per day for urinate of defecate. - Maintenance : clean the toilet, improve infiltration and drain septic tanks	- the existence of RT tents can hamper cleanliness - lack of users, - lack of road access from the seafront to portable toilets - there is a new toilet building, the Karya Mina Jaya toilet, so that the number of users becomes divided	

No.	Name	General Condition	Utilization	Management	Problems	Figure
		➢ a toilet guard place Physical condition; This portable toilet is poorly maintained where on the inside there is moss on the floor and the lower door and visible dirt and dust on the wall and portable toilet plastic door.	bathing and washing clothes			
3.	Karya Mina Jaya (KMJ) Public Toilet	Built in 2017 with assistance from the Housing and Settlement Agency: - Number of cubicle : 6 - Area per cubicle : 3 m^2 (2m x 1,5m) - Type of toilet : squat and sitting toilet - Facilities and infrastructure : ➢ clean water sourced from the wellbore, ➢ electric tokens, Physical condition; this toilet looks clean from the outside but the inside is quite dirty with mossy and crusty floors and the floor in the hallway is stained with rust from the septic tank lid.	- User Origin: the average user comes from Pesajen - The number of average users per day: 30 people - Form of community activity : urinating/defecating	- Manager : This toilet does not have an institution or management group and is only managed by one guard. - Toilet Utilization Fees : ➢ 500 IDR per day for urinate/defecate. - Income : 20.000 IDR per day - Maintenance : The cash collected will then be used for toilet maintenance such as buying 50,000 IDR electricity tokens and paying the wages of toilet guards of 300,000 IDR per month	- Salted well water - Poor sewage	
4.	Madin Public	built in 2017 with assistance from the Housing and Settlement Agency: - Number of cubicle : 8 - Area per cubicle : 3 m^2 (2m x 1,5m) - Type of toilet : ➢ Squat toilet : 6 unit ➢ Sitting toilet : 2 unit - Facilities and Infrastructure :	- User Origin: the average user comes from RT 03 RW 04 and fishermen including school children because they are located in one area with the	- Manager : management consisting of four people to manage and maintain toilets. - Toilet Utilization Fees: ➢ 2000 IDR per day for washing ➢ 1000 IDR per day for bathing	- Aluminum doors are often damaged due to ignorance of the user when opening the door lock, especially some users are school children so the door is forcibly opened and causes damage - lack of clean water - lack of users of public toilet	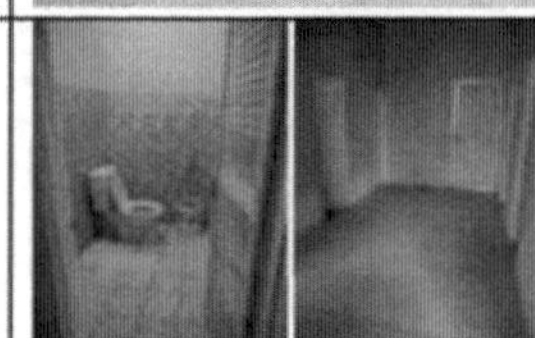

IOP Conf. Series: Earth and Environmental Science **396** (2019) 012023 doi:10.1088/1755-1315/396/1/012023

No.	Name	General Condition	Utilization	Management	Problems	Figure
		➢ Clean water sourced from dug well, ➢ Electricity token, ➢ Washing area 2 units ➢ A toilet guard place Physical condition : The condition in this toilet cubicle is quite clean, but the washing place looks dusty and mossy	Madrasah Diniyah (primary school) - The number of average users per day : 10 people - Form of community activities : bathing, washing clothes, urinating/defecating and washing motorbikes	➢ 500 IDR per day for urinaring/defecating. - Income : 7.000 IDR per day - Maintenance : buying electricity tokens, replacing broken lights and cleaning toilets with a total expenditure of 90,000 IDR per month		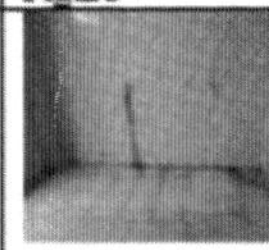 
5.	Floating Market Public Toilet	was built in 2012 from a government program called Noto Deso (ND) with a fund of 25,000,000 IDR and then added with self-help funds of 2,256,500 IDR: - Number of cubicle : 3 - Area per cubicle : 3,75 m^2 (2,5 m x 1,5 m) - Type of toilet : ➢ Squat toilet : 3 units - Facilities and infrastructures : ➢ lean water sourced from wells and regional drinking water companies, ➢ electricity token, Physical condition : the physical condition of the toilet is poorly maintained, dirty and a lot of moss sticks to both the door and on the walls and floors of the toilet	- User Origin: the average user comes from RT 03 RW 04 and traders and floating market visitors - The number of average users per day: 30 people - Form of community activities: bathing, washing, and urinating/defecating	- Manager : does not have an institution/management group, but is only managed by a pair of husband and wife appointed by the residents. - Toilet utilization fees : ➢ 2000 per day for bathing ➢ 1000 per day for urinating, defecating or washing. - Income : 10.000-20.000 IDR per day, or 200.000 – 300.000 IDR per month - Maintenance : pay electricity 130,000 IDR and the remain fund is used for RT and wages for managers including cleanliness	- Water tower is not big enough so that clean water often runs out - septic tanks are full fast - low awareness of users in paying toilet fees. Due to the limitations of the management personnel, this floating market toilet is not guarded every day, this is what is then used by users to use public toilets with fees not according to the agreement and even there are those who do not pay. This is compounded by the presence of people who pry up the contribution box and steal money in it when the toilet is not guarded. These conditions then have an impact on the decline in the amount of cash in public toilets including maintenance costs	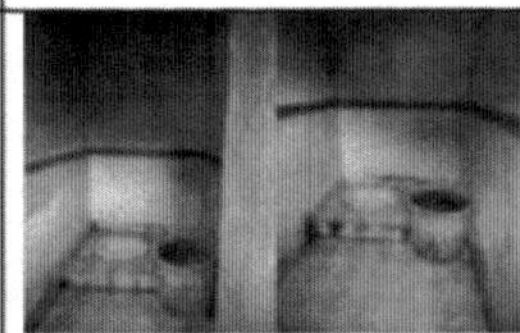 

Sources: Field Observation and Interview with Informant, 2019

Based on the results of studies in the field, it was also found that in the management and maintenance of toilets, public toilet managers were dependent on the fees paid by users when using the toilet. Whereas if public toilets suffer more serious damage and the repair fees are not sufficiently fulfilled from the toilet utilization fees, the manager/administrator is still dependent on assistance from the government. This happens because it is difficult to invite both users and community to collect money outside the fees for using the toilet.

From the general description of the toilet, it can be concluded that there are four main problems faced by public toilets in the study locations, namely (1) the absence of management institutions for several toilets such as floating market toilet and KMJ toilet, (2) minimal number of users such as Madin toilet, SLBM Barokah Toilet, and Portable toilet, so that the fund for maintenance and management is very limited, (3) the limitations and damage to facilities and infrastructure that occur in almost all public toilets, and (4) low level of community empowerment in the maintenance and management of public toilets.

4.3. Analysis of Use and Management of Public Toilets

The construction of public toilets in the slum area of Demaan in five locations is an effort to improve the quality of sanitation carried out by the government to replace helicopter latrines located above the sea. With the dismantling of helicopter latrines on the sea, people must switch to using public toilets as a result of government programs. Activities commonly carried out by the community in public toilets are not only defecating or urinating, but there are other activities such as bathing, washing clothes and washing motorized vehicles (motorbikes or cars). However, after several years of using public toilets, different problems emerged for each of the public toilets in the Demaan slum area.

The problem of public toilets in the form of the lack of users, so that there is an imbalance between income and expenditure and has an impact on the maintenance and management of public toilets. In the beginning the community was difficult to accept the use of public toilets. In the transition from utilization of helicopter latrines to public toilets this requires adaptation from the community because the utilization of these two types of sanitation facilities is far different. One example is the helicopter latrine has an open structure so that people can interact with other users while defecating/urinating, while public toilets on land have closed structures and more soundproof so that people cannot interact with other users. This was also a cause of the unwillingness of the community to use helicopter latrines at first.

But over time, people's behavior began to change and gradually began to accept the use of public toilets on land. According to Walgito [26], behaviors are formed in three ways, namely habits, understanding and use of models or role models. In this case, there are two ways that shape people's behavior, habits and understanding. People who were used to using helicopter latrines were then forced to switch to using public toilets on land because helicopter latrines were dismantled by the government. The use of public toilets which begins due to forced slowly then becomes a habit that forms new behavior for the community. Then when the community has felt the benefits and is accustomed to using public toilets on land, it raises their understanding and awareness within the community of the importance and benefits of proper and healthy sanitation facilities. With this understanding, coupled with an increase in ethics, knowledge, health, and comfort for the community, they began to think of greater convenience and benefits for themselves and finally decided to build a private toilet at home. Along with the increasing ownership of private toilets, the number of public toilet users has also diminished. Changes in the way of thinking of the community in the use of toilets and followed by actions in the construction of private toilets, this shows an increase in community capacity. Community needs in utilizing healthier, more efficient and creative toilets is an action taken by humans as individuals privately by changing the way of thinking to become smart way of thinking (smart people). The process of change in the utilization and management of public toilets is as expressed by the informant, namely the Chief of RW 04, Demaan Urban Village, as follows:

> *"From the helicopter latrine, the community moved to the public toilets that had been provided, a change that now the community has been able to build private toilets in their homes." (I/F/Chief of RW 04).*

The process of behavior change and community mindset also occurs in RT 02 RW 05. RT 02 RW 05 is an area outside the research that has successfully manage the public toilets and is still active today. This is in accordance with what was revealed by the public toilet manager in RT 02 Demaan Village which can be said to be successful, namely:

"Initially it was difficult to make people aware, little by little socialization was carried out. In the beginning even though all the helicopter latrines had been dismantled, there were still those who returned defecate in the sea, not in the helicopter latrines but on the outskirts of the sea. Eventually there was a sense of shame because of being seen by other people, so that eventually they wanted to use existing public toilets. .... At the beginning of the construction of public toilets, almost all residents were users, after time, now they have made their own toilets. ..... People are aware, that in the helicopter there are many obstacles, during the rainy season and the wave season. Adjustment of the community to be fully aware, reaches 1 year. It is not easy to change people's habits, there are many obstacles. After using public toilets, for those who experience economic improvement, start building private toilets" (I/MA/Manager of Communal Toilets in RT 02 RW 05)

Regarding smart city, Holland [19] argues that to lead to a progressive smart city, it should start with people and the human capital side rather than blindly believing in IT can itself transform and improve the city. In accordance with the literature, it was found that changes and developments in the behavior of people from open defecation above the sea, then forced to move to public toilets to land until finally changes in habits and community mindset about healthy living habits and fulfill personal sanitation needs could be a prefix good to lead to smart city through smart people.

The lack of public toilet users because there are many people who have private toilets shows that the government's goal in promoting improved quality of sanitation with stimulation of assistance in the form of public toilets has paid off. So that from the problem of the minimum number of users and the imbalance of operational funds, a step is needed to increase the economic value and value of benefits from the public toilet. McCreary, et al. [12] provides several options in planning the operational and financial sustainability of public toilets, one of which is the toilet design suggestion from Clara Greed [13] cited in [12] that is toilet design that can bring support partner such as co-locating public toilets with other public buildings such as fire stations or police stations. In addition, there are also options through sponsorship and advertising as done on Automatic Public Toilets (APTs) that are built and managed by multinational advertising firms and use them as advertising media [14] cited in [12]. The literature shows that public toilets can be integrated with other functions to support each other and can help in their operational and financial sustainability. Therefore, one of the things that can be done to add economic value and the value of the benefits of public toilets in the slums of Demaan is to add additional functions according to the needs of the community and the character of each toilet. Figure 2 is an analysis map of handling public toilet problems in Demaan slum.

:

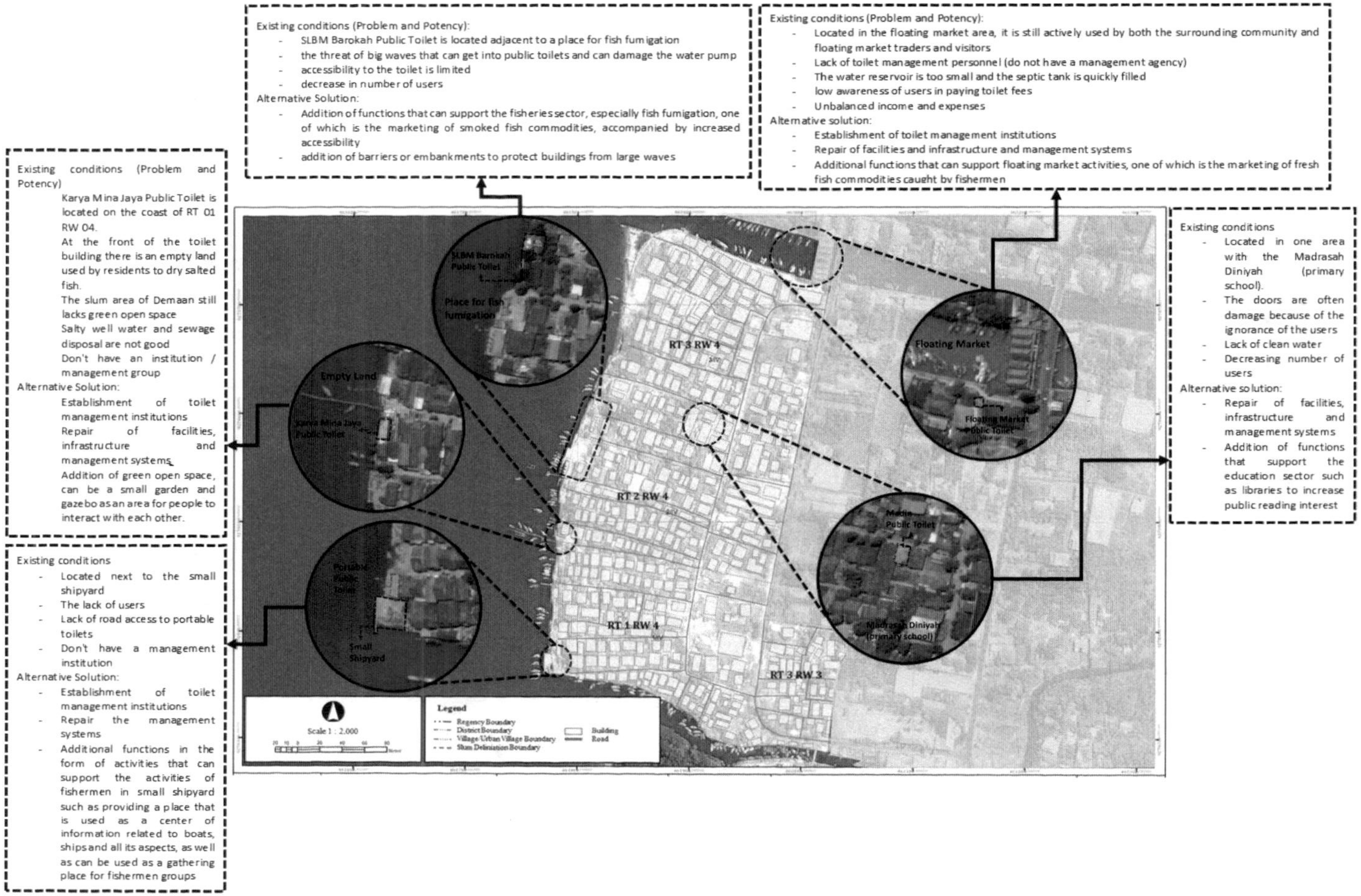

Sources: Researchers Analysis, 2019

Figure 2. Map of Public Toilet Problems Analysis in the Research Location

Public toilets that are spread in research location have different characteristics. Potential and problems in each public toilet can be developed both for the utilization and development of supporting activities. The first toilet, Portable Public Toilet, has a location adjacent to the small shipyard so that a potential additional function is activities that can support the activities of fishermen in boat repairs, such as a place that can be used for information centers about boats, ships and its aspects, as well as can be used as a gathering place or hold meetings for fishermen groups. The second toilet, the Karya Mina Jaya Public Toilet, has the potential for a large amount of vacant land in front of the toilet building. In 2010-2015, the regional government prepared a plan for the arrangement of the settlement environment of the Demaan urban village with one of its activities being to build a pavilion/gazebo in RW 04, where the location was planned to be located near the location of the KMJ Toilet. Based on the potential and plans previously initiated by the regional government, the potential additional function for KMJ toilets is the addition of green open spaces, in the form of small parks and gazebos as places where people interact with each other. The third toilet, the SLBM Barokah Public Toilet, has the potential in the form of a location adjacent to the fish fumigation place. So that a potential additional function is an activity that can support the activities of a fish smokers group, one of which is the marketing place for the smoked fish. The fourth toilet is the Floating Market Public Toilet, which has the potential to be located near the Floating Market. So that a potential additional function is activities that can support trade activities in the Floating Market, one of which is a place for marketing fresh fish caught by fishermen. The last toilet, Madin Public Toilet, which has a location located in the Madrasah Diniyah (primary school) area, has the potential to be developed with additional functions related to education such as library to increase reading interest in children in Demaan slum. Addition of additional functions to increase the economic value and benefit of each public toilet can be done by increasing the number of floors or expanding public toilet buildings.

Turning from the problem of reduced number of toilet users, public toilets in the slums of Demaan still have other problems, namely the absence of management institutions, the low level of community empowerment in their maintenance and management, and the limitations and damage to facilities and infrastructure. These problems can be solved by establishing a public toilet management institution that is accompanied by improvements in facilities, infrastructure and improvements to the existing management system. To solve the four main problems in public toilets in the slums of Demaan by adding other potential functions, establishing management institutions, improving facilities and infrastructure and improving the management system, can be done through community empowerment. In addition, through community empowerment it is hoped that it can encourage community participation in becoming a smart community not only in the management of public toilets but also in the maintenance of their settlement environment. This is in accordance with the opinion of Bartle [27] that one of the approaches that can be taken to increase community involvement is through community empowerment. This empowerment approach is supported by the potential of smart people owned by the community of Demaan Slum area as capital to carry out activities that are managed together.

4.4. Analysis of Empowermrnt Process for Public Toilets Management and Maintenance

Community empowerment is a method for strengthening communities where people are stimulated to develop themselves [27]. Whereas Lindskog [25] revealed that smart communities can be built from top down and bottom up steps but active involvement from every sector of society is very important. So that community empowerment is chosen to be one of the efforts that can be used to improve management, increase economic value and benefit from public toilets in the Demaan slums. Steiner and Farmer [29] states that community empowerment consists of three processes, namely engagement carried out by exogenous aspect, participation carried out jointly between endogenous and exogenous and empowerment which has been fully implemented by endogenous aspects. In the case of community empowerment in the management of public toilets and its additional functions/activities to create smart communities in Demaan slums, participation from both inside (endogenous) and outside (exogenous) is needed or in accordance with Lindskog [25], which requires active involvement from every sector of society. Exogenous aspects in this case are facilitators from local government and community self-reliance institution. Whereas endogenous aspects are capabilities owned by the community internally in the Demaan slums.

Community empowerment begins with an engagement process as expressed by Steiner and Farmer [29] This engagement process is a process carried out by exogenous aspects, namely facilitators from the local government and community self-reliance institution to invite the community, both the community as a user of public toilet and non-user, to disseminate the information about management and maintenance of the public toilets (Figure 3). After that, move to the participation process, in this process there is a transfer of power from the facilitator to the community. In this process the community is invited and given the power to identify the problems faced in the use and maintenance of public toilets and identify what they need for the management and development of public toilets. Then, with assistance from the facilitator, the community form public toilets management institution through member's recruitment. The selection of members of the management includes the entire community, both citizens in general, community leaders and youth. The next step is empowerment, Bartle [27] stated that one of the principles of empowerment is direct participation in decision making, in this case the community is given the power to make regulations on the use of public toilets, carry out development activities and utilization of public toilets and its supporting functions/activities and improve financial management. In this empowerment process, the community is not directly left independently but still needs assistance from the facilitator. Regarding the development and utilization of toilets with its supporting functions/activities, it is possible for the community to still need assistance from the government before they can stand independently. In accordance with the principle of empowerment mentioned by Bartle [27], the assistance offered is assistance that leads to partnerships, therefore local governments can provide assistance but only as a stimulant for the development of public toilets and its supporting functions/activities, and the rest of the community must seek their own resources. This is done to create community independence and increase their capacity. The following is a chart of community empowerment processes in the management and maintenance of public toilets and its supporting functions/activities:

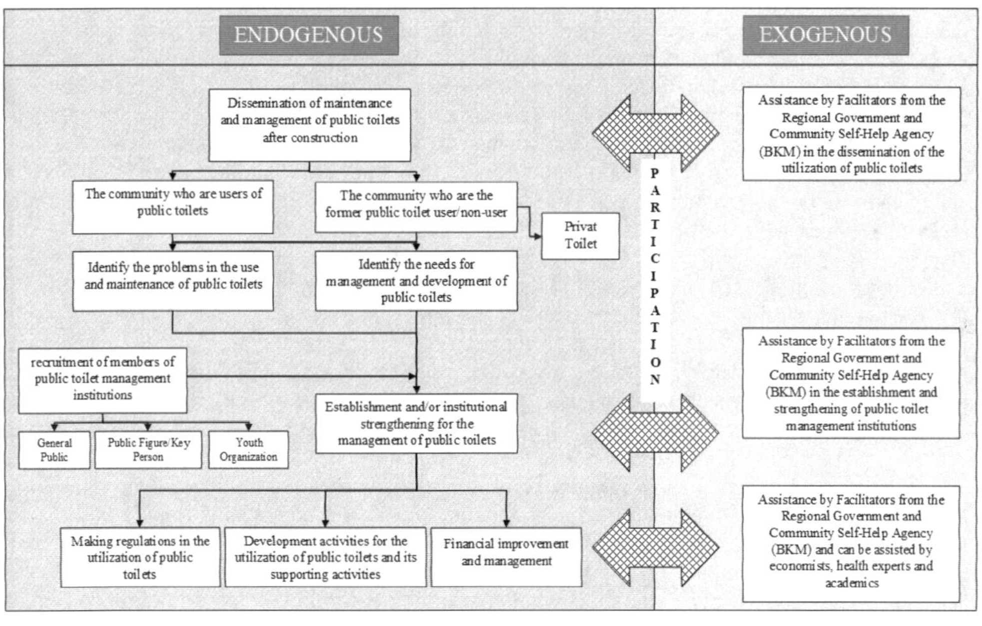

Sources: Researchers Analysis, 2019

Figure 3. Chart of Community Empowerment Flow in the Use and Management of Public Toilets

4.5. Smart People and Smart Community Form Analysis in Public Toilet Management

The process of changing behavior and mind set of the community in the use of helicopter latrines to public toilets on land, requires adjustment with a considerable amount of time and requires knowledge. So that the community can finally receive the public toilet and even increase their quality life to be more feasible and healthier. Holland [19] states that to lead to smart city starts with people and human capital. If seen from the conditions in the study area with the community change process, then the Demaan slum community is included in smart people with increased public awareness and understanding of the health and maintenance of the environment. However smart cities is closely related to building smart communities where in the process of social change must be accompanied by a sustainable social modernization process as expressed by Cifaldi and Serban [17]. Whereas in the conditions in the research area, the community has not led to smart communities with social modernization, public toilet utilization by users, management, maintenance and supervision by managers is still done conventionally. Therefore, after the behavior change and mindset of being smarter in the sanitation aspect, the use of information and communication technology is needed to create a smart society so that it can help increase economic value and the benefits of public toilets with its supporting function/activities. Information technology that can be applied in the utilization of public toilets and its supporting functions/activities is the use of the internet that is applied in accordance with the additional functions provided for each toilet.

Trencher [18] revealed that the concept of smart city 2.0 is centered on people where smart technology is used to overcome social problems, population needs and to encourage stakeholder collaboration. Regarding information technology, the internet is the most dominant tool or media in the new media era that can be used by anyone and for any interests such as politics, business (online business and marketing), education and socio-culture [23]. In accordance with the concept of Smart City 2.0 and the benefits of the internet as the new media, information technology in the form of the internet can be used to help meet the needs of the community, one of which is to increase economic value and the value of the benefits of existing public toilets. For portable public toilet with additional functions as information centers for fishermen, the use of the internet can help fishermen to gain information about boats and ships and information to develop the fisheries sector. Likewise for the SLBM Barokah public toilet with additional potential functions in the form of marketing places for smoked fish commodities. In this case the internet can be used to help with marketing activities both to find information on smoked fish market coverage and to market smoked fish through the website. Similar to the use of the internet for marketing smoked fish, the internet can also be used in the additional function namely places for marketing fresh fish in Floating Market Public Toilet. As for the additional function in the form of library in Madin public toilet, the internet can be used to increase knowledge for the community in addition to the books provided. The use of the internet to increase economic value and the value of benefits from public toilets can be done through intermediaries in the form of computer devices. In addition, for KMJ public toilet which have the potential for additional functions in the form of green open space, the internet can be provided in the form of WiFii network so that the community can learn and obtain various kinds of information that can be accessed through laptop, computer or smartphone. As stated by Setiyani [24] that through the internet, unlimited and actual sources of information can be accessed quickly including exchange of information or question and answer with an expert. From this benefit, in this case the internet is not only useful for public toilet's supporting function/activities but can also be used to help facilitate the management of public toilets, one of which is to buy electricity token that can be done instantly via an internet-connected smartphone, besides that it can help managers to gain information about management and maintenance of public toilets. To achieve maximum utilization of the internet, then in the process of community empowerment in the recruitment of the management institution members, are not only recruiting the community in general and community leaders but also the participation of the youth. The youth tend to be familiar with the utilization of the internet in their daily activities. Socialization and assistance in IT use will be given to young generation groups such as Karang Taruna or groups that have the desire and ability to increase their capacity to use technology so that they can help other communities in learning to use the internet for toilet management and its supporting function/activities.

The use of information technology in the form of internet is intended to encourage community creativity. The internet is expected to be a portal for people to get to know the wider world and get a lot of information that is useful for developing creativity. With the increase in community creativity, it is hoped that ideas, innovations and activities can have benefits and can improve the quality of life and the environment in which they live. As expressed by Nam & Pardo [21] that the main driver of smart cities is creativity, where people, education, learning and knowledge are of primary importance to smart cities. The following is a chart of the forms of smart people towards smart communities in the management of public toilets (Figure 4) and its supporting functions/activities as a first step towards smart city:

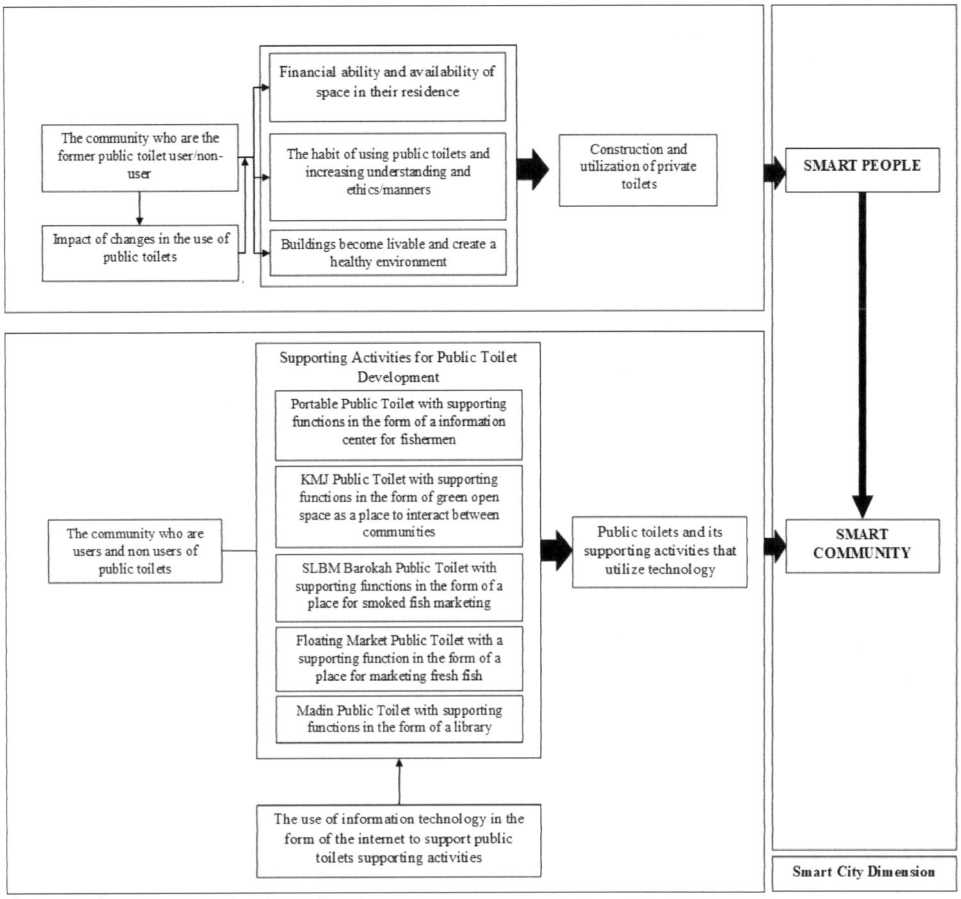

Sources: Researchers Analysis, 2019

Figure 4. Chart of the Forms of Smart People towards Smart Communities in the Management of Public Toilets

5. Conclusion

The form of public toilet management in Demaan Urban Village requires smart management with smart community. Smart cities is closely related to building smart communities where in the process of social change must be accompanied by a sustainable process of social modernization [17]. Holland [19] argues that to lead to a progressive smart city, it should start with people and human capital rather than blindly trusting IT can transform and improve cities themselves. A successful smart community is the result of a coalition of business, education, government and individual citizens. Successful smart communities can be built from the top down or bottom up, but active involvement from each sector of society is very important [25]. One approach that can be taken to increase community involvement is through community empowerment. Steiner and Farmer [29] states that community empowerment consists of three processes, namely engagement carried out by exogenous aspect, participation carried out jointly between endogenous and exogenous and empowerment which has been fully implemented by endogenous aspects. In addition, Bartle [27] stated that one of the principles of empowerment is direct participation in decision making. Based on those literatures, the form of management and maintenance of public toilets in the slums of Demaan village can be done by community empowerment using the EPE method, namely by optimizing endogenous aspects (from the community, toilet managers, chief of RT/RW) which are supported by exogenous aspects namely local government or related agencies and Community self-reliance agency. The results of community empowerment include establishing public toilet management institutions along with regulations and financial management, as well as developing economic values and the benefits of public toilets by adding other potential functions accompanied by the use of information technology in the form of the internet. To increase economic value and the value of benefits from public toilets is to add additional functions according to the character of each toilet and community needs. Additional function that can be added in public toilets in Demaan including marketing activities of smoked fish and fresh fish for fishermen, information center for the fishermen, educational activities namely by providing libraries and also green open spaces

In improving management and utilization of public toilets and its supporting functions/activities, it cannot work if the community has not yet become a smart people who understands the importance of sanitation for their health and environment. After individuals have an awareness and understanding the importance of proper and healthy sanitation, smart communities can occur through efforts namely the community empowerment and the utilization of technology. Novelty from this study is that for the utilization and management of public toilets preceded by the formation of smart people first, after that smart communities can be achieved. This is because the use of public toilets relates to the interests of each individual so that it starts from an individual awareness to increase to the willingness to sacrifice in forming a community in developing public toilets.

6. Acknowlegment

This article was presented at the 2nd International Conference on Smart City INNOVATION (ICSCI) 2019, jointly held by Universitas Indonesia and Universitas Diponegoro. ICSCI conferences have been supported by the United States Agency for International Development (USAID) through the Sustainable Higher Education Research Alliance (SHERA) Project for Universitas Indonesia's Scientific Modelling, Application, Research, and Training for City-centered Innovation and Technology (SMART CITY) Center for Collaborative Research, administered through Grant #AID-497-A-1600004, Sub Grant #IIE-00000078-UI-1

7. References

[1] Cohen B 2006 Urbanization in developing countries: Current trends, future projections, and key challenges for sustainability *Technol. Soc.* **28** 63–80

[2] Crow H 2010 *Factors influencing rural migration decisions in Scotland: an analysis of the evidence* (Scottish Government)

[3] Brøgger D 2019 Urban diaspora space: Rural-Urban Migration and the Production of Unequal Urban Spaces *Geoforum* **102** 97–105

[4] WANG S X and Benjamin F U Y 2019 Labor mobility barriers and rural-urban migration in transitional China *China Econ. Rev.* **53** 211–24

[5] Ezeh A, Oyebode O, Satterthwaite D, Chen Y-F, Ndugwa R, Sartori J, Mberu B, Melendez-Torres G J, Haregu T, Watson S I, Caiaffa W, Capon A and Lilford R J 2017 The history, geography, and sociology of slums and the health problems of people who live in slums *Lancet* **389** 547–58

[6] Ooi G L and Phua K H 2007 Urbanization and Slum Formation *J. Urban Heal.* **84** 27–34

[7] Santi, Bachrun R and Ornam K 2017 Typology of Slum Management in Coastal Settlement as a Reference of Neighborhood Planning in Konawe *J. Phys. Conf. Ser.* **846** 12018

[8] UN-Habitat 2006 State of the World's Cities 2006/7 *New York United Nations*

[9] Avvannavar S M and Mani M 2008 Retracted: A conceptual model of people's approach to sanitation *Sci. Total Environ.* **390** 1–12

[10] Chaplin S E 1999 Cities, sewers and poverty: India's politics of sanitation *Environ. Urban.* **11** 145–58

[11] Katukiza A Y, Ronteltap M, Niwagaba C B, Foppen J W A, Kansiime F and Lens P N L 2012

Sustainable sanitation technology options for urban slums *Biotechnol. Adv.* **30** 964–78
[12] McCreary C, Brown A, Holiman J and Sharma P 2016 Public Toilet Advocacy Toolkit *PHLUSH (Public Hyg. Lets Us Stay Human)*
[13] Greed C 2010 *Creating a Nonsexist Restroom" in Moloch and Noren, Toilet* (New York: New York University Press)
[14] Matier P and Ross A 2015 City's latrine team tries to keep public toilets tolerable *San Fr. Chron.*
[15] Thuzar M 2011 Urbanization in Southeast Asia: Developing Smart Cities for the Future? *Regional Outlook* ed M J Montesano and P O Lee ({ISEAS}{\textendash}Yusof Ishak Institute Singapore) pp 96–100
[16] Albino V, Berardi U and Dangelico R M 2015 Smart Cities: Definitions, Dimensions, Performance, and Initiatives *J. Urban Technol.* **22** 3–21
[17] Cifaldi G and Serban I 2017 Between a Smart City and Smart Society *Intelligent Human Systems Integration* (Springer International Publishing) pp 714–9
[18] Trencher G 2019 Towards the smart city 2.0: Empirical evidence of using smartness as a tool for tackling social challenges *Technol. Forecast. Soc. Change* **142** 117–28
[19] Hollands R G 2008 Will the real smart city please stand up? *City* **12** 303–20
[20] Gil-Garcia J R, Pardo T A and Nam T 2015 What makes a city smart? Identifying core components and proposing an integrative and comprehensive conceptualization *Inf. Polity* **20** 61–87
[21] Nam T and Pardo T A 2011 Conceptualizing smart city with dimensions of technology, people, and institutions *Proceedings of the 12th Annual International Digital Government Research Conference on Digital Government Innovation in Challenging Times - dg.o 11* (ACM Press)
[22] Eger J M 2009 Smart growth, smart cities, and the crisis at the pump a worldwide phenomenon *I-WAYS-The J. E-Government Policy Regul.* **32** 47–53
[23] Situmorang J R 2012 Pemanfaatan internet sebagai new media dalam bidang politik, bisnis, pendidikan dan sosial budaya *J. Adm. Bisnis* **8**
[24] Setiyani R 2010 Pemanfaatan Internet Sebagai Sumber Belajar *Din. Pendidik.* **5**
[25] Lindskog H 2004 Smart communities initiatives *Proceedings of the 3rd ISOneWorld Conference* vol 16pp 14–6
[26] Walgito B 2004 Pengantar Psikologi Umum Edisi IV *Yogyakarta ANDI Yogyakarta*
[27] Bartle P 2008 THE Human Factor and Community Empowerment. *Rev. Hum. factor Stud.* **14**
[28] Hollweck T 2016 Robert K. Yin. (2014). Case Study Research Design and Methods (5th ed.). Thousand Oaks, CA: Sage. 282 pages. *Can. J. Progr. Eval.*
[29] Steiner A A and Farmer J 2017 Engage, participate, empower: Modelling power transfer in disadvantaged rural communities *Environ. Plan. C Polit. Sp.* **36** 118–38
[30] KOTAKU 2017 *Prosedur Operasional Baku (POB): Penyusunan Baseline Data 100-0-100* (Jepara)

Smart Kampung: Characterisation of Surabaya Urban Coastal Settlements through Smart City Measurement

K P Nariratih[1], D Rahmawati[2]
[1]Department of Urban and Regional Planning, Institut Teknologi Sepuluh Nopember (ITS)
[2]Faculty of Architecture, Design, and Planning, Institut Teknologi Sepuluh Nopember (ITS)

E-mail: putrinariratih@gmail.com

Abstract. Surabaya is a metropolitan city in Indonesia, with the 4 million population. Therefore, it takes the smart city system to help The Government to optimize their service for citizen needs, and increasing its life quality, by putting interaction in the community based on ICT. Surabaya as a coastal city located on north coast of East Java Province make this city has several district in coastal area, one of them is Bulak District which is known as the fishing settlement integrated with tourism. Their character, totally different with midtown settlement. So do the given policy, there will be a difference treatments between coastal and midtown settlement. Therefore, it is necessary to explore the characteristic of coastal settlements that will support Surabaya as a smart city. This study aims to determine the criteria of urban settlements development in Surabaya based on smart city concept, focused on coastal urban settlement. To achieve this goal, it will use qualitative and quantitative approach (mixed method) which focuses on field conditions and community preferences in the location of settlement in the study area and must be through the 3 stage analysis. The first stage is identifying the smart potentials of coastal settlement using qualitative descriptive analysis. The second is identifying influential variables with using Confirmatory Factor Analysis feature on SmartPLS Application. The third was a determination of smart coastal settlement criteria using Pairwise comparison-triangulation. The result is there are 13 criteria of smart coastal settlement, spreaded on 6 major dimension. These criteria became a basic of dynamic measurement for smartkampung prototype which has so far resulted for identification and intelligence character on Bulak coastal settlement, included an entrepreneurship activities with local resources and uses ICT in its development; good accesibility which supported by the community's ability to follow transportation technology; ICT-based environmental management activities and community participation; creative community, open-technology-minded and able to preserve local coastal culture; sea tidal disaster free area that is equipped with mitigation facilities; developing an ICT-based tourist attraction area, and also achieving positive impact from the online Government's service.

Key Words: coastal settlement, smart city, Surabaya

1. Introduction

Surabaya is a metropolitan city with around 4 million population, causing more problems that faced by The Government and its people. To keep the standard of living, The Government uses a smart concept, called as smart city concept, which based on ICT (information and communication technology) that can make a livable, safe, comfortable and sustainable city. This concept can be seen as an urban planning model that can reduce and fix the urban problems nowadays, so the city will be a comfort place to live [1]. Smart city is a concept that uses ICT in 6 dimensions of life as follows: smart economy, smart people, smart living, smart environment, smart mobility and smart governance [2].

Surabaya has become a smart city based on its potential and characteristics, and also manage their resources effective and efficiently to solve several challenges with innovative, integrated and sustainable solution to provide infrastructure and service that can make quality of life up. It proved by 14 awards, achieved in the "Rating Kota Cerdas Indonesia 2017" event. However, this achievement has not completed all the dimensions yet, so it needs some efforts to make Surabaya completes all the dimension line of smart city that one of which is settlement [3].

Based on population, settlement is the most important and dominant dimension in Surabaya. There are distinguished characteristics of housing needs in Surabaya based on its process, formal housing and informal housing, so-called with the name kampung. Related to that, Surabaya is a coastal city located on north coast of East Java Province that makes this city has several districts in coastal area. This is mentioned with government policy on Detailed Urban Spatial Planning document [4,5], about the coastal settlement development located in Bulak District. To proving if a coastal settlement can support Surabaya as a smart city, needed a two-way perspective assessment through a top-down and bottom-up approach [6]. The top- down approach is used to find the smart potentials of coastal settlements based on smart city concept, and the bottom-up approach is used to build a smart perspective according to the people on the coast.

Bulak District has some smart potencies that can support Surabaya as a smart city as follows: there's a Fish Landing Base Infrastructure development plan in it; there's an embankment to anticipate the settlement from tidal flooding; there are several tourist attractions, one of them is a colorful settlement (kampung warna-warni); and there's economic development such as sea result processing into new products like crackers and smoked fish.

These potentials describe the relationship between the characteristics of coastal settlements and the development of smart city in Surabaya. It's mean that all problems in coastal settlements can be solved by the dimensions of smart city concept. Therefore, it is necessary to explore the characteristics of coastal settlements that will support Surabaya as a smart city [7], which can describe smart coastal settlements's criteria development. The question of this study is "*what are the development criteria that show coastal settlements as smart kampung*?".

2. Methods

Based on problems discussed, this study of exploration the smart coastal settlements's criteria is using qualitative and quantitative approach (mixed method) which focuses on field conditions and community preferences in the location of settlement in the study area. Qualitative methods are needed to produce descriptive data from interviewing people and something observed. While quantitative methods are needed to reduce the subjectivity of researchers in obtaining and interpreting data. In determining smart coastal settlement's criteria that can support Surabaya as a smart city, needed a two-way perspective assessment through a top-down and bottom-up approach [6].

There are 3 stages of analysis, the first is identifying the coastal settlements's potencies based on smart city concept. This first identification stage is done through observation and interviews (in depth interviews) to some people who are considered capable to representing coastal communities and have knowledge of coastal settlement in this study area. Then, data that have been obtained will be processed with qualitative descriptive analysis to take the conclusion about smart potentials of Bulak coastal settlement that are aligned with the related theories.

The second stage is identifying the influential variables according to preferences of people in coastal settlements based on smart city concept. This stage is done by distributing questionnaires to coastal communities. The questionnaires contain all the variables that are expected to influence the determination of smart settlement's criteria. The result will be obtained from assessments conducted by people on coast.

And then to sort out the variables, the results will be calculated and processed by using a Confirmatory Factor Analysis (CFA) method on SmartPls application. The research variables can be said valid with the following spesifications: loading factor >0.5; composite reliability >0.7; AVE >0.5; t-statistics ≥1.96 and p-value ≥0.05.

The last stage is determining the smart coastal settlement's criteria based on smart city concept. Different with the previous two stages, a top-down approach is used at this stage. This stage will combine some potentials from the uninfluential variables to influential variables using Pairwise Comparison method. And it will be continue with Triangulation method to combine the previous two analysis results with the policy and best practice, and also the opinions of experts who have settlement and smart city knowledge. Triangulation uses to correct the criteria by adjusting the plans that have been drafted by the government and theories that have been created by academic experts.

3. Result and Discussions

To determine the criteria of smart settlement in Bulak District, it is necessary to identify the smart potentials of the coastal settlement that will be explained by people in the Bulak coastal settlement. This first stage is done through observation and interviews (in depth interviews) with some questions about variables in each dimension of smart city concept. The variables used were obtained from several leading theories about smart city concepts, one of which was proposed by Cohen [8] who create the Smart City Wheel (Figure 1) and spreaded out the smart city in to 6 dimensions.

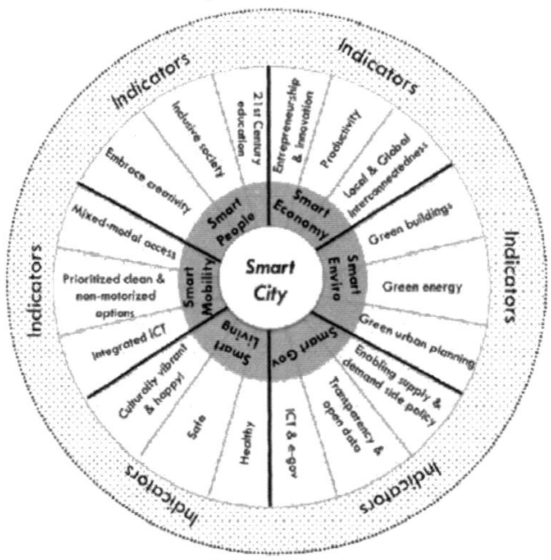

Figure 1. Smart City Wheel [8].

Other author [2] explained in more detail about 6 dimensions above. Smart governance emphasizes the conditions of good governance by establishing relations between the city government and its people. Smart economy which can be said as smart industries, which in its concept explains that this area is supported by various uses of information and communication technology including in the production process and carrying out its economic activities. Smart people are explained by people who have good skills and education levels, and who have good quality in integrating social life and can make a wide network of cooperation [9]. Smart living is cumulating several aspects of life to improve the quality of life of urban communities such as culture, health, security, housing, tourism, etc. Smart environment is an effort to protect the urban environment using technology. While smart mobility is a provision of access to new technology that can be used every day by the community, and the available infrastructure can share and process various information quickly from various places by users [10].

Based on the theory mentioned above, this study will distribute the dimensions of smart city to a smaller scope such as settlement in coastal area. Each variable has its own potential. The results of potentials identify from Bulak coastal settlement are described in Table 1.

Table 1. The smart potentials of Bulak coastal settlement in Surabaya.

No	Dimensions of Smart City	Variable	Potency
1	Smart Economy	Household-based economic avtivities	• This settlements has UKM-shaped economic activities with products output.
2		Entrepreneurship activities	• Another professions from fisherman families as entrepreneurs, because sea resources will be produced into new products by themselves.
3		Market range	• This settlements has become a brand image for the Surabaya people as a place to buy some sea resources processed products.
4		Coastal productivity	• The community uses local sea resources to support their ecoomic activities.
5		Quality of producing	• Seashell waste wlll be produced into new products such as paving, handicrafts and animal feeds. And another fish waste are easily decomposed.
6		Level of business innovation	• The level of business innovation can be seen from number of types product produced and also from other trades.
7		Economic contributions to the settlement	• Economic activities greatly helpful to increase the income of people in this settlements.
8		The citra of area	• This settlements is known as a coastal area that can produces sea resources processed products.
9		Internet usage in economic activities	• Some people have already sold their products through internet.
10		Technology usage in economic activities	• Processing of marine products by UKM used several modern machine technologies. Some of this marine products are sale on Bulak Fish Center, using e-payment system (OVO).
11	Smart Mobility	Road conditions	• Paving is a kind of road pavement in this settlements. The widht of the road around 2 metres with good condition. And there is 4 metre wide inspection road that can be passed by cars.
12		Kind of vehicle passes	• Just only 2 or 3-wheeled vehicles that can pass this road. For 4-wheeled vehicles can pass by inspection road.
13		Regulation of transportation	• For riders are required to turn off their engine and guide the vehicle in this settlements.
14		Transportation technology awareness	• The community is able to using and following the development of transport technologies such as taxibike online.

15		Connecting system for land - sea transportation	• There are no connecting facilities between sea and land transportations, society parked boats marked only by ordinary wooden stakes.
16	Smart Environment	Environmental condition	• The environment is more cleaner than before, and there is a little green space.
17		Level of polution	• There is only air pollution from fish fumigation activities in this settlements.
18		Building landscape	• The building in this settlements is quite neatly arranged.
19		Environmental management programs	• There are some environmental management programs such as garbage bank system (named as *bank sampah*), society-cooperation work (named as *kerja bakti*) and seashell waste processing into new products.
20		Technology usage in environmental management	• In seashell waste processing, the community uses a seashell crusher and there are also some WWTPs that are still functioning.
21		Community contributions in environmental management	• Society-cooperation work (named as *kerja bakti*) arranged by all settlements communities, held on unperiodic times, and the housewifes do the garbage bank system (*bank sampah*).
22		Avaibility of environment facilities	• There is a land fill and green open space around the settlements area.
23		Avaibility of environment infrastructures	• There is a trash can for each house, garbage carts for each neighborhood unit, and some WWTPs.
24	Smart People	Level of education	• The community has studied until high school and university.
25		Level of creativity	• The high level of creativity can be seen from several walls drawn by murals and some handicrafts made.
26		Openess	• People tend to be open about new things in their settlements development.
27		Community empowerment acivities	• There are empowerment activities such as routine recitation for adults to elder and drumband for children to adults.
28		Participation of community	• Many people participate in routine recitation and drumband activities.
29		Internet usage in daily	• Most of the people in this settlements, especially young people, have used smartphones.
30		Coastal cultural resilience	• There are coastal cultures that still preserved such as boat racing competitions, boat decorantion and istighosah on a boat using a torch.
31	Smart Living	Level of crime	• The crime rate is low because people know each other.
32		Security system	• There is no criminal security system in this settlements. But there is a security system for fire disaster such as their own fire station.
33		The impact of tidal flood	• There is no impact from tidal flood in this settlements.
34		Safety system for tidal flood	• There is no mitigation system in this settlements.
35		Health Facilities	• Health facilities have been fulfilled with an affordable and easily accessible.

36		Education Facilities	• Education facilities have been fulfilled for the playgroup level to elementary school and are easily accessible because it near from settlement.
37		Avaibility of internet etwork	• There are free Wi-Fi facilities with good quality.
38		The tidal flood mitigation infrastructure	• There are embankments, toas, and drone videos to prevent the impact of the tidal flood in this settlements.
39		Tourist attraction	• There are some tourist attractions such as Colorful Settlements (kampung warna-warni), Surabaya Bridge, Kenpark Atlantis, THP Kenjeran, coastal market named as Sentra Ikan Bulak and Surabaya Park.
40		IT usage in tourist attraction	• All tourist attractions have been exposed to internet.
41	Smart Governance	Online-based public service	• There is an online-based public services such as *e-Lampid* with *e-kiosk* machine.
42		Number of online access	• Most people who sble to use technology in daily, can use this online-based public service too.
43		The impact of online-based public service	• It is very helpful to make it easier and shorten the time of the community in taking care of letters and registration for medical treatment.

The description above is only based on data obtained from several community leaders who are representing the coastal communities. So it needs another identification that directly involves the community in determining criteria, namely identification of influential variables. The research variables can be said valid with the following specifications: loading factor >0.5; composite reliability >0.7; AVE >0.5; t-statistics ≥1.96 and p-value ≥0.05 (Table 2a, 2b, 2c, 2d). From the 43 initial variables formed, there are 22 variables that influence according to coastal settlement communities which will be described in Table 3.

Table 2a. The Result of Validity Test

No.	Dimensions of Smart City	Variable Code	TheInfluence Variables	Loading Factor Value
1	Smart Economy	SEc1	Household-based economic activities	0.855
2		SEc2	Enterpreneurship activities	0.619
3		SEc3	Market Range	0.731
4		SEc5	Quality of Producing	0.567
5		SEc8	The citra of area	0.553
6		SEc9	Internet usage in economic activities	0.869
7		SEc10	Technology usage in economic activities	0.831
8	Smart Mobility	SM3	Regulation of transportation	0.793
9		SM4	Transportation technology awareness	0.833
10	Smart Environment	SEn4	Environment management programs	0.849
11		SEn6	Community contributions in environmental management	0.612
12		SEn7	Avaibility of environment facilities	0.783
13		SEn8	Avaibility of environment infrastructures	0.606
14	Smart People	SP2	Level of creativity	0.679
15		SP3	Openess	0.620
16		SP6	Internet usage in daily	0.622
17		SP7	Coastal cultural resilience	0.890
18	Smart Living	SL8	The tidal flood mitigation infrastructure	0.948
19		SL10	IT usage in tourist attraction	0.516
20	Smart Governance	SG1	Online-based public service	0.868
21		SG2	Number of online access	0.915
22		SG3	The impact of online-based public service	0.918

Table 2b. The result of reliability test

Laten Variable	Composite Reliability	Average Variance Extracted (AVE)
Smart Economy	0.885	0.532
Smart Mobility	0.796	0.662
Smart Environment	0.808	0.519
Smart People	0.800	0.506
Smart Living	0.720	0.582
Smart Government	0.928	0.811

Table 2d. The result of P-Value test

Latent Variable	Original Sample	T-statistics	P-Value
Smart Economy	0.732	9,703	0.000
Smart Mobility	0.828	26,220	0.000
Smart Environment	0.426	4,547	0.000
Smart People	0.909	64,806	0.000
Smart Living	0.859	36,203	0.000
Smart Government	0.390	3,638	0.000

Table 2c. The result of t-statistics test

Latent Variable	Indicator	Loading Factor	T-Statistics
Smart Economy	SEc1	0.855	19,194
	SEc2	0.619	9,391
	SEc3	0.731	12,959
	SEc5	0.567	9,490
	SEc8	0.553	6,735
	SEc9	0.869	23,296
	SEc10	0.831	14,384
Smart Mobility	SM3	0.793	14,084
	SM4	0.833	24,586
Smart Environment	SEn4	0.849	20,806
	SEn6	0.612	5,494
	SEn7	0.783	10,675
	SEn8	0.606	4,566
Smart People	SP2	0.679	10,479
	SP3	0.620	9,132
	SP6	0.622	7,090
	SP7	0.890	55,817
Smart Living	SL8	0.948	39,739
	SL10	0.516	3,659
Smart Goverment	SG1	0.868	11,418
	SG1	0.915	11,622
	SG1	0.918	10,530

Table 3. Influential variables based on community coastal settlements preferences.

Smart City Dimensions	The Influence Variables
Smart Economy	Household-based economic activities
	Entrepreneurship activities
	Market range
	Quality of producing
	The citra of area
	Internet usage in economic activites
	Technology usage in economic activites
Smart Mobility	Regulation of transportation
	Transportation technology awareness
Smart Environment	Environmental management programs
	Community contributions in environmental management
	Avaibility of environment facilities
	Avaibility of environment infrastructures
Smart People	Level of creativity
	Openess
	Internet usage in daily
	Coastal cultural resilience
Smart Living	The tidal flood mitigation infrastructure
	ICT usage in tourist attraction
Smart Governance	Online-based public service
	Number of online access
	The impact of online-based public service

Source: Paramasatya, D.A., et al. [6]

Based on the two steps that have been done before, the criteria will be determined by pairwise comparison and triangulation. This stage will combine some potentials from the uninfluential variables to influential variables using Pairwise Comparison method (Figure 2). And it will be continue with Triangulation method to combine the previous two analysis results with the policy and best practice, and also the opinions of experts who have settlement and smart city knowledge. Triangulation uses to correct the criteria by adjusting the plans that have been drafted by the government and theories that have been created by academic experts. This last stage obtained 13 criteria of smart settlements according to Bulak Coastal Settlements in Surabaya are described in Table 4.

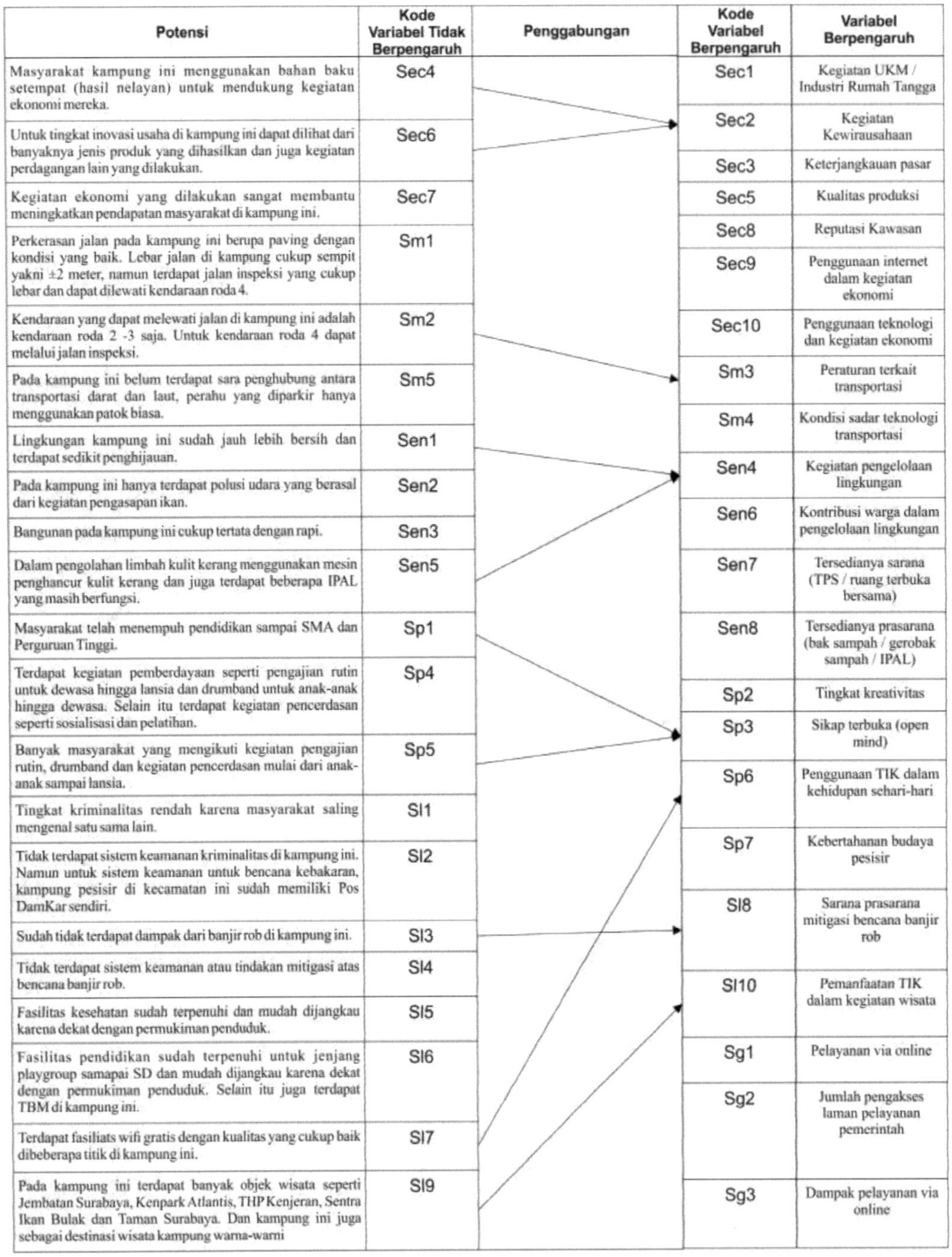

Figure 2. Pairwise Comparison Analysis, 2019

Table 4. The criteria of smart coastal settlements in Surabaya.

Dimension	Criteria
Economy	There are household-based economic activities or home-based entrepreneurships that use sea resources to make products with the environmental friendly processing and use ICT in its development;
	There are some facilities that can give a good image to sea resources market;
Mobility	There are good accessibilities with regulation that can improve comfort;
	The community is able to using and following the development of transport technologies;
Environment	There are based-technology environmental quality improvement programs by community contribution;
	There is implementation of waste water and waste management from sea economic activities;
	The quality of environment increased by the completeness of infrastructure and green space;
People	A creative and open-minded society about the development, one of which is education and able to preserve local coastal culture;
	There are skill-improve activites such as socialization and traine which followed by the community;
	The community is able to implement technology usage in daily activities and supported by free internet network;
Living	The area is free from tidal flooding with adequate mitigation infrastructures;
	There are tourist attractions with a synergy of maritime tourism concept which use ICT in its development;
Governance	The reacted of positive impact by public's ability to use the Online-based public service.

4. Conclusions

This research was conducted to determine the criteria that describe the characteristics of smart coastal settlement based on the smart city concept in Surabaya. To get criteria that are suitable with the characteristics of coastal settlements, there are 3 stages of analysis that must be carried out. The first stage is identifying the smart potential of coastal settlements using qualitative descriptive analysis. After identifying the potential, it is also necessary to identify which variables are affected according to the coastal settlement community. The identified variables refer to the 6 dimensions of smart city. Then to get the results of valid criteria, the formulation of criteria obtained from a combination of potential and influential variables will be confirmed to experts in the field of housing and smart cities through triangulation analysis.

The following are the result obtained from this study which are 13 smart coastal settlements development criteria:

1. There are household-based economic activities or home-based entrepreneurships that use sea resources to make products with the environmental friendly processing and use ICT in its development
2. There are some facilities that can give a good image to sea resources market.
3. There are good accessibilities with regulation that can improve comfort.
4. The community is able to using and following the development of transport technologies.
5. There are based-technology environmental quality improvement programs by community contribution
6. There is implementation of waste water and waste management from sea economic activities
7. The quality of environment increased by the completeness of infrastructure and green space.
8. A creative and open-minded society about the development, one of which is education and able to preserve local coastal culture.
9. There are skill-improve activites such as socialization and traine which followed by the community.
10. The community is able to implement technology usage in daily activities and supported by free internet network.
11. The area is free from tidal flooding with adequate mitigation infrastructures.

12. There are tourist attractions with a synergy of maritime tourism concept which use ICT in its development.
13. The reached of positive impact by public's ability to use the Online-based public service.

5. Acknowledgements

This article was presented at the 2nd International Conference on Smart City INNOVATION (ICSCI) 2019, jointly held by Universitas Indonesia and Universitas Diponegoro. ICSCI conferences have been supported by the United States Agency for International Development (USAID) through the Sustainable Higher Education Research Alliance (SHERA) Project for Universitas Indonesia's Scientific Modelling, Application, Research, and Training for City-centered Innovation and Technology (SMART CITY) Center for Collaborative Research, administered through Grant #AID-497-A-1600004, Sub Grant #IIE-00000078-UI-1. Authors wishing to acknowledge encouragement from Department of Urban and Regional Planning Institut Teknologi Sepuluh Nopember for the opportunity to cooperation with important stakeholder in Surabaya Smart City concept.

6. References

[1] Nam T and Pardo T A 2011 Conceptualizing smart city with dimensions of technology, people, and institutions *Proceedings of the 12th Annual International Digital Government Research Conference on Digital Government Innovation in Challenging Times - dg.o 11* (ACM Press)

[2] Giffinger R, Fertner C, Kramar H, Meijers E and others 2007 City-ranking of European medium-sized cities *Cent. Reg. Sci. Vienna UT* 1–12

[3] Kementerian Bappenas 2015 Pengembangan Kota Cerdas di Indonesia *Makalah disampaikan dalam Konferensi e-Indonesia Initiative dan Smart Indonesia Initiatives Forum ke-1* vol 15

[4] Bappeda Kota Surabaya 2018 *Detailed Spatial Planning of Surabaya City 2018 – 2038* (Surabaya)

[5] Bappeda Kota Surabaya 2014 *Spatial Planning of Surabaya City 2014 - 2034* (Surabaya)

[6] Paramasatya D A and Rahmawati D 2017 Penentuan Variabel Berpengaruh Terhadap Pengembangan Kampung Cerdas dalam Mewujudkan Konsep Smart City *J. Tek. ITS* **6**

[7] Susanti R, Soetomo S, Buchori I and Brotosunaryo P M 2016 Smart Growth, Smart City and Density: In Search of The Appropriate Indicator for Residential Density in Indonesia *Procedia - Soc. Behav. Sci.* **227** 194–201

[8] Cohen B 2012 What Exactly Is A Smart City?

[9] Sutriadi R 2017 *Media sosial dan perencanaan kota* (Penerbit ITB)

[10] Suryotrisongko H, Kusuma R C and Ginardi R V H 2017 Four-Hospitality: Friendly Smart City Design for Disability *Procedia Comput. Sci.* **124** 615–23

Towards Collaborative Batik Waste Management Model in Kampong Batik Semarang

G L Wungo[1], S P Dewi[1], H Susanto[2], R Desiriani[2], E F Karamah[3]

[1]Departement of Urban and Regional Planning, Diponegoro University, Semarang, Indonesia
[2]Departement of Chemical Engineering, Diponegoro University, Semarang, Indonesia
[3]Departement of Chemical Engineering, University of Indonesia, Depok, Indonesia

E-mail: grandywungo@live.undip.ac.id

Abstract. Kampong Batik is one of thematic urban kampongs located in Semarang inner-city with batik production and sales as the main activities. One of the important issues related to this kampong is batik waste management; there is not installation wastewater treatment plan. It caused by several reasons such as land availability, fund, government and community commitment. Therefore, this paper intended to analyses the batik waste management model through collaborative planning strategy. It used qualitative research method through interview and observation techniques. Likewise, secondary data has also used as the data collection method. The research objectives was break downed into several activities; first, identification of existing waste management. Second, analyze waste management gap that compare between the existing situations with the expected performance. Then, it can be structured the collaborative waste management model as the third objective. This study output is a collaborative model in batik waste management that emphasized on the managerial aspect rather than the physical aspect.

Keywords: collaborative, kampong batik, waste management

1. Introduction

Indonesia is a developing country that has challenges in developing the economy, Empowerment of Micro, Small and Medium Enterprises or Usaha Mikro Kecil Menengah (UMKM) is considered to be one of the effective ways to build a stable [1]. Micro, Small and Medium Enterprises (UMKM) have proven to be more resilient in the face of the economic crisis that had occurred in Indonesia in 1997 [2]. UMKM Batik becomes one of the UMKM that has an important role in developing Indonesia's economy, especially in the city of Semarang. One of the UMKM Batik that is fostered by the Semarang City Government is Semarang Batik 16. The development of Semarang Batik 16 has brought many positive impacts such as the existence of branding which is getting better, the amount of batik production that is increasing, increasing tourist visitors both local and foreign tourists.

The potential of Semarang Batik and its wide network attracts many investors who want to invest. One of the investments and cooperation were undertaken with Batik Semarang 16 is the provision of facilities for Wastewater Treatment Plant (WWTP) by the Center for Industrial Pollution Prevention Technology (BBTPPI). But the problem that is happening right now at Batik Semarang 16 is that activities are focused only on developing batik production. Production activities are carried out every day with a target number of approximately 50 (fifty) pieces of batik cloth per day. However, the amount of production also tends to be inconsistent, this condition has an impact on the amount of wastewater produced. The minimum amount of liquid waste produced per day is 10 (ten) liters per day. Batik Semarang 16, which already has Waste Water Management (WWTP) Installation with Aerob and Anaerobic WWTP, demands that the amount of waste and disposal time must be consistent because it will have an impact on the ability to decompose bacteria and the quality of wastewater produced. This condition is contrary to the amount and consistency of waste disposal time carried out in Batik Semarang 16 at this time. Besides, the activity of Batik Semarang 16 also focuses on expanding marketing networks, and developing cooperation in the field of tourism so that it hasn't maximized the wastewater management system. While the batik industry is an industry that has the potential to produce waste containing heavy metals and hazardous liquid waste that can cause environmental pollution [3]. Since nowadays sustainability emerge as high priority for the business world and all the key players in the various chains of production [4] thus to create a sustainable UMKM batik, quality improvement should not only be done in the marketing of products, the amount of production, and increasing in human resources production activities. But it must also pay attention to waste management, both from the WWTP, management, and also the location of disposal. Due to poor waste management will have an impact on the emergence of various problems both in terms of economic, environmental and social.

In addition, the unavailability of resources or experts in the field of wastewater management in Batik Semarang 16, creates a mismatch of wastewater management procedures stipulated in article 18 of Semarang City Regulation Number 13 2006 [5]. Based on article 18 section 2 of Semarang City Regulation Number 13 2006, states that each party responsible for a business and/or activity must have expert technical staff waste treatment. This requirement becomes one of the conditions that are not much attention by the Semarang Batik Party, currently for WWTP operations only carried out by Semarang Semarang 16 employees with a background in literary education thus there is no any experts for the relevant issue which is waste management.

Wastewater management which is considered not profitable, and asset ownership status with the service contract system causes a lack of attention by the Batik Semarang 16 to the wastewater management. This condition will have an impact on unsustainable wastewater management. Until now there has been no effort made to review the management of Batik Semarang 16 wastewater. Following up on this condition there needs to be a study of Batik Semarang 16 wastewater management to create sustainable wastewater management. The purpose of this study is to analyze the ideal of Batik Semarang 16 waste management using the concept of collaborative waste management.

2. Collaborative Waste Management Model

Wastewater management is one of the efforts made to create sustainability of an industry in terms of liquid waste expenditure and also its management system [6]. Waste management can be done by integrating social, management and economic aspects to create sustainability [7,8]. A good form of liquid waste management will reduce the risk of environmental pollution, environmental degradation [9]. To see the ideal liquid waste management, it can be seen as follows: social, management, and economy aspects [7,8].

According to Nururrohmah et al. [8], collaborative planning emphasizes the consensus among various stakeholders in making decisions to the implementation of the program. Collaborative planning (collaborative planning) is a decision making process where various stakeholders, who see the problem from various angles, sit together to explore their differences constructively, then search a solution, and to get more than what is obtained if only looking for a solution individually [10]. Collaborative planning is seen as important in developing a shared vision and goals that focus on ecological and social sustainability outcomes [11–13]. There are several important aspects that affect, namely as follows:

- Social aspects, the collaborative waste management social aspect is a variable that reviews the management of wastewater from two components, namely network power and also community power [8]. Network power is the ability to adjust liquid waste management by a group/business entity through cooperation. Network power can be grouped can be grouped into three classes, as follows [14]: strong, medium and weak. A good community power will start an empowerment management activity and increase the capacity of resources that will affect the capacity, knowledge, behavior, and degree of participation so that they can play an important role in achieving goals. The degree of participation is as follows [15]: Spontaneous participation, Participation induced, Participation depressed by habit, Participation depressed by socio-economic reasons, Participation depressed by regulation.

- The management aspect, in collaborative waste management, the emphasis on management variables leads to three indicators, namely the collaboration model, the application of team regulations and also stakeholder mapping. The principles in management activities, namely as follows: Planning (planning), Organizing (organizing), Actuating (implementation), Controlling (supervision) [16]. In addition, the pattern of cooperation in management aspect is how the relationship between two people or the relationship between two companies to create effectiveness, risk sharing, the existence of the company, cost savings, improvement or maintain service levels that encourage sustainability [17].
- Economic aspects, the collaborative waste management model emphasizes how the financing patterns for wastewater management activities can proceed sustainably without a deficit concept. This financing pattern aims to accumulate how operational costs, procurement costs do not cause losses that lead to the concept of deficit in a company. To encourage funding and assistance in operational activities a company must carry out the principle of cooperation [18].

3. Data and Methods

The research approach used in this study is a qualitative is one of the models invented by James Spradley in 1980 [19]. The qualitative approach used in this study is due to see the Batik Semarang 16 waste management system, as a social phenomenon of the development of the batik industry that still does not pay attention to the sustainability of waste management.

3.1 Data Collection

The method used in the sampling of this study was purposive sampling with the type of snowball. The use of snowball is used to make it easier for researchers to find sources that have not yet been identified and obtain sources that are relevant to the research. For data collection, the authors used 3 types of data collection:

a. Interviews in this study use open and in-depth interviews related to one's response to, perceptions, opinions, feelings and knowledge. Based on the type of snowball in sampling method, this research found 4 (four) resource person who have an understanding related to the management of Batik Semarang 16 wastewater. They are :
    - Resource person I is Semarang Batik 16 party who played an important role in the management of liquid waste generated from batik production activities.
    - Resource person II is the Central Institute of Industrial Pollution Prevention Technology (BBTPPI) which has an important role in the supply of liquid waste management tools
    - Resource person III is one of the leaders of the batik community in the Meteseh Village.
    - Resource person IV is the government in the Kelurahan Meteseh. This resource person has a role in the licensing process for the establishment of Batik Semarang 16.

b. Observation, carried out to see, the availability of waste management installations, waste management conditions, environmental conditions and see the impacts that may arise due to the lack of development carried out on the management of liquid waste from industrial waste Batik Semarang 16.

c. In this study, supporting documents such as Law Number 32 of 2009 concerning Environmental Protection and Management, Environmental Ministerial Regulation No. 5 of 2014 and Semarang City Regulation number 13 of 2006. By using these supporting documents, this research only focuses on management aspect, such as in article 18 section 2 that states each party responsible for a business and/or activity must have expert technical staff waste treatment.

3.2 Analysis Techniques

Domain analysis is performed to obtain a general and overall picture of the social situation under study or the object of research. The result of this analysis is a general description of the condition of the research topic that has never been known [20]. In this analysis the form of analysis is not exhaustive and can be classified into the recapitulation activities of an identification activity. In this study, the domain analysis is used to find out how the condition of waste management in Batik Semarang 16 by looking at several aspects namely management, social, economic aspects [7,8].

4. Findings and Discussion

4.1 Gap Analysis

In this gap analysis, three variables used to describe how the existing waste management will be compared with the ideal conditions of a collaborative waste management model. The following are the results of the gap analysis conducted.

4.1.1 Social

The lack of knowledge about the management of wastewater is influenced by the unavailability of experts on wastewater management in Batik Semarang 16 so that it impacts on the efficiency and effectiveness in managing wastewater in Batik Semarang 16 (Table 1). The achievement of efficiency and effectiveness in wastewater management is also influenced by weak networking power in current wastewater management. It is shown that profit tends to be felt by only one party (W2-J2-P10), (W1-J1-P10), has the same vision / different vision (W2-J2-P18) (W1-J1-P13), cooperation not sustainable (W2-J2-P15) (W1-J1-P15) (W1-J1-P13). Weak networking power also influences effectiveness, efficiency, innovation, and adaptability in wastewater management activities.

Table 1. Social Aspect Gap

Aspect	Existing Wastewater Management			Collaborative Waste Management		Gap
	Participation	Person	Knowledge	Community Power	Networking Power	
Social	• The actors involved in waste management are the outsourcer, BBTPPI, and Batik Semarang 16 (W2-J2-P23), (W4-J4-P1). • The current participation has not yet created transparency and accountability (W2-J2-P13), (W2-J2-P15), (W1-J1-P13), (W4-J4-P13).	• Semarang Batik 16 behavior is at the stage of caring for the environment and trying to preserve (W3-J3-P2), (W1-J1-P2), (W1-J1-P23), (W2-J2-P23). • No attempt has been made to provide understanding that waste that is disposed of must-be monitored, (W4-J4-P1).	• Not yet available human resources who understand wastewater management (W2-J2-P15), (W1-J1-P2). • Limited understanding regarding the characteristics of WWTP provided (W2-J2-P12), (W1-J1-P3). • There has not been any innovation related to waste water management (W1-J1-P13, (W2-J2-P13)	Transforming organizational systems to improve effectiveness, efficiency, innovation, and adaptability. The way that can be done by doing empowerment is the management of wastewater et al., 2016).	Network power can be done if both parties are equally benefited, have the same views and principles, and are interdependent. the stronger network power between actors will affect transparency, accountability and sustainability of management [14]	• Not yet achieved the effectiveness, efficiency in waste water management • There is no transparency and accountability in waste water management activities at Batik Semarang 16. • Participation involving only three actors needs to be broadened and increased participation.

The absence of transparency and accountability among actors in waste water management is influenced by weak community power, weak community power seen from the form of participation. The degree of participation that is formed tends to be Participatory pressure (W2-J2-P10) (W1-J1-P1)

(W1-J1-P15). Distressed participation is carried out for fear of losing social status or suffering losses due to not carrying out the assigned role.

4.1.2 Economy

The sustainability of a management system is inseparable from good financing. In the gap analysis specifically, the collaborative waste management model emphasizes that in good management activities all economic activities must be accumulated so as not to cause harm [8]. The results of the gap analysis of economic aspects can be seen in the Table 2.

Table 2. Economy Aspect Gap

Aspect	Existing Wastewater Management			*Collaborative Waste Management*	*Gap*
	Income	Source of Funding	Operational Cost	Financing Patterns	
Economy	• Revenues of 500 million per month (W1-J1-P4), (W3-J3-P4).	• Operations sourced from Batik Semarnag 16 (W1-J1-P8). • The cost of constructing WWTP is sourced from the government (W2-J2-P17).	• Operating costs incurred every month of Rp. 2.000.000-Rp. 3.000.000 **(W2-J2-P4).** • The operational costs are fully borne by the Batik Semarang **(W1-J1-P8).** • There were no deficits and difficulties in operational activities **W1-J1-P4).**	• Accumulate how operational costs, procurement costs do not cause losses [8]. • Carry out the principle of cooperation in operational and procurement activities [8]. • Emphasizing sustainability with innovations in technical management and also management in an organizational structure [8,18]	• There has not been any innovation in technical management and organizational structure

The results of the analysis of the economic aspects found a gap, where there has been no innovation in technical management and organizational structure. The purpose of innovation in technical management is that the results of the activities of treating wastewater can be developed not only directly disposed of. This concept is used to achieve sustainability from water management [14,18]. While innovation in organizational structure is how roles that have been divided can be increased to increase effectiveness and efficiency which can also have an economic impact [8,18]. The absence of innovation is since participation in the management of Batik Semarang 16 wastewater management is fairly minimal (W2-J2-P13), (W2-J2-P15).

4.1.3 Management.

The lack of communication among stakeholders to develop wastewater management is influenced by the nature of wastewater management which tends to be non-profit and is one of the reasons that Semarang batik is more developing activities that generate economic benefits. The management aspect emphasizes the cooperation and leading of regulations and actors that must be involved based on existing regulations. As for the gap analysis conducted, it can be seen in the Table 3.

Table 3. Management Aspect Gap

Aspect	Existing Wastewater Management		Collaborative Waste Management			Gap
	Cooperation Pattern	**Regulation**	**Cooperation Model (*Service Contract*)**	**Mapping Stakeholder**	**Regulation**	
Management	• WWTP facilities are assets owned by the government / public (W2-J2-P20). • Batik Semarang 16 has the status of usage rights and is responsible for the operation and maintenance of WWTP (W1-J1-P16). • Initial funding (construction, material, operational) is carried out by BBTPPI (W2-J2-P16, W2-J2-P17). • Loss / risk will be borne by the government / public (W2-J2-P17, W2-J2-P13) • The collaboration between Batik Semarang 16 and BBTPPI lasts for + 2 years consisting of the building and mentoring process (2016-2018). • Short cooperation time causes the lack of further development of waste management (W2-J2-P15, W2-J2-P13).	• There is no continuing communication between Batik Semarang 16 and BBTPPI as the organizer of WWTP (W2-J2-P13). • Batik Semarang already has a waste disposal permit (W1-J1-P1). • At present Batik Semarang 16 does not have the resources / expertise in wastewater management (W2-J2-P15), (W1-J1-P2).	• Characteristics of service contract cooperation models [17]: - Ownership by the public - Public and private operations and maintenance - Investment by the public - Risks borne by the public - Cooperation duration is 1-2 years • Collaborative Waste Management requires collaboration between business actors and other stakeholders that lead to the principle of public private partnership to create participatory engagement [18] • To create an ideal collaboration, the pattern of cooperation must have a long-term duration (Chong, 2015).	• Agencies that have the main task in controlling water pollution (Henri Subagyo, S.H., 2017): Local Government, Public Works & Public Housing Agency, Environmental Agency, Regional Development Planning Agency	• Testing the quality and measuring the discharge of liquid waste and reporting to the responsible government agency at least once a month • Meet quality standards in accordance with applicable laws and regulation and have a liquid waste disposal permit. • Must have permission to use the waste application and waste disposal • City / regional government The Regional Government is required to provide infrastructure and facilities for the management of wastewater generated from small businesses and / or domestic waste	• Lack of cooperation with other agencies / stakeholders that have influence and interest in waste management • Lack of communication between stakeholders for the development and sustainability of waste management. Lack of commitment from Good Semarang 16 in providing expert resources in waste management to do reporting and meet quality standards in accordance with regulations.

Currently, there are several actors involved in the management of Batik Semarang 16 wastewater management, while the stakeholder mapping in the existing management activities can be seen in the Table 4.

Table 4. Existing Stakeholder Mapping

Actor	Interest	The Ability of Actors in Mobilizing Resources (Technology, Human, Financial)	Tendency of Relationship Direction	Levels of Interest	Level of Influence
Batik Semarang 16	Business doers	The actor is one who can make decisions regarding available resources	Internal	The most critical interest	Very influential
BBTPPI	Government body that focuses on developing industrial waste management technology	The actor is one who can make decisions regarding available resources	Internal	Very important	Significant influence
Semarang City Government (Village)	Government bureaucracy at the level of output in the context of administration	Cannot Make Decisions Regarding Resources	External	Very small / unimportant	The effect is very small / none

The lack of cooperation has an impact on ideal management principles if seen from a management principle that consists of Planning, Organization, Actuating, and Controlling Analysis. The planning phase has been carried out by BBTPPI with the main objective to reduce hazardous liquid waste that is discharged into the environment through WWTP installations in Batik Semarang 16. Organization Currently the division of roles has been done in the management of Batik Semarang 16 waste. BBTPPI as a party that provides infrastructure and management facilities wastewater and Batik Semarang 16 also get a role as a manager. However in the management of wastewater Batik Semarang 16 has not been able to identify all the actors who must be involved in management activities, this condition causes several gaps to emerge such as on social aspects, there is no transparency, efficiency, and effectiveness, which also has an impact on the unavailability of a clear management organizational structure in waste management. Actuating is an implementation of planning and organizing. Due to the implementation of wastewater management activities in Batik Semarang 16 the involvement of the actors is minimal and the participation of actors who are depressed in their implementation (W2-J2-P13), encourages the implementation of management activities that are efficient and effective and are not formed accountability and transparency in wastewater management • Controlling, is the process of control or supervision to create dynamic, effective and efficient management activities. At the supervision stage, there is currently no form of controlling activities in the Semarang Semarang wastewater management 16.

4.2 Collaborative Waste Management of Batik Semarang 16

Based on the results of the gap analysis, the lack of communication and participation of other stakeholders causes waste management in Batik Semarang 16 to be less effective, efficient, accountable, and transparent. So it is needed to increase cooperation by involving several stakeholders who were not yet involved at all. Increased cooperation will have an impact on the three aspects used to analyze the condition of existing waste management on collaborative waste management. The stakeholders who need to be involved from the results of the gap analysis of the three aspects are contained in the stakeholder mapping of Semarang Semarang 16 waste management can be seen in the Table 5.

Table 5. Stakeholder mapping

Actor	Interest	The Ability of Actors in Mobilizing Resources (Technology, Human, Financial)	Tendency of Relationship Direction	Levels of Interest	Level of Influence
Batik semarang 16	Business doers	The actor is one who can make decisions regarding available resources	Internal	The most critical interest	Very influential
BBTPPI	Government body that focuses on developing industrial waste management technology	The actor is one who can make decisions regarding available resources	Internal	Very important	Significant influence
Semarang City Government (Village)	Government bureaucracy at the level of output in the context of administration	Cannot Make Decisions Regarding Resources	External	Very small / unimportant	The effect is very small / none
Industry Agency	Government Agency whose role is to develop procedures for industrial production.	The actor is one who can make decisions regarding available resources	Internal	Very important	Significant influence
Balai Besar Batik	Government Agency in research and development of quality, standards, institutional capacity and technology of the batik industry	Cannot Make Decisions Regarding Resources	Internal	Very important	Moderate influence
Environmental Agency	Government Agency whose role is to control the impact of development on environmental sustainability	Cannot Make Decisions Regarding Resources	External	Moderate interest	Moderate influence
Development and Financial Planning Agency	Government agencies that play a role in budgeting, implementation, operational supervision, and monitoring and evaluation of development implementation	The actor is one who can make decisions regarding available resources	Internal	Very important	Significant influence
Public Work Agency	Government Agency which has the role of developing environmental facilities and infrastructure	Cannot Make Decisions Regarding Resources	External	Moderate interest	Moderate influence

The results of stakeholder mapping will be used in the preparation of collaborative waste management models by emphasizing the consensus among various stakeholders in decision making, implementation, and monitoring the results of the program [8]. In developing collaborative waste management models of Batik Semarang 16, the principles of management activities consisting of Planning, Organization, Actuating, and Controlling (POAC) become a component that is used to separate the roles of each actor that also fixes the application of the principles of management activities when this is still not ideal. The collaborative waste management model in Batik Semarang 16 can be seen in Figure 1.

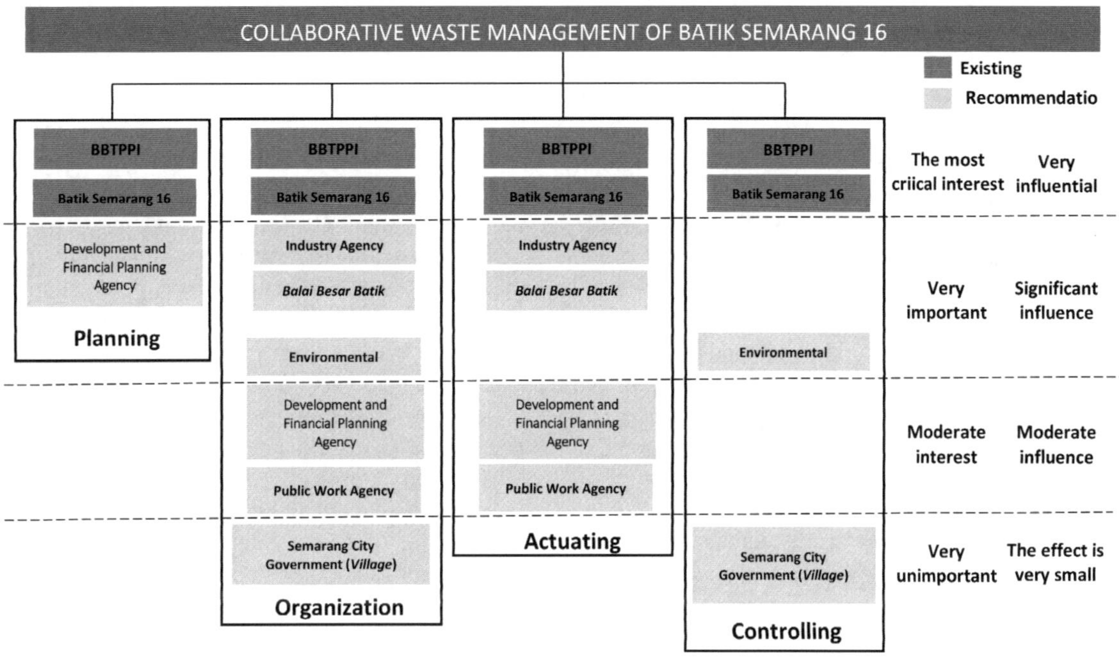

Figure 1. Collaborative waste management Batik Semarang 16

5. Conclusion

At the present social aspects do not yet have competence, efficiency, transparency, and accountability in wastewater management activities at Batik Semarang 16, Participation that only requires three actors needs to be increased and increase participation. In the social aspect, waste management is related to two indicators namely network power and community strength, the existing network strength in wastewater management is currently under weak conditions, while community strength is not yet ideal due to the influence of research involving depressed actors.

Economic aspects, there is no innovation in technical management and organizational structure. Folded innovation in terms of technical and organizational structure to encourage management in terms of economic sustainability and bring the economic benefits of waste management activities Batik Semarang 16.

Management aspects, currently there is still a collaboration with other agencies/stakeholders that have less influence and interest in waste treatment, as well as lack of communication and commitment among stakeholders in developing waste management. Batik Semarang 16 also still has limitations on the availability of resources that are experts in waste management. Seen from the four principles of management activities namely Planning, Organization, Actuating, and Controlling (POAC) there are still several conditions that are not specific to the conditions of waste management in Semarang Batik in the Organization, Actuating and Controlling sections.

6. Recommendation

To implement collaborative waste management in Semarang 16 waste management, to create sustainable wastewater management, the following recommendations can be given:

- Identify relevant actors who have a common interest in the context of liquid waste management.
- Increase collaboration by paying attention to the principle of community power that leads to spontaneous participation and also strong networking power.
- Implement provisions following applicable regulations especially in the provision of human resources.
- Make a clear organizational structure in the waste management activities of Batik Semarang 16
- Pay attention to the principles of management especially in the Organization, Actuating and Controlling sections.
- Increase attention to developing Batik Semarang wastewater management 16.

This research has a weakness in the study of the application of regulations related to wastewater management stipulated in Semarang City Regulation Number 13 of 2006 [5] which contains four policy items. Of the four policy items, it has not yet elaborated on the detailed process related to the waste disposal permit, whether the permit is still valid or has to be extended. In the third item, it is written that each business entity must provide experts, but in this study, the researchers have not provided a detailed explanation of the requirements that must be met by experts and the researchers have not yet explained how sanctions can be imposed on a company that does not have experts. On the fourth item reads that the government is obliged to provide facilities/infrastructure for waste treatment, but in this study, the researchers merely stated that the government had provided the infrastructure but had not yet identified who was involved in detail in the supply activities. Not only in the aspect of regulation in stakeholder mapping activities, but the author also has not described the advantages of each party when cooperating in wastewater management. The application of regulations and profit principles in cooperation can be a recommendation for further research especially in studying the management of liquid waste in similar industries or the same location. By specifying the application of regulations and the benefits of collaboration it is hoped that it will facilitate a more detailed identification of stakeholders.

7. Acknowledgement

This article was presented at the 2nd International Conference on Smart City INNOVATION (ICSCI) 2019, jointly held by Universitas Indonesia and Universitas Diponegoro. ICSCI conferences have been supported by the United States Agency for International Development (USAID) through the Sustainable Higher Education Research Alliance (SHERA) Project for Universitas Indonesia's Scientific Modelling, Application, Research, and Training for City-centered Innovation and Technology (SMART CITY) Center for Collaborative Research, administered through Grant #AID-497-A-1600004, Sub Grant #IIE-00000078-UI-1

8. References

[1] Gunartin 2017 Penguatan UMKM Sebagai Pilar Membangun Ekonomi Bangsa *EDUKA J. Pendidikan, Huk. dan Bisnis* **1** 59–74
[2] Anisyah, Yulianita; Atmanti H D Analisis Perkembangan Industri Batik Semarang **0**
[3] Purwaningsih R, Susanto N and Yudha M C 2016 Penilaian Keberlanjutan UKM Batik Kota Semarang dengan Metode Product Service System (PSS) *J. Tek. Ind.* **18** 31–42
[4] Sancha C, Wong C W Y and Gimenez Thomsen C 2016 Buyer-supplier relationships on environmental issues: A contingency perspective *J. Clean. Prod.* **112** 1849–60
[5] Pemerintah Kota Semarang 2006 *Peraturan Daerah Kota Semarang Nomor 13 Tentang Tata Cara Pengolahan Limbah*
[6] Subki N S R 2011 A Preliminary Study on Batik Effluent in Kelantan State: A Water Quality Perspective *Int. Conf. Chem. Biol. Environ. Sci.* 274–6
[7] M. Wawan Kurniawan; Purwanto S 2013 Kajian Pengelolaan Air Limbah Sentra Industri Kecil Dan Menengah Batik Dalam Perspektif Good Governance *Pengelolaan Sumberd. Alam Dan*

Lingkung. 501–8
[8] Nururrohmah Z and Suhirman 2016 Shared-power Governance in Managing Common Pool Resources Case Study: Collaborative Planning to Manage Thematic Parks in Bandung City, Indonesia *Procedia - Soc. Behav. Sci.* **227** 465–76
[9] Rizqon M A M, U D H and Taryana D 2010 Pengaruh Pencemaran Limbah Cair Industri Pengolahan Ikan Terhadap Kualitas Air Tanah di Kecamatan Muncar Kabupaten Banyuwangi
[10] Sufianty E 2014 Kepemimpinan dan Perencanaan Kolaboratif pada Masyarakat Non Kolaboratif *J. Perenc. Wil. dan Kota* **25** 78–96
[11] Moore M-L, Tjornbo O, Enfors E, Knapp C, Hodbod J, Baggio J A, Norström A, Olsson P and Biggs D 2014 Studying the complexity of change: toward an analytical framework for understanding deliberate social-ecological transformations *Ecol. Soc.* **19** 54
[12] Vacik H, Kurttila M, Hujala T, Khadka C, Haara A, Pykäläinen J, Honkakoski P, Wolfslehner B and Tikkanen J 2014 Evaluating collaborative planning methods supporting programme-based planning in natural resource management *J. Environ. Manage.* **144** 304–15
[13] Olsson P, Folke C and Hahn T 2004 Social-ecological transformation for ecosystem management: the development of adaptive co-management of a wetland landscape in southern Sweden *Ecol. Soc.* **9**
[14] Seadon J K 2010 Sustainable waste management systems *J. Clean. Prod.* **18** 1639–51
[15] Totok M and Poerwoko S 2013 Pemberdayaan Masyarakat Dalam Perspektif Kebijakan Publik *Bandung Alf.*
[16] Carpenter M, Bauer T and Erdogan B 2012 *Management principles*
[17] Fitriyah F 2016 Model Public Private Partnership Dalam Peningkatan Pelayanan Sumber Daya Air Bersih Di Pdam Kabupaten *Kebijak. dan Manaj. Publik* **4** 1–11
[18] Fuldauer L I, Ives M C, Adshead D, Thacker S and Hall J W 2019 Participatory Planning of the Future of Waste Management in Small Island Developing States to Deliver on the Sustainable Development Goals *J. Clean. Prod.* **223** 147–62
[19] Garrido N The Method Of James Spradley In Qualitative Research *Cuid. Humaniz.* 43
[20] Suryana 2012 Metodologi Penelitian: Metodologi Penelitian Model Prakatis Penelitian Kuantitatif dan Kualitatif *Univ. Pendidik. Indones.* 1–243

Usability of QR code in the design of Information system for recognizing historic buildings, Kota Lama, Semarang

D M K Nugraheni[1], D I K Dewi[2], A K Nugroho[3]

[1]Departement of Computer science/ Informatics, Faculty of Sciences and Mathematics, Diponegoro University Semarang, Semarang,Indonesia
[2]Urban Regional Planning Department, Faculty of Engineering, Diponegoro University, Jl. Prof Sudharto, SH Tembalang, Semarang, Jawa Tengah, Indonesia
[3]Electrical Department, Faculty of Engineering, Semarang University, Semarang, Indonesia

E-mail: dinar.mutiara@live.undip.ac.id

Abstract. Urban infrastructure or basic facilities such as museum is essential as the component of smart city. To support a natural museum for instances the exhibit of historical building in "Kota Lama Semarang "or The old city of Semarang, this paper used the QR code to recognising the information on the historical buildings, Kota Lama, Semarang. Leaflet map of Semarang Historic building commonly used to provide the location information. However, leaflet map has a limitations, it cannot provide enough information about the historic building. The limitations of leaflet maps is only provide little information which explaining the condition of the historic building. To expand the information about the historical building, prior research suggested the use of web technologies. However, some people suggested the usage of leaflet – map which can show the location and the web site can provide more additional information about the historical building. In this paper, the researcher purpose a design of Information System, which employs Quick Response (QR) code technologies as a link to combine the information between leaflet map and web technologies. Then, the design validated with identify the usability of design for using QR code for recognising historic building, Kota Lama, Semarang. The method that used to evaluate the design is SUS (System Usability Scale). The QR code usability test using SUS question showed the number that the design is in level usable from the user perspective.

1. Introduction

Smart city technologies are intended to improve citizens' lives. With the technology implemented in the smart city, the technology accelerates the pace of people searching of information. Urban infrastructures or basic facilities such as museum is essential to be supported by the technologies. "Kota Lama Semarang "or The old city of Semarang is a natural museum that exhibit historical buildings that tell stories the city of Semarang in the era of Dutch colonial rule. Many technologies has been used to engage the museum and the museum visitor. The engagement is essential because it can help the historical object in museum [1]. However, Rey et al. [2] noted that the used of information and communication technology (ICT) do not reached the usability of the technology from the users' perspective. The user prefer used traditional style when visiting a museum.

Many museum provides web site as a tool to attract the visitor to museum event, however, the validation that the usage of web site for attracting a museum success is need to be measure [3] Another research suggested for using a robot to replace the usage of human as guide tour in Museum [4]. But, the problem with the usage of technology in the museum to engages a visitor is financially problem.

With popularity of QR code, the usage of QR code for engage visitor for an art and museum exhibition is commonly usage. Evaluate the usage of QR code for help visitors personalize their visit to museum [5]. The result suggested that QR code is received positive respond from the museum staff, the used of QR code was assumed that generally young visitor owned smartphone and they familiar for using QR code.

Leaflet map of Semarang Historic building commonly used to provide the location information. However, leaflet map has a limitations, it cannot provide enough information about the historic building. The limitations of leaflet maps is only provide little information which explaining the condition of the historic building. To expand the information about the historical building, prior research suggested the use of web technologies. However, some people suggested the usage of leaflet – map which can show the location and the web site can provide more additional information about the historical building. In this paper, this paper purpose a new design for urban infrastructure facilities (natural museum), which employs Quick Response (QR) code technologies as a link to combine the information between leaflet map and web technologies. A usable interface is essential to allows mobility and ease of use of information using sharing system [6].

The originality or the value of the research is acceptable design for information system that combined map and web using QR code to access information about historic building in cultural space museum. And, the IS design has validated using usability testing that considered the user's prior knowledge for using QR code and SUS questionnaire and metric.

2. Validation Method

2.1 Grouping the user based on the QR code usage as respondents profile

It is expected that users will be familiar with technology and can return to the information system whenever necessary without prior training [7,8]. In usability, the user experience is influenced by subjective attitudes and feelings towards the system [9]. In 2010, ISO 9241-110 Ergonomics of human-system interaction: Human-centred design for interactive systems defined user experience as 'a person's perceptions and responses resulting from the use of a product, system, or service' [10,11]. The users' experience influences their perception of the system [8].

In the context of interactions with the system, the user's task also influences their perception of the system [12]. To use user perception as a measurement, it is essential that the user has an understanding about the technology. The users' perception is beneficial to the evaluation of technology that is reused to deliver public services; such as QR code or and website; because the user already uses the application. User's perception is also affected by the users' experience when they interact with the system [13].

User perception is a measurement commonly used when users are at the centre of the system evaluation and refers to users' opinions and judgements [14]. In addition, the user's perception of technology use is essential because it can influence the system information [15].

Table 1. Question for understanding the users' prior knowledge in the QR code's usage

No	Question	Scale
1	Smartphone ownership	- Smartphone - None
2	User experience for using QR code Have an experience for using QR code for payment or to get a link for an website.	- For Payment - To get link for a website
3	User's frequent of use for using QR code	- None - Beginner (1st time user) - Rarely used (1-3) - Regular user (more than 3 times)
4	User have experience for using Web to access information for historical building, Kota Lama, Semarang.	- None - Beginner (1st time user) - Rarely used (1-3) - Regular user (more than 3 times)
5	User have experience for using Map to get access as direction to explorer for historical building, Kota Lama, Semarang.	- None - Beginner (1st time user) - Rarely used (1-3) - Regular user (more than 3 times)
6	User have experience for using Map that contain with QR code to get access to a website for historical building, Kota Lama, Semarang.	- None - Beginner (1st time user) - Rarely used (1-3) - Regular user (more than 3 times)

To investigate prior knowledge and to assess the effectiveness of the QR code usage to provide information about the historic building, Kota lama, Semarang. This study used the user's Frequency of Use (FOU) of QR code. The user will be identified based on their ownership for smartphone, the experience for using QR code for payment and search for web link; the FOU of QR code usage; the FOU for using web as museum exhibit information; the FOU for using Map as guidance in museum exhibit and the FOU of QR code usage in map. Table 1 shows the user's prior knowledge for using QR code, map and web site. It essential for this study for grouping the participants, so the researcher can give a good recommendations about the design. As noted, to measure the effectiveness of museum information system design is not only about the user's acceptance. But, it also essential to understand the user's prior knowledge so the information system design can be usable for every visitor.

Trial and survey for this study used twelve students in urban regional planning, Universitas Diponegoro, Semarang, Indonesia. The reason for selecting the Urban Regional Planning department students due to that they familiar with the condition of Kota Lama, Semarang and they will be a potential user for using the Information System for study activities. The data of usability experiments should be collected from participants who are familiar with aim of the information system area being assessed [16]. The experiments conducted after the student finished a survey for recognising the historical building.

According to Krug [17] the ideal number of respondents is three to four for each test. Whereas according to Nielsen [18] the best number of respondents in each usability test is 5 people, respondents numbering 5 are considered to be able to represent most of usability problems without involving many resources.

And it suggested to use a real condition to trial the technology help the participants to feel the real environment when the technology used. In addition, it also to help the researcher with the condition of infrastructure such internet connection the area of Kota Lama, Semarang. Location and situation of the user when accessing the information system is included as one of factors in Physical Context in the

Context of Use in Human-Mobile Computer Interaction (CoU-HMCI model) [19]. In addition, Understanding a user's task includes the conditions and location in which the task is performed.

2.2 Method for understanding the users' usability

This study employed System Usability Scale (SUS) method to identify the user's design for using QR code. SUS is a tool that could quickly and easily collect a user's subjective rating of a product's usability[20]. The SUS, developed by Brooke (1996), reflected a strong need in the usability community for a tool that could quickly and easily collect a user's subjective rating of a product's usability. The System Usability Scale (SUS) is a reliable tool for measuring the usability. It consists of a 10 item questionnaire with five response options for respondents; from strongly agree to strongly disagree [21]. System Usability Scale (SUS) a reliable, low-cost usability scale that can be used for global assessments of systems usability [22].

SUS consist with ten question which described, below:

1. I think that I would like to use this system frequently.
2. I found the system unnecessarily complex.
3. I thought the system was easy to use.
4. I think that I would need the support of a technical person to be able to use this system.
5. I found the various functions in this system were well integrated.
6. I thought there was too much inconsistency in this system.
7. I would imagine that most people would learn to use this system very quickly.
8. I found the system very cumbersome to use
9. I felt very confident using the system.
10. I needed to learn a lot of things before I could get going with this system.

The questionnaire for this study adapt from the SUS question.

1. I think I would like to use a QR code to find information on ancient buildings when I visit Kota Lama, Semarang
2. I found that using QR code unnecessarily complex.
3. I thought the QR code was easy to use to find the story of the historical buildings when I visit Kota Lama, Semarang
4. I think that I would need the support of a technical person to be able to use QR code to find the story of the historical buildings when I visit Kota Lama, Semarang
5. I found the various functions in this system (QR code, map and web information) were well integrated to deliver information about the story of ancient building when I visit Kota Lama, Semarang.
6. I thought there was too much inconsistency in this system (when QR code used to integrate maps and the web) to get information about the ancient building at Kota Lama, Semarang.
7. I would imagine that most people would learn to use this system (when QR code used to integrate maps and the web) very quickly.
8. I found the system (QR code, map and web information) very cumbersome to use when I visit Kota Lama, Semarang
9. I felt very confident using the system (QR code, map and web information) to obtain information about ancient buildings without the other's help
10. I needed to learn a lot of things before I could get going with this system (QR code, map and web information) to obtain information about ancient buildings.

2.3 Analysing the result

To examine the result of the question, the SUS provide scale between one and five using Likert scale. For odd question (number r1, r3, r5, r7 and r9) using formula the scale – 1. As for the even number question (R2, R4, R6, R8 and R10) the participants respond should be count using formula scale -5. From the result of first and second equation the SUS final score can be reached by using the equation 1.

$$Score\ SUS: ((r1 - 1) + (5 - R2) + (r3 - 1) + (5 - R4) + \ldots\ldots + (5 - R10)) * 2.5 \quad (1).$$

Legend: *r1, r3, r5, r7, r9* *: result for odd number question*
R2, R4, R6, R8, R10 *: result for even number question*

The results of the SUS score for each respondent are then summed and divided by the number of respondents (Table 2). After calculating the average score of SUS, the score was compared with the range of values proposed [20].

Table 2. Score SUS

Score SUS	***Interprets***
< 50	Not Acceptable
50 – 70	Marginal
>70	Acceptable

3. Analysis and Result

3.1 User Prior Knowledge

User prior knowledge in this research is about, the technology ownership, user experience and user frequent of use QR code and web services is essential to identify the prior knowledge for user's usability perception in the usage of QR code to recognizing the historical building. This user prior knowledge is essential as precondition to measure the usability of design of Information System [21]. It as an aid to describe the result of SUS score.

Table 3 shows the user prior knowledge for QR code usage. From the Table 3 it can be defined all of the respondents has their own smart phone. The respondents with an experience for using QR code for payment is about 59.23% and 40.77% is to get link for a website.

Table 3. Users' prior knowledge in the QR code's usage

No	Question	Criteria	Frequency	Percentage
1	Smartphone ownership	Smartphone	27	100%
2	User experience for using QR code	For Payment	16	59.23%
		To get link for a website	11	40.77%
3	User's frequent of use for using QR code	Rarely used (1-3)	4	14,77%
		Regular user (more than 3 times)	23	85.12%
4	User have experience for using Web to access information for historical building, Kota Lama, Semarang.	None	4	14.82%
		Beginner (1st time user)	1	3.70%
		Rarely used (1-3)	9	33.33%
		Regular user (more than 3 times)	13	48.15%
5	User have experience for using Map to get access as direction to explorer for historical building, Kota Lama, Semarang.	None	2	7.41%
		Beginner (1st time user)	2	7.41%
		Rarely used (1-3)	3	11.11%
		Regular user (more than 3 times)	20	74.07%
6	User have experience for using Map that contain with QR code to get access to a website for historical building, Kota Lama, Semarang.	None	18	66.67%
		Beginner (1st time user)	4	14.82%
		Rarely used (1-3)	2	7.41%
		Regular user (more than 3 times)	3	11.11%

3.2 SUS result based on the user prior knowledge.

Table 4. Users perception on usability in the QR code's usage

<table>
<tr><th>No</th><th>User Profile</th><th>SUS Result</th></tr>
<tr><td>1</td><td>User experience for using QR code for Payment and to get a link for a website</td><td>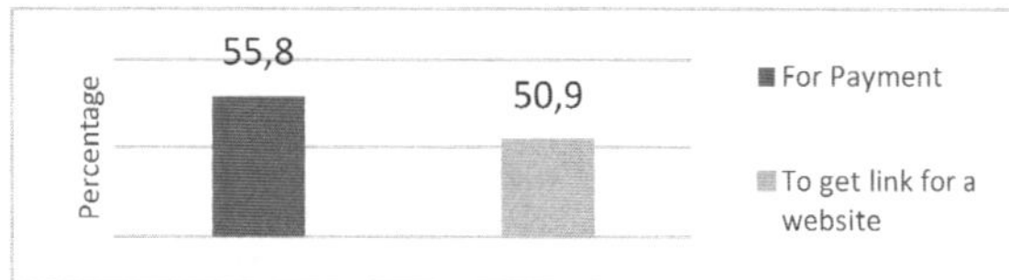
</td></tr>
<tr><td>2</td><td>User's frequent of use for using QR code</td><td>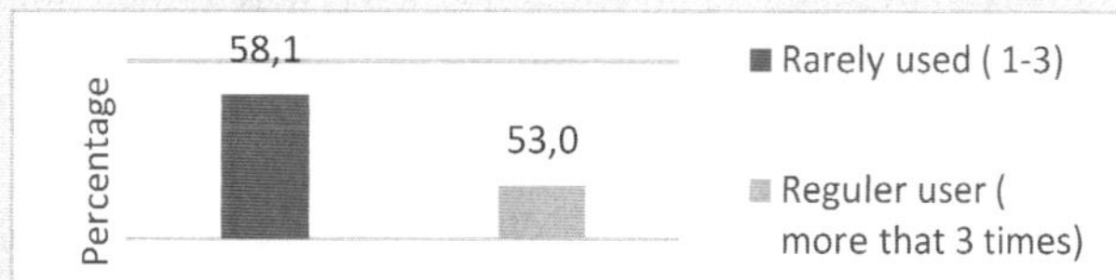
</td></tr>
<tr><td>3</td><td>User have experience for using Web to access information for historical building, Kota Lama, Semarang.</td><td>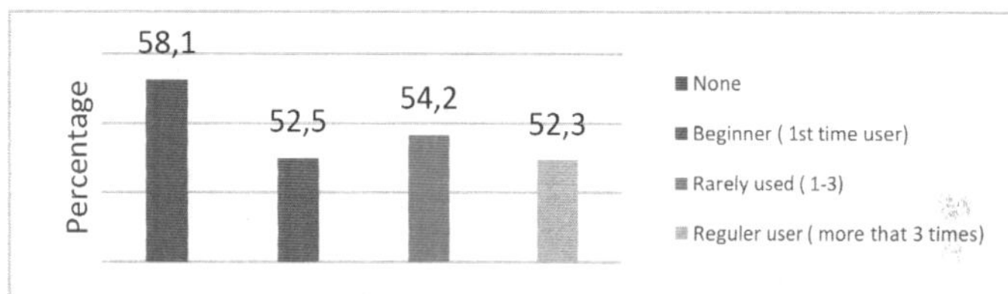
</td></tr>
<tr><td>4</td><td>User have experience for using Map to get access as direction to explorer for historical building, Kota Lama, Semarang.</td><td>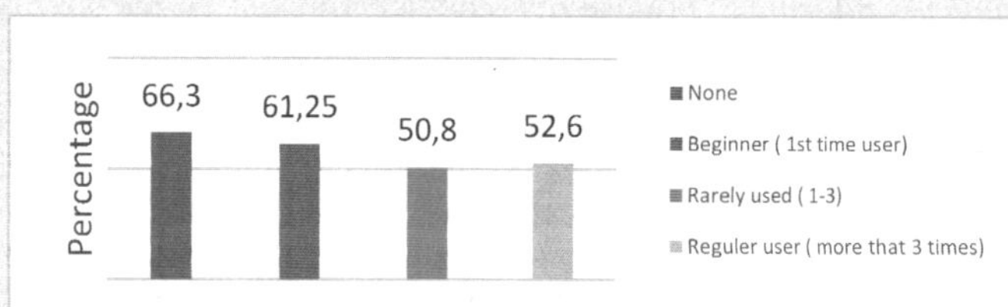
</td></tr>
<tr><td>5</td><td>User with experience for using Map that contain with QR code to get access to a website for historical building, Kota Lama, Semarang.</td><td>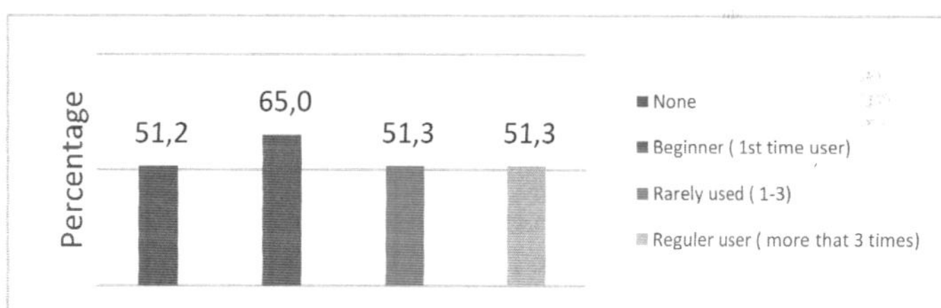
</td></tr>
</table>

Table 4 shows the result of the SUS score for each respondents that have identified in Table 3 is marginal. In Figure 1, it explains the position of marginal for an adjective rating for the user.

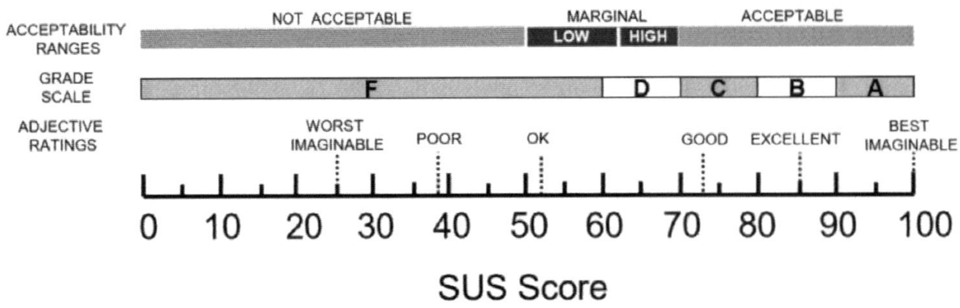

Figure 1. A comparison average SUS score

The usability testing in this research is to understand the Information System design using QR code for recognising the historical building has discover major problems in the user interface. It is suggested, that Usability testing has a number of possible goals and purposes. Certainly one of the most important is to discover major problems in the user interface that could result in human error, terminate the interaction, and lead to frustration on the part of the user [23].

Marginal according to Figure 1 with SUS score 50 can be describe as adequate, sufficient and acceptable for the user. It is due to that few of the respondent still not familiar with QR code technology (Table 3). With the result of the moderate. It is suggested that the design is still usable to be used, but is still need time for the responded to be familiar with technology. As noted in Nugrageni et al. [24] the user device and frequent of use of technology influence the ease of use of technology. Therefore, with the moderate result on usability by implementing QR code for recognising the historical building it can be applied in Kota Lama, Semarang.

Therefore, for each group that participate in the experiments examining the usability of QR code in the design of Information system for recognizing historic buildings, Kota Lama, Semarang is acceptable. Therefore, the design has a positive feedback from the participants who has an experience for visiting Kota Lama, Semarang. From the observation, many of the respondent still new with the QR code technology.

4. Conclusion

The result of SUS score is 50 which describe as the QR code as link between map leaflet and web are adequate, sufficient and acceptable for the user. Information access using QR code can overcome the problem of short explanation in map leaflet that commonly used a guide in cultural space museum.

The information of historical building using QR code is sufficient or the visitor to learn the building history. As, for he QR code usability test using SUS question showed the number that the design is acceptable. Using QR codes on map might help the cultural space museum enhance their accessibility, particularly as visitors increasingly arrive with their own smartphones or tablets.

5. Acknowledgment

This article was presented at the 2nd International Conference on Smart City INNOVATION (ICSCI) 2019, jointly held by Universitas Indonesia and Universitas Diponegoro. ICSCI conferences have been supported by the United States Agency for International Development (USAID) through the Sustainable Higher Education Research Alliance (SHERA) Project for Universitas Indonesia's Scientific Modelling, Application, Research, and Training for City-centered Innovation and Technology (SMART CITY) Center for Collaborative Research, administered through Grant #AID-497-A-1600004, Sub Grant #IIE-00000078-UI-1.

IOP Conf. Series: Earth and Environmental Science **396** (2019) 012026 doi:10.1088/1755-1315/396/1/012026

6. References

[1] Pérez-Sanagustín M, Parra D, Verdugo R, García-Galleguillos G and Nussbaum M 2016 Using QR codes to increase user engagement in museum-like spaces *Comput. Human Behav.* **60** 73–85

[2] Rey F B and Casado-Neira D 2013 Participation and Technology: Perception and Public Expectations about the Use of ICTs in Museums *Procedia Technol.* **9** 697–704

[3] Kabassi K 2017 Evaluating websites of museums: State of the art *J. Cult. Herit.* **24** 184–96

[4] Yamazaki K, Heath C, Yamazaki A, Okada M, Kuno Y, Kobayashi Y, Hoshi Y, Pitsch K, Luff P and vom Lehn D 2009 Revealing Gauguin *Proceedings of the 27th international conference on Human factors in computing systems - CHI 09* (ACM Press)

[5] Schultz M K 2013 A case study on the appropriateness of using quick response (QR) codes in libraries and museums *Libr. Inf. Sci. Res.* **35** 207–15

[6] Ozok A A, Gurses A P, Wu H, Nelson M, Moen D and Wei J 2009 Usability and User Acceptance for Personal Health Records: A Perspective from Healthcare Citizens *Online Communities and Social Computing* (Springer Berlin Heidelberg) pp 690–9

[7] Dishaw M T and Strong D M 1999 Extending the technology acceptance model with task-technology fit constructs *Inf. Manag.* **36** 9–21

[8] Nielsen J 1992 Finding usability problems through heuristic evaluation *Proceedings of the SIGCHI conference on Human factors in computing systems - CHI -92* (ACM Press)

[9] Lund A M 2006 Post-modern usability *J. Usability Stud.* **2** 1–6

[10] International Organization for Standardization 2006 *ISO 9241-110: Ergonomics of Human-system Interaction* (Dialogue Principles)

[11] Vermeeren A P O S, Law E L-C, Roto V, Obrist M, Hoonhout J and Väänänen-Vainio-Mattila K 2010 User experience evaluation methods *Proceedings of the 6th Nordic Conference on Human-Computer Interaction Extending Boundaries - NordiCHI -10* (ACM Press)

[12] Bradley N and Dunlop M 2005 Toward a Multidisciplinary Model of Context to Support Context-Aware Computing *Human-Computer Interact.* **20** 403–46

[13] Kuniavsky M 2007 User Experience and Hci *The Human-Computer Interaction Handbook* (CRC Press) pp 897–916

[14] Nielsen J and Levy J 1994 Measuring usability: preference vs. performance *Commun. ACM* **37** 66–75

[15] Dumas J and Fox J 2009 Usability Testing *Human Factors and Ergonomics* (CRC Press) pp 231–51

[16] Chowdhury G G and Chowdhury S 2011 *Information users and usability in the digital age* (Facet Publishing)

[17] Krug S 2014 Don't make me think, revisited: a common sense approach to Web usability *Choice Rev. Online* **51** 51–6218

[18] Nielsen J 2012 Usability 101: Introduction to usability **9** 35

[19] Jumisko-Pyykkö S and Vainio T Framing the Context of Use for Mobile HCI *Social and Organizational Impacts of Emerging Mobile Devices* (IGI Global) pp 217–48

[20] Bangor A, Kortum P and Miller J 2009 Determining what individual SUS scores mean: Adding an adjective rating scale *J. usability Stud.* **4** 114–23

[21] U.S. Department of Health and Human Services 2017 Usability

[22] Brooke J and others 1996 SUS-A quick and dirty usability scale *Usability Eval. Ind.* **189** 4–7

[23] Norman K L and Panizzi E 2006 Levels of automation and user participation in usability testing *Interact. Comput.* **18** 246–64

[24] Nugraheni D M K and de Vries D 2016 Profile of a typical mobile SMS user in emergency situations (empirical study in an urban flood prone area) *2016 2nd International Conference on Science in Information Technology (ICSITech)* (IEEE)

Traffic Management for Jalan Kayu Ayu in Seminyak Area, Badung Regency

I N K Mataram[1], A A N A J Wikrama[1], I G A M Suartika[2]
[1]*Civil Engineering, Universitas Udayana*
[2]*Arcitecture Engineering, Universitas Udayana*

E-mail: nym.karnata@unud.ac.id

Abstract: The usage of land for commercial purpose along the side of the road often causes conflicts. One of the areas in Badung Regency, namely along the road called Jalan Kayu Aya, is an attractive tourism area, the passageway is quite large, while on the other hand the effective width of the road is only 5 meters. The purpose of this study is to analyze the performance and the current condition of the road as well as to analyze alternative solutions for improvement. Based on the results of the analysis, the peak hours at Jalan Kayu Aya Badung Regency is at 23.15-00.15 WITA, with a traffic volume of 1,078 pcu / hour, road capacity of 1236.51 pcu / hour, actual speed of 19, 26 km / h and a saturation level is 0.87 and an F for the level of road service. The analysis results of alternative 1 with the arrangement of side barriers shows that the traffic volume becomes 1,078 pcu / hour. The road's capacity is 1404.43 pcu / hour, the actual speed is 27.23 km / h and the saturation level is 0.77 with a C for the level of road service. While the analysis of alternative 2 with the application of one-way roads results in a traffic volume of 550.2 pcu / hour, the capacity of the road's passage is 1407.06 pcu / hour, the actual speed is 27.23 km / h and saturation level is 0.77 with a road service level of B.

Keywords: road performance, side obstacle

1. Introduction

Badung Regency is among the Regencies in Bali Province which is active in the trade, agriculture and tourism sectors. The tourism sector is the main economic activity in the Southern Badung region, mainly in Kuta District. One of the locations with a high density of visits from local and foreign tourists is Jalan Kayu Aya. There are many facilities that support the tourism activities in this area such as hotels, restaurants, cafés and art shops, which have a positive impact on the increase in the number of tourist who visits.

The increase in economic activity in the areas of Jalan Kayu Aya has lead to an increase in the road traffic volume. The utilization of land for commercial purposes along the sides of the road creates various problems[1]. The land along the sides of the road is generally used by pedestrians (both walking on and crossing), non-motorized vehicles, access for vehicles from the road side, and there are also vehicles that stop and park on the street. The various activities has caused side friction conflicts [2,3].

The main problem in Jalan Kayu Aya Badung Regency is the tall barriers on the road side. While the width of the road is narrow, it only has an effective width of 5m. So that traffic delays often occur due to the high side barriers on these roads.

Based on the background of the traffic problems in Jalan Kayu Aya, this study will examine the extent of the influence of side barriers on the road performance along with alternative solutions. This research is the first study to analyze side barriers in the areas of Jalan Kayu Aya.

2. Materials And Method

2.1. Side Obstacles

Many transportation problems are linked to the problems of delays that occur during the trip. [4,5]These problems are caused by vehicles entering and exiting the road, vehicles that park on the roadside, pedestrians using the road bodies, and non-motorized vehicles [6]. These are what are called side obstacles. Side obstacles in certain circumstances may result in a temporary travel delay and also cause traffic jams (Table 1, 2).

Table 1. Side Barriers based on the Classifications of City Size

Total Population of The Regency (Million People)	Classification of City Size (CS/ City Size)
CS < 0.1	Very Small
0.1 ≤ CS < 0.5	Small
0.5 ≤ CS < 1.0	Mediun
1.0 ≤ CS < 3.0	Large
3.0 ≤ CS	Very Large ,

Source: Departemen Pekerjaan Umum, 1997 [7]

Table 2. Factor Weightage for each type of events on the side barrier

Types of Side Obstacles	Symbol	Factor Weightage
Pedestrian	PED	0.5
Parking and Stop Vehicle	PSV	1.0
Enter and Escape Vehicle	EEV	0.7
Unmotorized	UM	0.4

Source: Departemen Pekerjaan Umum, 1997 [7]

2.2. Traffic Volume

In MKJI, the traffic flow value (Q) is expressed in passenger car units (pcu). All traffic flow values (per direction and total) are converted into passenger car units by multiplying the number of vehicles with passenger car equivalents (pce) (Table 3) which are derived empirically based on vehicle type [7]. According to Putranto, the definition of traffic flow / volume is the number of vehicles passing a point on a cross section of a road within a certain time frame [8].

The research results volume can be calculated using the following equation :

$$Q = \frac{n}{T}$$

The equivalence of passenger cars (imp) for each type of vehicle depends on the road type and total traffic flow expressed in the vehicle / hour.

Table 3. Passenger Car Equivalent (PCE) on undivided city roads

Type: Undivides Road	Total Two Ways Traffic Flow (Vehicle/ Hour)	pce		
		HV	MC	
			Width of the Traffic Path (m)	
			≤ 6	> 6
Undivided Two Paths	0 – 1800	1.3	0.5	0.40
(2/2 UD)	≥ 1800	1.2	0.35	0.25
Undivided Four Paths	0 – 3700	1.3	0.40	
(2/2 UD)	≥ 3700	1.2	0.25	

Source: PU Department, 1997

2.3. Basic Capacity

Capacity is expressed in passenger car units (pcu). The basic equation to determine capacity is:

$$C = Co \times FCw \times FCsp \times FCsf \times FCcs$$

2.4. Free Flow Velocity

Free flow velocity is defined as the velocity in which there are no obstacles on the road, depending on the speed that the driver choose driving a motorized speed without obstruction of other motorized vehicles on the road (i.e. when the current = 0). The free flow velocity of passenger cars is usually 10-15% higher than other types of vehicles. The equation to determine the free flow velocity on urban roads is as follows:

$$FV = (FVo + FVw) \times FFVsf \times FFVcs$$

The road side performance is quantitative measure to improve the operational conditions of the traffic facility, such as what was evaluated by Bina Marga [7]. The road side performance is measured using five indicators, namely volume, capacity, travel speed, degree of saturation and level of service.

2.5. Speed

Time Mean Speed is the average speed of all vehicles that cross a point on the road within a certain period of time, while Space mean speed is the average speed of all vehicles occupying a piece of road within a certain period of time. In this study, the mean space speed is used to calculate the average velocity value. Space Mean Speed is calculated using the following formula:

$$Vs = \frac{n.d}{\Sigma ti}$$

2.6. Degree of Saturation

Degree of Saturation (DS) is defined as the ratio of the current (Q) to the capacity (C), used as the main factor in determining the level of performance of intersections and road segments. The DS value indicates whether the road segment has a capacity problem or not.
Formula to calculate the Degree of saturation:

$$DS = Q/C$$

2.7. Level of Service

The concept of service level is used to measure the quality of the road service. Measures that are suitable for determining the level of service can be by identifying the speed of vehicles passing a highway and / or the volume of vehicles on the road. The level of road service is a quantitative measure that reflects the driver's perception of the road quality related to the vehicle. Level of road service can be classified from service level A to F, which is measured from the V / C ratio as shown in Table 4 where V is the flow or current (pcu / hour) and C is the capacity (pcu.hour).

Table 4. The Relationship between Service Level

Level of Service	**Actual Conditions**	**V/C Ratio**
A	Free flow allowing drivers to choose any speed they want without any delay	$0 < A \leq 0.20$
B	Stable flow, speed may be limited due to traffic conditions, driver has enough liberty to choose their speed	$0.20 < B \leq 0.45$
C	Stable flow, but speed and vehicle movement may be limited due to traffic conditions, driver's ability to choose there are limited	$0.45 < C \leq 0.75$
D	The flow is close to unstable, speed is still controlled by traffic conditions, V/C ratio is still tolerable	$0.75 < A \leq 0.85$
E	Traffic volume is near its capacity, unstable flow, speed occasionally goes down to zero	$0.85 < A \leq 100$
F	Traffic flow congested, low speed, long lines of vehicles and large amount of obstacle	$F \geq 100$

Source: Morlok, 1984

3. Research Design

The steps to conduct the research are shown in Figure 1 as follows:

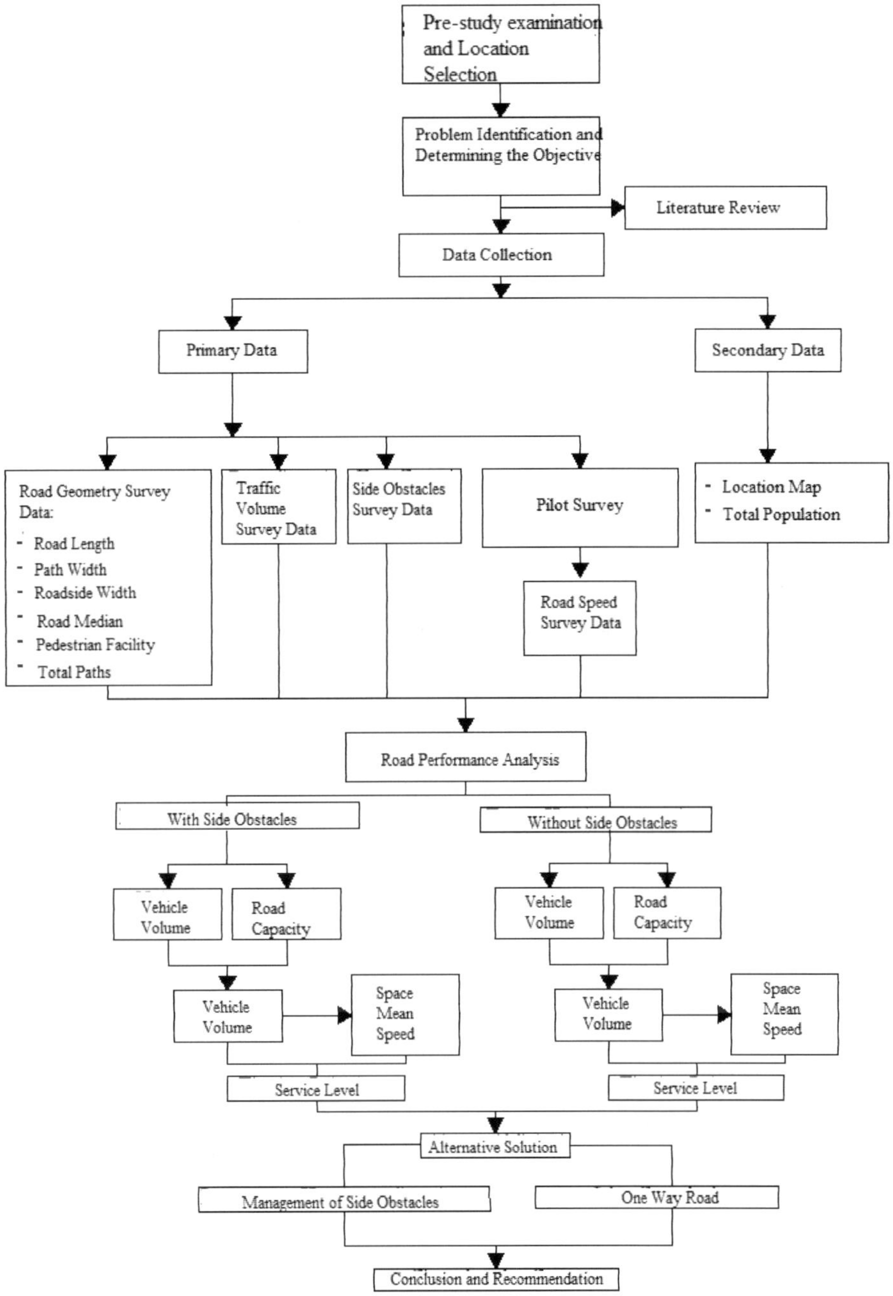

Figure 1. Research Design

The criteria considered in choosing the location of this study are: there must be pedestrians that walk on and cross the road, vehicles are able to enter and exit the road from the shopping area along the sides of the road, there are vehicles that stop or park on the road, there are non-motorized vehicles, and there are no signs or information regarding parking or stop restrictions on the road despite being a shopping district with high traffic activity.

Collecting primary data in this study was done using manual methods and digital methods. The manual method is by collecting data directly at the study location that involves several people to record the needed data. The data needed are road geometry data, side obstacle data, volume data, traffic speed

IOP Conf. Series: Earth and Environmental Science **396** (2019) 012027 doi:10.1088/1755-1315/396/1/012027

data, and side obstacle data. Digital methods are used to determine the traffic volume data. Data collection is conducted on days when tourism is busy, which was on weekends. The time in which the traffic volume survey on Jalan Kayu Aya in Badung Regency was carried out was on Saturday, January 19, 2019 for 16 hours without pause. The survey was conducted in conjunction with the survey regarding side barriers and speed but by using different methods.

4. Results And Discussion

4.1. Total Population of People

The total population in Badung Regency in 2018 is 643.500 people and this is used as a reference for analysis [9]

4.2. Side Obstacle Analysis

The peak hours of the side obstacles in Jalan Kayu Aya, Badung Regency, adjusted for the peak hours of the traffic on this road, is from 23.15-00.15. The results of the side obstacle analysis conducted during the peak hours are shown in Table 55.

Table 5. Results of the Analysis on the Side Obstacles

Types of Side Obstacles	Symbol	Factor Weightage	Frequency of Occurences in the Field (per hour/200m)	Weighted Frequency (per hour/200m)
Pedestrian	PED	0.5	205	102
Parking and Stop Vehicle	PSV	1	218	218
Enter and Escape Vehicle	EEV	0.7	264	184,8
Unmotorized Vehicle	SMV	0.4	1	0.4
	Total Occurrence (per hour/200m)			505.7

Source: Analysis Results, 2019

4.3. Spot Speed Analysis

Spot speed is the speed of a vehicle at a certain time measured starting from a specified place. Spot speed is among the various types of speed in the science of traffic. There are two types of velocities obtained in the study of traffic flow velocity, namely Time Mean Speed (Vt) and Space Mean Speed (Vs) where both are part of Spot Speed. The speed calculation used is the average space speed (space mean speed), in which the vehicle speed at Jalan Kayu Aya in Badung Regency is calculated using Equation 2.4.

The space mean speed of vehicles at Jalan Kayu Aya Badung regency is calculated as follows:

$$Vs = \frac{n.d}{\Sigma ti} = \frac{128x(25m/1000)}{(598,43\ dt/3600)}$$

$$= 19,26\ km/jam$$

The average speed of light vehicles at Jalan Kayu Aya Badung Regency is 19,26 km/hour.

4.4. Analysis on the Current Road Performance Traffic Volume

From the fluctuation in the graph, the peak hour is at 23.15-00.15 where the total volume is 1575 vehicles / hour and Q = 1078 pcu / hour. The next analysis conducted is the direction separator analysis, in order to calculate the percentage of direction separators. The calculation is displayed in Table 6.

Table 6. Direction Separator

Direction	Type				Arus (Kend/Jam	Vehicle Type			Total	Flow (pcu/hour)
	MC	LV	HV	Total		0.35 x MC	1.0 x LV	1.2 x HV		
East to West	392	413	0	805	1575	137.2	413	0	550.2	1078
West to East	373	397	0	770		130.55	397	0	527.55	
Percentage of the Direction Separation	East to West				0.511			**50%-50%**		
	West to East				0.489					

Source: Analysis Results,2019

4.4.1. Current Capacity

To calculate the capacity during the traffic peak hours at jalan Kayu Aya, Badung Regency, the following equation is utilized:

C = Co x FCw x FCsp x FCcs x FCsf
= 2900 x 0,56 x 1,00 x 0,94 x 0,81
= 1236,51 pcu/hour

4.4.2. Degree of Saturation

After conducting the traffic volume survey and the results of road capacity analysis is obtained, the degree of saturation (DS) analysis can afterwards be conducted. To find the degree of saturation of Jalan Kayu Aya in Badung Regency, the following equations are used:

$$DS = \frac{Q}{C}$$

$$DS = \frac{1078}{1236.51}\ SMP/Jam = 0{,}87\ SMP/Jam$$

4.4.3. Speed

The method used to analyze the speed of vehicles at Jalan Kayu Aya in Badung Regency is the space mean speed method. The space mean speed of vehicles at Jalan Kayu Aya in Badung Regency is 19.26 km/hour. After the speed is calculated, the next step is to determine the level of service.

4.4.4. Level of Service

Jalan Kayu Aya, Badung Regency, is a local road which has a width of 5 meters. The initial plan was so that the speed on the flat terrain of this road is 40 km / h [10]. The relationship between planned speed, actual speed, degree of saturation, and level of road service is shown in Figure 2.

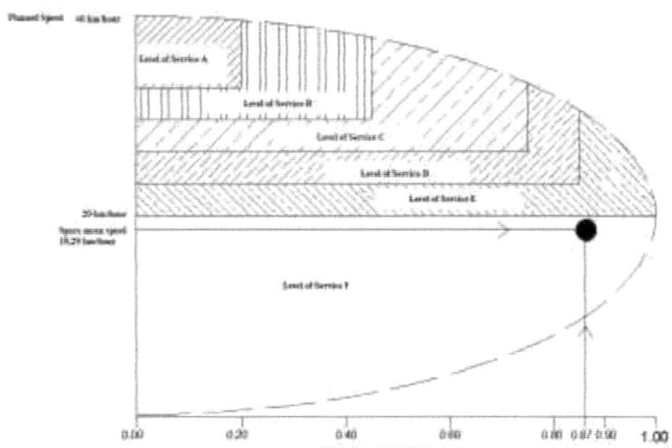

Figure 2. Existing Level of Service

4.5. *Alternative Performance Analysis 1: Managing the Side Obstacles*

In assessing the solutions, there are some side obstacles which can be assumed to have no weightage value.

Table 7. Side Obstacle Reduction Analysis

Types of Side Obstacles	Symbol	Factor Weightage	Frequency of Occurences in the Field (per hour/200m)	Weighted Frequency (per hour/200m)
Pedestrian	PED	0.5	0	0
Vehicle, Parking and Stop	PSV	1	0	0
Vehicle Enter and Exiting	EEV	0.7	264	184.8
Unmotorized Vehicle	SMV	0.4	0	0
	Total Occurrence (per hour/200m)			184.8

Source: Analysis Results, 2019

Based on the results of the analysis above, the level of the road side barriers is low (L) with an FCsf value = 0.92. Even by managing the side obstacles, vehicles coming in and out cannot be significantly reduced because the road does not allow for slow lanes to be constructed.

4.5.1. *Traffic Volume*

The traffic volume utilized to measure the performance of the road is the traffic volume during the peak hours. From the fluctuation graph, the peak hour is found to be from 23.15 –
00.15 n which the total volume is 1575 vehicles/hour and the Q = 1078 pcu/hour.

4.5.2. *Capacity*

C = Co x FCw x FCsp x FCcs x FCsf
= 2900 x 0,56 x 1,00 x 0,94 x 0,92
= 1404,43 pcu/hour

4.5.3. *Degree of Saturation*

Below is the equation used to analyze the degree of saturation due to the reduction of side obstacles:

$$DS = \frac{Q}{C}$$

$$DS = \frac{1078}{1404.43}\ SMP/Jam = 0{,}77\ SMP/Jam$$

4.5.4. *Speed*

Afterwards, the speed of vehicles at Jalan Kayu Aya is analyzed. The FVo in this study is not from the by the PU Department, 1997, but it is the planned speed which 40 km/hour [10].

FV = (FVo + FVw) x FFVsf x FFVcs
FV = (40+(-9,5)) x 0.95 x 0.94
FV = 27,24 km/jam

Based on the calculations above, the speed of vehicles at Jalan Kayu Aya is calculated to be 27,24 km/hour.

4.5.5. *Level of Service*

The service level of the road after assuming that the side obstacles has decreased with a DS = 0.77 and Vs = 27.23 km/h, is at the service level C, and this can be seen in Figure 3.

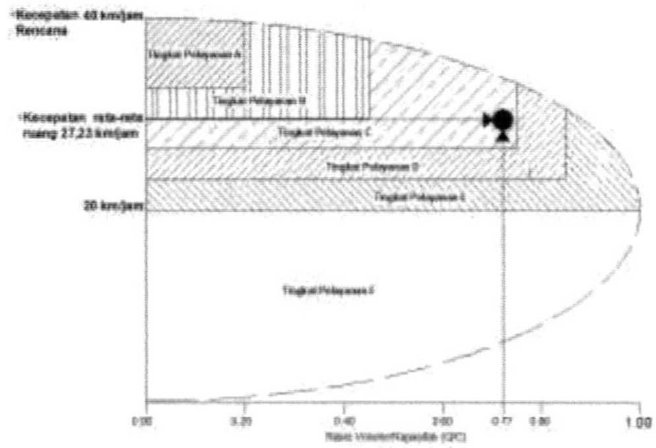

Figure 3. Level of Service for Alternative 1

4.6. Analysis of Performance Alternative 2: One Way Road System

It is recommended for Jalan Kayu Aya Badung Regency to be a one-way road from East to West. This can be seen from the results of the analysis of the direction separation factors which is 50% from the east and 50% from the west. Viewed from the volume of vehicles, a larger proportion are from the east so that one direction movement is determined to be from the east to the west.

4.6.1. Traffic Volume

During the highest peak hours of Jalan Kayu Aya in Badung Regency, from east to west, which is the at 23: 15-00: 15, the total volume of cars is 805 (vehicles / hour) with a One Direction Q = 550.2 pcu / hour.

4.6.2. Capacity

In the performance analysis of the one-way road here, the correction factor for the side obstacle is used for the correction of the existing condition. To determine the basic capacity, in this calculation Jalan Kayu Aya, Badung Regency was changed to a one-way road, because currently it has two lanes. The capacity was multiplied by the number of lanes. The calculation is as follows: Co = 2x 1650 pcu / hour.

Calculation of the One Way Road Capacity: C
= Co x FCw x FCsp x FCcs x FCsf
= (2x1650) x 0,56 x 1,00 x 0,94 x 0,81
= 1407,06 pcu/hour

4.6.3. Speed

In the speed analysis of the one-way road version of Jalan Kayu Aya in Badung Regency, the FVo is not derived from the tables of the Department of Public Works, 1997, but the planned speed of 40 km/h (Saodang, H, 2005) is utilized:

FV = (FVo + FVw) x FFVsf x FFVcs
FV = (40+(-9,5)) x 0.95 x 0.94
FV = 27,23 km/jam

Based on the calculations above, the free flow speed at Jalan Kayu Aya, Badung Regency is 27,23 km/hour.

4.6.4. Degree of Saturation

The following equation is used in the analysis of the one-way road's degree of saturation:

$$DS = \frac{Q\ puncak\ 1\ arah}{C}$$

$$DS = \frac{550.2}{1407.06}\ SMP/Jam = 0{,}39\ SMP/Jam$$

$$= 0{,}39$$

After analyzing the one way road performance alternative, the degree of saturation is 0,39. This means that the road can still accommodate 61% more vehicles than the road's capacity. In contrast with the current capacity of Jalan Kayu Aya, Badung Regency, which can only accommodate vehicles at 11% of the road's capacity.

4.6.5. Service Level

The level of road service has a purpose to determine the degree of the road's performance that is calculated based on the level of road usage, speed, density, and obstacles that occur. To calculate the level of service, some data are needed, namely the optimum speed, travel speed, and degree of saturation. The service level of the road segment, after assuming that the road is a one-way, is service level B with a DS = 0.39 and Vs = 27.23 km/h. This can be seen in Figure 4.

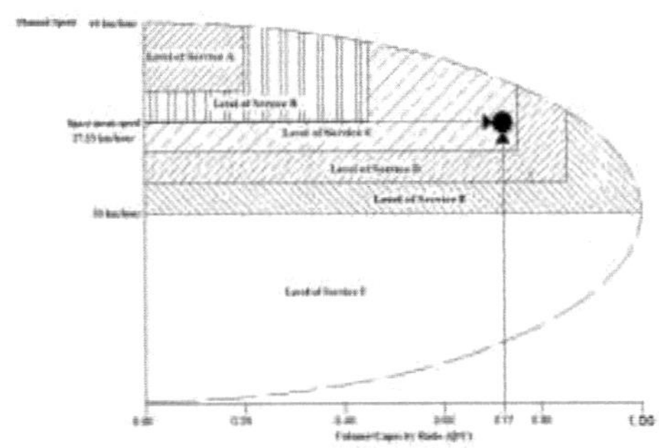

Figure 4. The Service Level of the One-Way Road

After obtaining the results of the road performance analysis, the comparison table was constructed and is displayed in Table 8 as follows.

Table 8. Comparison of the Road Performance Analysis Results

Road Peformance	Road Perfomance Analysis Result		
	Existing/ Current	Alternative I	Alternative II
		SO Management	One Way Road
V (Volume)	1078 pcu/hour	1078 pcu/hour	550.2 pcu/hour
C (Capacity)	1236.51 pcu/hour	1404,43 pcu/hour	1407,06 pcu/hour
DS (Degree of Saturation)	0.87	0,77	0,39
Vs (Space man Speed)	19.26 km/hour	27.23 km/hour	27.23 km/hour
Level of Service	F	C	B

Source: Analysis Results, 2019

5. Conclusion And Recommendation

5.1. Conclusion

From the results of the study regarding Jalan Kayu in Badung Regency, the conclusions are as follows:

1. The current performance of Jalan Kayu Aya in Badung Regency is:
   1. Volume = 1078 pcu/jam
   2. capacity = 1236,51 pcu/jam
   3. Degree of Saturation = 0,87
   4. Speed = 19,26 km/jam
   5. Level of Service = F
2. From the two alternative proposed for Jalan Kayu Aya in Badung Regency, the results are as follows:
   a. The road performance of alternative 1, which is by managing the traffic, is:
   - Volume = 1078 pcu/hour
   - capacity = 1236,51 pcu/hour
   - Degree of Saturation = 0,77
   - Speed = 27,23 km/jam

- Level of Service = C

b. The road performance using alternative 2, which is the one-way road system, is as follows:
- Volume = 550,2 pcu/hour
- capacity = 1407,06 pcu/hour

- Degree of Saturation = 0,39
- Speed = 27,23 km/jam
- Level of Service = B

5.2. Recommendation

The recommendations based on the study results are as follows:

1. The management of the traffic flow in the study location should be done at the earliest.
2. To solve the issue of cars parking on the road body of Kayu Aya Road, a central parking area should be constructed as there are still vacant land which can be used for a parking area.
3. If alternative 2 is selected, namely the one way road system, the road traffic flow circulation must be modified.

6. Acknowledgement

This article was presented at the 2nd International Conference on Smart City INNOVATION (ICSCI) 2019, jointly held by Universitas Indonesia and Universitas Diponegoro. ICSCI conferences have been supported by the United States Agency for International Development (USAID) through the Sustainable Higher Education Research Alliance (SHERA) Project for Universitas Indonesia's Scientific Modelling, Application, Research, and Training for City-centered Innovation and Technology (SMART CITY) Center for Collaborative Research, administered through Grant #AID-497-A-1600004, Sub Grant #IIE-00000078-UI-1.

7. References

[1] Morlok E . 1991 *Pengatur Teknik dan Perencanaan Transportasi* (Jakarta)
[2] Brindle R E 1996 *Local street traffic management-is European practice really relevant?*
[3] Pemerintah Republik Indonesia 2004 *Undang-Undang Nomor 38 Tahun 2004 Tentang Prasarana dan Lalu Lintas Jalan*
[4] Pemerintah Republik Indonesia 2009 *Undang-Undang Nomor 22 Tahun 2009 Tentang Lalu Lintas dan angkutan Jalan*
[5] Pemerintah Republik Indonesia 2006 *Peraturan Pemerintah Nomor 34 Tahun 2006 Tentang Jalan.*
[6] Pignataro L J, Cantilli E J, Falcocchio J C, Crowley K W, McShane W R, Roess R P and Lee B 1973 Traffic engineering: theory and practice
[7] Departemen Pekerjaan Umum 1997 *Manual Kapasitas Jalan Indonesia* (Dirjen Bina Marga)
[8] Putranto L . 2008 *Rekayasa Lalu Lintas, Indeks* (Jakarta)
[9] BPS Kabupaten Bandung 2018 *Bandung Dalam Angka 2017* (BPS Kabupaten Bandung)
[10] Saodang H 2005 *Kontruksi Jalan Raya* (Bandung)

Angle Characterization Radiation Detection of Microstrip Antenna for Short Range Terahertz Communication System

C Apriono, G Adriandi

Antenna Propagation and Microwave Research Group (AMRG), Department of Electrical Engineering, Faculty of Engineering, Universitas Indonesia, Depok, Indonesia

E-mail: catur@eng.ui.ac.id

Abstract. Communication technologies are crucial components to develop smart city systems related to low latency and massive data transfer. The near field communication (NFC) technology has been becoming more popular for close-range data transmissions, such as multimedia data transfer and non-contact payment, and still increasing in term of capacity and data transfer. Increasing carrier frequencies is a solution to fulfill those demands. In wireless communication technology, an antenna is a crucial element to convert electrical signals into electromagnetic waves or vice versa. Theoretically, an antenna dimension is comparable to the considered wavelength. This condition indicates that higher working frequency systems have a smaller antenna size due to its wavelength shorten. Terahertz (THz) spectrum band is promising to apply higher carrier frequencies. However, the antenna design and fabrication are becoming issues because of the tiny size, electronic components availability, and practical complexities. We have designed a microstrip antenna for working frequency of 0.35 THz by combining ground modification to cover distance and data rates up to 10 cm and more than 20 Gbps, respectively. By considering the previous design, this paper discusses different distances and rotation angles to analyze the effectiveness of wave transmission by simulation using CST Microwave Studio. Analysis use received power parameters to identify accepted signals detection possibility. The results show that the received power exceeded the sensitivity boundary from -50° to 40°, which show that the angles also limit the allowable accepted power level. Therefore, THz NFC systems should consider the antenna placement for an optimum condition application.

Keywords: wireless, Terahertz, Near Field Communication (NFC), communication, antenna

1. Introduction

Communication technologies are crucial components to develop smart city infrastructure systems, especially related to the demand for low latency and massive data transfer. These requirements have been coming from new emerging human communication behaviors, such as video-on-demand services, video conferences, and cloud computing. Those technologies need wide bandwidth and high-speed data transfer to obtain high quality of service (QoS) for various both fix and mobile multimedia applications. They have been becoming more popular and have shifted previous telecommunication technologies usage and the way people communicate. The current technologies have been in a stage which should

IOP Conf. Series: Earth and Environmental Science **396** (2019) 012028 doi:10.1088/1755-1315/396/1/012028

find other alternatives to catch up the tremendously near future demands, such as the upcoming 5G standards.

Among various communication technology solutions, the Terahertz (THz) waves, which is lying in a frequency band between 0.1 - 10 THz, offers broad bandwidth for wireless communication systems. Even though the frequencies are not practical for long transmission, due to the atmospheric attenuation, it is promising for short distance, point to point communication, and high rate data transfer [1–3].

A Near Field Communication (NFC) system has been an attractive technology for near communication systems, which combines card, reader/writer, and peer-to-peer communication [4]. These three provide intuitive, simple, and safe connection. The example use of an NFC system embedded in a mobile phone device is to increase data exchange, fast and secure methods for payments [5]. Another example of NFC advanced application is a multimedia data exchange vending machine for purchasing multimedia data. It can work more efficiently by increasing data rates into far higher, such as the THz wave spectrum band [6,7].

In such very high frequencies, propagation is a challenging issue of THz NFC systems development affected by the high attenuation in the THz spectrum band. Various research groups have researched the THz waves challenge for communication system applications. A recent study has used the link budget and gain analysis for indoor communication analysis [2]. It shows a promising configuration to obtain high-speed data transfer. The Non-line-of-sight (NLOS) propagation phenomena have been analyzed and investigated in detail to observe more constraint in real applications [8–11]. Another research has shown the possibility of reaching a multi-gigabit THz communication channel using wireless data exchange at 300 GHz [12]. These researches indicate that the THz waves are coming to a possible stage for supporting the inevitable demand of a massive data transfer.

In a wireless communication system, including a THz wireless network, an interface between the propagation medium and electronic device mostly is an antenna as a transducer. In the THz waves spectrum band, design and fabrication of THz antennas have many challenging issues, such as the tiny antenna size in a few micrometer scales and high gain required to compensate high attenuation in the propagation. We have been researching antenna designs for THz wave radiation source and detection based on quasi-optical lens antenna and microstrip antenna [13–16]. A lens antenna is a solution to gain antenna radiation power density in an expected direction. However, the challenge is to tackle radiation efficiency affected by a boundary condition between a dielectric and free space medium [17]. Meanwhile, the use of a microstrip type antenna is because of its low-profile characteristics, robustness, and possibility to fabricate. A proper design of a microstrip THz antenna can provide wide bandwidth [18]. In our previous work, THz planar antenna design is to offer working frequencies of 0.35 THz for short wireless communication purposes [19]. In this research, we study wave propagation between transmitter and receiver antenna by varying angles for NFC applications. This research is conducted with electromagnetic simulation software of CST Microwave Studio. The receiver antenna is expected to detect a minimum level for the received signal, namely a sensitivity threshold. This present paper aims to study the impact of angle variations of a THz microstrip antenna for several short-range communication scenarios to understand the best configuration in a short-range THzcommunication.

2. Propagation simulation between two antennas

Figure 1 shows an angular scheme of a transmitting antenna (Tx) rotated horizontally and radiates THz waves. The receiving antenna (Rx) receives the radiated power as a function of the parameterized angles. We consider our previous microstrip antenna design with a modified ground antenna and the resonant frequency of 0.35 THz to simulate the scenario [13]. Propagation simulation between the transmitter and receiver antenna for NFC applications is conducted with electromagnetic simulation software of CST Microwave Studio. The power emitted by the transmitter antenna is 500 mW. The received power is compared with the sensitivity level, which is around -40 dBm, to find out angles whether the received power is enough for data transmission. The considered distances between the transmitter and the receiver are 0.1, 1, 5, and 10 cm, which are the possible ranges in a short-range NFC system. The rotation is conducted from 180o to -180o with 10o spacing angle.

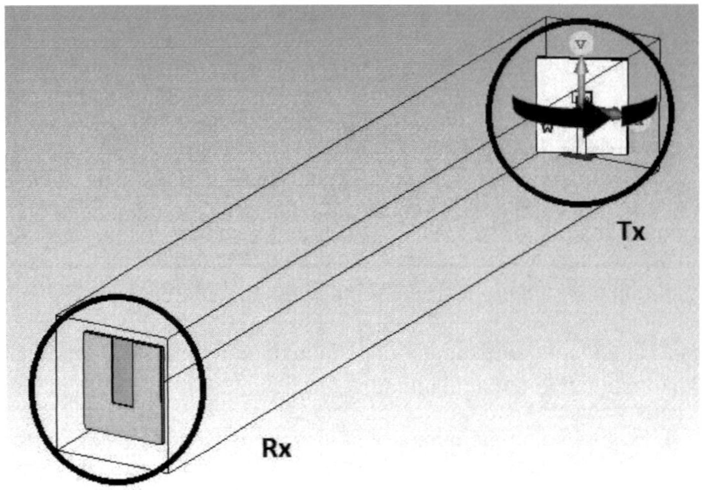

Figure 1. Schematic of angle changes simulation

3. Propagation simulation results

Figure 2 shows received power at a distance of 0.1 cm as a function of rotation angles. The line which has square marker shows the received power. The other line with a circle marker shows the sensitivity level of the receiver. Another shows the half-power beam width (HPBW). It shows that the entire received power is above the sensitivity level. These results indicate that the configuration can work well on the simulated angle variations.

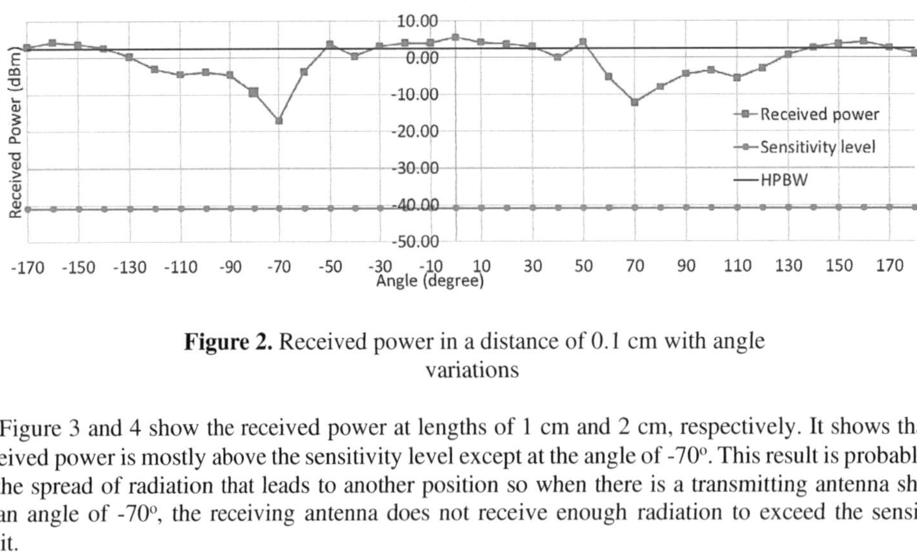

Figure 2. Received power in a distance of 0.1 cm with angle variations

Figure 3 and 4 show the received power at lengths of 1 cm and 2 cm, respectively. It shows that the received power is mostly above the sensitivity level except at the angle of -70°. This result is probably due to the spread of radiation that leads to another position so when there is a transmitting antenna shifting to an angle of -70°, the receiving antenna does not receive enough radiation to exceed the sensitivity limit.

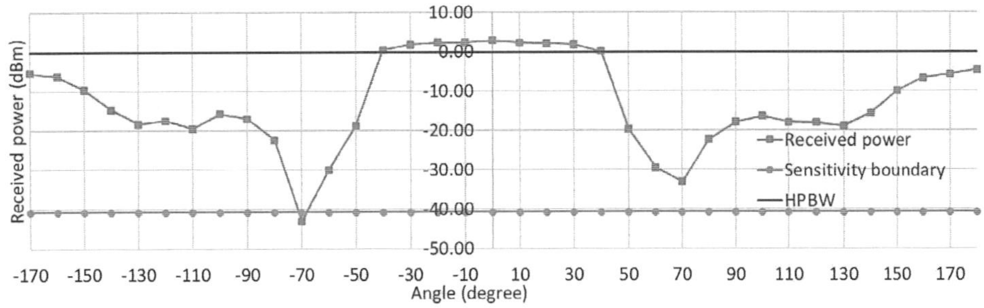

Figure 3. Received power in a distance of 1 cm with angle variations

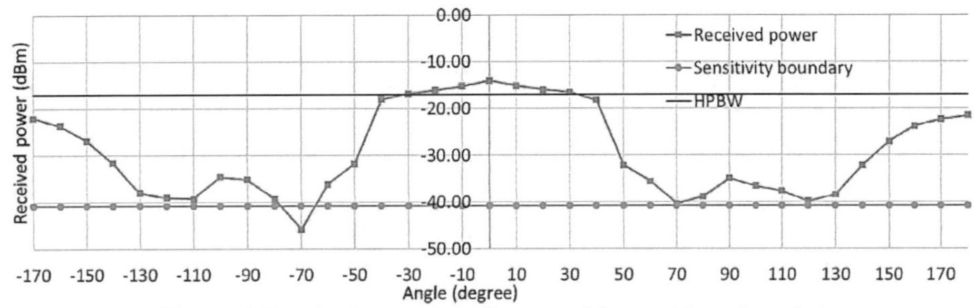

Figure 4. Received power in a distance of 5 cm with angle variations

Figure 5 shows the received power at a distance of 10 cm. It shows that the received power only exceeded the sensitivity level if the transmitting antenna directions are from angles of -50° to 40°. The received power for other scenarios does not exceed the sensitivity level. This condition is affected by the primary beam radiation direction, which is directive towards 0°. In the case of an NFC application, this scenario seems the last position for covering in the surrounding of the main beam direction except for correctly point to the point of zero degree condition. From the three scenarios, the covered area is decreasing at the given antenna input power. Higher antenna input power is necessary for more distances and covered area, which depends on the application requirements.

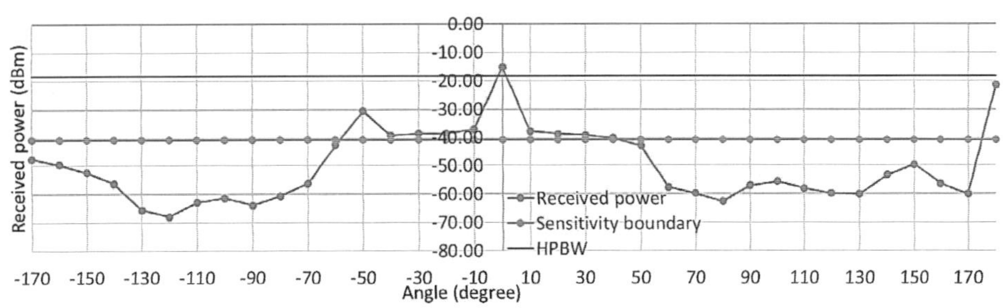

Figure 5. Received power in a range of 10 cm with angle variation

4. Conclusion

This paper has simulated and analyzed the THz wave propagation between a transmitting antenna and a receiving antenna in various near distances and rotation angles. By increasing the range, the received power levels do not exceed the sensitivity threshold. Received power in the furthest considered distance of 10 cm, exceeded the sensitivity boundary from -50° to 40°. Because the angles also limit the allowable accepted power level, therefore, NFC systems should consider the antenna placement.

5. Acknowledgments

This article was presented at the 2nd International Conference on Smart City INNOVATION (ICSCI) 2019, jointly held by Universitas Indonesia and Universitas Diponegoro. ICSCI conferences have been supported by the United States Agency for International Development (USAID) through the Sustainable Higher Education Research Alliance (SHERA) Project for Universitas Indonesia's Scientific Modelling, Application, Research, and Training for City-centered Innovation and Technology (SMART CITY) Center for Collaborative Research, administered through Grant #AID-497-A-1600004, Sub Grant #IIE-00000078-UI-1. Universitas Indonesia supports this research through Q1Q2 Grant, the year 2019, contract number: NKB-0292/UN2.R3.1/HKP.05.00/2019.

6. References

[1] Federici J and Moeller L 2010 Review of terahertz and subterahertz wireless communications *J. Appl. Phys.* **107** 111101

[2] Piesiewicz R, Kleine-Ostmann T, Krumbholz N, Mittleman D, Koch M, Schoebei J and Kurner T 2007 Short-Range Ultra-Broadband Terahertz Communications: Concepts and Perspectives *IEEE Antennas Propag. Mag.* **49** 24–39

[3] Mann C M 2001 Towards Terahertz Communications Systems *Terahertz Sources and Systems* (Springer Netherlands) pp 261–7

[4] Lesas A-M and Miranda S 2016 *The Art and Science of NFC Programming* (John Wiley & Sons, Inc.)

[5] Al-Tamimi M and Al-Haj A 2017 Online security protocol for NFC mobile payment applications *2017 8th International Conference on Information Technology (ICIT)* (IEEE)

[6] Risolia J R 2004 Multi-media vending machine with digital docking station

[7] Grobler B F 2001 Data Vending System

[8] Piesiewicz R, Jansen C, Mittleman D, Kleine-Ostmann T, Koch M and Kurner T 2007 Scattering Analysis for the Modeling of THz Communication Systems *IEEE Trans. Antennas Propag.* **55** 3002–9

[9] Jansen C, Priebe S, Moller C, Jacob M, Dierke H, Koch M and Kurner T 2011 Diffuse Scattering From Rough Surfaces in THz Communication Channels *IEEE Trans. Terahertz Sci. Technol.* **1** 462–72

[10] Jansen C, Piesiewicz R, Mittleman D, Kurner T and Koch M 2008 The Impact of Reflections From Stratified Building Materials on the Wave Propagation in Future Indoor Terahertz Communication Systems *IEEE Trans. Antennas Propag.* **56** 1413–9

[11] Jacob M, Priebe S, Dickhoff R, Kleine-Ostmann T, Schrader T and Kurner T 2012 Diffraction in mm and Sub-mm Wave Indoor Propagation Channels *IEEE Trans. Microw. Theory Tech.* **60** 833–44

[12] Jastrow C, Priebe S, Spitschan B, Hartmann J, Jacob M, Kürner T, Schrader T and Kleine-Ostmann T 2010 Wireless digital data transmission at 300 0.25em GHz *Electron. Lett.* **46** 661

[13] Aji A P, Apriono C, Zulkifli F Y and Rahardjo E T 2018 Double Crossed Planar Bow-Tie on a Lens Antenna at Terahertz Frequency for Imaging Application *2018 Progress in Electromagnetics Research Symposium (PIERS-Toyama)* (IEEE)

[14] Hidayat M V and Apriono C 2018 Design of 0.312 THz microstrip linear array antenna for breast cancer imaging application *2018 International Conference on Signals and Systems (ICSigSys)* (IEEE)

[15] Apriono C, Nurfitri I, Aji A P and Rahardjo E T 2019 Geometrical Parameterization of Bowtie Terahertz Receiver Antenna Connected with Conductor Bar as Bolometer *2019 IEEE International Conference on Industry 4.0, Artificial Intelligence, and Communications Technology (IAICT)* (IEEE)

[16] Apriono C, Aji A P and Rahardjo E T 2019 Study of Impedance Matching Effect between Terahertz Bowtie Antenna and Bolometer to Enhance Signals Detection *2019 7th International Conference on Smart Computing & Communications (ICSCC)* (IEEE)

[17] Apriono C, Rahardjo E T and Hiromoto N 2015 Ray tracing method with Fresnel's transmission to calculate polarized radiation power distribution focused by a terahertz Silicon lens antenna

Infrared Phys. Technol. **69** 102–6

[18] Wahyudi T, Apriono C, Zulkifli F Y and Rahardjo E T 2017 Broadband planar bow-tie antenna on high resistivity silicon substrate for terahertz application *2017 15th International Conference on Quality in Research (QiR) : International Symposium on Electrical and Computer Engineering* (IEEE)

[19] Adriandi G and Apriono C 2018 Modified ground microstrip antenna design for near field communication application at frequency 350 GHz *2018 International Conference on Signals and Systems (ICSigSys)* (IEEE)

PROVISION OF A SUSTAINABLE PUBLIC SPACE: LEBIH COASTAL AREA IN THE AFTERMATH OF RAMPANT ABRASIONS

G A M Suartika, N P D A Permanasuri, and K E Saputra

Masters Program in Architecture - Udayana University, Bali

E-mail: ayusuartika@unud.ac.id

Abstract. The study aims to examine sustainable uses of public spaces of Lebih's coastal line, in the aftermath of uncontrolled abrasion. It implements a qualitative research approach. Data was collected through a series of field observations, historical studies, and unstructured interviews. This study relates functions with spatial quality. While function is analysed based on social, economic and environmental roles accommodated by a public space, quality is assessed in terms of three conditions that a public space is expected to have, which are being responsive, democratic, and meaningful. Discussion over these two issues is based on the examination of both physical and non-physical attributes. The physical attribute is discussed in terms of spatial design and supporting facilities available to support the functions of Lebih as a public space, and the non-physical attribute is examined based on the capacity to be a responsive, democratic, and meaningful space to the public who uses it. This study alarmingly shows that abrasion has eroded Lebih coastal area to a level of 3.2% per year, which is equal to a 4m reduction annually. This condition directly influences the quality of Lebih Coast and consequently its capacity to accommodate the three inherent functions. The first two functions, to certain extents, remain well served, but the last one is feared to disappear along with the erosion of Lebih cost by continuous abrasion. This fear is especially underlined by the role of this coast as a buffer to protect the adjacent mainland and its associated ecosystem.

Keywords: abrasion, public space; public infrastructure, sustainability

1. Introduction

A public space often functions as a spatial unit dedicated for various communal interactions, social activities, and recreations. In acting on these roles, a well-planned public space, on the one hand, usually has a clear design as to how it is expected to be used by members of the public. An unplanned open space on the other hand, happens accidentally often on an unoccupied space or even on an unclearly defined space [1]. Many local governments refer to public space as an open space that is accessible to urban residents for free. Its provision is part of duties a city government has to fulfill as part of its public services, especially of those concerning to the maintenance of the public life as well as the associated social interactions. This public (open) space can take shape in many forms, such as parks, sport venues, children's playgrounds and so on.

Being classified as a public space, coastal area has its uniqueness. Its potential in attracting the public is built by untiring waves, sandy terrain, panoramic sunrises and sunsets, the scenic coming in and going out of ships, calming daily fishing and its associated activities, and many more inviting sceneries. On top of these, the coast is a protected area for its natural resources and biota that are environmentally important as well as strategic to meet people's basic needs. The beach is a spatial unit that accommodates the daily living of fishermen as well as salt and seaweed farmers. In the case of Balinese tradition, many processions and rituals are carried out along the beach which supposedly aims at maintaining a balance between nature, human beings, and the belief system. Following this tradition, various rituals/ceremonies are performed along beaches, including *melukat* (cleansing ritual dedicated for individual), *nganyut* (to wash away body ashes as part of cremation ceremony), *nyegara-gunung* (sea to mountain ceremony, a demonstration of gratitude to all blessing) and *melasti* (cleansing rituals dedicated for deities).

Taking Lebih Beach of Gianyar Regency - Bali Province of the Indonesian State as its case study, the research documented in this article also observes that the aforementioned conditions are also found at this Beach. Lebih Beach has been well known being a recreational venue, mainly, for the surrounding communities. This is one of many destinations for culinary experiences available in Gianyar Regency. This position is preceded by its former role being the central venue for *melasti* ceremony in the Regency. This role however has been disappearing for years in the aftermath of a rampant abrasion that has continuously hit Lebih's waterfront for the last two decades. Another prominent development is the role of Lebih Beach as a destination for sport activities, including jogging, soccer, yoga and kite playing exercise. In lining its role in governing development activities within its spatial administration, Gianyar Local Government then assigns this waterfront as one of prime recreational zones in its Regency. This stamp has consequently placed more weight on the development and the use of Lebih waterfront and its surrounding areas. One can observe several villas start to sprout following this Local Government's Regulation.

However, the fear over the impact of the rampant and continuous abrasion on the beach as a form of important public space that is important to the public is more of a concern. At certain timing of the year, the waterfront is a dangerous place to be in. Members of the surrounding communities even anticipate that at not in far distance the abrasion will hit their communities and their homes. Several endeavors that were aimed at preventing further damages caused by this natural phenomenon have been put in place. These include the use of bamboo as a water breaker in 1992 and the construction of rock revetments in 2010. The latter has since been functioning as barrier that reinforces and therefore protects the shoreline from impacts brought by the strong waves and sea current. As a result many functions, that were used to be accommodated by Lebih as a public space, are disappearing. To date, many sporting activities can no longer take place. Place for social and cultural activities have been also vanishing. Lebih's fishermen who had used the coast to embark on their fishing activities also find it hard to maintain their job. Being grounded by circumstances mentioned above, this research aims at investigating the sustainability of Lebih public space and its uses in the aftermath of continuous abrasion.

2. Methods

This research implements a qualitative research. Discussions within focus on events, social activities, attitudes, beliefs, and perceptions of either individual/s or a group of people [2]. The research starts with an existing phenomenon pertaining to a deep concern over the impact of coastal abrasion on the quality and ability of Lebih Beach to accommodate activities that usually take place in it as a public space. The research then progresses with a demonstration of the overlapping utilization of Lebih public space in the aftermath of abrasion. This spatial utilization is discussed in three different categories embracing social, economic and environmental functions. This discussion will be followed with an analysis of the sustainable use of Lebih Public Space, again, in terms of these three groups of function. Data presented in this paper has been collected by conducting a series of interviews, physical observations and secondary data available at many government institutions and provided by previous studies relevant to the discussed topic.

3. Overview of the Location

Lebih Beach is located in the Banjar Lebih Beten Kelod, village of Lebih Gianyar. The boundaries of the research location are as follows, the northern part is Tohpati-Kusamba Bypass, the southern part is Lebih Beach, the eastern part is rice field, and the western part is rice field (Figure 1).

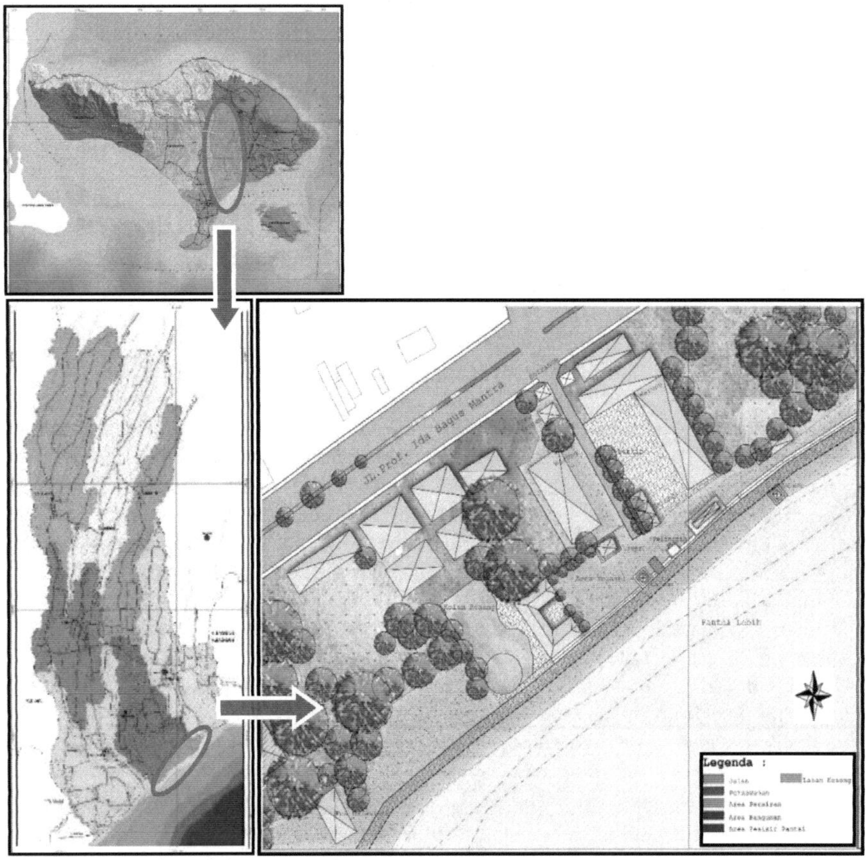

Figure 1. Location of the case study
Source: [3]

4. Lebih Beach as a Public Space

Lebih village is located on the coast of Lebih. The coastal area is a meeting area between land and sea, which in land direction covers parts of the land, both dry and submerged in water, and towards the sea covering parts of the sea that are still affected by natural processes. Public space that occurs in the coastal area is space that is used for the common interest, both in the form of social, economic, and environmental activities. Lebih Beach as a tourism area continues to run from year to year. According to an interview with one of the local fishermen, Mr. I Made Ana, as far as he concerns, Lebih Beach has been used as a place to carry out religious ceremonies, such as *melasti* activity, *nangluk* and *nganyud* activities. Like coastal tourism objects in general, Lebih Beach is also a recreational place for visitors who intend to enjoy the beautiful panorama of the beach and its exotic black sand.

Besides visited as a destination for recreational purposes, Lebih Beach also provides marine products, including fish caught by local fishermen, which are then processed by local *warung* (stall). The catches of the fishermen are served in various form of dishes including satay *lilit* (grounded fish, wrapped around a skinny bamboo stick and then grilled), *pepes* (steamed fish wrapped in banana leaf), fish meatballs, and grilled fish. Sate *lilit*, first popularized in 1987 by a resident of Banjar Lebih Beten Kelod named Bapak Lodig. He works as a fisherman and the pioneer in processing fish into satay. Satay and fish dishes become a special icon of Lebih Beach up to now.

The existence of Lebih Beach is a publically shared place that accommodates public activities. The establishment of Lebih Beach as one of the tourism areas is set in accordance with Local Government Regulation number 16 year 2009 [4] concerning Bali Spatial Plan for Bali Province and Local Government Regulation Number 16 year 2012 [5] concerning Spatial Plan for Gianyar Regency. These policies stipulate Lebih Beach as a strategic region of Gianyar Regency. Subsequently, both of these laws increase number of visitors using Lebih as a public space. This position is intensified in line with the completion of another prominent public infrastructure of Tohpati-Kusamba Bypass.

5. Community and Lebih Beach as a Public Space

Community who uses Lebih coast areas can be grouped into 4 (four) categories. Details of each category group are explained as follows:

1. Food vendors who sell seafood based dishes in many culinary stalls exist at the surrounding areas of Lebih Beach. They usually start 7:00 am to 11:00 pm. There are 18 permanent culinary stalls in the research location. Each culinary stall is managed privately by the owners who are generally relatives of fishermen who supply them with various catches.
2. Fishermen who form Lebih coastal community. In 2017, there were 95 fishermen operating at the Lebih Beach.
3. Visitors who visit the beach. There is an average of 100 to 150 visitors visiting the beach daily. This figure however can reach three times as much (300 people) per day during public holidays.

6. Abrasion in Lebih Coastal Public Space

Abrasion is a process of eroding the coast by destructive ocean waves and currents [6]. Damage to the coastline due to abrasion is triggered by disruption of the natural balance of the coastal area. Based on interviews conducted in July 2017 with the Secretariat of Conservation and Environmental Rescue Workers (SKPPLH), I Made Mangku that the abrasion is caused by changes in currents due to physical development and the consequences of human activities namely sand and stone brush mining.

No one knows for sure when abrasion at Lebih Beach actually started. According to I Made Ana, chairman of fishermen in Lebih Village, in the 1980s the coastal was far from the present situation. According to him, more than 200 meters of coastline has been eroded. In the past, between Masceti Beach and Watu Klotok Beach was a straight line, there was no bay like now, moreover since the reclamation of Serangan Island in 1994 exacerbated the abrasion at Lebih Beaches. The same thing was also reaffirmed through a direct interview to Mr. I Wayan Weta (55 years) who is a food stall owner at Lebih Beach, he recalled that abrasion in Lebih Beach has occurred since 1985. But it did not take place as rampant as it is recently. This was reinforced by the Gianyar Regency Public Works Office. According to satellite imagery and Spatial Plan for Gianyar for the period of 2010-2017, there was a significant change in Lebih coastline within this period. The abrasion affects the physical existing of public spaces in the Lebih Coastal Area. The changes in the existing physical abrasion that occurred from 2010 to 2017 appear as follows:

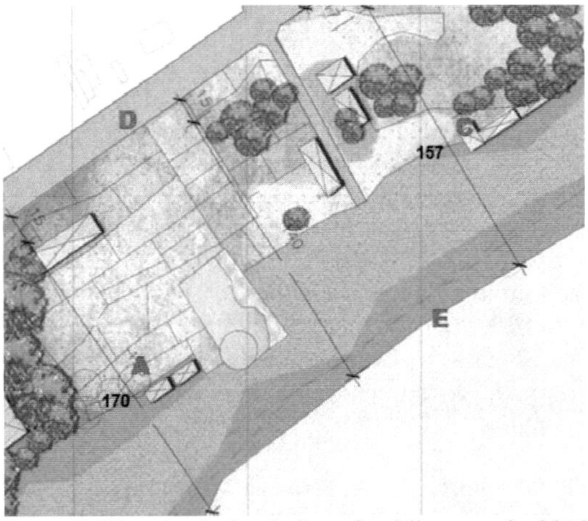

A. The western side of the selected coastal line - as a public space
B. The middle part of the selected coastal line - as a public space
C. The eastern side of the selected coastal line - as a public space
D. The Tohpati-Kusamba Bypass
E. The high water mark in 2010

Figure 2. The physical existing of public space in 2010
Source: Diah, 2017

The Spatial Plan for the year 2012 and 2010 satellite imagery show physical existing on the Lebih coast in which A is the western side distance from the public space which has a distance of 170 meters from the edge of Tohpati-Kusamba Bypass. then B is the middle side which is 155 meters from the edge Tohpati-Kusamba Bypass, and C is the east side with a distance of 170 meters. Overall the length of the coastal public space area reaches 323 meters. Conditions in 2010 showed that the street is one-lane road with a width of 7 meters. The coastal area in 2010 was 267,520 m² and the area used as a public space reached 9,700 m², with a percentage of the coast over public space reaching 3.6% (Figure 2).

In 2012 the physical condition (width) of public space decreased from the previous year. A (west side) decreased to 11.7%, then B (middle side) decrease by 13.5%, and C (east side) which is the eastern side the distance between the by-pass to the coast has decreased by 14%. In the same year, the Tohpati-Kusamba Bypass started to be constructed as a into two lanes road of ± 15 meter. The notation of E is the highest water mark in 2010. It can be seen how Lebih public space is narrowed down physically by the effect of abrasion. The with of notation A narrowed into 150m, B into 135m and C into 135m (Figure 3). The scale of public space available in the early 2010 reached 9,700 m² then in 2012 it was reduced into 8,500 m². It is a decrease of 12.37% (Figure 3).

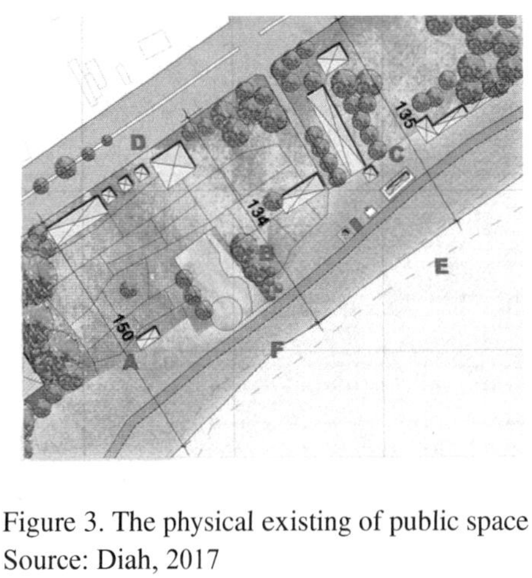

Notes:
A. The western side of the selected coastal line - as a public space
B. The middle part of the selected coastal line - as a public space
C. The eastern side of the selected coastal line - as a public space
D. The Ida Bagus Mantra Road
E. The high water mark in 2010
F. The low water mark in 2012

Figure 3. The physical existing of public space in 2012
Source: Diah, 2017

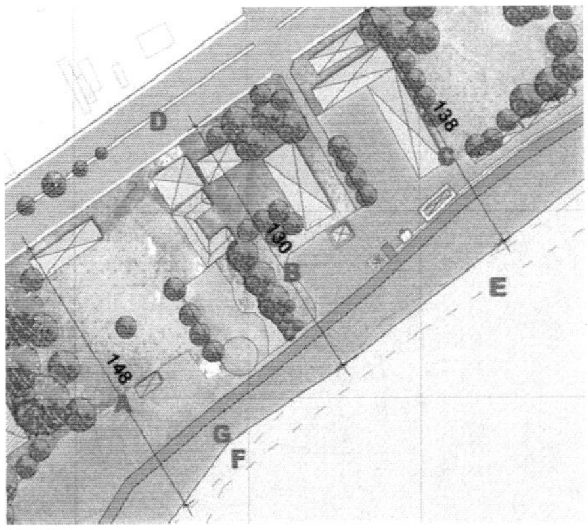

Notes:
A. The western side of the selected coastal line - as a public space
B. The middle part of the selected coastal line - as a public space
C. The eastern side of the selected coastal line - as a public space
D. The Ida Bagus Mantra Road
E. The high water mark in 2010
F. The high water mark in 2012
G. The high water mark in 2014

Figure 4. The physical existing of public space in 2014
Source: Diah 2017

In 2014 there were some constructions taking place including the making of revetment, jogging track, hall and shrine. In this year, the physical condition of public space again reduced to to abrasion. The width of C (east side) reduced by 2.2%, then B (middle side) experienced a reduction of 2.98%, and A (west side) was reduced by 1.3 % (Figure 4). The scale of public spaces in 2014 (Figure 4) was also reduced. For safety reason, government installed several forms of coastal security facilities. And to develop Lebih as a more friendly public space, several facilities to accommodate social uses were also built in this year. There were also culinary stalls and parking lot being built in area noted as C in the map, which used to be a paddy field. The physical percentage of Lebih public space over the total coastal area in 2014 reached 3.35%.

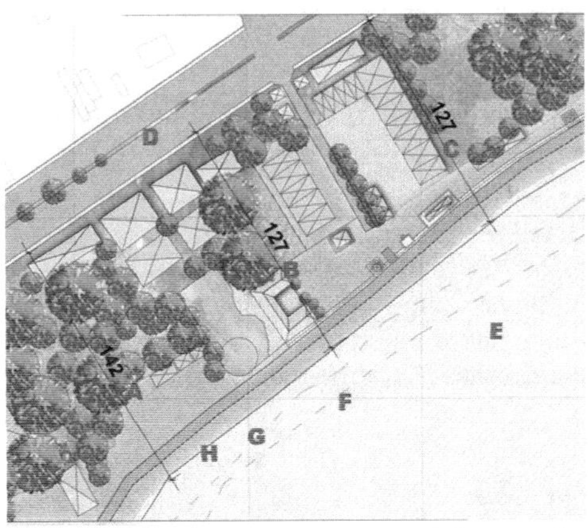

Notes:
A. The western side of the selected coastal line - as a public space
B. The middle part of the selected coastal line - as a public space
C. The eastern side of the selected coastal line - as a public space
D. The Ida Bagus Mantra Road
E. The high water mark in 2010
F. The high water mark in 2012
G. The high water mark in 2014
H. The high water mark in 2016

Figure 5. The physical existing of public space in 2016
Source: Diah, 2017

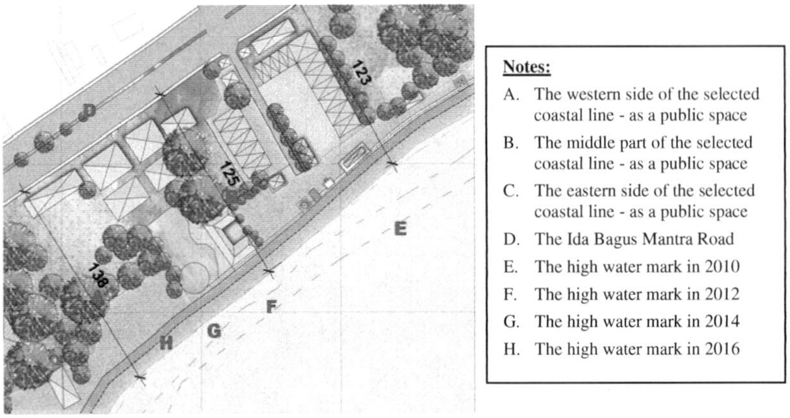

Figure 6. The physical existing of public space in 2017
Source: Diah, 2017

In 2016, the area of public space noted as C (east side) decreased by 3.78% from 2014, B (middle side) reduced by 2.3% and the area noted as A (west side) was also reduced by 4.05% compared to the 2014's data. Data also shows that there were more activities taking place at Lebih public space in 2016, accommodating more social, economic and environmental functions. However, the physical area of public spaces in 2016 decreased by 10.25% (Figure 5) compared with the same data recorded for 2014. Based on this data, abrasion had eroded the coast as much as 3.2% annually. The physical area of public space in 2016 reached 7,000m² (Figure 5).

In 2017 C (east side) decreased by 3.14% since 2016, B (middle side) decreased by 2% and A (west side) decreased by 2.8% since 2016 (Figure 6). In this year, the abrasions rate reached an annual average of 4-5 meters with an averaged percentage of 3.2%. In 2017 the area of public space reached 3.15% of the total coastal area available in the same year. The scale of area left for public space after abrasion in this year was 6,500m² (Figure 6).

7. Coastal Utilization as a Public Space Based on Social, Economic and Environmental Functions in a period of 2010-2017

The definition of spatial use according to Law No. 26 year 2007 [3] concerning spatial planning is an effort to realize the structure of spatial patterns in accordance with spatial plans. Withis this regulation, the utilization of public space is expected to deliver three important attributes, including: it has meaning (meaningful), it has the capacity to accommodate the needs of the public (responsive), it accommodates public activities without discrimination (democratic) [7]. In this case public space is seen as a place for social interaction and communication conducted either by individuals or groups.

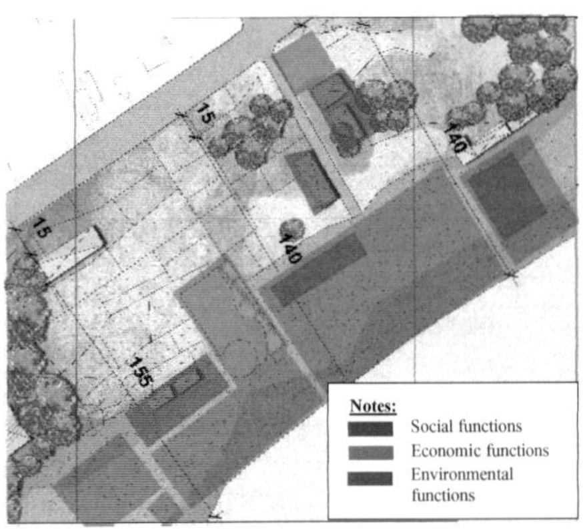

Figure 7. Land use in 2010
Source: Diah, 2017

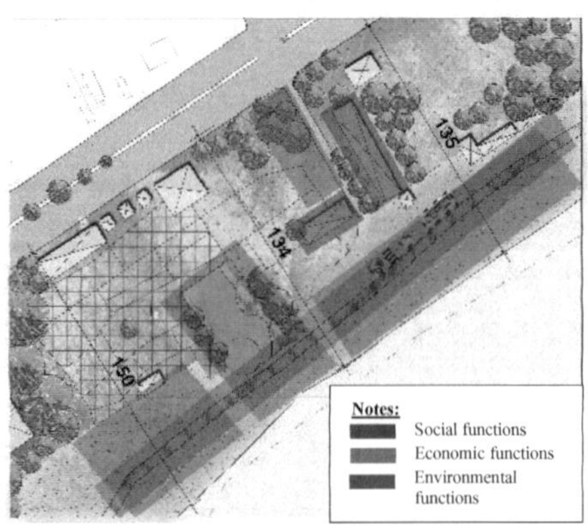

Figure 8. Land use in 2012
Source: Diah, 2017

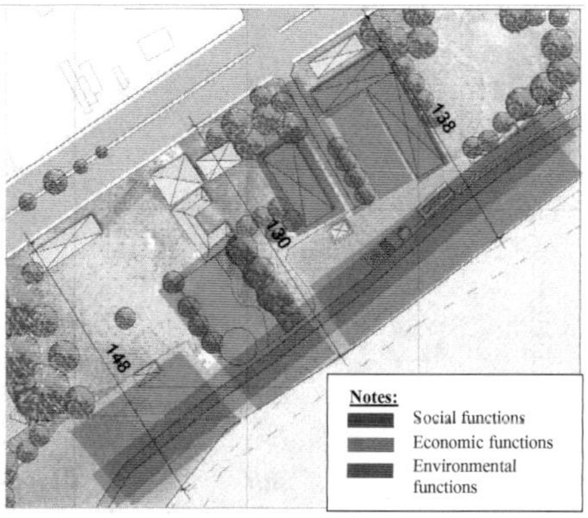

Figure 9. Land use in 2014
Source: Diah, 2017

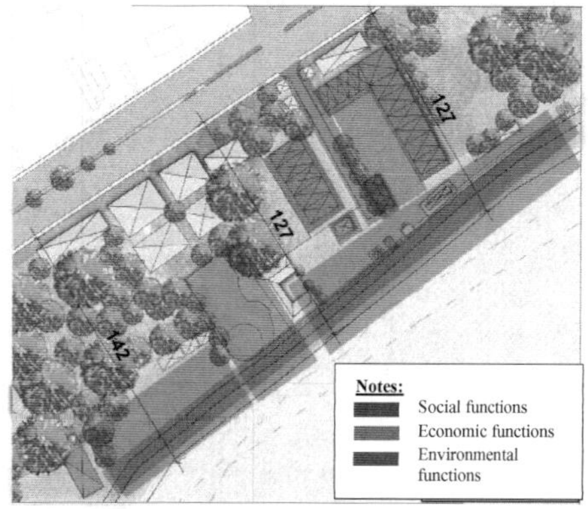

Figure 10. Land use in 2016
Source: Diah, 2017

The use of a space is inseparable from the sustainable availability of the space. Sustainability in general means the ability to maintain the balance of a process or condition of a system, which is related to biological and guided systems. In the context of ecology, sustainability is understood as the ability of ecosystems to maintain processes, functions, productivity, and ecological diversity [8]. The use of space based on sustainability functions is related to economic, environmental and social sustainability.

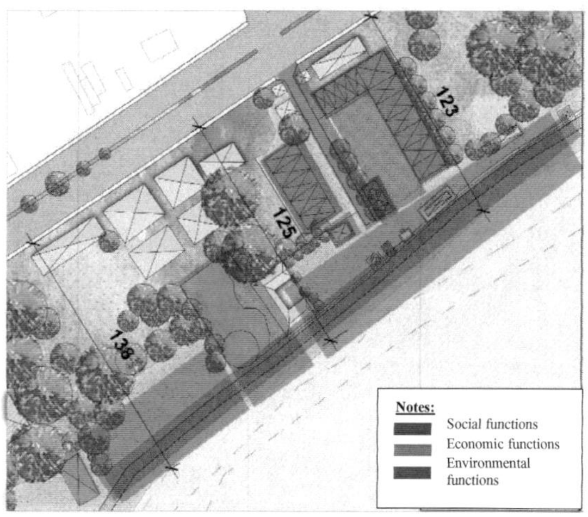

Figure 11. Land use in 2017
Source: Diah, 2017

The scale of Lebih public space used for social function in a period of 2010 to 2017 experienced several stages of change. In 2010 (Figure 7), the area available for this use was 9.700m². Then in 2012 (Figure 8) this figure decreased by 12.37% which again reduced by 8% in 2014 (from 2012's data). In 2016, the decrease was recorded as 5.8% (based on 2014 data). Then in 2017 (Figure 11) the decrease was 1.5%, based on the 2016 figure 10. These reductions are the immediate result of the narrowing scale of the public space due to abrasions. Every year the land is eroded by sea water and decreases by ± 4m, this has an impact on the narrowing of the social function area in Lebih coastal public spaces.

Utilization of Lebih public space to accommodate economic function, as a whole has also undergone an increase in changes from 2010 to 2017. In 2010 a number of culinary stalls are still located on the beach (south) adjacent to the sandy area of Pantai Lebih. Then in 2012 the shops were moved to the east of the two-wheeled parking area (eastern culinary stalls). And in 2014 the eastern culinary stalls experienced a rearrangement and an addition of several stalls. Then in the period of 2016-2017 the stall arrangement was further enhanced, in which the stalls' front area which used to be a parking area, then made into a more organized area of the same use.

Environmental functions of Lebih public space in a period of 2010 to 2017 is explained as follows. In order to overcome the physical impact of the abrasion to the natural environment, Gianyar Local government decided to build coastal safeguards in the form of revetments in 2010. Then in 2012 the same government installed safety facilities along the coast and also the constructed jogging tracks. These all were aimed at maintaining the safety of the public members when they are using the area. Then in 2016 to 2017, there were constructions of buildings to further accommodate social functions, including the provision of public hall, *padmasana* (shrine) and several gazebos. Gazebos were placed along the coastline parallel to the coastal security facilities. Further detailed land utilization in the period of 2010 and 2017 is presented as follows.

8. Utilization and Quality of Public Spaces

The quality of a public space can be viewed from two aspects, namely physical and non-physical aspects. Physical aspects include regional facilities, design, and conditions. Then the non-physical aspects are how responsive, democratic, meaningful and accessible a public space is [7]. The impact of abrasion has changed as to how public space of Lebih Beach of Gianyar is utilized. This is mainly due to the narrowing of the available spaces. The following section (Table 1) discusses how rampant abrasion has affected the quality of Lebih public space and its uses in accommodating social, economic and environmental functions.

Tabel 1. Utilization of Public Spaces in 2010 - 2017

No	Year	Social	Economy	Environment	Remarks
1	2010	- Sports (soccer, volley, raw) - Religion (*melasti, nangluk, nganyud, melukat*) - Recreation - Social service - *Metanem* (sand therapy)	- Culinary stall - Stone craftsmen - Fishermen	- Parking	
2	2012	- Sports (Soccer, volley, raw, jogging) - Recreation - Religion (*melasti, nangluk, nganyud, melukat*) - *Metanem* - Social service	- Culinary stall - Fishermen	- Parking	Starts to reduce brush stone craftsmen to look for stones in the public space area
3	2014	- Sports (Soccer, volley, raw), jogging - Recreation - Ritual (*melasti, nangluk, nganyud, melukat*)	- Culinary stall - Fisherm en	- Parking - Beach lifeguard (revetment)	The construction of coastal security/lifeguard
4	2016	- Sport (jogging) - Recreation - Ritual (*melasti, nangluk, nganyud, melukat*)	- Culinary stall - Fishermen	- Parking - Beach lifeguard (revetment) - Monitoring tower	Construction of monitoring towers
5	2017	- sport (jogging) - Recreation - Ritual (*melasti, nangluk, nganyud, melukat*)	- Culinary stall - Fishermen	- Parking - Beach lifeguard (revetment) Monitoring tower	

9. Sustainability of Social, Economic and Environmental Functions of Public Spaces

Public space is interesting and never-ending topic to discuss. Government and urban planner in their effort to enhance the quality of its urban area have been trying to find ways in which the need for an adequate public space can be met. This attempt is always challenged with the fact, that there is a limited scale of land available to fulfill this objective. In some cases, this availability is even zero. In the case of Lebih coastal public space, the situation is worsened by the rampant abrasion that gets worse year by year. It does not only reduce the physical scale of this public space, it also affects the sustainability of the functions accommodated within (socially, economically, and environmentally) and the meanings imbedded within (meaningful, responsive, and democratic). As discussed, there are certain functions or activities that can no longer be contained by Lebih public space. However, we are challenged by the need to prolong both the existence and the use of this public space for future generation [9,10]. The following sections discuss further about this issue (Table 2).

Table 2. Utilization and Quality of Public Spaces

No	Aspect	Social Function	Economic Function	Environmental Function
1	Responsive	-	√	-
2	Democratic	√	√	-
3	Meaningful	√	-	-
4	Accessible	-		-
5	Area for facilities	-	√	-
6	Design	-	√	-
7	Condition	-	-	√
	Note:			
	√	Exist		
	-	Not exist		

10. Sustainability of Social Functions

Based on the analysis grounded by the trend of how Lebih public space is used in accordance with its social functions, it is concluded that: (i) interactions that would no longer be possible to be contained in the future include activities pertain to sports (except jogging) and religious uses; (2) social interactions for recreational purposes would be prolonged. These remarks drawn here are based on interviews held to a group of respondents who visit Lebih Beach in a regular mode. Jogging will remain possible, especially after the installment of facilities that secure the safety of the people who use the beach as a public space. However, there will be no more football games, kites playing, and volleyball. People have their anxiety to be in Lebih Beach at certain timing of year, but overall the relaxing feeling and the rejuvenating impacts brought by being at the beach overtakes this anxiety. Most respondents to this study view, the meaning of Lebih Beach as a place for recreation will stay for much longer. Religious activities such as melasti, nganyud, nangluk and melukat may still be possible now and for the next few years. But in a long run, there will be no adequate space available to accommodate a large number of crowds, as is often required by these activities. However, ritual activities organized individually will be possible to be accommodated for a longer time, since they don't involve a large crowd.

11. Sustainability of Economic Functions

In terms of the sustainability of the economic functions, this study finds that it will go ahead for quite sometimes to come. This is inseparable from image of Lebih beach area as a destination for culinary experience. Existence of food stalls and restaurants will be continuing since local government has installed safety measures as well as the construction of physical barrier to protect the beach from further impacts inflicted by the rampant abrasion. This study also believes that fishing activities will be thriving just that further investment will have to be put in by the fishermen. This includes investment to buy technology aided boats and investment to mediate the constructed revetment between the beach and the mainland. This analysis relates to earlier prediction stating that the seafood stalls located at the surrounding Lebih Beach will remain exist for some times to come. These premises will need catches provided by the fishermen. Therefore, as long as the demand is there the fishing community and their catches will subsequently be sustained.

12. Sustainability of Environmental Functions

In principal the sustainability of environment refers to the capacity of Lebih public space to maintain the existence and the quality of its biodiversity. Naturally at present, Pantai Lebih has no substantial biodiversity related resources to be protected. Instead, the beach has been a target of natural threats - abrasion. As has been explained, attempt to protect the coast has been done by revetments. This action is not merely aimed to protect the beach but also in consequence to protect the function of this zone from disappearing. The beach is a buffer to protect the mainland from erosion and the impact of abrasion. As a public space, Lebih Beach has been also used as a place to educate member of the public to enhance their environmental awareness. Many informal activities have been taking place to educate kids to properly dispose rubbish that they produce daily and to recycle them as much as possible. Action to free the beach from rubbish is also a routine activity held by groups of the community that involved many wider groups of different ages. This study finds that the use of Lebih beach being an environmentally friendly public space will be sustained. This is basically drawn from the fact that concern over the sustainability of our natural environment nowadays is a business of every single one of us, every single spatial unit, and every land development.

13. Conclusion

As one of important public spaces in Gianyar Regency, Lebih Beach has gone through a massive physical change. It is largely due to the impact of rampant abrasions. No one knows for sure when this natural phenomenon started but it has been estimated taking place for (at least) more than two decades now. Abrasions has significantly diminishes the physical scale of the Lebih Beach public space to a level that affect its capacity to act as a spatial unit, important for public uses. In discussing the sustainability of a Lebih public space and its roles, the social function has been the most that affected by the rampant abrasion, while economic and environmental related uses are thriving. Parts of Lebih beach that used to be utilized for numerous kinds of ritual procession are diminishing. This is due to the limited space left to accommodate a large number of people involved in each procession. The thriving of Lebih beach public space in accommodating economic and environmental related functions is due to the image of the place being a destination for seafood-based culinary experience and being part of Gianyar tourist zone. In this context, the creation of Lebih beach public space being an environmentally friendly tourist destination is promoted at all levels. This objective subsequently encourages the accommodation of activities that focus on sustaining our nature and its associated environment. Nevertheless, the impact of abrasion over Bali coastal zone has been a major concern right now. This is a natural phenomenon that will be going away, but it will continue to hit our waterfronts. By conducting this preliminary study, it is expected more researches will be carried out in this field in the near future.

14. Acknowledgement

This article was presented at the 2nd International Conference on Smart City INNOVATION (ICSCI) 2019, jointly held by Universitas Indonesia and Universitas Diponegoro. ICSCI conferences have been supported by the United States Agency for International Development (USAID) through the Sustainable Higher Education Research Alliance (SHERA) Project for Universitas Indonesia's

Scientific Modelling, Application, Research, and Training for City-centered Innovation and Technology (SMART CITY) Center for Collaborative Research, administered through Grant #AID-497-A-1600004, Sub Grant #IIE-00000078-UI-1.

15. References

[1] Darmawan E 2003 Teori dan Kajian Ruang Publik Kota *Semarang Badan Penerbit Univ. Diponegoro*

[2] Sukmadinata N S 2009 *Metode Penelitian Pendidikan* (Bandung: PT. Remaja Rosdakarya)

[3] Ruang D J P 2007 Undang-Undang Republik Indonesia Nomor 26 Tahun 2007 tentang Penataan Ruang *Dep. PU. Yakarta*

[4] BALI R T R W P 2009 Peraturan Daerah Provinsi Bali Nomor 16 Tahun 2009 *Tentang Rencana Tata Ruang Wil. Provinsi Bali Tahun* **2029**

[5] Bappeda 2012 *Rencana Tata Ruang Wilayah Kabupaten Gianyar Tahun 2012-2032*

[6] Triatmodjo B 1999 Teknik pantai *Beta Offset, Yogyakarta* **397**

[7] Carr S, Stephen C, Francis M, Rivlin L G and Stone A M 1992 *Public space* (Cambridge University Press)

[8] Wardhono F and Hesti N 2012 Pembangunan Kota Berkelanjutan *Present. sebagai bagian dari Lap. akhir berjudul "Kajian Upaya Perwujudan Kota Jakarta yang Berkelanjutan*

[9] Suartika G A M 2015 Sand, sea and ceremony: conflict over the littoral public realm in Sanur, Bali *Procedia-Social Behav. Sci.* **179** 128–40

[10] Suartika G A M and Saputra K E 2019 The Creation of Harmonious Living Spheres-Landscape, Capital, and the Balinese Way *IOP Conference Series: Earth and Environmental Science* vol 248 p 12065

IOP Conf. Series: Earth and Environmental Science **396** (2019) 012030 doi:10.1088/1755-1315/396/1/012030

Try and Find it - Understanding Public Space and Civil Society in Bali

N M Swanendri, G A M Suartika

Doctoral Program in Engineering Faculty - Udayana University

E-mail: ayusuartika@unud.ac.id

Abstract. An ethnographic study using qualitative research methodology underwrites the following study. Withinthe current urban planning practice, the concept of public space exists in contradiction to private ownership. Within the global capitalist system, land is commodified like any other goodand is subject to ownership via the market mechanism. The endgame is for all land to have this status. In this context, there isbusiness. This paper, however, offers to renege on this perspective. It tries to understand public space within a system where use-value is the dominant concept. Taking Bali as its laboratory, this paper examines a hypothetical typology of public space that brings use values to the fore and denies the primacy of land as solely a profit-seeking medium for land rent, surplus value extraction from labor and profit on built form. While no doubt these processes will still take place, they must be assessed in the context of environmental ethics and fundamental human needs. In conclusion, this paper seeks a benevolent and humane planning practice that places people before profits. It implements this idea by saying that the creation of spaces whose use values can be accessed by all members of society must become the central planning strategy.

Keywords: communal infrastructure; public space, planning practices, land values, public/rights over land

1. Introduction

The ensuing overview of public and private realms lays a conceptual framework for ensuing discussions focussed on the built environment. The remaining three sections respectively discuss communal space and civil society in Bali; adopted forms of communal space; and necessary connections to a sustainable living environment for Bali.

In a globalized capitalist world, privatization is recognized as an essential part of development. It involves a conflictual process whereby public rights to space are converted to private interest and ownership. It frequently happens to various elements of a public realm. This process invariably focuses on the conversion of public space and assets, to private control, and not the reverse- for example littoral zones including beaches, forests, sacred sites, culturally important landscapes and settings,and natural features. This has a palpable effect, particularly on the public space of a city and human settlement, such as streets, open spaces, plazas, green open spaces, and public facilities. This privatization of the public realm is often facilitated and accelerated by the closing public functions, redesigning existing public space and the construction of business-focussed commodity architecture that is privately operated and where digital monitoring becomes the norm [1, 2, 3].

The use of the word 'public' in either 'public realm' or 'public space' has a correlation with certain historical phases and social conditions. It is also associated with the word 'common' as in 'the commons' a right to space established as common grazing land in the Magna Carta [4]. Throughout the Middle Ages, the term 'public' and 'private' and of the public sphere were understood as *res publica* (public matters). This understanding is embedded in the definition of Roman Law. We call something 'public' when it is open to all andcontains an element of both public recognition [5], as well as shared interests. It also embraces the need to live together or live under the same rights [6].

Under the umbrella of the built environment, the term public also embraces spaces contained within settlements that exist for sole use of the public and are often protected in law [7]. They representplaces for all sorts of social interactions in streets, playgrounds, parks, and areas designated for conservation that are, in theory at least, out of reach of private sector avarice[8].These are spaces that can be accessed by everyone without restriction regardless of age, gender, ethnicity, physicalhandicap, or other characteristics. Problematically, state governments are usually forced to charge entry fees to offset maintenance costs and are frequently exported to the private sector to manage.

These are the placeswhere people come together for social and/or economic purposes [2, 7, 9]. In general, public and private spacesare distinguished by attributes pertaining to the level of access, control and behavior sanctions associated with their use [10]. Some territorial forms may be considered publically accessible with certain terms and conditions. In this regard, Altman views that*"the public does not necessarily relate to ownership, but rather to use. Some privately owned places and spaces are accessible to the public and some publicly owned areas are not"* [7, p.1]. If this was not the case, capitalism would grind to a halt since the entire commercial sector would exist out of bounds for the general population. Capital is forced by its own logic to encourage interfaces with the general population so commodity circulation can take place.

In our attempt to comprehend public space in a deeper manner, one has to understand the matrix of power that mediatesits existence within the built environment. The key element is the state and it is important to recognize why the state arose in the first place.As one might expect, there is a variety of interpretation [11]. For the sake of argument we adopt a capital logic position that proposes that the state arose to serve the interests of capital. In the process of capital accumulation, social reproduction and its development were viewed as an unnecessary diversion and a debt, thus reducing profits, by giving money to the poor. Capitalists focussed on creating more capital through trading commodities. As capitalism became more advanced, social reproduction had to become organized as a healthy, educated workforce was necessary to guaranteed production and advancing technology.

Capital brought the state into existence to administer the overall process, one that focuses not on profit, but the social consumption of resources that were funded through taxation of those who were being reproduced. Jonas Brodin (2007) distinguishes the terminology of public space into two references. The first is symbolic(referring to an abstract mental construct) and the second is literal (referring to a collection of physical objects) [9]. Both necessarily combine in designating specific states of being. So the symbolic attributes of a public space rely literally on the specific forms that are adopted and the social meanings they incorporate. This does not necessarily work in reverse since no symbolic meaning might be present in the built form.

In the symbolic conception of public space, we find the expression of power in various forms, usually manifested in monumental architecture which emphasizes the social hierarchy in the form of gods, emperors and kings, heroes, the power of the state and the manifestation of conflict such as war. Public space then manifests itself as the limits to capital, which capital will obviously want to resist. It is also clear that most monumental architecture is located in public spaces, a clear reminder of the power differential between the people, organized religion, the state, and private capital. In many ways this can be considered an oppressive context since the public does not even dictate who their real heroes are.

Public space, therefore,relates to our concept of civil society that in principle requires the existence of public awareness of shared values and equality, rooted in democratic politics. Hegel [12] maintained that it is only through the state that civil society can prevail, in contradiction to Adam Smith [13] who considered that the innate rationality of civil society would lead to the general good. Civil society can also be seen as a necessary transitional stage between feudalism and capitalism.

Hence the state cannot be seen merely as a form of government since it also incorporates elements of civil society in its functioning.This is a totally different concept of social relations that operate on a principle of domination [6]. Arguably the concept of a civil society arosein the 18th Century which differentiated society from the state and operated on the basis of a self-organizing social life.

There are five visions of civil society, which are: 1) social-ability; 2) market-based which is shifted from a confined economy into a digitally-driven market supported by a sophisticated computerized information system; 3) collective choice, organized in groups and associations; 4) concerted public actions; and 5) focuses on communal cultures [14]. From this point of view, public space seems to be a representation of civil society on the one hand, and representation of a state on the other, although as we have argued above, a more sophisticated analysis would envision a degree of overlap between them [8].

Nowadays the existence of public space is not merely about people's rights. The discussion is, however, more about various interests competing to use the very public space. Many facts have demonstrated thatmany public spaces which are supposedly open for public needs, have their uses confined by private sectors who hold rights over spaces adjacent to these public territories. Shopping centers, malls, access between buildings, sports facilities, etc which are open to the public during the day are invariably shut off at night. Hong Kong Central is a perfect example of how an entire city center can be partially closed at night and even weekends since the private sector controls pedestrian access between most buildings.

Apart from such expropriation, practice, the encroachment of business and economic interests into the public realm are also of an alarming and challenging concern in this context. State and local governments now hire out a diversity of public spaces for rent to raise revenues by catering to private sector interests. In line with Altman's views, the public does not have to own a right to use a public space but they are born with such a right since they are members of a community and part of civil society. Paradoxically and in contrast with this idea, there is no constitution in any developed society that guarantees any individual 'rights' to space whatever. This is a utopian ideal that is denied in reality. The only functional right is that of ownership as the basic tenet of the capitalist system.

Contextualized by the preceding considerations, this paper aims at emphasizing the use-values of a public space that conveys certain meanings and attributes that are of functional and symbolic value. The spaces do not belong to the public unless we also consider that the public and the state are in a coalition against capital. Otherwise it functions for both parties, increasingly in the private interest, so a more sophisticated view of the state apparatus is required.

2. Method

The following text is based on an ethnographic study. Data collection was carried out by a series of interviews and the examination of relevant human settlementsin Bali and their associated public spaces. It attempts to investigate the existence of public space as a communalevent where use-valuesdominate. The above overview of public and private realms lays a conceptual framework for ensuing discussions focussed on the built environment. The remaining four sections respectively discuss built form and the Public interest, adopted forms of communal space in Bali, concept, and forms of communal space in Bali; and communal space and sustained living environment.

3. Built Form and the Public Interest

In discussingthe public versus the private realms and how the two alter each other, Michael Brill suggests using the term 'intrusion' in order to bring a more neutral sense regarding the possibility of each (private or public) bringing benefitsto one another [15]. In the context of human settlement, the intrusion of the public into the private realm occurs when this settlement is seen as a public entity in which every private place (home/dwelling) owner is required to follow a series of building traditions, codes, and practicesto not interrupt public consensus in regard to built forms. On the other hand, discussionover the intrusion of the private into the public realm can be demonstrated by a place that is owned, created and managed privately but used by members of the public.

This often occurs in many counties and is encouraged and facilitated by the government's policy and done with a certain mechanism. Tax incentive and corporate social responsibility (CSR) is the most common one implemented in the context of urban planning practices in Indonesia. In contradiction to this circumstance, the intrusion of the private interest into public space leads to what has been stated as the privatization of public space. Popular among common practices are the exclusion of public access to certain spaces; the private claims and uses of public spaces for private needs; and privatized development of trendy gated communities [3, 7, 16]. In order to contextualise these ideas, three case studies as examples of privatization of public space are given below

Firstly, limited access of public space -in a city park for example - occurs when an organization has acquired a permit to take over theuse of the park for business purposes, either temporarily or on a permanent basis. Such a practice is legal under government consent. On different occasions, limited public access is also due to badly designed public space. A park often lack facilities and features that are suitable and fulfilthe need of the public as users.This is inline with the concept of 'public' that bears the idea of equality. Public parks are a pivotal element in determining the liveabilitylevel of a city. This is dependent on its location, accessibility, vegetation, facilities, maintenance and other features [17].

LapanganPuputanBadungIGustiNgurahAgung is an example of a central and successful public park in Denpasar. In terms of access, this park is open to the public. The park is also well equipped with all sorts of facilities dedicated to sports activities, games, kids, art performances and other temporal uses. Figures 1 – 5 demonstrate the current situation of this park as an example of a well used and people-friendly public space.

Figure 1. The central part of the park
(Source: Swanendri, Survey 2019)

Figure 2. Kid corner
(Source: Swanendri, Survey 2019)

The central part of the park (Figure 1) is well used for a variety of dynamic activities. It has no physical boundary that segregates one part of the park to another. The most common activities organized in this park namely: soccer, cycling, kiting, meditating, picnic and even parenting.A kid corner (Figure 2) exists on it's peripheral, precisely on the northwest corner. It is a space dedicated especially for children to have fun. It is literallya children's playground, equipped withan assorted of popular playing equipment. Here, parents can observe and control their beloved ones from a close distance dan while socializing with each other in a clustered park bench.On the otherside, the northeast corner is known as a chess corner since the main attraction is the provision of giant chess grounds (Figure 3). Apart from being used for chess exercises - which is dominantly by senior citizens-, the area is often used for chess competitions (Figure 4). Others part of the park is dedicated to teenagers, oftento hold art-related activities(Figure 4).

Figure 3. ChessCorner
(Source: Swanendri, Survey 2019)

Figure 4. Dominantly by senior citizens.
(Source: Swanendri, Survey 2019)

Figure 5. Art-related activities
(Source: Swanendri, Survey 2019)

Secondly, privatization is commonly experienced by public spaces located in areas with high economic value, such as city main streets or coastal areas with potential tourist development. Paved pedestrian on main streets is often treated as a private area(see Figures 6-8), used especially for commercial activities. It is claimed as part of a local shop to exhibit all sorts of merchandise. At the end, these pedestrians can not accommodate people passing by.

Figure 6-8. Paved pedestrian as a private area
(Source: Swanendri, Survey 2019)

Suartika on her study denoted Sanur as one of the famous littoral public realms in Bali.She found that excessive tourist development at Sanur coast has failed to provide transitional space to separate the coastand tourist facilities constructed along Sanur Beach. This has incurred limited access to the beach by the public. She suggests that the physical development of Sanur should be suspended so the public needs and meanings embedded in the coast (littoral) can be retained[16]. Under the influence of global tourism, the intrusion of private interest into the public realm is not improving. The most current pictures of this area are displayed in Figure 9-12.Theseimages show that there has been an occupancy of public spaces[18]and the weakness of formal and social institutions to prevent such occupancy.

Figure 9-10. The occupancy of the beachside paved pedestrian
(Source: Swanendri, Survey 2019)

On Sanur beach, the pavement pedestrian along the beachoriginally is meant to be a delineation between public realm (beach) and private realm/private property (hotel, restaurant, shop), thus an intrusion among two of them are becoming obvious(Figure 9). The occupancy of public spaceson the beachside of the paved pedestrian has blocked the view to the beach which one is the truly genuine 'attraction' for this area, even though it's only occurred in the busiest part of the area (Figure 10).

Figure 11-12. Flags, umbrellas and semi-permanent enclosureas territorial marking
(Source: Swanendri, Survey 2019)

In some parts of the area, certain flags and umbrellas are becoming a kind of territorial marking (boundaries) for the part of the public realm/public space that has been claimed by the private interest(Figure 11). Furthermore, these public space occupancies are intensified by adding a more physical semi-permanent enclosure and wooden flooringon a 'reclaimed' public realm/public space to increase the quality of the space (that often used as dining area) as well as define the territory(Figure 12).

The third mechanism of privatization of the public place is internationally known as gated communitiesthatpromote socio-spatial segregation of the city. Gated communities first emerged in the US in the 1980s[3, 8], quickly followed by others around the globe including Indonesia when on 1990 BumiSerpongDamai emerged in Jakarta as the prototype of gated communities [19].In Bali the characteristic of a gated community is alike: a restricted access neighborhood which enclosed by circumferentialwalls. They are equipped with high-quality facilities and occupied by middle classes and upwards.Hence, social classes are segregated from each other and privatization of housing expands.The phenomenon has been seen merely as a mechanism of supply and demand for the property market [20].

4. Adopted Forms of Communal Space in Bali

The conception of civil society has a long history and can be traced back to Aristotle (quote – ref?). Nowadays, civil society is now understood as the intermediate realm between the family and the state which is the realm of culture, ideology, and politic. To enlighten it, Mary Kaldor emphasized that *"civil society is the process through which individuals negotiate, argue, struggle against or agree with each other and with the centers of political and economic authority. Through voluntary associations, movements, parties, unions, the individual is able to act publicly"* [21, p.585]. Thus, the main basis of civil society is unity and social integration referring to shared values; guaranteeing a balance between individual freedom and the stability of society based on shared life values.

In the context of Bali, civil society will be strongly related to the existence of *desaadat*(adatvillage) which has grown over the centuries into a system of self-government due to its inherent autonomy and traditions. Bali is not only well-known for the beauty of its natural landscape but also and religious traditions that characterize Balinese community life. The basic philosophy is *Tri Hita Karana* -three causes of happiness- that ensures the harmony of the universe and human life, namely harmonization of human to God, harmonious relationship within member of the society, and harmonization of human life with the environment [22].Thus Balinese structure their lives in relation to the gods and nature, expressed in their daily life as well as in the landscape of village planning and building that nowadays reflects the ongoing conflict between a proto-feudal society and globalization [23].

The term of *desaadat*was reaffirmed in"*PeraturanProvinsi Bali No. 4 Tahun 2019 tentangDesaTradisional di Bali*(Provincial Regulation for Bali Number 4 Year 2019 in regard to Traditional Villages)." Previously, the entity was called call as *desapakaraman* based on "*PeraturanProvinsi Bali Nomor 3 Tahun 2001 tentangDesaPakraman* (Provincial Regulation for Bali Number 3 Year 2001 in regard to *Desa Pakraman*)" which was then amended by "*Peraturan Daerah Provinsi Bali Nomor 3 Tahun 2003* (Provincial Regulation for Bali Number 3 Year 2003 in regard to *Desa Pakraman.*"*Desaadat* is a communal association at a village level whose operation is based on local codes, tradition, and culture. It usually has one or more *banjar adat* (neighborhood associations) working under its command. Parallel to these institutional practices, there are state-imposed institutions working at both the village and neighborhood level called *desa dinas* and *banjar dinas*. Further explanation of how these four associations work and cooperate can be explored further from Suartika (2010)[24].

Desaadat and *banjaradat* play an important role in the life of each community member which is based on agreement and shared values, including social capital of the community. Lisa Wood [25]summarized that social capital is often characterized as the glue that holds society together, with core aspects: networks, norms, and trust, mutual goals/actions. These traditional forms maintain social capital and organize and facilitate the dynamic activities of the community all of which have spatialimplications. To contain all of the related activities, the existence of sustainable public space is vital in Balinese life,not only for sacred activities but also for those of the non-sacred form of interactions. In Balinese communal space, use value is the dominant concept that promotes a series of social interactions. So,such a space has great social, cultural, and economic importance.' Itsexistencealsoexpresses liveability of the *desaadat* as an institution as well as a community. Thus in the perspective of civil society, *desaadat* along with its inherent characteristic has been proven as an intermediate realm between the Balinese (an individual) and the state.

5. Conceptand forms of communal space in Bali

Spatial practices in Balinese society are generally influenced by belief systems, natural conditions, and economic potential that lies in these settlements. Therefore, basically, humans build (architecture and built environment) mimic the good nature of the universe (macrocosm) and the human realm (microcosm) and try to achieve balance and harmony with the environment through the utilization of local potential such as natural resources, human resources as well as the surrounding physical pattern of architecture.The main characteristics of villages in Bali (in terms of rural and urban areas) are communal life and the existence of liveability communal space that not only accommodates communal activities but also becoming a container for social capital on daily lives of a community member.

The form of these communal spaces can be taking place in the form of open spaces and confined spaces. In some (traditional) settlements in Bali, the existence of a communal space can easily be identified. It is usually placed at the center of the settlement and most of the time is the core territory for both sacred and profane activities. It is also usually grand on a physical scale that makes it obvious compared with the surrounding structures [26, 27]. Such a practice can be observed in many traditional villages of Karangasem Regency, namely Timbrah, Asak, Bungaya, Tenganan, Kastala, see Figure 13.

a. Timbrah Village

b. Asak Village

Figure 13. Maincommunal space situated in the middle of its human settlement structure
(Source:Swanendri [28])

This communal space has become a center for a community that can be shifting in accommodate between sacred and profane activities in regard to a certain time and boundary, as demonstrated in Figure 14-20.

Usaba ceremony drying peanut *chatting*

Figure 14-16. Communed space in Timbrah Village is used for profane and scared activities
(Source: Swanendri, Survey 2015, 2018)

Figure 17-18. Communal space in Tenganan Village during *Mekare-kare* Festival
(Source: Swanendri&Mantra, Survey 2019)

Figure 19-20. Communal space in Tenganan Village during *Mekare-kare* Festival
(Source: Swanendri&Mantra, Survey 2019)

One of confined communal spaceexistsin Bali, such as *bale banjar*, - hall for the neighbourhood association. There are many kinds of activities can be performed in the *bale banjar*, both in term of sacred-profane. These includeactivities relating to preparation for religious ceremonies, centerfor learning traditional dances and musical instruments, *samua/sangkep* - community meetings, the conduct of community welfare issues, and public election. The type and the size of *banjar*hall can be varied from a simple semi-open building (Figure 21) which is exist most common in a rural area or less densely populated area, or a huge *wantilan*(an open hall) type of building so it can be utilized to accommodate various kinds of activities which is exist most common in an urban area or densely populated area(Figure 22).

Figure 21. A simpler type of the *bale banjar*
(Source: Swanendri, Survey 2018)

Figure 22. A *wantilan* type of the *bale banjar*
(Source: Swanendri, Survey 2019)

6. Communal space and sustained living environment

It is widely understood that the relationship between humans and their environment is basically dynamic and reciprocal where it is not only that the environment that influences behavior but also in vice versa, (human) behavior can perform series of adaptations to change and develop their environment to fulfill the certain behavior.Regarding that, one of the important issues in successful public spaces design is the chance for the community to adjust and improve their relationship with designed space.

In achieving communal space as a sustained living environment,the quality and its liveability should be addressed in the way of participation, sense of attachment and ownership. This quality is simply considered from the way people experience using the place and right to control the use, or in other words how a communal space as a form of sustainable public space consent to its use-value and fulfill the concept of 'public'. And all of these,from the perspective of civil society, promote and ensure a dynamic life and lesser control nor pressures from the state as well as the economic forces.

7. Conclusion

Consequent tothe global capitalist system, both in an urban and rural, context, the privatization of land occurred.In the process, itoften leads to a conflict in interests between the public realm and its need to fulfill the demands of civil society for public space. Learning from Balinese traditional settlements, where use-value is a dominant concept that promotes the opportunity for the community members to meet and interact, participate and supportthe formation of social networks. Another important feature of (Balinese) communal space is the existence of control by public or community itself that implies communal space as spatial form that is to be accommodated as a part of central planning strategy in order to achieve a liveable and sustained living environment.

8. Acknowledgement

This article was presented at the 2nd International Conference on Smart City INNOVATION (ICSCI) 2019, jointly held by Universitas Indonesia and Universitas Diponegoro. ICSCI conferences have been supported by the United States Agency for International Development (USAID) through the Sustainable Higher Education Research Alliance (SHERA) Project for Universitas Indonesia's Scientific Modelling, Application, Research, and Training for City-centered Innovation and Technology (SMART CITY) Center for Collaborative Research, administered through Grant #AID-497-A-1600004, Sub Grant #IIE-00000078-UI-1

9. References

[1] Vatn A 2018 Environmental Governance–from Public to Private? *Ecological economics***148** 170-77
[2] Punter J V 1990 The Privatisation of the Public Realm *Planning Practice & Research***5** 9-16
[3] Low S 2013 *How Private Interests Take over Public Space: Zoning, Taxes, and Incorporation of Gated Communities* in: Low S and Smith N (Eds.) The Politics of Public Space (Routledge) 81-104
[4] Suartika G A M 2013 *Lost in Translation: The Balinese Vernacular Open Space Tradition* in: Suartika G A M (Ed.) Vernacular Transformations (Pustaka Larasan and Udayana University's Program) 9 131-49
[5] Habermas J and Habermas J 1991 *The Structural Transformation of the Public Sphere: An Inquiry into a Category of Bourgeois Society* (MIT press)
[6] Supriadi Y 2017 Relasi Ruang Publik Dan Pers Menurut Habermas *Kajian Jurnalisme***1**
[7] Altman I and Zube E H 2012 *Public Places and Spaces* (Springer Science & Business Media)
[8] Smith N and Low S 2013 *Introduction: The Imperative of Public Space* in: Low S and Smith N (Eds.) The Politics of Public Space (Routledge) 1-16
[9] Brodin J 2007 Public Space: Exploration of a Concept *California Politics & Policy***11** 47-65
[10] Low S and Smith N 2013 *The Politics of Public Space* (Routledge)
[11] Clark G L and Dear M J 1984 *State Apparatus: Structures and Language of Legitimacy* (Allen & Unwin)

IOP Conf. Series: Earth and Environmental Science **396** (2019) 012030 doi:10.1088/1755-1315/396/1/012030

[12] Hegel G W F 1991 *Hegel: Elements of the Philosophy of Right* (Cambridge University Press)
[13] Smith A 2010 *The Wealth of Nations: An Inquiry into the Nature and Causes of the Wealth of Nations* (Harriman House Limited)
[14] Calhoun C 2011 *Civil Society and the Public Sphere* in: Edwards M (Ed.) The Oxford Handbook of Civil Society. Oxford Handbooks in Politics & International Relations (Oxford University Press) 311-23
[15] Brill M 1989 *Transformation, Nostalgia, and Illusion in Public Life and Public Place* in: Altman I and Zube E H (Eds.) Public Places and Spaces (Springer) 7-29
[16] Suartika G A M 2015 Sand, Sea and Ceremony: Conflict over the Littoral Public Realm in Sanur, Bali *Procedia-Social and Behavioral Sciences***179** 128-40
[17] Sepe M 2017 Placemaking, Livability and Public Spaces. Achieving Sustainability through Happy Places *The Journal of Public Space***2** 63-76
[18] Setiawan B B 2006 Ruang Publik Dan Modal Sosial: Privatisasi Dan Komodifikasi Ruang Di Kampung *Jurnal Fakultas Hukum UII***29** 28-38
[19] Widhyharto D S 2009 Komunitas Berpagar: Antara Inovasi Sosial Dan Ketegangan Sosial (Studi Kasus Komunitas Berpagar Di Propinsi Di Yogyakarta, Indonesia) *Jurnal Ilmu Sosial dan Ilmu Politik***13** 204-30
[20] Sueca N P and Fitriani L R D, Profile Gated-Community Di Denpasar, in: Wibowo A S, Widiastuti I, Kusuma H E and Larasati D (Eds.) TEMU ILMIAH IPLBI 2012, Ikatan Peneliti Lingkungan Binaan Indonesia (IPLBI), Bandung, 2012, pp. 45-48.
[21] Kaldor M 2003 The Idea of Global Civil Society *International affairs***79** 583-93
[22] Bali G, Peraturan Daerah Provinsi Bali Nomor 4 Tahun 2019 Tentang Desa Adat Di Bali, Denpasar, 2019.
[23] Suartika G and Saputra K, The Creation of Harmonious Living Spheres-Landscape, Capital, and the Balinese Way, IOP Conference Series: Earth and Environmental Science, IOP Publishing, 2019, pp. 012065.
[24] Suartika G A M 2010 *Morphing Bali: The State, Planning and Culture* (Lambert Academic Publishing)
[25] Wood L and Giles-Corti B 2008 Is There a Place for Social Capital in the Psychology of Health and Place? *Journal of Environmental Psychology***28** 154-63
[26] Parimin A P, Fundamental Study on Spatial Formation of Island Village : Environmental Hierarchy of Sacred-Profane Concept in Bali, University of Osaka, Japan, 1986.
[27] Swanendri N M and Setiada I N K, Pola Spasial Permukiman Masyarakat Bali Aga Di Desa Pakraman Timbrah, Desa Pertima, Kecamatan Karangasem, Kabupaten Karangasem, Universitas Udayana, 2016.
[28] Swanendri N M, Social Capital of Human Settlement: Learning from Bali Aga Villages in Karangasem, in: Suartika G A M and Saputra K E (Eds.) Reframing the Vernacular: Politics, Semiotics, and Repesentation (the First International Conference on Cultural Communication and Space (ICCCS) and 9th International Conference on International Society for Vernacular Settlement (ISVS)), Udayana University Press, Denpasar, Bali, 2018, pp. 4.10-17.

The Implication of Smart Environment on Old Palembang Cultural Heritage Places

W F F Anwar

Departement of Architeture, Universitas Sriwijaya. Jl Raya Prabumulih KM.32 Inderalaya Ogan Ilir Sumatera Selatan 30118

E-mail: widyafransiska@ft.unsri.ac.id

Abstract. Palembang as old city has many historical places in various conditions. Some of these places are still easy to be found, well maintained, abandoned as well as leaved as named without any physical evidences. The cultural heritage deals with modern live and the threat of extinction. In the internet era, the term smart environment for cultural heritage becomes another approach to conserve the historical artefacts. However, the conservation should also consider the non-physical aspect such as the relationship between people and the environment in the heritage place. This paper elaborate the use of people place relationship in determining the smart implication at the at Palembang cultural heritage area. To achieve this aim, study have two objectives; (1) examining how people of historical places attached to their living environment as well as their city, (2) determining the level of potency of historical places to find the appropriate smart strategy for conservation. Place attachment was measured through questionnaires filled by the 136 residents from seven old urban kampongs in Palembang. Then, results from this were confirmed by respondents understanding on the history of place and their visiting experience to determine the potency of historical places. Those statistical analyses used PASW v.19. The result shows three groups of heritage places at Palembang cultural heritage area; the ones with high, medium and low attachment in urban scale. The smart strategy can be applied at medium and low attachment to rebirth and strengthen the attachment to the places. The use of virtual reality on these places is the appropriate strategy for smart heritage conservation. The combination between Virtual Reality and multimedia technologies will support the development at historical places of Palembang

1. Introduction

Palembang is one of old cities in Indonesia with long history. It was an international hub, a capital of Palembang Darussalam Sultanate. In the past time, the center of city was located at riverside area. Therefore, the historical places can be found along the Musi River. Musi river divide the city into two riverside area; *Ulu* at southern and *Ilir* at northern. The historical places are the old riverside settlements', historical buildings and old public facilities. In present, these places are still a part of city center and influenced by modern development. Generally, their conditions are varying. Some of them are well maintained and easy to find. The others are abandoned or no longer found.

Conservation of historical place is important for a society. The globalization is spread all over the world brings some disadvantages to the issue of identity. The international architecture style is easily to be found in any modern city. The general style brings similar expression of modern thought in architecture. However, the loss of identity becomes threat in modern society. In fact, the identity is important intangible assets for any society. The identity differentiates one society to another and become competitive advantage. Further,the society can learn their locality from the historical places.. The local wisdom expressed at the historical places is the important thought for dealing with future life. Now, the historical places are threatened by the modern development. Therefore, to conservation of historical places is a must for city like Palembang.

Historical places reflect identity. Relph (1976) coins the place represents the identity of physical and social environment [1]. The place has physical elements. The elements reflect the uniqueness of the place. In other words, the specific element and its setting differentiate place from other places. Therefore, the place can be a specific character that generates identity for its surrounding as well as urban reminder in bigger scale [2,3]. The historical places (along with its architecture) were the expression of identity. In order to sustain the identity, the distinctiveness should be well maintained and recognized by the residents of places and citizen.

Effort to sustain the identity must take people into account. Conserving the historical place means not only protect and maintain the place but also brings the people to get in the effort. In the internet era, the term smart environment for cultural heritage becomes another approach to conserve the historical artifacts. However, the conservation should also consider the non-physical aspect such as the relationship between people and the environment in the heritage place. This paper elaborate the use of people place relationship in determining the smart implication at the at Palembang cultural heritage area.

2. Literature review

2.1. Place Attachment in Conservation

Place attachment is one of concept that relates people and their physical environment. It defines as emotional bonding between an individual and his living environment. The scale of this relationship can be in living environment, neighborhood, urban, region and country [4,5]. The bonding brings the memory and perception toward the living environment. When the experience towards the place is continuing, the attachment will become stronger and lead to identity [6,7].

Place attachment has four dimensions; place identity, social bonding, place dependence and nature bonding [8–10]. Place identity is a person definition on his self, caused by the setting and the meaning of the place. One's interaction with the society and the natural environment directs to the social and nature bonding. The ability of a place to accommodate one's activities leads to place dependence. Further these four dimensions is complemented by the place attachment predictors caused by one's preference towards the place. The predictors are place value, special place and response on development [9]. The predictors and dimension are used to measure the place attachment according to the scale of place [11].

The existence of historical place results the people perception. It is one of attachment toward the place. The perception of historical places in a city implies its history. How the history of the city appreciated by the people is influenced by the perception towards the historical places. If the perception is positively appreciated, then the identity of the place strengthens city identity. Moreover, the existence of historical places in the city reflects the culture of city society. It contributes to city distinctiveness [12,13]. Those places are tangible assets of the city and the distinctiveness is the intangible asset.

The place attachment and historical place are strongly tied. People and the environment have related since a long time ago. Therefore in conserving the historical places, the planner should consider the people who lived there. The cultural significance of certain historical place will be lasting if people were engaged in the conservation effort [12,14,15]. Place attachment is the antecedent to the people's loyalty [16]. Level of attachment determines the support for conservation. The stronger quality of attachment is the stronger support for conservation. At the end, it creates more sustained identity of the place [17].

2.2. Smart Tools in Heritage Conservation

Historical place is also named as cultural heritage. Cultural heritage refers to a locality, natural landscape, settlement area, architectural complex, archaeological site or standing structure that is recognised and often legally protected as a place of historical and cultural significance (ICOMOS 2007 in [18]). Physical setting of historical place can promote identity in two contexts; the built heritage and innovative design [2]. The built heritage accommodates the tradition in a certain place. The innovation in places, elements and supporting social wisdom contribute to the identity. Conserving the built heritage along with offering new innovation design are the basic factor in conservation. Dealing with internet technology, the term smart environment for cultural heritage becomes another approach to conserve the historical places.

The trend of smart city leads to the application of smart technology in conservation. The application of smart technology can be in form of restoration techniques, data bases, user interactive technology and the cultural heritage [19]. Restoration techniques include the smart metode of archiving and authentic digital surrogates. Smart in data base includes photogrammetric, reality-based 3D documentation and perpetual digital conservation [20,21]. The interactive use includes social media, 3d visualization and virtual reality technology [22–24]. The application of each form of smart application requires certain condition as mentioned in previous studies. The restoration techniques should apply latest friendly technologies, equipment, software, authenticity assurance and spatial data [22]. Database should be efficient in using the data, analysis and presentation opportunities [20]. When users apply the digital in cultural heritage, it brings the literature on digital engagement, interactive and participation of user in heritage place. The digital tool should enhance and not replace the artifacts or historical evidences [25].

Since the early of 2000s, the virtual heritage has become important term in preservation, conservation and interpretation of culture and history [26,27]. As many historical places are threaten to be destroyed or lost, then the emergence of long lasting documentation must be formed to sustain the culture and history. The advanced technology is needed to overcome the threat on cultural heritage assets [28] such as 3D reconstruction, hybrid 3D models, virtual reality devices and software and web-based virtual tour system [24,29,30]. In virtual heritage, the historical places were examined and preselected as the potential digital conserved built heritage. The virtual heritage is applied for the historical places that no more exist, unreachable, geographically remote or has limited access caused by some constraints such time, distance, scale, safety even cost [27]. The best scenario of virtual heritage is the places or collections are chosen by the public [21], presented in interesting presentation, assisted young generation to learning and have strong motivational impact [27].

The use of virtual reality in conservation has been studied in many cases. For some international cases, the VR has also been applied in form of interactive simulation museum for Hagia Sophia and Pompeii, visualization of environment for Ancient Malacca and PC webbased application for Campeche old Mexican City, Pompei. Further, the presentation of VR use some tools such as PC engine for Notre Dame, virtual overlay seen in certain kiosk at Ename, Belgium, or projection in certain wall surface like Salzburg, Austria and Sala Della Volta Dorata, Roma [26]. Studies on implication of VR in Indonesian virtual heritage have also been developed for documentation. The documentation by using VR GIS allows the 3D model georeferencing to put the coordinate to the model with in the of Sewu Temple area. Further, VR is also used for promotion the cultural tourism destination such as 3D walk through of Lawang Sewu in Semarang virtual museum at Istana Siak Sri Indra Pura (Supardi, Danuri and Prayitno, 2016) and at Istana Basa Pagaruyung Batusangkar. The aforementioned case of VR implementation in heritage places are conducted both within the object of heritage, area of conservation and heritage site. Hagia Sophia, Salzburg, Sala Della Volta Dorata, Lawang Sewu, and at Istana Basa Pagaruyung Batusangkar are the case that use VR for visualization within the deteriorated built heritage. The built heritages are still existed. Campeche old Mexican City, Pompei, Notre Dame and museum of Istana Siak Sri Indra Pura are cases that use PC for VR presentation of the built heritage in museum. Sewu temple is the case that uses VR for visualization in a wide area of conservation. Ename and Ancient Malacca are the practices that use VR for visualization of environment of heritage site where the built heritages are no longer found.

3. Methodology

This paper elaborate the use of people place relationship in determining the smart implication at the at Palembang cultural heritage area. To achieve this aim, study have two objectives; (1) examining how people of historical places attached to their living environment as well as their city, (2) determining the level of potency of historical places to find the appropriate smart strategy for conservation. In order to examine the place attachment, the survey questionnaire was done to 136 respondents. It measured the attachment towards 23 historical places in seven urban kampongs along Musi riverside area. The data was analyzed by using factor analysis and cross tabulation. Factor analysis measured the dimensions of attachment; place identity, place dependence and place value. To determine the level of potency of historical places, the result form place attachment measurement was crosstabulated with the respondents' understanding on the history of places and visit. This statistical analysis used PASW v.19.

The urban kampongs are the old settlements at Musi riverside area which have historical as well as cultural value. The kampongs are named as Kapiten, Klenteng, Bahrak, Munawar, Suro, Sekanak and Kuto. The kampongs are located at two side of riverside area (Figure. 1). Historically, those kampongs had been established since pre sultanate era. They were addressed for certain ethnics in Palembang. Kapiten and Klenteng were the Chinese settlements (*klenteng* means the temple). Bahrak, Munawar and Kuto were Arabian settlements. Except Kuto, these four old settlements are located at Ulu. Meanwhile, Suro and Sekanak were Malay Palembang settlements and located at Ilir. All of these kampongs have old buildings that full of history.

A total of 136 questionnaires were distributed in seven kampongs (Table 1). The respondents are the residents of those kampongs. Generally, each kampong contributes more than 16 % to the total 136 valid questionnaire.

Table 1. Distribution of questionnaires

Kampong Name	F	%
Kapiten	25	18.4
Klenteng	16	11.8
Bahrak	18	13.2
Munawar	16	11.8
Sekanak	23	16.9
Suro	20	14.7
Kuto	18	13.2
Total	136	100

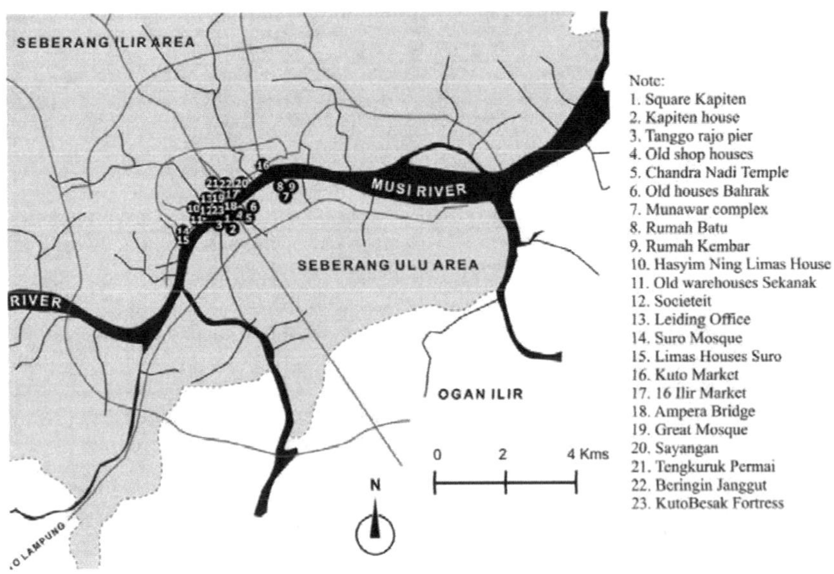

Figure 1. Location of 23 historival places in Palembang city map

The respondents are relatively equal between male and female. About 86 % of respondents are more than 20 years old, with 10.3% senior residents. More than 80% of them have stayed for more than 20 years in the kampong. The demographic profile of respondents shows in Table 2. The questionnaire consists of 15 questions on place attachment dimensions, one question on respondents' understanding on the history of the places and one question on respondent visiting experience.

Table 2. Demographic profile N= 136

Description		F	%
Gender	Male	67	49.3
	Female	69	50.7
Age	<20years old	18	13.2
	21-30 years old	60	44.1
	31-50 years old	44	32.4
	> 50 years old	14	10.3
Length of stay	<5y	5	3.7
	6-20y	20	14.7
	21-30y	62	45.6
	31-50y	36	26.5
	>50y	13	9.6

4. Result and Discussion

4.1. Place Attachment and potency of historical place

Place attachment in seven kampongs was measured in forms of 15 questions. The questions were derived from three dimension of place attachment which are place identity, place dependence and place value. Respondents gave their response within 1 to 5 Likert scale, namely strongly agree, agree, neutral, disagree and strongly disagree. The questions were translated as component in factor analysis. The Kaiser-Meyer-Olkin (KMO) value is 0.91 showing the sample is adequate. From the factor analysis running, the 15 components were extracted into three components, as shown in Table 3 as the highest value for each component. They are love (0.847), satisfaction (0.753) and economic value (0.869). Respectively, place identity, place dependence and place value contribute the most influence factor of attachment. This result represents the historical value is not the main factor in forming the attachment. Factor analysis show historical value (0.67) is lower than recreational value (0.831).

To confirm the results on historical value, respondents also answered their understanding on history of 23 historical places at seven kampongs. The places had been existed since sultanate era. Places are shown in Table 4. At kampong kapiten and klenteng, the square, the kapiten houses and temple were part of Chinese settlements that have long history and unique architecture setting. The tangga rajo pier was used by the sultan when visit this territory. Similar to them, Kuto, Bahrak and Munawar also have unique architecture and history as Arabian settelement. Suro and Sekanak were the local malay settlements that are mixed with colonials buildings. At city center, there are many places with history of Palembang sultanate. The *beringin janggut* was an area of sultanate palace named *kuto beringin* or *kuto cerancang* before it was burnt by the Dutch colonial and moved to *kuto* kecik, the place near Kuto Besak Fortress and the Great Mosque. Tengkuruk Permai and Sayangan were the other places and settlement area that had been existed since sultanate era. Unlike the other places, the physical heritage at these three places is no longer existed anymore. The area becomes modern trading area and the names were given to name street to memorize their history (toponim).

Table 3. Factor analysis results on place attachment

Rotated Component Matrix[a]

Place Attachment Dimension	Component		
	1	2	3
Place Identity			
A part of me	.771	.399	.211
Love	**.847**	.329	.214
Define who I am	.765	.412	.191
Meaningful	.609	.599	.233
Being accepted	.729	.390	.255
Not a part of it	-.686	-.156	.001
Place dependence			
The best place	.600	.572	.076
Doing	.524	.699	.113
Not going to move	.252	.734	-.002
The only one	.366	.727	.241
Satisfy	.448	**.753**	.234
Place value			
Scenic value	.002	.581	.579
Economic value	-.008	.151	**.869**
Historical value	.493	-.027	.671
Recreational value	.242	.138	.813
Kaiser-Meyer-Olkin Measure of Sampling Adequacy.			.918

In general, the statistic shows that the understanding on history of kampong is relatively lower than in city center. In average, less than 50% of residents understand the history of mentioned places. It reflects that the residents have limited understanding on history of other kampongs. He could understand his kampong history, but not so well on other kampongs. For respondents, their understanding on Munawar complex history is relatively mediocre (51.5%). Among 23 places, Munawar complex history is relatively well-known for residents of all kampongs.

Statistic also shows that the highest value goes to Kuto Besak Fortress (82.4%), Ampera Bridge (81.6%) and Great Mosque (79.4%). All of these three places are not located at kampongs. They are at city center, near the Musi River. However, some of historical places are not well known by the respondents which are Sayangan, Tengkuruk Permai, Kuto Market, old shop houses at Klenteng, Societeit at Sekanak. Extremely, the Beringin janggut history has lowest statistics (2.9%). Residents almost do not know the history of Beringin Janggut.

Table 4. Respondent's understanding of history (N=136)

Kampung	Historical places	F	%
Kapiten	Square	46	33.8
	Kapiten house	62	45.6
	Tanggo rajo pier	41	30.1
Klenteng	Old shop houses	22	16.2
	Chandra Nadi Temple	53	39
Bahrak	Old houses	50	36.8

Munawar	Munawar complex	**70**	51.5
	Rumah Batu	48	35.3
	Rumah Kembar	52	38.2
Sekanak	Hasyim Ning Limas House	37	23.8
	Old warehouses	35	25.7
	Societeit	9	6.6
	Leiding Office	66	48.5
Suro	Suro Mosque	33	24.3
	Limas Houses	41	30.1
Kuto	Kuto Market	26	19.1
City center	16 Ilir Market	75	55.1
	Ampera Bridge	**111**	**81.6**
	Great Mosque	**108**	**79.4**
	Sayangan	27	19.9
	Tengkuruk Permai	26	19.1
	Beringin Janggut	4	2.9
	KutoBesak Fortress	**112**	**82.4**

To confirm the attachment on 23 historical places at seven kampongs, the factor analysis were cross tabulated with the respondent's history understanding. To simplify paper presentation, only the results on history understanding and positive response on attachment (agree and strongly agree) were shown in Table 5. Statistic shows consistency on places that are well known with the love, satisfaction and economic value. Ampera Bridge has highest results, followed by Kuto Besak fortress and Great Mosque respectively. Societeit and Beringin Janggut have very low results which is in line with previous statistical analysis.

Table 5. Cross tabulation of attachment on historical understanding

No	Place			Love	Satisfaction	Economic value
	Kapiten					
1	Square	agree	count	12	14	19
			%	26.1%	36.8%	27.9%
		strongly agree	count	31	27	26
			%	47.0%	52.9%	46.4%
2	Kapiten house	agree	count	15	15	20
			%	32.6%	39.5%	29.4%
		strongly agree	count	42	33	37
			%	63.6%	64.7%	66.1%
3	Tanggo rajo pier	agree	count	10	10	18
			%	21.7%	26.3%	26.5%
		strongly agree	count	26	24	22
			%	39.4%	47.1%	39.3%
	Klenteng					
4	Old shop houses	agree	count	3	6	7
			%	6.5%	15.8%	10.3%
		strongly agree	count	17	15	15
			%	25.8%	29.4%	26.8%
5	Chandra Nadi Temple	agree	count	15	16	21
			%	32.6%	42.1%	30.9%
		strongly agree	count	32	27	30
			%	48.5%	52.9%	53.6%
	Bahrak					
6	Old houses	agree	count	19	20	21
			%	41.3%	52.6%	30.9%
		strongly agree	count	27	24	24
			%	40.9%	47.1%	42.9%
	Al Munawar					
7	Al Munawar complex	agree	count	22	20	28
			%	47.8%	52.6%	41.2%
		strongly agree	count	41	34	34
			%	62.1%	66.7%	60.7%
8	Rumah Batu	agree	count	13	11	17
			%	28.3%	28.9%	25.0%
		strongly agree	count	32	28	26
			%	48.5%	54.9%	46.4%
9	Rumah Kembar	agree	count	14	14	17
			%	30.4%	36.8%	25.0%
		strongly agree	count	32	27	29
			%	48.5%	52.9%	51.8%
	Sekanak					
10	Limas of Hasyim Ning	agree	count	7	6	16
			%	15.2%	15.8%	23.5%
		strongly agree	count	14	9	11
			%	21.2%	17.6%	19.6%
11	Old warehouses	agree	count	8	9	18
			%	17.4%	23.7%	26.5%
		strongly agree	count	20	14	16
			%	30.3%	27.5%	28.6%
12	**Societeit**	agree	count	1	1	2
			%	**2.2%**	**2.6%**	**2.9%**
No	Sekanak			Love	Satisfaction	Economic value
12	**Societeit**	strongly agree	count	8	7	7
			%	12.1%	13.7%	12.5%
13	Leideng office	agree	count	21	20	29
			%	45.7%	52.6%	42.6%
		strongly agree	count	42	33	36
			%	63.6%	64.7%	64.3%
	Suro					
14	Suro Mosque	agree	count	9	9	16
			%	19.6%	23.7%	23.5%
		strongly agree	count	18	14	16
			%	27.3%	27.5%	28.6%
15	Limas houses	agree	count	12	13	18
			%	26.1%	34.2%	26.5%
		strongly agree	count	18	12	18
			%	27.3%	23.5%	32.1%
	Kuto					
16	Kuto market	agree	count	6	7	10
			%	13%	18.4%	14.7%
		strongly agree	count	17	15	14
			%	25.8%	29.4%	25%
	City Center					
17	16 Ilir Market	agree	count	21	22	28
			%	45.7%	57.9%	41.2%
		strongly agree	count	47	37	42
			%	71.2%	72.5%	75.0%
18	**Ampera Bridge**	agree	count	32	28	52
			%	69.6%	73.7%	76.5%
		strongly agree	count	58	45	49
			%	**87.9%**	**88.2%**	**87.5%**
19	**Great Mosque**	agree	count	34	29	52
			%	73.9%	76.3%	76.5%
		strongly agree	count	56	42	47
			%	**84.8%**	**82.4%**	**83.9%**
20	Sayangan	agree	count	9	9	13
			%	19.6%	23.7%	19.1%
		strongly agree	count	15	14	13
			%	22.7%	27.5%	23.2%
21	Tengkuruk	agree	count	6	7	10
			%	13.0%	18.4%	14.7%
		strongly agree	count	17	15	14
			%	25.8%	29.4%	25.0%
22	**Beringin janggut**	agree	count	1	1	2
			%	**2.2%**	**2.6%**	**2.9%**
		strongly agree	count	2	2	2
			%	**3.0%**	**3.9%**	**3.6%**
23	**Kuto Besak Fortress**	agree	count	39	33	56
			%	**84.8%**	**86.8%**	**82.4%**
		strongly agree	count	57	44	48
			%	**86.4%**	**86.3%**	**85.7%**

Table 6. Response on place visit

Places	**Visiting the place**	**%**
Kapiten	**93**	68.4
Klenteng	**79**	58.1
Bahrak	61	44.9
Munawar	70	51.5
Sekanak	**83**	61.0
Kuto Besak Fortress	**115**	84.6
Suro	75	55.1
Kuto	19	14.0

Table 7. Cross tabulation of attachment on place visit

Crosstab value			Love	Satisfaction	Economic value
Kapiten	Agree	Count	26	25	40
		% within	56.5%	65.8%	58.8%
	Strongly agree	Count	52	40	45
		% within	**78.8%**	**78.4%**	**80.4%**
Klenteng	Agree	Count	19	23	33
		% within	41.0%	60.5%	48.5%
	Strongly agree	Count	46	36	40
		% within	69.1%	**70.6%**	71.4%
Bahrak	Agree	Count	22	23	27
		% within	47.8%	60.5%	39.7%
	Strongly agree	Count	32	22	28
		% within	48.5%	43.1%	50%
Munawar	Agree	Count	25	21	29
		% within	54.3%	55.3%	42.6%
	Strongly agree	Count	36	27	32
		% within	54.5%	52.9%	57.1%

Crosstab value			Love	Satisfaction	Economic value
Sekanak	Agree	Count	21	22	37
		% within	45.7%	57.9%	54.4%
	Strongly agree	Count	43	32	43
		% within	65.2%	62.7%	**76.8%**
Kuto Besak Fortress	Agree	Count	38	32	57
		% within	**82.6%**	**84.2%**	**83.8%**
	Strongly agree	Count	56	43	49
		% within	**84.4%**	**84.3%**	**87.5%**
Suro	Agree	Count	21	23	36
		% within	45.7%	60.5%	52.9%
	Strongly agree	Count	40	28	34
		% within	60.6%	54.9%	60.7%
Kuto	Agree	Count	8	6	12
		% within	17.4%	15.8%	17.6%
	Strongly agree	Count	8	7	7
		% within	12.1%	13.7%	12.5%

4.2. Smart heritage conservation implication

The previous discussion shows how people relate towards historical places and the factor that drive the attachment. The different result on two cross tabulation shows that the historical value does not stimulate the attachment and willingness to visit the place. Economic value, satisfaction and love are more influencing to the attachment and visit. Regarding the conservation effort, study determined the smart implication that should be applied to 23 historical places (Table 8). The implication is applied to all places, in terms of restoration, database and sustained maintenance. For places with strong attachment and not physically threatened, the Virtual Reality is an optional. The VR can be used within the built heritage for visualization of history. For places with weak attachment, VR is still applied to increase the attachment. Virtual Reality become a must to be applied at places with weak attachment, physically threatened even no longer exist. VR can be implemented for visualization of environment of heritage site where the built heritages are no longer found. In this case, Sayangan, Tengkuruk Permai and Beringin Janggut are the sites that has no built heritage. The physical evidences of built heritage are no longer existed. Therefore, these sites can use virtual overlay seen in certain kiosk or projection in certain surface. By doing so, their existence will be not as toponim only, but also the virtual visualization. Therefore, the future generation will have more understanding on city history, strong attachment and support the conservation of historical places in their city.

Table 8. The smart heritage implication based on people place relationship

Places		Square	Kapiten house	Tanggo rajo pier	Old shop houses	Chandra Nadi Temple	Old houses at Bahrak	Munawar complex	Rumah Batu	Rumah Kembar	Hasyim Ning Limas House	Old warehouses	Societeit	Leiding Office	Suro Mosque	Limas Houses	Kuto Market	16 Ilir Market	Ampera Bridge	Great Mosque	Sayangan	Tengkuruk Permai	Beringin Janggut	KutoBesak Fortress
No more exist																					v	v	v	
Physically threatened		v	v	v	v		v														v	v	v	
Constraint	Time				v		v	v	v	v	v	v	v	v		v	v							v
	Scale	v	v	v																	v	v	v	v
Attachment	Weak											v	v								v	v	v	
	Medium				v		v	v	v	v	v			v	v	v	v	v						
	Strong	v	v	v		v													v	v				v
Smart Heritage conservation		RES Db VR SM	RES Db VR SM	RES Db VR SM	RES Db VR SM	RES Db SM	RES Db VR SM	RES Db SM	RES Db SM	RES Db SM	RES Db SM	RES Db VR SM	RES Db VR SM	RES Db SM	RES Db SM	RES Db SM	RES Db SM	RES Db SM	RES Db SM	RES Db SM	RES Db VR SM	RES Db VR SM	RES Db VR SM	RES Db SM

RES: restoration; Db: Data Base; VR: Virtual Reality; SM: Sustained Maintenance Roussou, 2002).

5. Conclusion

This paper elaborate the use of people place relationship in determining the smart implication at the at Palembang cultural heritage area. The study shows that the measurement of place attachment can support the effort to conserve the historical places. Attachment toward the place can be used to determine what smart tools to be applied as a part of conservation. By combining the attachment and smart technology, the smart environment at cultural heritage places can be implemented. Further, the conservation become people based effort, and it will be sustained by the support from the people.

6. Acknowledgement

This article was presented at the 2nd International Conference on Smart City INNOVATION (ICSCI) 2019, jointly held by Universitas Indonesia and Universitas Diponegoro. ICSCI conferences have been supported by the United States Agency for International Development (USAID) through the Sustainable Higher Education Research Alliance (SHERA) Project for Universitas Indonesia's Scientific Modelling, Application, Research, and Training for City-centered Innovation and Technology (SMART CITY) Center for Collaborative Research, administered through Grant #AID-497-A-1600004, Sub Grant #IIE-00000078-UI-1.

7. References

[1] Relph E 1976 1976: Place and placelessness. London: Pion.

[2] Gospodini A 2004 Urban morphology and place identity in European cities: built heritage and innovative design *J. Urban Des.* **9** 225–48

[3] Lewicka M 2008 Place attachment, place identity, and place memory: Restoring the forgotten city past *J. Environ. Psychol.* **28** 209–31

[4] Lewicka M 2010 What makes neighborhood different from home and city? Effects of place scale on place attachment *J. Environ. Psychol.* **30** 35–51

[5] Lewicka M 2011 Place attachment: How far have we come in the last 40 years? *J. Environ. Psychol.* **31** 207–30

[6] Low S M and Altman I 1992 Place Attachment *Place Attachment* (Springer {US}) pp 1–12

[7] Smaldone D 2007 The role of time in place attachment *In: Burns, R.; Robinson, K., comps. Proceedings of the 2006 Northeastern Recreation Research Symposium; 2006 April 9-11; Bolton Landing, NY. Gen. Tech. Rep. NRS-P-14. Newtown Square, PA: US Department of Agriculture, Forest Service, Northern Research Station: 47-56.* vol 14

[8] Williams D R and Vaske J J 2003 The measurement of place attachment: Validity and generalizability of a psychometric approach *For. Sci.* **49** 830–40

[9] Brown G and Raymond C 2007 The relationship between place attachment and landscape values: Toward mapping place attachment *Appl. Geogr.* **27** 89–111

[10] Raymond C M, Brown G and Weber D 2010 The measurement of place attachment: Personal, community, and environmental connections *J. Environ. Psychol.* **30** 422–34

[11] Hernández B, Hidalgo M C, Salazar-Laplace M E and Hess S 2007 Place attachment and place identity in natives and non-natives *J. Environ. Psychol.* **27** 310–9

[12] Tweed C and Sutherland M 2007 Built cultural heritage and sustainable urban development *Landsc. Urban Plan.* **83** 62–9

[13] Handal J 2006 Rebuilding city identity through history: the case of Bethlehem-Palestine *Des. Sustain. Cities Dev. World* 51–68

[14] Samant S 2004 Manifestation of the urban public realm at the water edges in India{\textemdash}a case study of the ghats in Ujjain *Cities* **21** 233–53

[15] Hanh V T H 2006 Canal-side highway in Ho Chi Minh City ({HCMC}), Vietnam {\textendash} Issues of urban cultural conservation and tourism development *GeoJournal* **66** 165–86

[16] Prayag G and Ryan C 2011 Antecedents of Tourists' Loyalty to Mauritius *J. Travel Res.* **51** 342–56

[17] Walker A J and Ryan R L 2008 Place attachment and landscape preservation in rural New England: A Maine case study *Landsc. Urban Plan.* **86** 141–52

[18] Wessels S, Ruther H, Bhurtha R and Schroeder R 2014 Design and creation of a 3D virtual tour of the world heritage site of Petra, Jordan *Proc. AfricaGeo* 1–3

[19] Graziano T 2014 Boosting innovation and development: the italian smart tourism, a critical perspective *Eur. J. Geogr.* **5** 6–18

[20] Toz G and Duran Z 2004 Documentation and analysis of cultural heritage by photogrametric methods and GIS: A case study *XXth ISPRS Congress. Istanbul, Turkey* pp 438–41

[21] Taylor J and Gibson L K 2016 Digitisation, digital interaction and social media: embedded barriers to democratic heritage *Int. J. Herit. Stud.* **23** 408–20

[22] Mudge M, Ashley M and Schroer C 2007 A digital future for cultural heritage *AntiCIPAting the Future of the Cultural Past, Proceedings of the XXI International CIPA Symposium* pp 1–6

[23] Zabulis X, Grammalidis N, Bastanlar Y, Yilmaz E and Cetin Y Y 2008 3D Scene Reconstruction Based on Robust Camera Motion Estimation and Space Sweeping for a Cultural Heritage Virtual Tour System *2008 3DTV Conference: The True Vision - Capture, Transmission and Display of 3D Video* (IEEE)

[24] Remondino F and Rizzi A 2010 Reality-based 3D documentation of natural and cultural heritage sites{\textemdash}techniques, problems, and examples *Appl. Geomatics* **2** 85–100

[25] King L, Stark J F and Cooke P 2016 Experiencing the Digital World: The Cultural Value of Digital Engagement with Heritage *Herit. Soc.* **9** 76–101

[26] Addison A C 2000 Emerging trends in virtual heritage *{IEEE} Multimed.* **7** 22–5

[27] Roussou M 2002 Virtual heritage: from the research lab to the broad public *Bar Int. Ser.* **1075** 93–100

[28] Refsland S T, Ojika T, Addison A C and Stone R 2000 Virtual Heritage: Breathing new life into our ancient past *{IEEE} {MultiMedia}* **7** 20–1

[29] Bastanlar Y, Grammalidis N, Zabulis X, Yilmaz E, Yardimci Y and Triantafyllidis G 2008 3D reconstruction for a cultural heritage virtual tour system *Int. Arch. Photogramm. Remote Sens. Spat. Inf. Sci. Beijing* **37** 1023–8

[30] Barsanti S G, Caruso G, Micoli L L, Rodriguez M C and Guidi G 2015 3D Visualization of Cultural Heritage Artefacts with Virtual Reality devices *{ISPRS} - Int. Arch. Photogramm. Remote Sens. Spat. Inf. Sci.* **{XL}-5/W7** 165–72

IOP Conf. Series: Earth and Environmental Science **396** (2019) 012032 doi:10.1088/1755-1315/396/1/012032

Desktop-Based Population Data Information System to Support The Sumbawa Electronic Government in Rhee District

Muhammad Hidayatullah[1], Fery Hendrawan[2], Titi Andriani[1], Shinta Esabella[2], Nurhairunnisah[3]

[1]Department of Electrical Engineering, Faculty of Engineering, Universitas Teknologi Sumbawa, Indonesia, 84371
[2]Department of Informatics, Faculty of Engineering, Universitas Teknologi Sumbawa, Indonesia, 84371
[3]Department of Educational Technology, Faculty of Education, Universitas Samawa, Indonesia, 84316

E-mail: mhhidayat07@gmail.com

Abstract. This research aims to improve the population service system which is optimal in supporting electronic government in Sumbawa Regency. In software development using the waterfall model and desktop-based Programming with Visual Studio and SQL Server as a Database Management System. The system testing method uses the Black Box Method, a test that focuses on the functional requirements of the software. The results of the research can be used easily, quickly, and useful for processing population data so that it can provide optimal service. This system has been used at the Luk Village Office (Rhee District) in improving service to the community optimally.

Keywords: population service system, waterfall, Sumbawa e-government.

1. Introduction

Village or Village Level Government is the closest government to the community. Change continues along with the development of increasingly sophisticated science and technology [1]. The emergence of various technological products that can be utilized in various fields provides convenience to the government apparatus to improve the quality of services to the community, both at the village or urban level government services to the central level. Luk Village is one of the villages in the District of Rhee, Sumbawa Regency, located at Jln. Lintas Sumbawa - Tano KM.27 Desa Luk. The Luk Village office is a government agency that provides services to take care of the needs of the village of Luk. But in providing services to the community, the village government of Luk still uses the process of processing population by manual method, such as making books by hand writing and using semi-computerization with Microsoft Excel data processing software [2,3].

Desktop-based Population Data Management Information System in Petir Village, Rongkop Subdistrict by Waspodo Dumadi in 2011 was made to make it easier to help search for population data checking. The system is designed using the System Development Life Cycle (SDLC) which includes the design of system requirements, manufacturing and application to system maintenance. There are deficiencies in system maintenance which results in disruption of the system as a whole and there is a repeated restart of its application, so that it is necessary to update the method using the Waterfall method which is tested to be more structured and collapsed to the stage of system maintenance without significant disruption [4].

Based on the above problems, the authors built a Village Population Information System that can process and regulate the flow of the process of managing population data and information stored digitally in a SQL Server base. The system is built using the desktop programming language, Visual Studio 2015 [5,6]. The results of the research become the Village Population Information System which can be used easily, quickly, and is useful for Population processing so that it can provide optimal services to the Luk Village community in order to support electronic government in Sumbawa Regency [7].

2. The Material and Method

The software development method used in this study is the waterfall method, because the waterfall software development process model implies a systematic and sequential approach to software development [8,9].

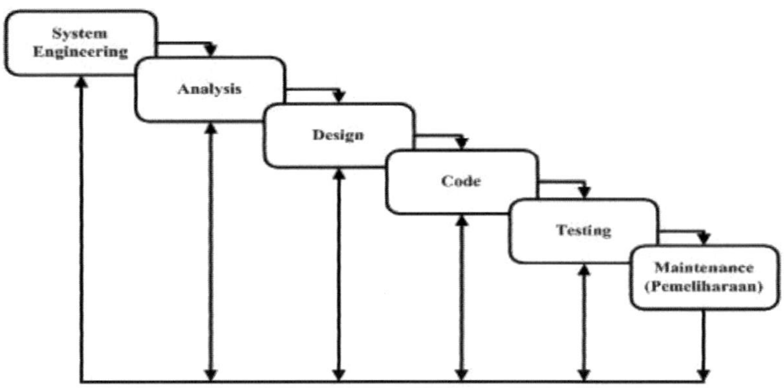

Figure 1. Flowchart of the waterfall method

Following are the stages in software development using the waterfall model (Figure 1):

a. System Engineering: In this step is the stage in forming the needs of all elements. System engineering is very necessary, because software is usually part of a larger system. Making a software can be started by looking at and looking for what is needed by the system. Of the system requirements will be applied to the software created.

b. Analysis: In this step is the process of gathering software requirements. To understand the basis of the program that will be created. In this stage the researcher looks for data and information at the Luk Village office to find out the scope of information, the functions needed, the performance capabilities to be produced and the design of the software user interface

c. Design: This step is a stage that focuses on four important parts, namely: Data structure, software architecture, procedure details, and user interface characteristics.

d. Program Writing (Coding): In this step the researcher writes the program code so that it can be understood and executed by the machine. If the design is correct and detailed, then writing the program can be achieved efficiently.

e. Testing: After writing the program code is complete, the researcher tests the program code that has been created by focusing on the part in the software. The aim is to ensure that all statements have been tested and also ensure that the inputs used will produce the appropriate output.

f. Maintenance: At this stage the researcher gives changes to the software that is new to the customer. If later an error can be encountered when it is run in the customer's environment. Or customers ask for additional new functions.

3. Result and Discussion

This section discussion about how the system built in desktop language by using visual studio 2015 based-on waterfall method.

3.1. System Design Result

The following is a system design carried out in the manufacture of the Population Information System of the District of Luk Village in Rhee:

3.1.1. Context Diagram. Context diagram serves to define the beginning and end of incoming data and output on a system [10,11]. The context diagram below explains that the user enters birth and death cover letters into the system, then the user will issue output in the form of birth and death letters. Then the birth and death letter will be submitted to the Village Head to be approved and the Village Head will hand it back to the user. the following is a context diagram of the information system of the Village Government (Figure 2):

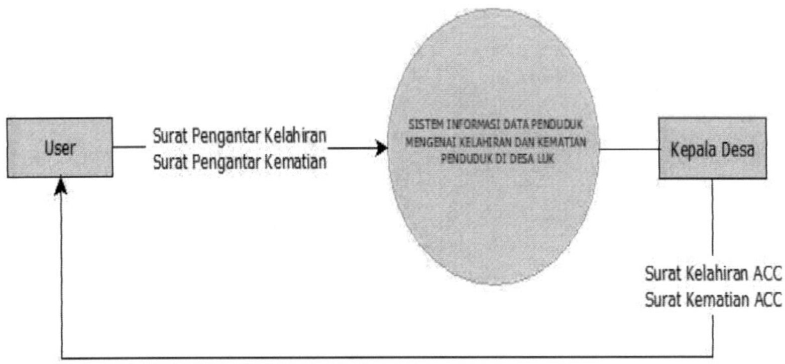

Figure 2. Context Diagram System

3.1.2. Input Design

Form Login Design

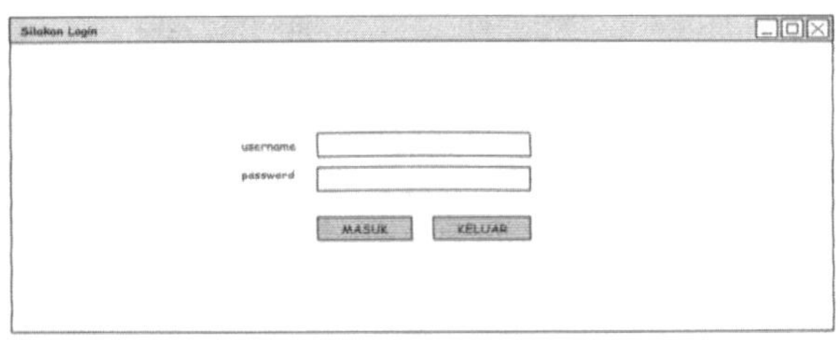

Figure 3. Form Login Design

Figure 3 show a draft login form that displays the design process for inputting username and password to determine the system user access rights. In the design view above there are controls used in making the login form, namely the text box that functions as a place to enter the username and password and the command button is used for the login process and exit the application.

Form of Input Identity Card Data Design

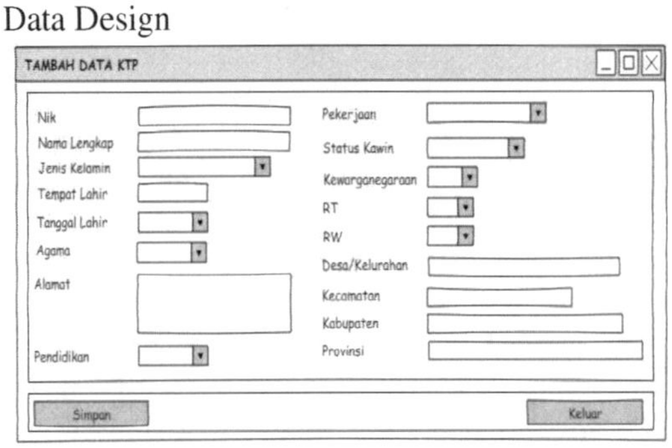

Figure 4. Form of Input Identity Card Data Design

Figure 4 show the design of the Identity Card (KTP) data input form that displays the design process that allows users to fill in the text box control, full name, place of birth, address, village / sub-district, district, district, province and combo box gender, date of birth, religion, education, employment, marital status, citizenship, household, and RW to be saved to the database.

Design of Initial Family Card Menu Form

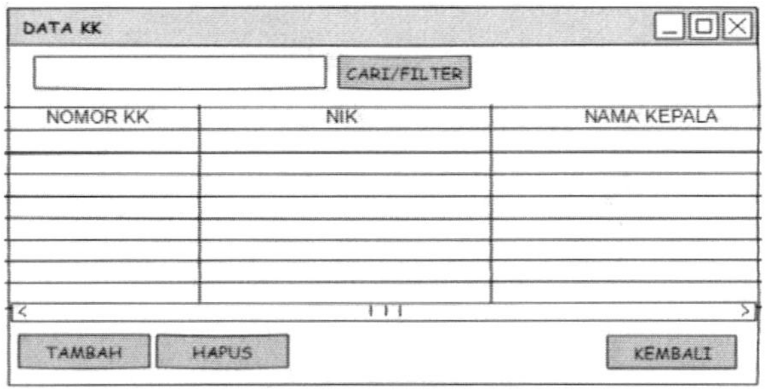

Figure 5. Design of Initial Family Card Menu Form

Figure 5 show a draft form of the KK initial menu that displays the design process that allows users to view, add, delete and perform KK data searches. In the initial Family Card menu form design there are controls used for the process of processing KK data, namely ADDED, DELETE and SEARCH / FILTER data for Family Card control.

Design of Family Card Data Input Form

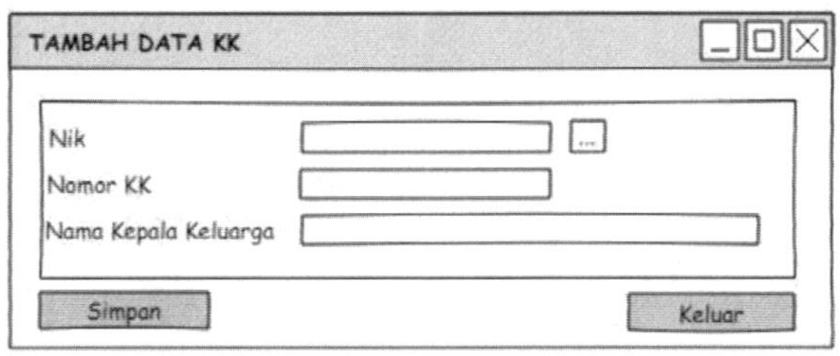

Figure 6. Design of Family Card Data Input Form

Figure 6 show a design for the family card data input form (in Bahasa called "Tambah Data Kartu Keluarga" form) that displays the process design that allows the user to fill in the text box control, number ("Nomor KK"), and name of the family head (in Bahasa "Nama Kepala Keluarga") to be saved to the database. There are two bottom "Simpan" (Save) and "Keluar" (Exit).

Form of the Initial Birth Data Menu Design

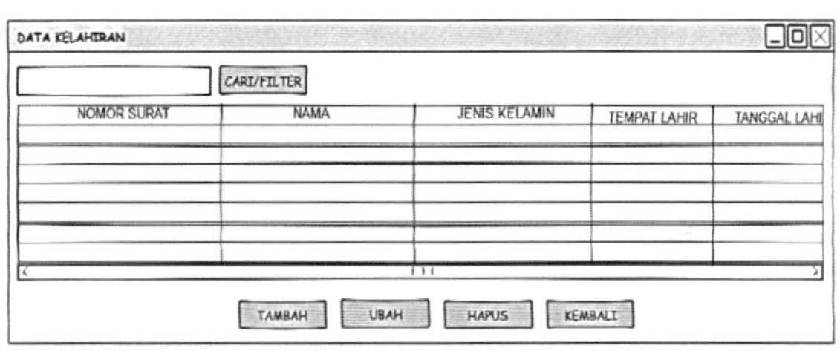

Figure 7. Form of the Initial Birth Data Menu Design

Figure 7 show a draft birth data (in Bahasa called "Data Kelahiran") menu form that displays the process design that allows users to process KK data. In the design of the birth data initial menu form there are several function keys that are used in the process of processing birth data, namely

ADDITIONAL ("Tambah"), CHANGE ("Ubah"), DELETE ("Hapus") and SEARCH / FILTER birth data functions.

Design the Form of Input Data Birth Menu

(a) (b)

Figure 8 (a) Design the Form of Input Data Birth Menu, and (b) Design the Form of Early Death Data Menu

Figure 8(a) show a draft birth data input form that displays the design process that allows users to fill in the control text box birth number, card number, head name, full name, place of birth, o'clock, baby's weight, baby length, father's name, mother's name and combo box gender, date of birth, type of birth, birth to birth and birth attendants to be saved to the database. Figure 8(b) show the design of the initial death data menu form that displays the design process that allows users to be able to see, add, change, delete and search death data in the design of the initial data death menu form there are several function keys used in the process of processing death data, namely functions ADDITION, CHANGE, DELETE and SEARCH / FILTER Birth data.

Print a birth certificate (Figure 9)

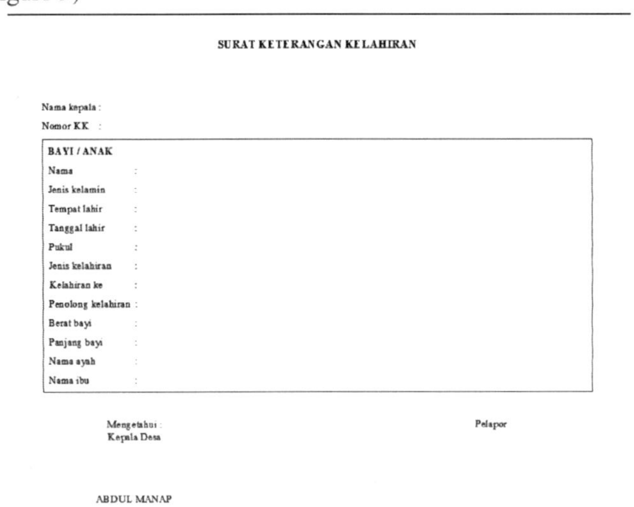

Figure 9. Design of Print a birth certificate

IOP Conf. Series: Earth and Environmental Science **396** (2019) 012032 doi:10.1088/1755-1315/396/1/012032

Print a Death certificate (Figure 10)

PEMERINTAH KABUPATEN SUMBAWA
KECAMATAN RHEE
KANTOR KEPALA DESA LUK
Alamat : Jalan Lintas Sumbawa - Tano KM.27 Desa Luk

SURAT KETERANGAN KEMATIAN
Nomor :

Yang bertanda tangan dibawah ini Kepala Desa Luk Kecamatan Rhee Kabupaten Sumbawa merangkan bahwa :

Nama :
Jenis Kelamin :
Agama :
Alamat :

Bahwa benar telah meninggal dunia pada :

Tanggal Meninggal :
Jam :
Tempat :
Sebab :

Demikian surat keterangan kematian ini dibuat untuk dapat diketahui seperlunya.-

SEKRETARIS

ABDURRAHMAN

Figure 10. Design of Print a Death certificate

Print a Population Report (11)

PEMERINTAH KABUPATEN SUMBAWA
KECAMATAN RHEE
KANTOR KEPALA DESA LUK
Alamat : Jalan Lintas Sumbawa - Tano KM.27 Desa Luk

REKAPITULASI DATA PENDUDUK DESA LUK

Nik	Nama	Tempat lahir	Tanggal Lahir	Jenis kelamin	Agama	Alamat	RT	RW	Pendidikan	Pekerjaan	Status kawin	Nomor KK	Nama Ayah	Nama Ibu	Status
5204210107400010	A GANI BANI	LUK	Kamis, 01 Januari 1940	Laki-Laki	Islam	DUSUN LUK KARYA	006	003	SD/SEDERAJAT	PETANI/PEKEBUN	Kawin	5204211206090003	ABDUL	SADE	AKTIF
5204210403810002	SUDIRMAN	LUK	Rabu, 04 Maret 1981	Laki-Laki	ISLAM	DUSUN LUK A	001	001	SD/SEDERAJAT	PETANI/PEKEBUN	KAWIN	5204212604110001	CUNGANG	EMBO	AKTIF
5204211002020001	ARKAN SUYADI	SUMBAWA	Minggu, 10 Februari 2002	LAKI-LAKI	ISLAM	DUSUN LUK KARYA	006	003	SLTA/SEDERAJAT	PELAJAR/MAHASISW	Belum Kawin	5204210102080208	M.SIDIK	MARYAMA	AKTIF
5204211604650001	M.SIDIK	ALAS	Jumat, 16 April 1965	Laki-Laki	Islam	DUSUN LUK KARYA	006	003	SLTA/SEDERAJAT	KARYAWAN SWASTA	Kawin	5204210102080208	ISMAIL	SAPIYAH	AKTIF
5204214107590010	MASITA	LUK	Rabu, 01 Juli 1959	PEREMPUA	ISLAM	DUSUN LUK KARYA	006	003	SD/SEDERAJAT	PETANI/PEKEBUN	Kawin	5204212604110001	MUHAMMAD	SENI	AKTIF
5204214807860001	SUKMAWATI	RHEE LOKA	Jumat, 08 Juli 1983	Perempuan	Islam	DUSUN LUK A	001	001	SD/SEDERAJAT	PETANI/PEKEBUN	Belum Kawin	5204211206090003	KARI	MINA	AKTIF

Figure 11. Design of Print a Population Report

3.1.3. Program Implementation.

Implementation is a continuation of the design of Population Information System in Luk Village. It has been created and carried out using the Visual Studio 2015 programming language with the database used is SQL Server 2014. Visual Studio 2015 can be run on Microsoft Windows operating system platforms and various hardware devices, but the implementation is only done on Notebook hardware with Microsoft Windows operating system.

3.1.4. Application Display.

The program display in the Population Information System Application in Luk Village, Rhee-Based Desktop District is as follows:

a. Login Form Display

The display of Login Form for Population Information System Application Desktop-Based in Luk Village, Rhee District is as follows:

Figure 12. Login Form Display

The Figure 12 show is the initial display and will appear when the user first starts the application.

b. Main Menu Display

The display of the Application Main Menu Display is as follows:

Figure 13. Main Menu Display

Figure 13 show the main display form menu of the Information System. Of the six menus contained in the main menu have their respective functions such as the home menu, when the click process is carried out it will display the added sub-menu of the user, where the user sub-menu adds when the user clicks it, the user will add the form. Then the data menu, when the click process is done, it will display sub-menus, namely the KTP sub menu and KK sub-menu, where when the user clicks on the KTP sub-menu, it will display the KTP data form and when the user processes click on the sub menu KK, it will display the KK data form. Then menu residents, when the click process is done it will display the Population form. Then the birth menu, when the click process is done it will display the birth data form. Then the death menu, when the click process is done it will display the death data form.

c. Initial Identity Card Data Form

The display of the Initial Form of Identity Card Data on the Population Information System is as follows:

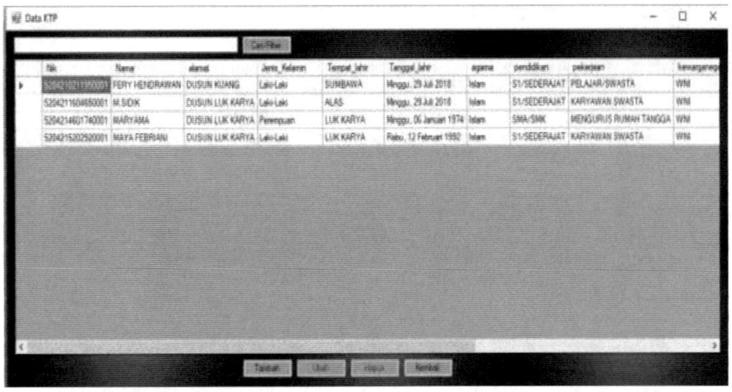

Figure 14 Initial Identity Card Data Form Display

Figure 14 show a display of the initial KTP data form. In the display above there are several function keys used in the process of processing user data, namely the function added, change, and delete.

d. Add Identity Card Form
Display of Add KTP Form on Population Information System is as follows:

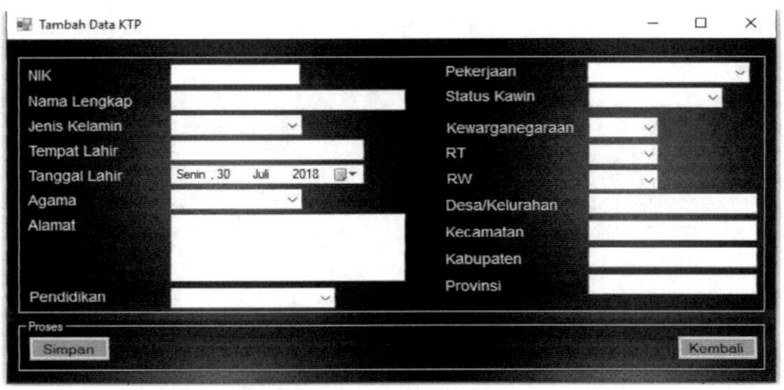

Figure 15. Add Identity Card Form Display

The display in Figure 15, there are several function keys that are used in the process of processing the new Identity Card data plus the save and return functions. The save function button functions to add new KTP data and the function key is used again to return to the initial KTP menu form.

e. Family Card Data Initial Form
The display of Family Card Data Initial Form in the Population Information System is as follows:

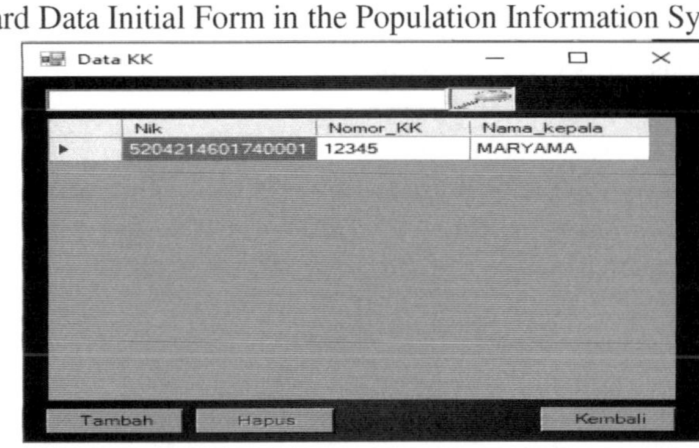

Figure 16. Family Card Data Initial Form Display

The display showed in Figure 16 has several function keys that are used in the process of processing KK data, namely the function added, delete, search or filter and return.

f. Initial Form of Birth Data

The display of Initial Form of Birth Data in the Population Information System is as follows:

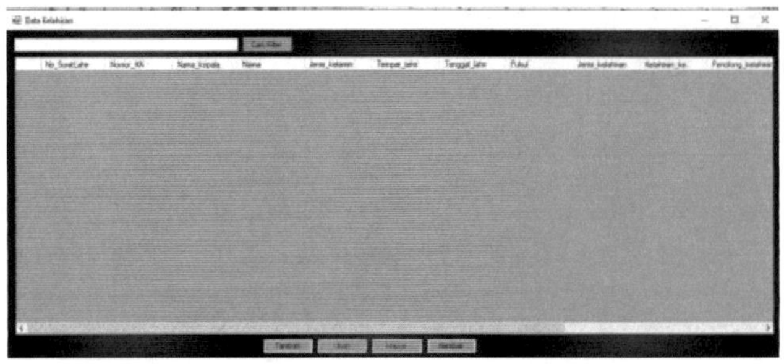

Figure 17. Initial Form of Birth Data Display

Figure 17 show a display of the initial form of birth data from the Population Information System in the Desktop District of Luk Rhee District. In the above display there are several function keys that are used in the process of processing birth data, namely the function of adding, changing, deleting, searching or filtering and returning.

g. Form Initial Death Data

The display of Form Initial Death Data in the Population Information System is as follows:

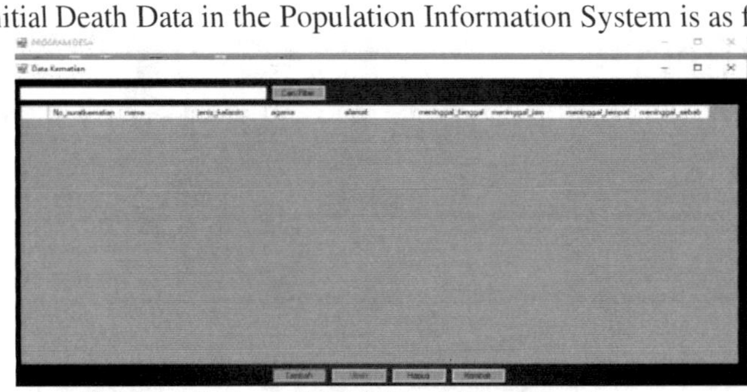

Figure 18. Form Initial Death Data Display

Figure 18 show several function keys that are used in the process of processing death data, namely the function added, change, delete, search or filter and return.

3.1.5. Testing.

Testing is an important part of the software development cycle. Testing is done to guarantee the quality and also to know the weaknesses of the software. The purpose of this test is to ensure that the software that is built has a reliable quality, which represents the principal study of specifications, design analysis and coding of the software itself.

3.1.6. Test Plan.

Testing the software on the Population Information System application on the Village of Luk District based on Desktop Rhee is by using black box testing. Black box testing focuses on the functional requirements of the software created.

4. Conclusions

With this Population Information System, it has become one form of service to the community (electronic government). Where, the processing that was initially carried out conventionally has now become computerized, such as the process of input, update and search for Population, which makes birth certificates and death certificates. Digitizing with the data has been able to minimize errors in the process of adding data and simplifying the process of finding Population.

5. Acknowledgment

This article was presented at the 2nd International Conference on Smart City INNOVATION (ICSCI) 2019, jointly held by Universitas Indonesia and Universitas Diponegoro. ICSCI conferences have been supported by the United States Agency for International Development (USAID) through the Sustainable Higher Education Research Alliance (SHERA) Project for Universitas Indonesia's Scientific Modelling, Application, Research, and Training for City-centered Innovation and Technology (SMART CITY) Center for Collaborative Research, administered through Grant #AID-497-A-1600004, Sub Grant #IIE-00000078-UI-1.

6. References

[1] Sofya N D, Esabella S and Rodianto R 2017 Rancang Bangun Aplikasi Kamus Bahasa Sumbawa Berbasis Android *MATRIK J. Manajemen, Tek. Inform. dan Rekayasa Komput.* **17** 36–45

[2] Surantalla R 2017 PERANCANGAN SISTEM INFORMASI DAN PENDATAAN IKATAN PELAJAR MAHASISWA SUMBAWA-YOGYAKARTA (IPMSY) BERBASIS WEB *J. TAMBORA* **2**

[3] Setiawan P 2015 *TA: Rancang Bangun Aplikasi Pengolahan Data Evaluasi Proses Belajar Mengajar Berbasis Web Pada Stikes Yayasan RS. Dr. Soetomo Surabaya* (Institut Bisnis dan Informatika Stikom Surabaya)

[4] MZ M M K 2016 Pengujian Perangkat Lunak Metode Black-Box Berbasis Equivalence Partitions Pada Aplikasi Sistem Informasi Sekolah *MIKROTIK J. Manaj. Inform.* **6**

[5] Imam 2015 *Aplikasi WPF (WindowsPresentation Foudation) dengan Visual Basic* (Jakarta: Elex Media Komputindo)

[6] Nurhayani 2016 Implementasi Bahasa Pemrograman Visual Basic pada Pengolahan Data Reservasi , , Juli 2016. *J. IPTEK* **7** 47–57

[7] Anwar T and Utomo Y W 2017 IMPLEMENTASI PAPERLESS OFFICE PADA SISTEM MONITORING DAN EVALUASI PROGRAM KERJA ORGANISASI MAHASISWA *MATRIK J. Manajemen, Tek. Inform. dan Rekayasa Komput.* **17** 56–65

[8] Pressman R S 2002 Rekayasa Perangkat Lunak Pendekatan Praktisi (Buku Satu) *Yogyakarta Andi*

[9] Esabella S 2017 PERANCANGAN APLIKASI SURAT KETERANGAN TANDA LAPOR KEHILANGAN PADA KANTOR KEPOLISIAN RESOR SUMBAWA BERBASIS ANDROID *J. TAMBORA* **2**

[10] Wardhana H and Hasanah B D U 2016 APLIKASI MONITORING PENERIMA BEASISWA BIDIKMISI BERBASIS WEB, ANDROID DAN SMS GATEWAY *MATRIK J. Manajemen, Tek. Inform. dan Rekayasa Komput.* **16** 22–32

[11] Astuti T, Kusumastuti G and Fitriyanto R 2017 PEMANFAATAN ANALYTICAL HIERARCHY PROCESS (AHP) PADA E-VOTING PEMILIHAN KETUA OSIS *MATRIK J. Manajemen, Tek. Inform. dan Rekayasa Komput.* **17** 95–104

The Analysis of Barriers for Implementation of Sustainable Construction in Indonesia

B Susanti[1], S F H Filestre[1], I Juliantina[1],

[1]Department of Civil Engineering, Faculty of Engineering, Universitas Sriwijaya, Palembang, Indonesia

E-mail: bettysusanti0401@gmail.com

Abstract. The sustainable development requires the provision of adequate infrastructure, but the construction activities on infrastructure development itself still have been considered as a sector that has a negative impact to the environment. The implementation of the sustainable construction becomes a necessity, but the implementation of the sustainable construction in developing countries is still very limited, since there are still many factors that inhibit the implementation of this concept. This study aims to analyse the barriers in the implementation of sustainable construction in the infrastructure development projects in Palembang, Indonesia. The barriers are identified based on the results of literature studies. The survey using a questionnaire was conducted to state-owned and private companies that were carrying out infrastructure development projects in Palembang and its surrounding areas. The survey confirms the various barriers that may occur based on the perceptions of respondents representing state-owned and private companies. The mean item score was used to determine the ranking of each barrier. This study indicate that the following factors are the main barriers to the implementation of sustainable construction, namely factors related to the limited number of trained and/or certified workers and lack of communication between parties involved in the project.

Keywords: barriers, infrastructure development, sustainable construction

1. Introduction

Sustainable development has become a major issue in global development today. One effort to achieve sustainable development is through the provision of physical infrastructure facilities for society. The construction industry plays a significant role in the development of physical infrastructure [1]. The infrastructure development is believed to be able to provide benefits to the social and economic aspects of the community. Nevertheless, many infrastructure construction activities are claimed to have the potential to cause negative impacts on the environment [2]. Global data shows that the construction industry uses the most natural resources. As an illustration, world construction activities consume an average of 40% of energy, 25% of wood, and 16% of water from the total available resources on earth [3]. In terms of pollution, construction activities contribute to produce 35% of CO2 gas emissions for the planet [4]. Efforts are needed to overcome the negative impacts caused by construction activities. One concrete effort to overcome the negative impact of construction activities is through the application of the concept of sustainable construction.

Sustainable construction is a concept that aims to balance ecological, social, and economic aspects in construction activities [5]. The application of this concept ultimately aims to improve the quality of life of the people [6], so that it becomes the main goal to be achieved by the construction industry in the future [7]. Until now, the implementation of the concept of sustainable construction is still limited, especially in developing countries [2]. Another problem is that many countries have failed to implement the concept of sustainable construction due to various barriers. Various factors related to operations, financial capability, quality and quantity of human resources, experience, technology, regulations, and many other factors are thought to be the barriers in implementing the concept of sustainable construction in general [7].

Palembang is one of the big cities in Indonesia which is actively conducting infrastructure development. Various strategic infrastructure projects are being developed to improve the local, regional and even national economy. The identification and analysis of the barriers toward the implementation of sustainable construction can provide an accurate picture of the factors inhibiting the implementation of this concept in Indonesia. This study aims to analyze the barriers in the implementation of sustainable construction in Indonesia based on the perceptions of representatives of state-owned and private companies that involved in infrastructure construction projects.

2. Sustainable Construction

Sustainable construction is a construction process that aims to balance the ecological, social, and economic aspects [5]. Many studies equate the concept of sustainable construction with the concept of green construction, but there are fundamental differences between the two [8]. Green construction is a terminology that only focuses on environmental aspects, while sustainable construction covers environmental, social and economic aspects simultaneously, as shown in Figure 1.

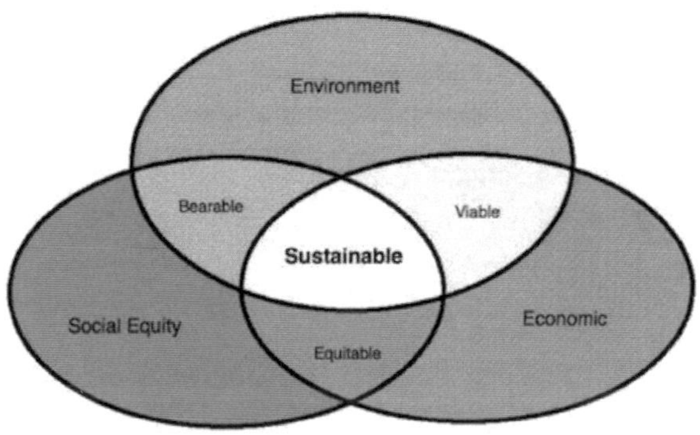

Figure 1. Three issues of sustainable construction concept, adopted from [9]

The study conducted by Ref [9] showed that the construction industry in developing countries generally faced difficulties to adapt and apply the concept of sustainability. Several ASEAN countries have started to implement the concept of sustainable construction in various construction projects. Some agenda in the form of training, seminars, and discussions related to the environment in the construction sector has done frequently [10]. The same thing has been implemented in Indonesia, as shown in Ref [11]. In general, it can be concluded that the implementation of sustainability concept in the construction sector only touches environmental aspects, yet it does not touch the social and economic aspects. Applying this concept is increasingly difficult in developing countries due to various inhibiting factors, such as economic, social, and technical [12].

The previous studies revealed that there were various barriers in the implementation of sustainable construction, as shown in Table 1. There are 16 factors that have the potential to inhibit the implementation of sustainable construction. These factors are grouped into economic/financial, regulatory, management, technical, and understanding aspects regarding the concept of sustainable construction itself.

IOP Conf. Series: Earth and Environmental Science **396** (2019) 012033 doi:10.1088/1755-1315/396/1/012033

Table 1. Identification of barriers for the implementation of sustainable construction

No.	Barriers	Ref [7]	Ref [13]	Ref [1]	Ref [6]	Ref [10]
X1	**Economic/Financial Aspect**					
X11	Potential increase in total project costs	√	√	√	√	√
X12	Potential increase in the project duration	√	-	√	-	-
X13	Limited capacity of the country's economy	√	-	√	-	√
X2	**Regulation Aspect**					
X21	Limited regulations and/or standards that encourage the application of the concept of sustainable construction	√	√	√	√	√
X22	The low implementation of regulations that encourage skills/expertise training to support sustainable construction	√	-	√	-	√
X3	**Management Aspect**					
X31	Lack of preparation and preparedness of the company in facing the demands of the implementation of sustainable construction	√	-	-	√	√
X32	Lack of communication between parties involved in the project	√	-	√	-	-
X4	**Technical Aspect**					
X41	The limited number and types of equipment and materials that meet the criteria of sustainable construction	√	√	√	-	√
X42	The limited amount of trained/certified workers	-	√	√	-	√
X43	Limited guidelines/modules regarding the concept of sustainable construction.	√	-	√	-	√
X44	Limited rating tools or benchmarks to assess the level of sustainability	√	√	√	-	√
X5	**Social-Culture Aspect**					
X51	The conventional construction method is preferable compared to sustainable construction method	√	√	√	√	√
X52	Limited demand for projects that use the concept of sustainable construction	√	√	-	-	√
X6	**Understanding aspect**					
X61	Limited knowledge regarding	-	√	√	√	√

	the advantage of sustainable construction					
X62	A perception that sustainable construction is a similar concept to the green construction	-	-	√	√	√
X63	The implementation of sustainable construction concept is more risky than conventional concept.	-	-	√	√	√

3. Methodology

This study identifies barriers to the implementation of sustainable construction from various previous studies. Based on the factors that have been identified, a questionnaire is then prepared to confirm whether these factors are also contributed as barriers to the implementation of sustainable construction in Palembang. Validity and reliability tests were performed on each factor included in the questionnaire. This questionnaire is also intended to obtain respondents' perceptions on how much they agree with each inhibiting factor. The assessment of respondents' perception uses Likert scale with a range of perceptions such as strongly disagree, disagree, agree, and strongly agree. The data was collected at a contractor company in Palembang City that was carrying out infrastructure construction projects and was recorded in the Directory of Construction Companies in South Sumatra Province.

Non-probability sampling technique was used in this study because of the limited number of contractor companies implementing the infrastructure construction projects in Palembang and its surrounding areas. As many as 35 respondents came from state-owned and private company contractors with the company classification of medium to large. Respondents' positions include project managers, site managers, site engineers, and field supervisors. Mean item score is used to determine the average value of respondents' perceptions of each inhibiting factor. Based on the mean item score of each inhibiting factor, ranking is then performed to obtain the main barrier to the implementation of sustainable construction.

4. Results and Discussion

This study employed a questionnaire as a data collection instrument. The questionnaire contained 16 questions related to the barriers for the implementation of sustainable construction. The validity and reliability of the questionnaire were tested before the questionnaire was distributed to respondents. The validity test is conducted using product moment correlation. Each question is valid if the product moment correlation shows that r arithmetic greater than r table. System analysis is used in testing the validity of each question, by correlating the score of each question with the total score. Having 35 respondents as sample for a significance level of 5%, a product moment correlation value of 0.334 was obtained. The result of the validity test of factors inhibiting the implementation of sustainable construction is shown in Table 2. This validity test showed that all questions used in the questionnaire were declared valid because the value of r arithmetic greater than r table, so that the questionnaire could be used to determine the barriers for implementation of sustainable construction.

Tabel 2. The result of validity test

Question item	$r_{calculate}$ Value	r_{table} Value	Description
Economic/Financial aspect			
X11	0,780	0,334	Valid
X12	0,859	0,334	Valid

X13	0,624	0,334	Valid
Regulation Aspect			
X21	0,851	0,334	Valid
X22	0,779	0,334	Valid
Management Aspect			
X31	0,793	0,334	Valid
X32	0,879	0,334	Valid
Technical Aspect			
X41	0,758	0,334	Valid
X42	0,656	0,334	Valid
X43	0,821	0,334	Valid
X44	0,694	0,334	Valid
Social-Culture Aspect			
X51	0,868	0,334	Valid
X52	0,848	0,334	Valid
Understanding Aspect			
X61	0,585	0,334	Valid
X62	0,589	0,334	Valid
X63	0,839	0,334	Valid

The Reliability tests were also carried out to ensure the consistency of the assessment. This reliability test used Cronbach Alpha [14]. In this study, the questionnaire was declared reliable if the Cronbach Alpha value for each question item was greater than 0.7. The reliability test results of the questionnaire are shown in Table 3. The reliability test shows that the Cronbach Alpha for each aspect has a value between 0.80 - 0.87 so that the questionnaire is declared reliable and it can be used to determine the barriers for implementation of sustainable construction.

Tabel 3. The result of reliability test

Criteria	Cronbach's Alpha value	Description
Barriers related to economic aspect (X_1)	0,83	Reliable
Barriers related to regulation aspect (X_2)	0,83	Reliable
Barriers related to management aspect (X_3)	0,83	Reliable
Barriers related to technical aspect (X_4)	0,80	Reliable
Barriers related to social-culture aspect (X_5)	0,85	Reliable
Barriers related to conscious aspect (X_6)	0,87	Reliable

Data collection resulted in 35 responses from 42 questionnaires distributed to contractor companies. A total of 7 questionnaires were not filled because the company was not willing to be a research respondent. Overall, the reasons of refusal were related to respondents' lack of understanding of the sustainable construction concept. The profile of respondents is shown in Table 4. Out of the 35 respondents, 62.86% came from private companies and the rest of the respondents came from state-owned companies. In terms of company grade, 51.43% came from medium grade companies and 48.57% of companies came from large grades.

In general, respondents had a sufficient understanding regarding the sustainable construction concept and also had adequate practical experience in the construction field. More than 82% were less than 40 years old and they had a tertiary level of education, thus the respondents are quite updated with the concept of sustainable construction. As many as 42% of respondents had at least work experience in construction projects for more than 5 years, even 22% more respondents had work

IOP Conf. Series: Earth and Environmental Science **396** (2019) 012033 doi:10.1088/1755-1315/396/1/012033

experience of more than 10 years. In this way, the respondents are considered to be quite knowledgeable related to the problems and potential barriers on the implementation of the sustainable construction concept. This is also supported by the roles and positions of respondents in the project.

Tabel 4. Profile of respondents

Category	Respondents	
	Number	%
Ages		
20 – 30 years	24	68,57
30 – 40 years	5	14,29
40 – 50 years	4	11,43
> 50 years	2	5,71
Latest Academic		
Highschool	6	17,14
Undergraduated	27	77,14
Post Graduate	2	5,71
Working experience in construction project		
< 5 years	20	57,14
5 – 10 years	7	20,00
> 10 years	8	22,86
Respondent's Profile		
Project Manager	2	5,71
Site Manager	3	8,57
Site Engineer	16	45,71
Others	14	40,00
Company's Status		
BUMN	13	37,14
Private	22	62,86
Company's Grade		
Grade-4 (Middle 1)	3	8,57
Grade-5 (Middle 2)	10	28,57
Grade-6 (Middle 3)	5	14,29
Grade-7 (Big1)	5	14,29
Grade-8 (Big 2)	12	34,29

The result of the survey related to the respondents' perception to the factors inhibiting the sustainable construction concept in Palembang is shown in Table 5. The respondents' perception related to the inhibiting factor is scored using likert scale ranging from 1 to 4. Scale 1 to 4 each shows a perception of strongly disagree, disagree, agree, and strongly agree to the statement that the factors reviewed are barriers for the implementation of sustainable construction.

IOP Conf. Series: Earth and Environmental Science **396** (2019) 012033 doi:10.1088/1755-1315/396/1/012033

Table 5. The respondents' perception regarding the barriers for sustainable construction implementation

Respondent	The Respondents' Score															
No	**X11**	**X12**	**X13**	**X21**	**X22**	**X31**	**X32**	**X41**	**X42**	**X43**	**X44**	**X51**	**X52**	**X61**	**X62**	**X63**
R1	4	4	3	3	4	3	4	4	4	4	3	4	4	4	3	3
R2	3	3	4	2	4	4	4	3	4	2	3	3	4	3	3	3
R3	2	2	2	3	2	2	4	2	3	2	3	3	4	1	3	1
R4	3	3	3	3	3	2	2	2	4	3	3	2	2	3	3	3
R5	4	3	2	3	3	4	4	4	4	3	2	2	2	3	2	2
R6	4	4	3	4	3	3	4	3	3	3	3	4	2	4	3	3
R7	3	3	3	3	3	3	3	3	4	3	3	3	3	3	3	3
R8	3	3	3	2	3	3	3	3	3	3	3	3	3	3	3	2
R9	3	3	4	3	4	3	3	3	3	4	4	2	3	4	3	2
R10	3	3	3	2	3	3	2	3	4	2	2	4	2	3	3	2
R11	3	4	3	4	4	4	4	4	4	4	4	4	3	3	3	3
R12	1	1	1	1	2	1	1	1	1	1	1	1	1	3	3	1
R13	2	3	3	4	3	3	4	2	4	3	3	3	3	3	3	2
R14	3	4	3	4	4	4	4	4	4	4	4	4	3	3	4	3
R15	3	3	3	3	3	3	3	3	3	3	3	3	3	3	3	3
R16	3	3	3	3	3	3	3	3	3	3	3	3	3	3	3	3
R17	3	2	3	3	3	3	3	3	4	3	3	3	3	4	4	3
R18	3	3	4	4	4	3	4	4	3	3	3	3	3	3	3	2
R19	3	3	4	4	4	3	4	4	3	3	3	3	3	3	3	2
R20	2	2	4	3	3	3	3	2	3	3	3	3	3	3	3	2
R21	2	3	3	3	4	3	4	3	4	4	3	3	4	4	2	3
R22	2	3	3	2	3	3	3	3	4	3	3	4	4	3	4	3
R23	3	2	2	3	4	4	3	3	4	3	3	3	3	4	2	3
R24	1	2	3	3	3	4	2	3	3	3	3	3	3	3	2	2
R25	3	2	3	3	3	3	4	2	2	3	4	3	3	4	1	2
R26	3	3	4	3	3	3	3	3	3	3	3	3	3	3	3	3
R27	4	3	2	4	3	3	4	4	3	3	3	4	4	3	3	2
R28	3	3	3	3	4	3	4	3	3	4	3	3	3	3	3	2
R29	3	3	3	3	3	3	3	3	3	3	3	4	3	2	2	2
R30	3	3	3	3	3	3	2	3	3	3	3	3	2	3	3	3
R31	3	3	3	3	4	3	3	2	4	3	2	3	3	4	2	2
R32	2	2	3	3	4	3	4	2	3	4	4	2	3	3	3	2
R33	3	3	3	4	3	3	3	3	3	3	4	3	3	3	4	3
R34	3	3	2	3	3	3	4	3	4	3	3	2	3	3	3	3
R35	3	2	3	3	3	2	3	2	3	3	2	2	2	3	2	2

The data from barriers assessment in the above table are then analysed using the mean item score. Mean Item Score is employed to rank the barriers, thus the dominant factor that inhibits the implementation of sustainable construction based on the respondents' perception is found. The mean item score analysis is employed using the equation as follows:

$$MIS = \frac{1n_1 + 2n_2 + 3n_3 + 4n_4}{\Sigma N};$$

Where n_1 = The number of respondents who answered strongly disagree (score on a scale of 1)
n_2 = The number of respondents who answered disagree (score on a scale of 2)
n_3 = The number of respondents who answered agree (score on a scale of 3)
n_4 = The number of respondents who answered strongly agree (score on a scale of 4)
N = The total number of respondents

The result of the mean item score analysis and the ranking of inhibiting factors are shown in Figure 2. There are 9 factors that are generally assessed as barriers for the implementation of sustainable construction with a mean item score greater than 3. A total of 7 other factors are generally regarded as factors that do not become barriers to the implementation of sustainable construction, with a mean item score of less than 3. Factors related to the limited number and types of equipment and materials that meet the criteria of sustainable construction (X41), limited demand for projects that use the concept of sustainable construction (X52) and factors related to understanding aspects regarding the sustainable construction concept (X62 and X63) are not considered as barriers for the implementation of sustainable construction. Factors related to economic/financial aspects are also assessed as factors that does not become a barriers to the implementation of sustainable construction in infrastructure development projects. This can be understood as the financial and economic aspects are not a burden for the contractor, but rather a burden for the owner.

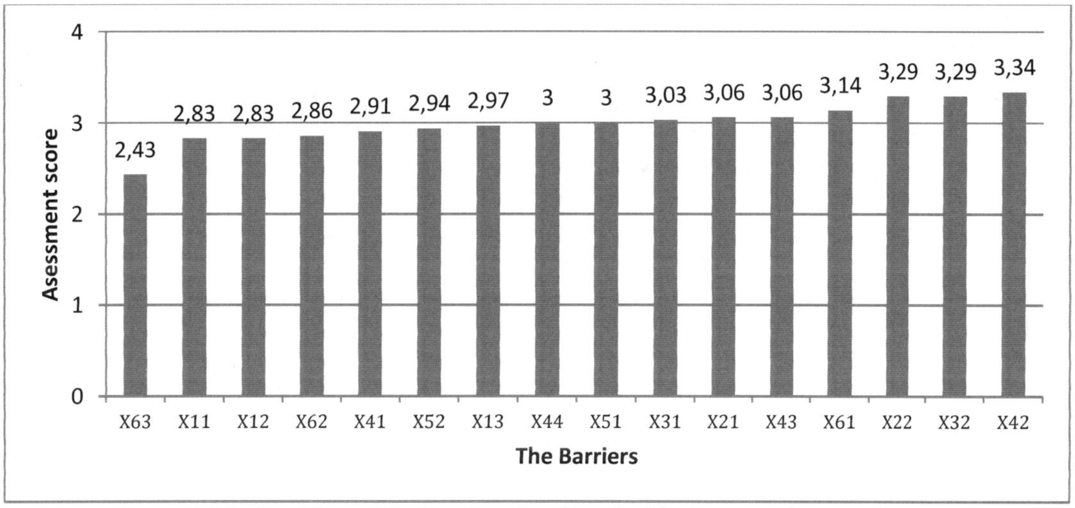

Figure 2 The Ranking of the Barriers for Sustainable Construction Implementation

There are 3 factors that are generally considered by respondents as barriers to the implementation of sustainable construction in Palembang. The factor with the highest mean item score is at X42, stating the limited number of trained or certified workers in construction project. Based on the understanding of the majority of respondents, the skills and expertise of workers at the company and project level will greatly influence the process and the final result of the work. Workers' skills and experience have an impact on decreasing defects, reworks, waste, duration of work, project costs, and work safety. This is consistent [14] and [15], and [16] the inhibiting factor named X42 is closely related to the low implementation of regulations that encourage training for workers, as mentioned in factor X22.

The limited number of workers who have the skills and expertise to support the implementation of sustainable construction in Indonesia is due to the uniqueness in the procurement of workers themselves. Some field workers in the construction sector are seasonal workers who come from the agricultural sector. In addition, there is a local government policy that requires contractor companies to involve field workers who come from residents around the project site. The limited number of skilled workers is also due to the limited information obtained by workers about the concept of sustainable construction itself. This problem reflects the need to improve skills and expertise for all workers in the construction sector. Efforts that can be proposed to overcome these barrier are through training, workshops, seminars, and formal education to improve the skills and expertise of construction project workers regarding the concept of sustainable construction.

Another factor that has a high mean score as a barrier to the implementation of sustainable construction is related to the lack of communication between the parties involved in the project, as referred to factor X32. Based on respondents' perceptions, good communication and coordination between project team members is an important aspect to respond to the complexity of the work and answer the specific requirements for implementing sustainable construction in infrastructure projects. The various levels of understanding of the project team regarding the concept of sustainable construction also require good communication, so that ideas and practices of sustainability can be understood by all parties involved in the project. Ref [17] and [18] also show that good communication between all project teams can prevent problems related to sustainable construction implementation. Efforts to overcome these inhibiting factors are through the assignment of project managers who have the awareness and ability to convince all team members to jointly achieve sustainable construction.

5. Conclusion

This research has provided an overview of various factors that are considered as the barriers in implementing the sustainable construction for infrastructure development projects in Indonesia. The results of this study can illustrate the barriers in implementing the concept of sustainable construction in Indonesia, bearing in mind that this case study was conducted not only at medium and large grade contractor companies, but also at state-owned companies. Based on the perception of the contractor, as many as 7 factors are assessed as insignificant barriers for implementation of sustainable construction. These factors are dominantly related to financial and economic aspects. A total of 9 other factors are considered as high barriers in implementing the sustainable construction. The factors related to the limited skill and expertise of workers and the lack of communication between parties involved in the project have the highest value as a barrier to the implementation of sustainable construction in infrastructure development projects.

This research determines the barriers in implementing sustainable construction in various infrastructure projects. Given the unequal complexity of projects, further analysis is needed to determine the barriers in implementing the sustainable construction concepts that are specific to certain types of construction projects. Determination of barriers to the implementation of sustainable construction in this study was carried out only from the perspective of the contractor. It is also necessary to identify barriers from the perspective of the owner and project designer, so that a comprehensive picture of the barriers regarding the implementation of the concept of sustainable construction in infrastructure construction projects in Indonesia can be obtained.

6. Acknowledgement

This article was presented at the 2nd International Conference on Smart City INNOVATION (ICSCI) 2019, jointly held by Universitas Indonesia and Universitas Diponegoro. ICSCI conferences have been supported by the United States Agency for International Development (USAID) through the Sustainable Higher Education Research Alliance (SHERA) Project for Universitas Indonesia's Scientific Modelling, Application, Research, and Training for City-centered Innovation and Technology (SMART CITY) Center for Collaborative Research, administered through Grant #AID-497-A-1600004, Sub Grant #IIE-00000078-UI-1.

7. References

[1] AlSanad S 2015 Awareness, Drivers, Actions, and Barriers of Sustainable Construction in Kuwait *Procedia Eng.* **118** 969–83

[2] Abidin N Z 2010 Investigating the awareness and application of sustainable construction concept by Malaysian developers *Habitat Int.* **34** 421–6

[3] Pulselli R M, Simoncini E, Pulselli F M and Bastianoni S 2007 Emergy analysis of building manufacturing, maintenance and use: Em-building indices to evaluate housing sustainability *Energy Build.* **39** 620–8

[4] Son H, Kim C, Chong W K and Chou J-S 2009 Implementing sustainable development in the construction industry: constructors{\textquotesingle} perspectives in the {US} and Korea *Sustain. Dev.* **19** 337–47

[5] Kibert C J 2016 *Sustainable construction: green building design and delivery* (John Wiley & Sons)

[6] Saleh M S and Alalouch C 2015 Towards Sustainable Construction in Oman: Challenges & Opportunities *Procedia Eng.* **118** 177–84

[7] Osaily N 2010 *The Key Barriers to Implementing Sustainable Construction in West Bank – Palestine* (UK: Dissertation, University of Wales)

[8] Berardi U 2011 Sustainability Assessment in the Construction Sector: Rating Systems and Rated Buildings *Sustain. Dev.* **20** 411–24

[9] Opoku A and Fortune C 2013 Implementation of Sustainable Practices in {UK} Construction Organizations: Drivers and Challenges *Int. J. Sustain. Policy Pract.* **8** 121–32

[10] Ametepey O, Aigbavboa C and Ansah K 2015 Barriers to Successful Implementation of Sustainable Construction in the Ghanaian Construction Industry *Procedia Manuf.* **3** 1682–9

[11] Ervianto W I 2014 Kendala Kontraktor Dalam Menerapkan Green Construction Untuk Proyek Konstruksi Di Indonesia *Seminar Nasional X Inovasi Struktur dalam Menunjang Konektivitas Pulau di Indonesia Teknik Sipil ITS*

[12] Plessis C Du 2007 A strategic framework for sustainable construction in developing countries *Constr. Manag. Econ.* **25** 67–76

[13] Djokoto S D, Dadzie J and Ohemeng-Ababio E 2014 Barriers to Sustainable Construction in the Ghanaian Construction Industry: Consultants Perspectives *J. Sustain. Dev.* **7**

[14] Taber K S 2017 The Use of Cronbach's Alpha When Developing and Reporting Research Instruments in Science Education *Res. Sci. Educ.* **48** 1273–96

[15] Lill I 2008 Sustainable Management of Construction Labour *{ISARC} 2008 - Proceedings from the 25th International Symposium on Automation and Robotics in Construction* (International Association for Automation and Robotics in Construction ({IAARC}))

[16] Oke A, Aigbavboa C and Khangale T 2017 Effect of Skills Shortage on Sustainable Construction *Advances in Human Factors, Sustainable Urban Planning and Infrastructure* (Springer International Publishing) pp 303–9

[17] Hwang B-G and Tan J S 2012 Sustainable project management for green construction: challenges, impact and solutions *World Construction Conference* pp 171–9

[18] Robichaud L B and Anantatmula V S 2011 Greening Project Management Practices for Sustainable Construction *J. Manag. Eng.* **27** 48–57

Comparison Analysis Operastional Cost of Vehicle (VOC) Between Kayu Agung-Palembang-Betung Toll Road Plan with Existing Road

E Kadarsa[1], Hanafiah[1], B B Adhitya[1], M Pataras[1], A Azari[2]

[1]Lecturer in Civil Engineering Study Program, Faculty of Engineering, Sriwijaya University
[2]Student in Civil Engineering Study Program, Faculty of Engineering, Sriwijaya University

E-mail: aedikadarsah@gmail.com

Abstract. South Sumatra is one of the provinces whose development is very rapid in Indonesia. With the increasing development of centers of economic activity, politics, socio-culture and education make it an attraction for residents outside the city to travel to Palembang City. However, the condition of the eksisting highway which has become the driver's mainstay for travelling are not free of congestion. This causes travel time and Vehicle operational costs increase for every vehicle that passes the road. To overcome this problem, the government brought investors to build a toll road. The results of preliminary research show that people are still confused about the benefits through toll roads which are built besides travel time faster because they have to spend money to pay tolls. The main objective of this study was to determine the vehicle operating costs (VOC) using both roads with the Pacific Consultant International Method (PCI) and Industrial and Affiliate Research, Bandung Institute of Technology (LAPI ITB). Variable of the equation to determine the cost components is travel speed determined the Floating Car Method. The results show that the operational costs of vehicles (VOC) through toll roads are smaller when compared with existing roads. VOC of passenger cars that pass through the existing road in the morning Rp. 3395.45, trucks 2 axles loaded Rp. 5570.55 and trucks 2 axles with no payload of Rp. 3815.00, while if through toll road, passenger cars Rp. 2618.82 and trucks 2 axles Rp. 2918.71. In addition, information is obtained, more than 90% of drivers only consider fuel costs, they do not consider the cost of other variable variables in choosing a route.

Keyword: vehicle operating cost, toll road, eksisting road

1. Introduction

Toll road is one alternative to overcome the increasing congestion in the Province of South Sumatra. Stable pavement of toll roads, free of side barriers, free of u-turn cause the travel time and vehicle operating costs will be lower [1]. However, every vehicle through the toll road has to pay according to the class of the vehicle and the distance from the trip. For most people in South Sumatra, the toll roads are a new transportation infrastructure. Before the existence of the toll road, the drivers were free through the existing road. The results of the preliminary research show that the public still confuses about other advantageous the toll roads besides lower travel time bacause they have to spend money to pay tolls.

The purpose of this study was to determine the space speed (space mean speeds) for several vehicle types (passenger cars and trucks) and operational costs of vehicles (VOC) through existing roads with planned toll roads built.

2. Literatur Review

2.1. Toll Road

According to the Law of the Republic of Indonesia No. 38 of 2004 concerning roads, toll roads are public roads that are part of the road network system and as national roads whose use is permitted to pay tolls. While the toll itself is a certain amount of money paid for toll road use. Toll road concessions are carried out by the government and / or business entities that meet the requirements [2]. Toll roads as part of the public road network system are cross-alternative, although in certain circumstances it is not an alternative crossing, has higher specifications and services than existing public roads, and has certain toll rates that must be paid by users if they want to use the toll road [3].

2.2. Vehicle Operating Costs (VOC)

Vehicle operating costs are the total costs incurred by road users using vehicles from the origin zone to the destination zone. Vehicle operating costs consist of two components, namely fixed costs and variable costs. Fixed costs are costs that do not change (do not change according to the volume of production volume of services to a certain extent) including depreciation, salary of drivers, overhead, insurance and permits [4,5]. Medium variable costs (costs) are costs that change based on the volume generated from fuel services, lubricant, tires, vehicle maintenance costs. Some of the most common methods in Indonesia are PCI 1988, LAPI ITB 1996 and Road User Cost Model (RUCM 1992).

2.3. Previous Researches

- **Saputro D S C et al (2014)**
  Saputro et al conducted a comparative analysis of vehicle operational costs (VOC) on primary arterial roads with planned toll road sections: ungaran - Salatiga using the Pacific Consultant International (PCI) method. The vehicle speed in this study is the average velocity along the road which is the object of research [6].
- **Caesariawan I et al (2015)**
  Caesariawan et al. Conducted a study of the effect of time value on vehicle operational costs (VOC) for passenger cars on the existing road and Ambarawa ring road. Calculation of travel time values uses the Income Approach Method, while the calculation of vehicle operating costs (VOC) uses the Pacific Consultant International (PCI) Method [7].
- **Transport Canada Economic Analysis Directorate (2016)**
  Develop a model for estimating the operating costs of heavy vehicle perkilometers. The costs of recorded vehicle operating components are related to vehicle speed and road conditions. This study only calculates vehicle operating costs (VOC) and the value of travel time regardless of road pavement conditions. In this study, the calculation of vehicle operating costs (VOC) uses the Pacific Consultant International Method (PCI) and LAPI ITB. Both of these methods are used with consideration of data availability. The Road User Cost Model (RUCM) method is not used because the latest data roughness is not available. Vehicle speed which is the variable in equations to determine the cost components of the VOC using the floating car method considering the resulting data is quite accurate and easy to implement [8].
- **Kresnanto N C (2016)**
  Kresnanto conducts research on the analysis of VOC and the time value of several vehicle types of urban. The method for calculating vehicle operating costs uses pacific consultant International (PCI) and development of income approach for time value. The measurement of travel time and travel speed using the floating car method is by following each vehicle surveyed on the road. The road segment is divided into several segments, then the speed and travel time of each segment is averaged [9].
- **Ranawaka S et al. (2017)**

Ranawaka attempts to calculate the operating costs of vehicles, namely passenger cars and medium trucks by including the influence of roughness using HDM IV which is commonly used in Indonesia. This approach can be used to publish recommended values of VOC for various road conditions to be used in economic evaluation of feasibility studies [10].

3. Research Method

Structured and systematic stages are needed in conducting research. This will affect the effectiveness of time and work and avoid the occurrence of repetitive work and not needed. The flow chart of the research is presented in Figure 1

3.1 Primary Data

- Average Space Speed and Vehicle Travel Time
  The following will be explained the vehicle speed survey using the vehicle test method (Floating car method). This method is carried out with a test vehicle driven on a traffic flow following the following operating conditions:
  1. The driver tries to make a floating test vehicle in the traffic flow, which is to make sure that the number of vehicles that follow and are followed by the test vehicle are the same.
  2. Drivers regulate the speed of the test vehicle as expected speed of traffic flow.
  3. The test vehicle accelerates in accordance with the speed limit unless obstructed by traffic conditions.

  With this method can be obtained the total travel speed and location and the length of obstacles along the route. The study was conducted several times for each type of vehicle (passenger car, loaded and empty truck) for each direction of movement at the time of crowded and medium traffic flow. Based on the survey manual speed from the Ministry of Public Works, for each type of vehicle, a test vehicle survey is conducted 6 times for each direction.

- VOC Component Cost
  VOC component costs include fuel, engine oil, tires, costs maintenance, depreciation, interest rates and insurance. Fuel, engine oil, tires and maintenance costs were obtained through interviews with fueling stations, oil change shops, tire shops and workshops around the research site. Depreciation, interest and insurance costs are obtained from car dealers in the Palembang City Center

3.2 Secondary Data

Secondary data is obtained from the competent authority in the determination transportation policies such as the Transportation Agency and the Regional Government. Following secondary data needed:

- Minimum and average speed on toll roads
  The minimum speed set on the toll road depends on the location of the road located, both in the city and outside the city. Minimum toll road speed limit is 60 km / h or 80 km / hr. While the maximum limit is usually shown on toll road signs, which can be different on each road section.
- Road Geometry
  The road geometry includes the length and width of the road and the radius of the bend is obtained from authorized institutions, namely the Ministry of Public Works and Toll Road Managers. If the required data is difficult to obtain or not available, geometric measurements of the road are carried out directly in the field.

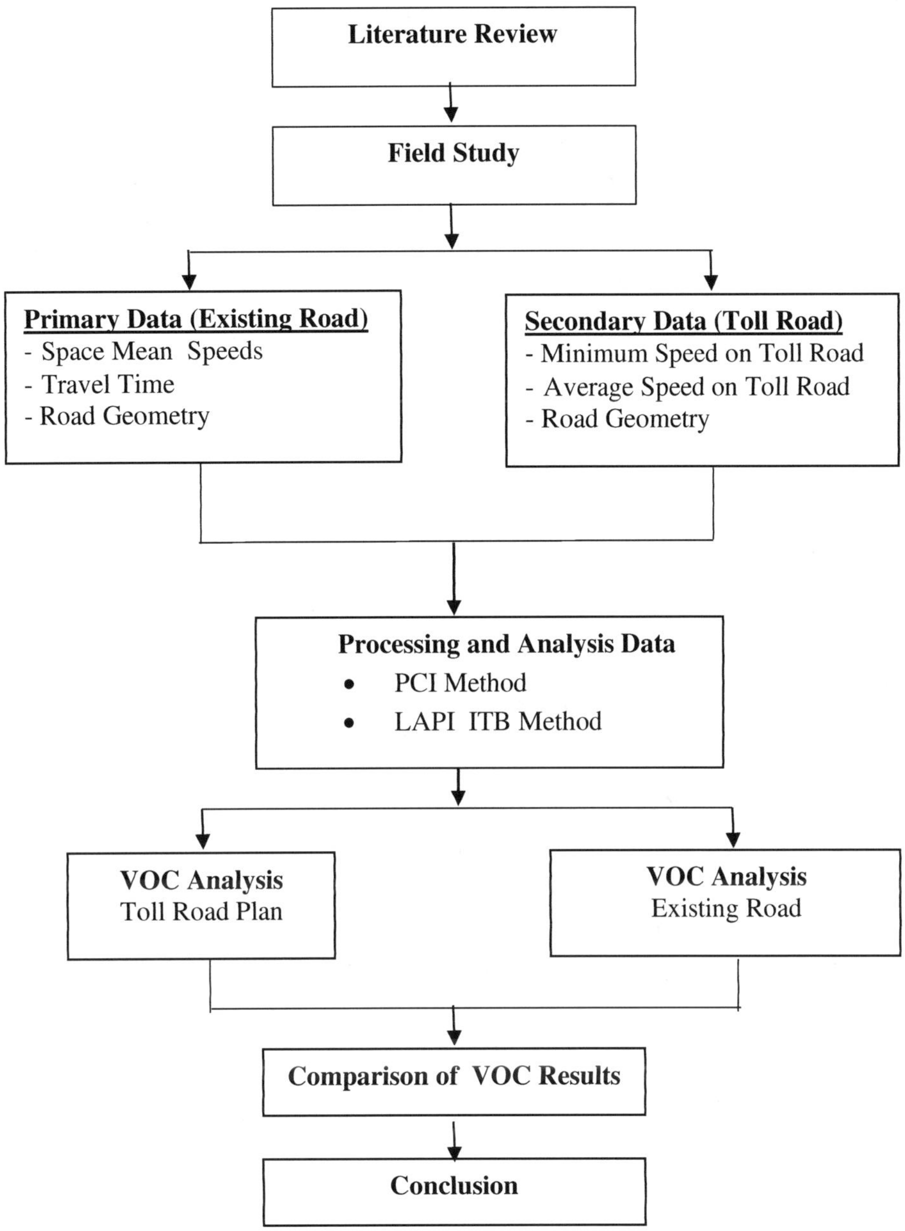

Figure 1. Research Flow Diagram

4. Results and Disscussion

The following will discuss tables and histogram curves as a result of data processing

4.1. Travel time and Travel Speed

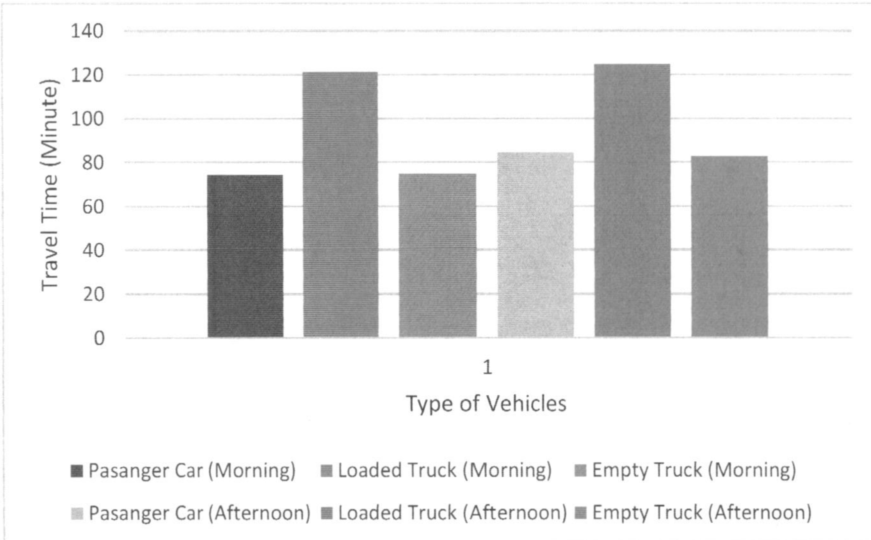

Figure 2. Travel Time Comparison for Passenger Cars and Trucks

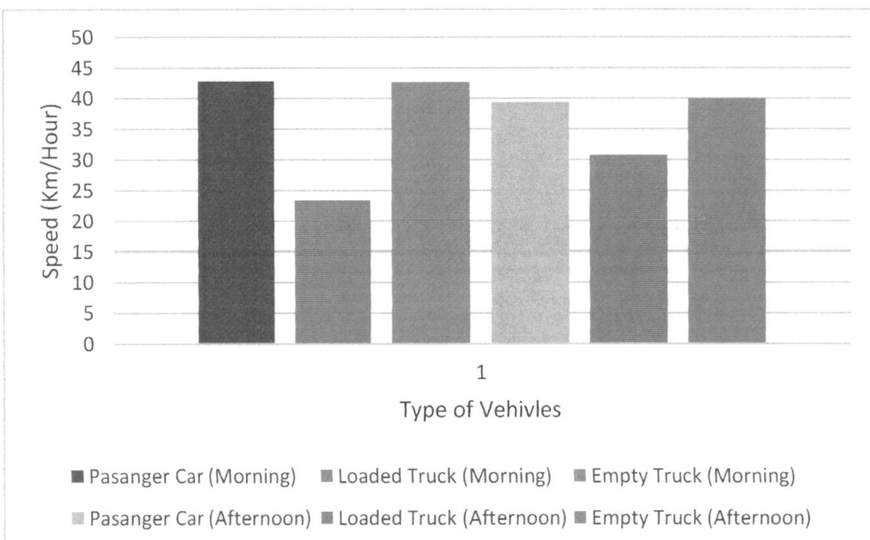

Figure 3. Comparison of Space Speed Curves

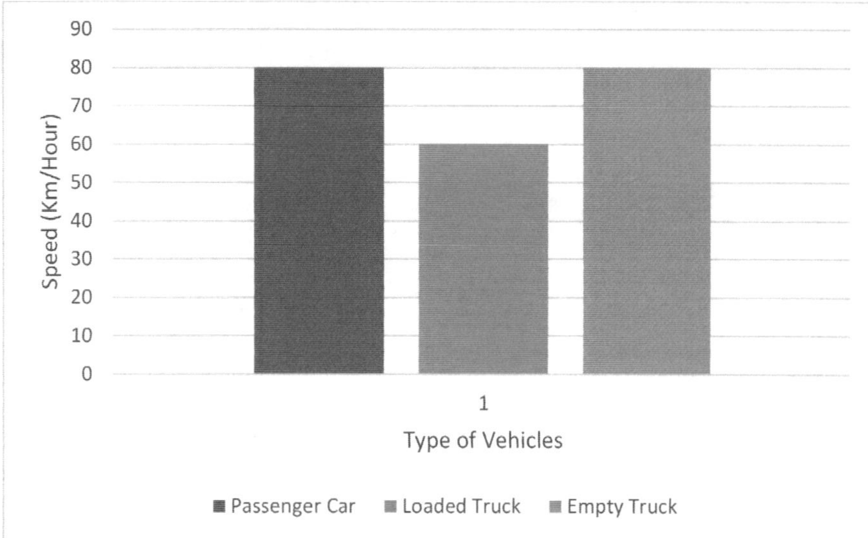

Figure 4. Minimum Speed Curve on Toll Road

Figures 2 and 3 show the average speed of passenger car space l higher or travel time shorter compared to loaded or empty trucks on the existing road. This is due to the car passengers are more efficient when going through existing roads. From the curve above, it can be seen that the average speed of the empty truck space is higher than the loaded truck and the speed of the empty truck is almost the same as the passenger car (Figure 4). This shows that when it is empty the truck can drive almost as efficiently as a passenger car. In the afternoon the space mean speeds of various vehicles are lower than in the morning. This is because in the afternoon the vehicle is denser than in the morning.

4.2. Vehicle Operating Costs

Table 1. Comparison of Vehicle Operating Costs Between Existing Roads With Toll Road Using the PCI Method

Variabel	Existing Road						Toll Road	
	Passenger Car		Loaded Truck		Empty Truck		Passenger Car	Medium truck
	Morning	Afternoon	Morning	Afternoon	Morning	Afternoon		
Fuel Costs	786,8	835,66	2572,64	2767,77	1591,45	1677,05	545,1	878,55
Lubricant Costs	59,16	61,03	316,87	331,17	242,11	249,07	44,71	251,47
Tire Charges	33,37	30,27	31,8	27,61	62,87	58,64	48,52	91,06
Maintenance Costs	247,13	241,17	593,64	578,73	701,07	686,15	279,85	802,47
Depreciation Costs	713,03	744,03	419,49	438,62	316,46	294,38	590,5	257,58
Interest Rate Fee	1033,38	1124,9	549,01	618,19	301,74	322,34	738,13	214,9
Insurance Fee	522,59	569,83	447,45	503,39	247,28	262	372,02	173,68
Travel Time	0	0	639,63	719,42	352,03	375	0	249
Total	3395,45	3606,9	5570,55	5984,91	3815	3924,64	2618,82	2918,71

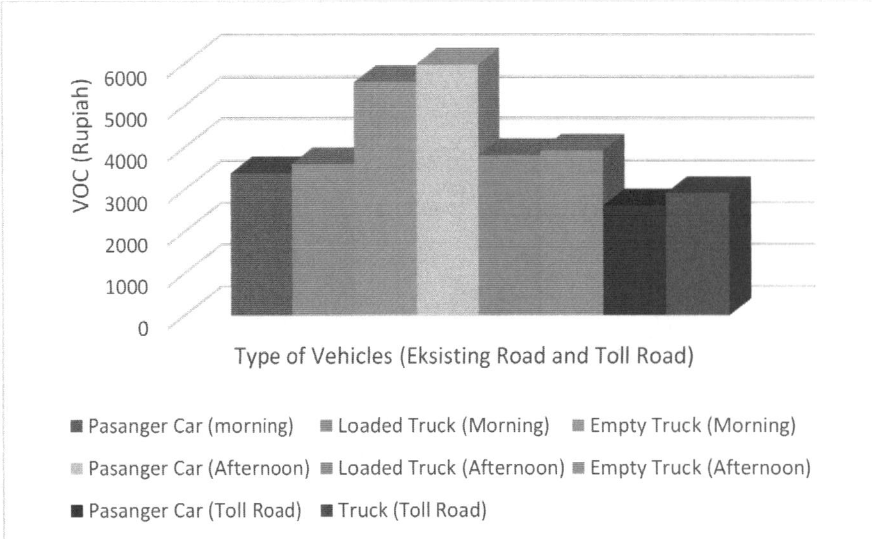

Figure 5. Comparison curve of Vehicle Operating Costs Between Existing Roads and Toll Roads

From the results of calculations in Table 1 and Figure 5 it can be concluded that operational costs of toll road vehicles are smaller than existing roads. This matter due to the condition of toll road pavement better than road existing, there are no intersections and motorbikes. In addition, there are no side barriers such as markets or vehicle parking on the road so that the driver can drive the vehicle faster.

Table 2. Comparison of Vehicle Operating Costs Between Existing Roads With Toll Road Using the LAPI Method

Variabel	Existing Road						Toll Road	
	Passenger Car		Loaded Truck		Empty Truck		Passenger Car	Medium truck
	Morning	Afternoon	Morning	Afternoon	Morning	Afternoon		
Fuel Costs	786,8	835,66	2572,64	2767,77	1591,45	1677,05	480,45	899,41
Lubricant Costs	59,25	61,12	316,87	331,17	242,11	249,07	4,78	7,61
Tire Charges	33,37	30,27	31,8	27,61	62,87	58,64	48,53	91,05
Maintenance Costs	247,43	240,76	587,97	573,4	697,2	682,33	280,06	796,29
Depreciation Costs	635,9	660,81	334,03	346,25	264,89	272,57	536,82	265,06
Interest Rate Fee	0	0	0	0	0	0	0	0
Insurance costs	523,54	570,1	446,88	502,4	245,83	261,87	373,98	174,58
Travel time	0	0	639,93	719,42	352,03	375	0	250
Total	2286,28	2398,72	4930,13	5268,02	3456,38	3576,54	1724,62	2484,01

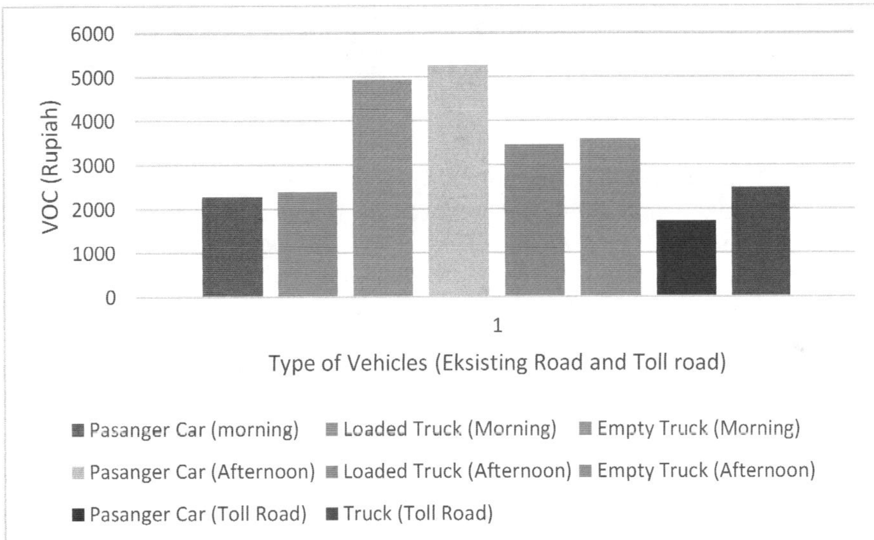

Figure 6. Curve Comparison of Vehicle Operating Costs Between Roads Existing with Toll Road

The results of the comparison of Tables 1 and 2 or Figures 5 and 6 can be concluded that the calculation of vehicle operating costs with the LAPI and PCI Method is not much different.

4.3. Vehivle Driver Perception of VOC Variabels

In this study an interview survey was conducted on vehicle drivers to see the effect of the VOC variable on consideration in choosing a route. The survey results show that more than 70% of motorists only consider fuel, while other VOC variable variables are not considered. Based on the interview, information was obtained that fuel was a consideration because it was directly purchased when the trip took place, while other variable variables were not. Thus this research can be input for motorized motorists that there are other costs to be incurred when traveling other than fuel.

5. Conclusion

Based on the discussion and the results on the calculation can be concluded as follows:

1) Space mean speed for direction of Palembang - Betung on the existing road for each vehicle and the departure times both morning and evening are as follows: Passenger car speed in the morning = 42.86 km / hr and afternoon = 39.36 km / h, truck 2 axles loaded in the morning = 23.44 km / hr and afternoon = 20.85 km / hr, trucks 2 axles empty in the morning = 42.62 km / hr and afternoon = 40 km / hour. This shows that the traffic flow is more crowded in the afternoon than in the afternoon so that the vehicle speed in the afternoon is lower than the morning
2) From the results of the comparison of vehicle operating costs (VOC), it is found that the cost of passing the toll road is smaller if compared with the existing road. Where is the vehicle in the morning for passenger cars, the cost is Rp 3395.45; loaded truck 2 axles Rp. 5570.55; and empty truck 2 axles Rp. 3815.00. Whereas if passing a toll road for passenger cars, that is Rp. 2618.82; for truck 2 axles Rp 2918.71.

6. Acknowledgement

This article was presented at the 2nd International Conference on Smart City INNOVATION (ICSCI) 2019, jointly held by Universitas Indonesia and Universitas Diponegoro. ICSCI conferences have been supported by the United States Agency for International Development (USAID) through the Sustainable Higher Education Research Alliance (SHERA) Project for Universitas Indonesia's Scientific Modelling, Application, Research, and Training for City-centered Innovation and Technology (SMART CITY)

Center for Collaborative Research, administered through Grant #AID-497-A-1600004, Sub Grant #IIE-00000078-UI-1.

7. References

[1] Walpole E R 1999 *Probability and Statistics for Wngineers and Scientists 9th Edition*
[2] Indonesia D G of H R 1997 *Indonesian Highway Capacity Manual*
[3] Direktorat Jenderal Bina Marga 1990 *Panduan Survai dan Perhitungan Waktu Perjalanan Lalu Lintas*
[4] Khisty C J and Lall B K 2006 *Transportation Engineering: An Introduction 2rd Edition* (Jakarta: Erlangga)
[5] Subandriyo E, Marpaung R R and Ismiyati 2014 *Comparative Analysis of Vehicle Operating Costs (VOC) Ambarawa Ring Road and Existing Roads* (Civil Engineering Masters Program UNDIP Semarang)
[6] Saputro D S C and Wirayoga R A 2014 *Comparative Analysis of Vehicle Operating Costs (VOC) on Primary Arterial Roads with planned toll road sections: Ungaran – Salatiga* (Semarang: Civil Engineering Study Program UNDIP)
[7] Caesariawan I, Rizky D N, Ismiyati I and Yulipriyono E E 2015 Pengaruh Nilai Waktu Pada Biaya Operasional Kendaraan (Bok) Mobil Penumpang Dalam Pemilihan Rute Jalan Eksisting Dan Jalan Lingkar Ambarawa *J. KARYA Tek. SIPIL* **4** 304–12
[8] Directorate T C E A 2006 *Estimation of Cost of Heavy Vehicle Use Pervehicle* (Kilometre in Canada Final Report)
[9] Kresnanto N C 2016 *VOC Comparison Analysis and Time Value of Several Types of Urban Modes, Civil* (Yogyakarta: Engineering Masters Program, UNJAN)
[10] Ranawaka S and Pasindu H R 2017 Estimating the Vehicle operating cost used for economic feasibility analysis of highway construction projects *2017 Moratuwa Engineering Research Conference (MERCon)* pp 347–50

Erosion Potential Based on Erodibility and Plasticity Index data on Cilengkrang, Bandung, West Java, Indonesia

N Khoirullah[1], IJ Mufti, I Sophian[1], T Yan W. M. Iskandarsyah, and D Muslim[1]
[1]Department of Applied Geology, Faculty of Geological Engineering, Universitas Padjadjaran

E-mail: Khoirullah @unpad.ac.id

Abstract. Cilengkrang is a sub-village in the north of Ujung Berung and Cijambe sub-District Bandung Regency, Indonesia. It is located on the west of Cikeruh sub-watershed. This subwatershed is part of greater upstream Citarum watershed. This area has been developed rapidly by new settlement and villa. However, the available data of erosion potential in this area is less convenient. The erosion is a severe problem that has a strong relation to flood potential. This study starts with collected engineering geological data from 14 samples. These samples were taken from different soil types from its engineering characteristics. Based on the field and laboratory observation, this area is form by fine-grained soils as a result of residual soils from young volcanic rocks — the silt and very fine sand content range from 49 to 70 % of the total soil. The erodibility value (K) in this area is from 0.03 to 0.06. These values are indicating moderate soil erodibility class. The plasticity index (PI) also shows value from 9 to 48 %. This value is pointing low to high plasticity behavior of this soil. The regression linear shows strong correlation between PI and K with r = 0.7. This correlation is interpreted that the cohesion controls the soil characteristic in Cilengkrang area and affect the moderate erodibility value. This study hopefully giving new insight that soil mechanic data is can also support data on erosion potential.

Keywords: Erodibility, Erosion Potential, Plasticity Index, Cilengkrang

1. Introduction

The erosion is a severe problem that has a strong relation to flood potential. Land degradation from forest to housing complex development, are some factors that lead an accelerated erosion. This accelerated erosion will also affect the decreasing water infiltration area during the rainy season. One of the well-known methods to predict this soil erosion is USLE (Universal Soil Loss Equation) proposed by Wischmeier and Smith in 1978 [1]. However, soil erosion prediction is a complex and multifaceted process that is affected by a host of factors. Soil erodibility is regarded as a critical parameter for evaluating the soil's susceptibility to erosion and is essential for predicting soil loss and assess its environmental effects. Soil erodibility is usually thought of as the amount of soil loss per unit of erosive force, whether it is rainfall, surface flow, or seepage [2]. The soil erodibility is expressed by the K factor. This K factor is obtained by the unit plot measurement on 22.1 m in length with a slope of 9 %, and it should remain uncovered, with conventional tillage along the slope [1,3]. While the Plasticity Index (PI) is a measure of the plasticity of a soil. It is also the difference between the liquid limit and the plastic limit of soil [4]. The higher the value of the PI, the more soil is plastics. The soil has potential to adsorbed water in significant number, by its cohesion due to the clay mineral. However, the lower value of PI, indicating that the soil has less cohesion and more silt characteristic. Since the silt is the high erodible material [5–7], the lower PI value would correlate with the erodibility factor.

The area of the study is located in Cilengkrang area. This area is a sub-village situated on the north of Ujung Berung, Bandung City, and close to Cijambe sub-District Bandung Regency, West Java Indonesia. This area is developed rapidly in housing complex development because of the strategic location near to Bandung capital city. Cilengkrang is located on the western part of Cikeruh sub-watershed. This subwatershed is part of greater upstream Citarum watershed. Several studies about erosion have done in Citarum watershed. However, most of them are studied in the southern part of this watershed, such as Subardja et al. [8] and Sukiyah et al. [6] focused on the south of basin of Bandung. However, there is also research on erosion potential in Bandung Basin, such as Saptari et al. [9] and Apandi [10] in Citarik sub-watershed. However, this research is studied by remote sensing analysis. Thus this paper will give a new insight by using field samples to determine the erodibility class. A previous study is did in a similar weathered volcanic material in Jatinangor area, Sumedang West Java by Khoirullah and Sophian [11]. This study showed that Jatinangor is consist of a high silty material in weathered volcanic material. This study also showed that the erodibility class in Jatinangor is categorized medium to high. Since weathered volcanic materials also form the Cilengkrang area, this study will present a wider result of erodibility data in weathered volcanic rocks. The rapid development in Cilengkrang area is present by many the housing complex built that changing the previous agriculture land. The decreasing green open space would lead a high erosion potential [12].

However, erodibility data is less available. Whereas by identify the erodibility data, we find out that an area would have higher erosion potential if left barely. Since the soil also has physical and mechanical properties, it would be indicated that these both parameters would have a significant correlation to the engineering geological data. This paper aims to show the relationship between the soil erodibility and the plasticity index value. Moreover, this paper will add new data collaboration. If we had PI data, it also could be used to identify the erodibility factor. Furthermore, the developer on the infrastructure in Cilengkrang will even think of the solutions to the impact of land clearing. The omission of this problem is feared could increase erosion rate that causes flash flood potential in East Bandung.

2. Method and Materials

This study started with field mapping to identify the surface condition of the soil. After that, the hand auger or hand boring test is done to obtain the shallow subsurface characteristic of soil. By the depth of 0.5 meters, the undisturbed sampling is done. There are 14 sites chosen by random sampling after the soil mapping result came. Later, the samples are tested in the laboratory to obtain grain size analysis, hydrometer analysis, and also Atteberg Limit test. The grain size analysis resulted in the percentage particle size, such as sand, silt, and clay. And the Atteberg limit tests were Liquid Limit test and Plastic limit test. These both test will result in plasticity index (PI) formula [13] shown as

$$\text{Plasticity Index (PI)} = \text{Liquid Limit (LL)} - \text{Plastic Limit (PL)} \quad \dots\dots\dots\dots (1)$$

While the soil erodibility is using Wischmeier and Smith formula [1] shown as

$$100\,K = \{\, 2{,}1\, M^{1{,}14} (10^{-4}) (12 - a) + 3{,}25 (b - 2) + 2{,}5 (c - 3)\} \quad \dots\dots\dots (2)$$

where
K = Soil erodibility
M = Particle Percentage (% of very fine sand + % of silt x (100 - % clay)
a = percent organic matter content
b = soil structure code used in soil classification
c = the soil profile permeability class

However, this K unit of short US Dimension (0,1-ton acre hour/ acre foot-ton inch). The K factor values obtained by the equation or nomograph constructed with the parameters must be multiplied by a factor of 0.1313 to be able to express the results in the International System of Units (metric ton hectare hour/hectare megajoule centimeter) [3]. The M factor is obtained by grain size and hydrometer analysis test. While the a, b, are collected by field observation. And the c factor is achieved by the permeability test.

3. Results and Discussion

Based on the field observation Figure 1, the high plasticity fine-grained soil is dominating in Cilengkrang area. This soil has brown to reddish-brown color, soft strength, and medium to high plasticity. This soil is interpreted as the weathered product of young and old volcanic materials, according to Djadja and Hermawan previous research published on the Engineering Geological Map of Bandung Quadrangle [14].

Figure 1 The Tuff rock and its weathering zone to brown soil (photo taken by Arif F. Sobari and team)

After the field observation, the samples are taken and test on the laboratory. In order to obtain the quantitative grain size distribution, the samples are tested with sieve analysis. The percent pass of mess 200 is categorize to fine grain soil. To determine the silt and clay distribution, the sample are test with hydrometer analysis. The summary of the distribution percentage of soil can be seen in the Figure 2.

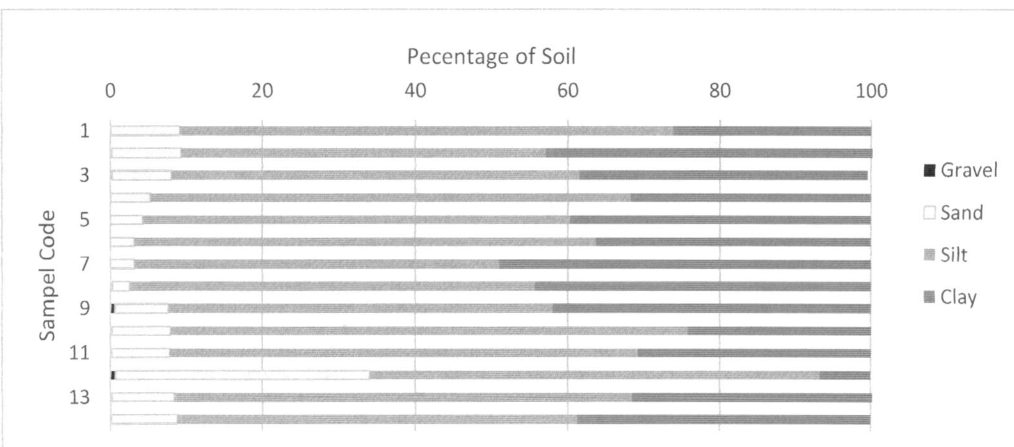

Figure 2 The distribution of soil percentage from 14 samples taken in Cilengkrang area

From figure 2, it can be seen that the silt and clay are dominating in all 14 samples with range 66 %— 92 %. This is indicating that there are no significant differences between residual soil source from breccia nor tuff. Likewise, there is no differences between silt and clay percentage. Except for sample number 12, that shows the sand percentage is more significant than the clay percentage. This grain size distribution is used to determine the K factor. the soil later analyzed by the erodibility characteristic. the summary of the erodibility can be seen below. The a factor is based on field observation and assumption, if the soil is left in bare condition with less organic materials. The b factor is based on the field observation, the soil condition is likely blocky and massive. While the c factor is based on the falling head of permeability test on the laboratory. The summary of erodibility factor is described on the Table 1.

Table 1 The summary of erodibility in Cilengkrang

Sample Code	Very Fine Sand (%)	Silt (%)	Clay (%)	a	b	c	K erodibility in US unit	K erodibility in the SI unit
	0.64	64.705	26.155	3	4	4	0.32	0.055
2	2.12	47.74	43.16	3	4	3	0.19	0.033
3	2	53.52	38	3	4	4	0.24	0.042
4	1.8	63.205	31.595	3	4	4	0.30	0.051
5	1.4	56.13	39.67	3	4	4	0.25	0.042
6	1.26	60.675	36.205	3	4	3	0.25	0.042
7	1.26	47.955	48.925	3	4	4	0.20	0.034
8	0.56	53.195	44.325	3	4	4	0.22	0.038
9	2.3	50.4	42	3	4	4	0.22	0.039
10	1.82	67.915	24.286	3	4	4	0.34	0.059
11	2.82	61.47	30.73	3	4	4	0.30	0.051
12	6.1	59.35	6.67	3	3	4	0.35	0.061
13	1.48	60.17	31.67	3	4	4	0.28	0.049
14	2.8	52.49	38.67	3	4	4	0.24	0.042

Based on Table 1, the erodibility in Cilengkrang are range from 0.03 to 0.06 t ha h ha-1 MJ-1 mm-1 in SI units. From this results, the erodibility class according to Foster et al.[15] in Cilengkrang is categorized into moderate soil erodible except on sample 12 that categorize into high soil erodible. This means that the soil in Cilengkrang soil has susceptibility to detachment and transport by erosion process, especially by rain [1,3,16].

After testing the grain size analysis, the samples then test with Atteberg limit to obtain the plasticity characteristic. The summary can be seen on the Table 2.

Table 2 The summary of Atteberg Limit test

Sample Code	Liquid Limits %	Plastic Limits %	Plasticity Index %	Description
1	85	50	35	High Plasticity
2	82	41	41	High Plasticity
3	74	35	39	High Plasticity
4	76	34	42	High Plasticity
5	83	35	48	High Plasticity
6	81	46	35	High Plasticity
7	90	46	43	High Plasticity
8	88	55	33	High Plasticity
9	72	29	43	High Plasticity
10	50	34	16	Medium Plasticity
11	47	29	18	High Plasticity
12	89	80	9	Medium Plasticity
13	95	62	33	High Plasticity
14	67	36	31	High Plasticity

Based on the Table 2, the plasticity characteristic of 12 samples are categorized into high plasticity, and only two samples are medium plasticity feature. The greater value of the plasticity index (PI) indicates that the soils have clayey behavior instead of silty expression. It is also concluded the clay mineral type form this soil has high cohesion value like illite mineral. While the samples on code 10 and 12 indicated the silty clay characteristic. This soil still has cohesion due to the clay mineral. However, the clay mineral type could be dominated by kaolinite.

Based on both erodibility and plasticity index data, the relationship is made between the plasticity index and erodibility factor can be seen on Figure 3.

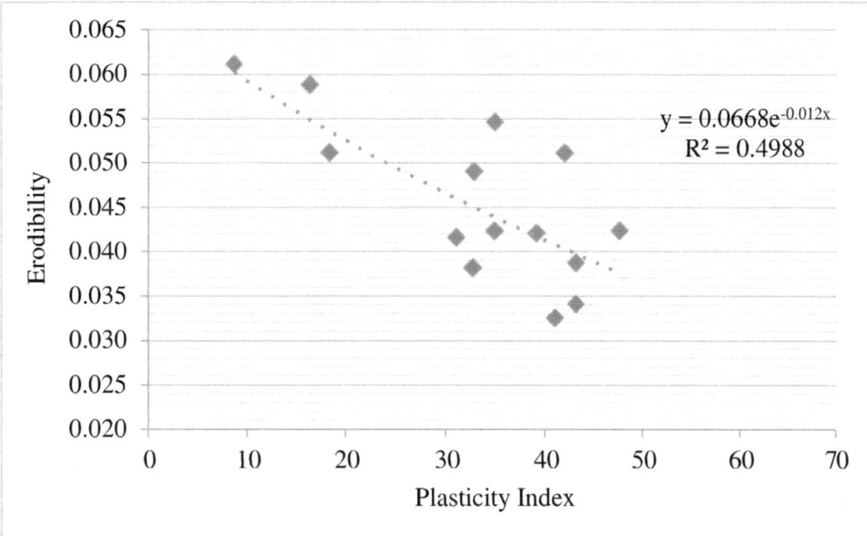

Figure 3 The correlation between Plasticity Index and Erodibility Factor

Based on the graphic above, it can be concluded that the plasticity index has a strong correlation to erodibility factor. This is interpreted as similarity behavior of the soil. The high erodible soil will have a lower value of PI. This would be a useful information for geotechnical or civil engineer before developed an area for a housing complex. If the erodible data are not available, the plasticity index can be a solution to estimate the soil characteristic. If an area has a lower PI value, that means the soil characteristic on that site is highly erodible. To prevent the erosion potential of the soil loss, the geotechnical engineer co-work with agriculture scientist by several solutions, e.g. with the movement of planting erosion-retaining plants. The simple prevention on the upstream management would prevent a flood disaster on the downstream

4. Conclusion

The Cilengkrang area is consist of medium to high plasticity fine-grained soil it is strongly correlated to the erodibility factor, from moderate soil erodible to high soil erodible. It can be concluded that the geotechnical data like plastic limit (PI) can be used to identify the erodibility factor. The developer needs to concern for not left the land barely especially during the rainy season, since the Cilengkrang soil has prone to erosion. Further work should be adding more samples to obtain a more accurate analysis

5. Acknowledgments

The author also would like to thank Cilengkrang team, Arif F. Shobari, Annisa Ayuingtyas A.P., Iqbal J. M. and Aldi Rexy for assisting the field data. This article's publication is partially supported by the United States Agency for International Development (USAID) through the Sustainable Higher Education Research Alliance (SHERA) Program for Universitas Indonesia's Scientific Modeling, Application, Research and Training for City-centered Innovation and Technology (SMART CITY) Project, Grant #AID-497-A-1600004, Sub Grant #IIE-00000078-UI-1. The research was funded by Indonesian Ministry of Research, Technology and Higher Education (Ristekdikti) under the scheme of University Basic Research Excellent 2019 Contract No. NKB-1636/UN2.R3.1/HKP.05.00/2019.

6. References

[1] Wischmeier W H and Smith D D 1978 Predicting rainfall erosion losses: a guide to conservation planning *U.S. Dep. Agric. Handb. No. 537*

[2] Wang B, Zheng F and Römkens M J M 2013 Comparison of soil erodibility factors in USLE, RUSLE2, EPIC and Dg models based on a Chinese soil erodibility database *Acta Agric. Scand. Sect. B Soil Plant Sci.*

[3] Cassol E A, da Silva T S, Eltz F L F and Levien R 2018 Soil erodibility under natural rainfall conditions as the K factor of the universal soil loss equation and application of the nomograph for a subtropical Ultisol *Rev. Bras. Cienc. do Solo*

[4] Das B M and Sobhan K 2018 Principles of Geotechnical Engineering, 9th Edition *Cengage Learning*

[5] Morgan R P C 2005 *Soil Erosion and Conservation, 3rd edition.* (Oxford: Blackwell Publishing)

[6] Sukiyah E, Hirnawan R F and Muslim D 2010 The Erosion Model Based on Grainsize Distribution Ratios of Weathering Product of Quaternary Volcanic Deposits (4218) *FIG Congr. 2010*

[7] Soemitro R A A and Asmaranto R 2016 The effect of drying - wetting cycles to the Wischmeier soil erodibility factors *Japanese Geotech. Soc. Spec. Publ.*

[8] Subardja A, Djuwansyah R and Wibawa S 2015 Prediksi Erosi Lahan Di Kabupaten Bandung, Bagian Selatan *Pros. Geoteknologi Lipi*

[9] Saptari A Y, Supriadi A, Wikantika K and Darmawan S 2015 Remote Sensing Analysis In RUSLE Erosion Estimation *Indones. J. Geospatial* **4** 34–45

[10] Apandi T 2003 Kajian Karakteristik Curah Hujan di Sub DAS Citarik, DAS Citarum Jawa Barat *Bionatura* **5**

[11] Khoirullah N and Sophian I 2019 Erodibility Factor in Weathered Volcanic Rock in Jatinangor, West Java, Indonesia *IOP Conf. Ser. Earth Environ. Sci.* **248** 012055

[12] Kusumandari A 2014 Soil Erodibility of Several Types of Green Open Space Areas in Yogyakarta City, Indonesia *Procedia Environ. Sci.* **20** 732–6

[13] Ishibashi I and Hazarika H 2015 *Soil Mechanics Fundamentals and Applications*

[14] Djadja and Hermawan 1996 *Engineering Geological Map Bandung Quadrangle, Jawa Barat, Indonesia* (Bandung, Indonesia)

[15] Foster G R, Mccool D K, Renard K G and Moldenhauer W C 1981 Conversion of the universal soil loss equation to SI metric units *J. Soil Water Conserv.*

[16] Renard K G, Foster G R, Weesies G A, McCool D K and Yoder D C 1997 Predicting soil erosion by water: a guide to conservation planning with the Revised Universal Soil Loss Equation (RUSLE) *USDA Agric. Handb.*

Evaluation and study of minimarket potential locations in BWK I, Semarang City

D Pramukti[1], W P Tyas[2]

[1] Departemen of Urban and Regional Planning, Diponegoro University, Semarang, Indonesia
[2] Departemen of Urban and Regional Planning, Diponegoro University, Semarang, Indonesia

E-mail: danangpramukti989@gmail.com

Abstract. Nowadays the population in Semarang City is increasing rapidly, this is also followed by the development of economic activities in the trade and services sectors. The minimization of the Minimarket to the lower middle trade sector can be seen from the increase in quantity. In 2010 the number of minimarkets in Semarang City had 244 buildings and continued to increase every year. This has an impact on land use change and will affect the existence of traditional markets. This research uses descriptive quantitative method with emphasis on spatial perspective. Aims to identify the needs, distribution patterns, service coverage (trade area), and suitability of the existing minimarkets. This study shows that the number of existing minimarkets exceeds the limits of what they should be. Based on the analysis of the suitability of the location of the existing minimarkets, as many as 11 minimarket buildings are included in the appropriate category from a total of 92 buildings.

Keywords: Minimarkets, Trade area, Service area

1. Introduction

The growth of the city is marked by an increase in population, usually followed by the development of economic activities in the trade and services sector. It is also characterized by changes in the use of agricultural land to more profitable and productive sectors such as trade and services. Semarang City is one of them, in 2010 the number of minimarkets reached 244 buildings and continues to increase every year [1]. In line with the increasing focus on customer value offered by minimarkets, especially in services, food safety guarantees, convenience, information, variety and comfort [2]. This is reinforced by Indonesia's gross domestic product (GDP) per capita of US $ 3,500 with an annual household budget of around US $ 5,000-US $ 15,000 expected to expand from 36% at present to 58% in 2020 [3]. Going Over 60 million low-income residents are projected to join the middle class population and drive higher consumer demand [3].

In determining the modern retail location in this study, especially the minimarket has a tendency to be located in the city center, this is often referred to as a symptom of the centralization of retail locations [4]. Richardson (1969) suggested that economic or corporate activity tends to be located at the center of activity in an effort to reduce uncertainty in decisions taken to minimize risk [5]. As a research location BWK I Semarang City is a city center that has offices, trade and services. Pearson's

[6] research suggests that in determining retail locations using demographic data as potential consumers. Berman & Evans [7] explained that the importance of identifying consumers according to their demographics, lifestyles, needs and desires, attitudes and shopping behavior, their perceptions of retailer actions and the environment. Site selection is very important because a minimarket can succeed or fail based only on its location [8].

Location analysis is very important for retail [9]. Some previous studies related to minimarkets in general are still focused on the determinants of minimarket locations, such as the most dominant geographical factors in determining new locations [10]. The distribution pattern of minimarkets and the reach of minimarket services to consumers and the impact on traditional markets [11]. This study bases on the saturation of minimarkets which has an impact on consumer reach. Both between minimarkets and traditional markets. The importance of structuring business space in urban areas (especially minimarket), this condition requires a structured problem solving through control and evaluation. In the future so that the development of minimarkets does not cause problems, this research will discuss the needs of minimarkets, the pattern of distribution of service areas (trade areas) and focus on evaluating the suitability of existing minimarket locations. So the results of this study are expected to be able to provide an overview and recommendations on the challenges of sustainable minimarket.

2. Data and Methods

2.1. Research data

Data requirements in this study use secondary and primary data. These data are grouped based on objectives and variables or attributes used in research. Primary data sources in this study were conducted by questionnaire (questionnaire) and field observations. As for secondary data, it is conducted through literature review and document review from books, laws, journals, previous studies, articles, internet, archives and data from government or private agencies. Secondary data in this study include population, number and location points of trade facilities (minimarkets and traditional markets) as well as spatial data such as land use maps, road networks and disaster-prone maps.

2.2. Research analysis method

This research uses descriptive quantitative method with emphasis on spatial perspective. Church states that the success of many future applications of the location of retail outlet sites may be closely linked to Geographical Information Systems (GIS) due to the fact that these systems are responsible for working with spatial information [12]. The analysis in this study mostly uses GIS software including the needs analysis and distribution patterns of the minimarket (nearest neighbor analysis), the trade areas analysis (services range) [13] of existing minimarkets and traditional markets, and the suitability analysis of the location of the existing minimarkets [14]. Sampling in this study using the technique of Probability sampling. This sampling is to provide equal opportunities for each member of the population to be chosen as a sample. With stratified random sampling the classification is distinguished through the street level of the minimarket location. Then the number of samples is determined by a random selection system. By using the formula from Slovin [15] the number of samples in the study obtained 100 respondents. then each sub-district in BWK I Semarang City has 33 respondents and subdivided into 3 road classes namely arterial roads, collector roads and local roads.

3. Results and discussion

3.1. Analysis of Minimarket Needs and Distribution Patterns

This minimarket needs analysis refers to the Indonesian national standard [16] each minimarket building can serve up to 6000 consumers Table 1. Overall in BWK I Semarang City, the total needs of suitable minimarkets are 36 minimarkets, while the existing minimarkets are 92 buildings, resulting in an excess of 56 minimarkets. So that the total number of existing minimarkets is over capacity or more than the needs of minimarkets.

Table 1. Minimarket needs per district

No	District BWK I	Total population (2018)	Exsisting minimarket	Minimarket needs	Add new minimarket	Over capacity
1	Central Semarang	69.711	40	11	-	29
2	South Semarang	79.162	28	13	-	15
3	East Semarang	76.608	24	12	-	12
	Total	**225.481**	**92**	**36**	**-**	**56**

But it is known from the results of research questionnaire data that there are several minimarkets that have consumers who live outside the city of Semarang. Minimarkets with consumers outside the study area are located along arterial roads, collector roads and minimarkets which are located around areas adjacent to other districts such as Kendal Regency and Demak Regency Figure 1. So this is in accordance with the theory developed by Losch [17] based on the landscape model, in this case the trade location (minimarket) is located in the city center and is related to the consumer movement sector. This is also reinforced by the land value theory by R.V.Retcliff [5] that the retail zone (minimarket) is located in the city center because business continuity requires the greatest degree of accessibility in order to obtain maximum benefits when viewed from an entrepreneur's perspective.

Figure 1. Map of existing minimarket locations distribution and analysis diagram of average nearest neighbors in BWK I Semarang City

It can be seen that in 2019 almost all parts of the research area have been covered by Alfamart and Indomart minimarkets. The distribution points of minimarkets are located in each class of roads, arterial roads, collector roads and local roads. Based on the analysis of the average nearest neighbor, the result of the minimarket distribution patterns found in the study area is to form an irregular random or random pattern. The Observed mean number values indicate that the average distance between the nearest minimarket is 203 meters. This shows that the distance between the existing minimarkets in the study area is closer than the service coverage limit if viewed from SNI No. 03-1733-2004 as far as 500 meters.

3.2. Analysis of minimarket and traditional market reach services (trade area)

Minimarkets with a distance of less than 500 meters with other minimarket locations produce a range of services that overlap with one another Table 2. So that it can be said the density between the locations of the minimarket results in competition in getting consumers. consumers will usually use more than one place to shop, especially in densely populated urban areas where shopping options are greater [18]. Whereas traditional markets are divided between services that overlap and some do not. when viewed from the perspective of consumers this has a positive impact on the variety of choices in meeting shopping needs.

Table 2. Distance, transportation and shopping frequency of minimarket consumers

No	Location distance		Transportation		Shopping frequency	
	Distance	**Respondents**	**Modes of transportation**	**Respondents**	**Shopping frequency**	**Respondents**
1	<500 Meter	16	Walk	8	Daily	35
2	<1 Km	23	Motorcycle	82	Weekly	10
3	>1 Km	61	Car	8	Monthly	2
4			Other	2	Uncertain	53
	Total	100	**Total**	100	**Total**	100

The diversity of the reach of minimarket consumers in the study area shows that the reach of minimarket services is not limited to the surrounding population Figure 2. This can be shown from the number of respondents between the location of residence and the destination minimarket which is more than 1 Km as many as 61 people. The most frequently used mode of transportation to the minimarket is a motorcycle with 82 respondents with an average answer that is more flexible and comfortable.

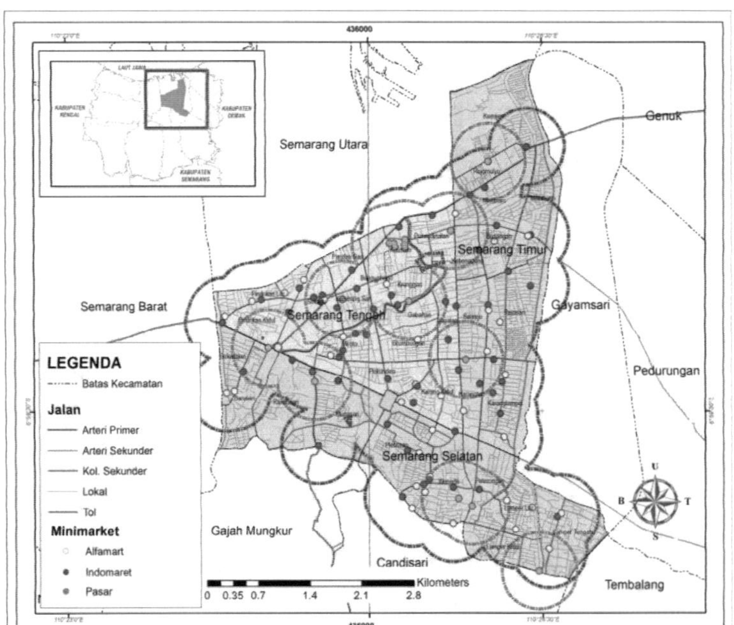

Figure 2. Map of the reach of minimarket and traditional market services at BWK I Semarang City

3.3. Analysis suitability of the existing minimarkets

The criteria chosen as a reference for selecting the suitability of minimarkets are minimarkets located in residential and commercial areas, minimarkets are located in every road network system, not located in the area or range of traditional market services, finally the existing minimarkets are not in disaster prone areas. By using the model builder in the Spatial Analysis, each geographic information system is realized in a spatial map [19]. The process of scaling and weighting of land use variables, road networks, the distance between existing minimarkets and traditional markets and to disaster prone locations.

The total locations of the existing minimarkets according to the category correspond to the total value of the value between 17-21 in BWK I Semarang City of 11 buildings. Average minimarkets are in land use in trade and service areas, arterial roads, located outside the area or range of services of other existing minimarkets and traditional markets and located outside flood-prone areas. Furthermore, more than half of the total minimarkets in the Semarang City BWK I are dominated by existing minimarkets with a category that is quite in accordance with the total value of between 13-16 as many as 65 buildings. While the location of the existing minimarkets in the category does not match the total value of the range between 7-12 as many as 16 buildings with an average land use in residential areas, located along the collector road and local roads, located within the area or range of other existing minimarket services and traditional markets and located in flood prone areas Figure 3.

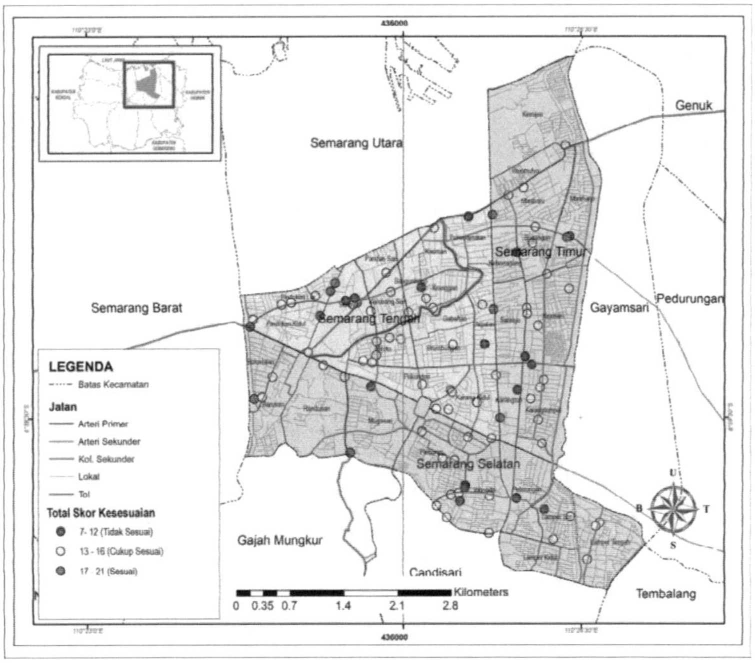

Figure 3. Map of the suitability analysis of the existing minimarket in BWK I Semarang City

4. Conclusion

The city of Semarang especially BWK I is a growth center for the surrounding area. This region is a strategic location for investment, especially in the trade sector, especially minimarkets. Based on the results of minimarket needs analysis that the number of existing minimarket buildings is disproportionate or more than their needs. Over capacity of up to 56 minimarkets out of a total of 92 existing buildings. This certainly affects the income of traders in traditional markets and grocery stores in the study area. But it is known from the results of questionnaire data that there are several minimarkets that have consumers or visitors residing outside the study area. Minimarkets with consumers outside the study area are located along arterial roads, collector roads and minimarkets which are located around areas adjacent to other districts such as Kendal Regency and Demak Regency. This shows that the minimarket is not only to meet the needs of residents in the Semarang City BWK I region, but also includes consumers outside the study area. Through analysis of the distribution patterns of minimarkets it is known that in 2019 almost all parts of the research area have been covered by minimarkets (Alfamart and Indomart) and form irregular random or random patterns. Based on the analysis of the nearest nearest average or average nearest neighbor, the average distance between the nearest minimarket is 203 meters.

In the analysis of the reach of minimarket services, it is known that all minimarkets in the study area overlap with one another. In accordance with the analysis of the nearest nearest neighbor, the average nearest neighbor shows that the average distance between the nearest minimarket is 203 meters. Even at some point the location of the minimarket is next to each other and facing a distance of less than 50 meters. So that it can be said the density between the locations of the minimarket results in competition in getting consumers. But based on the results of the questionnaire showed the diversity of consumers marked by as many as 61 respondents the distance of the location of their residence with a minimarket of more than 1 km. The average consumer minimarket is a private employee with the most commonly used mode of transportation is a motorcycle. Based on the analysis of the suitability of the location of the existing minimarkets, it is known that as many as 11 minimarket buildings are included in the appropriate category, as many as 65 buildings in the category are quite appropriate and with the category of incompatible as many as 16 buildings.

5. Acknowledgment

This article's publication is partially supported by the United States Agency for International Development (USAID) through the Sustainable Higher Education Research Alliance (SHERA) Program for Universitas Indonesia's Scientific Modeling, Application, Research and Training for City-centered Innovation and Technology (SMART CITY) Project, Grant #AID-497-A-1600004, Sub Grant #IIE-00000078-UI-1. The research was funded by Indonesian Ministry of Research, Technology and Higher Education (Ristekdikti) under the scheme of University Basic Research Excellent 2019 Contract No. NKB-1636/UN2.R3.1/HKP.05.00/2019.

6. References

[1] BAPPEDA 2009 *Masterplan Pengembangan Pola Perpasaran Kota Semarang* (Kota Semarang: Badan Perencanaan Pembangunan Daerah (BAPPEDA))

[2] Dyck J, Woolverton A and Rangkuti F 2012 Indonesia's modern retail food sector: Interaction with changing food consumption and trade patterns *USDA-ERS Econ. Inf. Bull.*

[3] Duniaindustri 2016 Data Industri Minimarket, Supermarket dan Hypermarket di Indonesia

[4] Yunus H S 2004 *Struktur Tata Ruang. Cetakan IV* (Jogjakarta: Pustaka Pelajar Offse)

[5] Tarigan R 2007 *Ekonomi Regional Teori dan Aplikasi (Ed. Revisi)* (Jakarta: Bumi Aksara)

[6] Pearson J 2007 A comparative business site-location feasibility analysis using geographic information systems and the gravity model *Pap. Resour. Anal.* **9**

[7] Berman B and Evans J . 2006 *Retail Management : A Strategic Approach (Tenth ed.)* (Pearson Prentice Hall)

[8] Mendes A B and Themido I H 2004 Multi-outlet retail site location assessment *Int. Trans. Oper. Res.* **11** 1–18

[9] Hernández T and Bennison D 2000 The art and science of retail location decisions *Int. J. Retail Distrib. Manag.* **28** 357–67

[10] Erbiyik H, Özcan S and Karabouga K 2012 Retail Store Location Selection Problem with Multiple Analytical Hierarchy Process of Decision Making an Application in Turkey *Procedia - Soc. Behav. Sci.* **58** 1405–14

[11] Önden .Ismail, CSen C G and CSen A 2012 Integration of Integer Programming with GIS Analyzing Abilities for Determining the Convenience Levels of Retail Stores *Procedia - Soc.*

Behav. Sci. **62** 1144–9

[12] Widaningrum D L 2015 A GIS-Based Approach for Catchment Area Analysis of Convenience Store *Procedia Comput. Sci.* **72** 511–8

[13] Murad A A 2003 Creating a GIS application for retail centers in Jeddah city *Int. J. Appl. Earth Obs. Geoinf.* **4** 329–38

[14] Suárez-Vega R, Santos-Peñate D R and Dorta-González P 2012 Location models and GIS tools for retail site location *Appl. Geogr.* **35** 12–22

[15] Werang B R 2015 Pendekatan Kuantitatif dalam Penelitian Sosial *Yogyakarta: Calpulis*

[16] SNI 2004 *Standar Nasional Indonesia (SNI) 03-1733-2004 concerning Procedures for Planning a Housing Environment in the Urban Area*

[17] Rustiadi E 2011 *Perencanaan dan Pengembangan Wilayah* (Jakarta: Kerjasama crestpent Press dan Yayasan Pustaka Obor Indonesia)

[18] Amaya F L, Gonzales T A, Hernandez E C, Luzano E V and Mercado N P 2012 Estimating Point and Non-Point Sources of Pollution in Biñan River Basin, the Philippines *APCBEE Procedia* **1** 233–8

[19] Roig-Tierno N, Baviera-Puig A, Buitrago-Vera J and Mas-Verdu F 2013 The retail site location decision process using GIS and the analytical hierarchy process *Appl. Geogr.* **40** 191–8

Community Perception of Cheese-Making Wastes Utilization and its Sustainability

G L Utama[1,2,3,4], I Dinika[4] and R L Balia[5,6]

[1]Sustainability Science Study Program, Postgraduate School, Universitas Padjadjaran
[2]Centre for Environment and Sustainability Science, Universitas Padjadjaran
[3]Centre for Environmental Innovation and Technology, Postgraduate School, Universitas Padjadjaran
[4]Faculty of Agro-Industrial Technology, Universitas Padjadjaran
[5]Faculty of Animal Husbandry, Universitas Padjadjaran
[6]Veterinary Study Program, Faculty of Medicine, Universitas Padjadjaran

E-mail: g.l.utama@unpad.ac.id

Abstract. Cheese-making wastes utilization has been done in KPBS (The Dairy Farmer Cooperatives of South Bandung) and affect the surrounding communities. The research aimed to determine the perception of the community towards the utilization of cheese-making wastes that has been done. Cluster random sampling was taken on three groups of respondents, namely KPBS employees, KPBS member farmers and the affected communities around KPBS. Randomized numerical tables used to determine the respondent which resulted 24 respondents of KPBS employees, 41 respondents of KPBS member farmers and 28 respondents of affected communities around KPBS, with the total of 93 respondents. Questionnaires used to obtain the community perceptions of cheese-making wastes utilization and its sustainability that covered social, economic, environmental and technological dimension with five levels Likert scale to express the respondent's preference. Results showed that the communities perception of cheese-making wastes utilization on social, economic and ecological dimension shown neutral to positives perception with average Likert score of 3.61, 3.82 and 3.35 respectively, while technological knowledge dimension shown neutral to negative perception (2.99) which could be a threat towards cheese-making sustainability .

Keywords: cheese-making wastes, perception, sustainability, utilization

1. Introduction

Perceptions about the ecology will be related to physical space and disturbances that occur in it [1]. Disposing of cheese-making wastes into the environment improperly can cause perception disruption in the community both directly and indirectly. The proper utilization of cheese-making wastes results in a reduction in pollution disturbances that trigger the emergence of comfort and assurance for environmental hygiene and health, thus affecting people's perceptions towards the cheese producers environment [2,3].

Cheese whey, which is a waste of cheese-making process, needs to be managed through a sustainable approach that generated from the community engagement so that positive community perceptions are resulted [4,5]. This approach is higher than community participation because it promotes authentic partnerships, including mutual support, respect, trust, equality, "win-win solutions" in the collaboration initiative [6]. The concept of community engagement in sustainable waste management means working together collaboratively with related partners who have common goals and interests in reducing the negative impacts caused.

Community perceptions of a process's sustainability can be used as an approach in managing food industry waste. This approach is able to ensure that the management carried out is socially acceptable and practically applicable. In addition, the approach taken can reduce the risk of interference from the community, avoid the risk of exploitation through equitable distribution of benefits, and be able to calculate the social hazards that may arise from the management carried out so that sustainability is achieved [7]. In addition to the calculation of social hazards, achieving sustainability from a business can also be determined by economic aspects perception. However, some companies that integrate this are only able to achieve economic performance indicators without being able to integrate sustainability with other performance [8].

Various developing companies around the world are trying to integrate sustainability in their business practices [9]. It is important to note that sustainability is not only achieved by one or two dimension. The economic dimension is not only limited to short-term performance indicators such as how much profit is earned, but must be able to cover various elements that support long-term financial success such as reputation and public trust in the company and company relations with the community [10]. In line with the various environmental and economic benefits produced, various social benefits are also expected to emerge.

Pollution load reduction will trigger the emergence of comfort and assurance for cleanliness and environmental health which ultimately impacts on changes in people's perceptions around cheese producers. This can create community engagement in the form of support from the community for various activities carried out by cheese producers. As a result, a sustainable cheese-making business will be generated. The research done to determine the community perception towards cheese-making wastes utilization by KPBS Pangalengan and its sustainability.

2. Materials and Methods

Cluster random sampling was taken as the methodology by directing random election towards three cluster of respondents (KPBS plant staff, KPBS farmer member and influenced communities). The survey had been conducted on April to May 2019, around KPBS Pangalengan, Bandung District, and West Java Province, Indonesia. Respondents' information were gotten from 93 respondents and inspecting utilizing randomized numerical tables [11]. The consequences of randomization are as per the following, 24 plant staff, 28 farmer members, and 41 influenced communities (Figure 1).

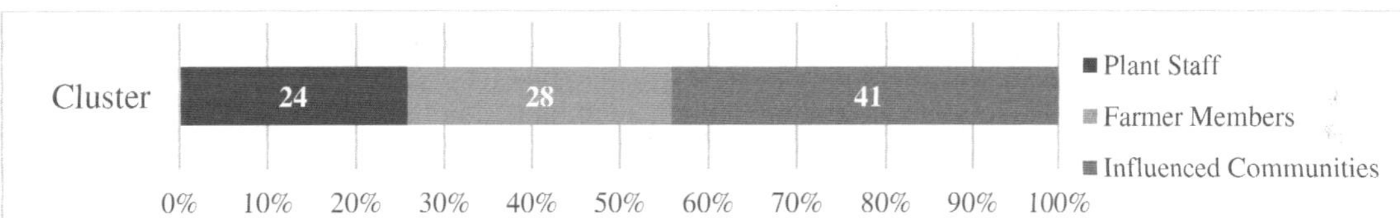

Figure 1. Respondents Compositions

The research variables are respondents' perception of cheese-making wastes utilization represented by social, economic, ecological and technological dimension. Community perception were taken utilizing questionnaires which contained 24 question related to the research variables which divided to 6 question of every dimension including social, economic, ecological and technological dimension. Data analysis was done by clustering the evaluations into five classification according to the Likert scale and outlined as radar graphs. All of the data were averaged, then analyzed descriptively.

3. Results and Discussions

The respondents' profiles plays an important role in the sustainability of the production process and utilization of cheese-making waste in the KPBS environment. Profiles of respondents showed the 88 (94.62%) of respondents were male and only 5 (5.38%) respondents were female. The largest proportion of respondents aged between 20-40 years and the last highest education is elementary school, which is which is 32 (34.41%) respondents (Figure 2).

Production activities that have been carried out by KPBS have had an impact on the surrounding community. The resulted cheese-making waste creates various perceptions either directly or indirectly. Waste generated from the production process is able to bring out various possible perceptions in the community [12]. Perceptions of social, economic and ecological impacts supported by the technological knowledge mastery can be useful in ensuring the continuity of production activities. One of the perceptions sampled was the social dimension which included six statements related to the social impact of cheese production and cheese-making wastes utilization (Figure 3).

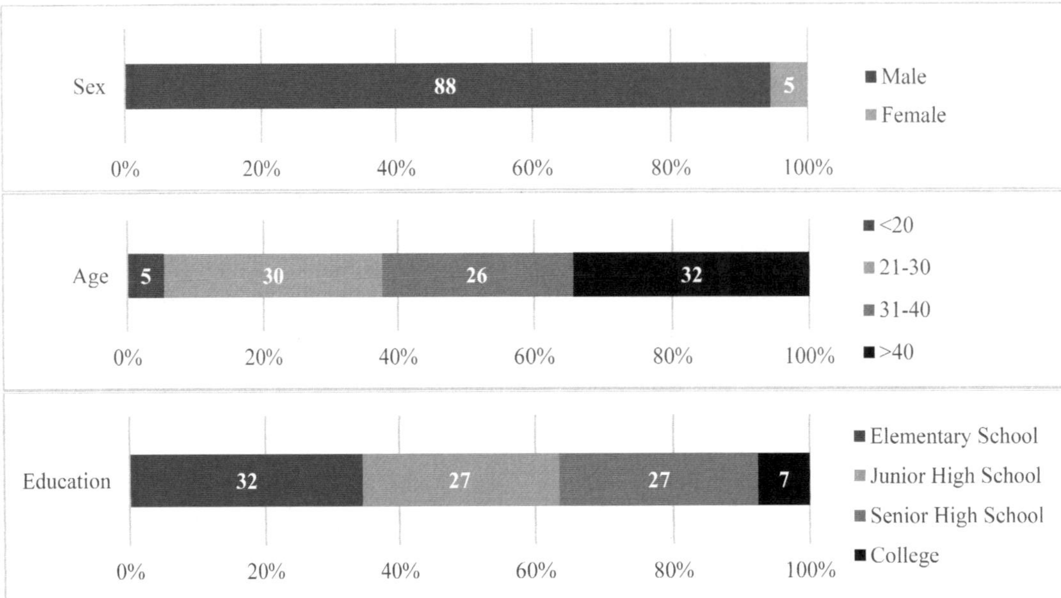

Figure 2. Respondents Profiles

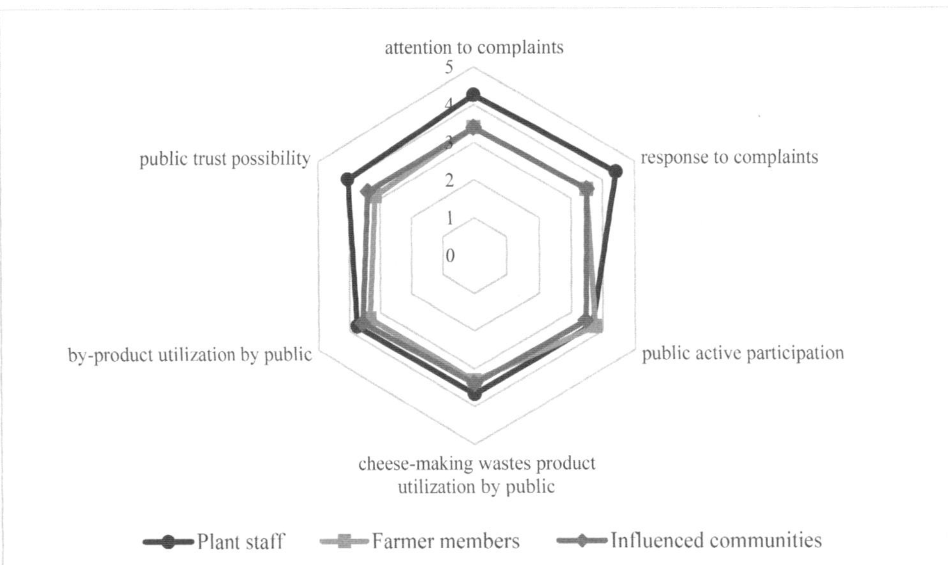

Figure 3. Community perception of social dimension of cheese-making wastes utilization

Three cluster of respondents had a variety of perceptions of the social dimensions that were generated. The three groups of respondents have the same perception of the cheese-making utilization and will actively participate in the utilization activities carried out. Meanwhile, there are differences in the three statements related to plant staff attention to public complaints, responsiveness towards complaints and the possibility of increasing public trust in KPBS.

Increased public trust can be generated through the quality of service [13]. The two groups of external respondents (influenced communities and farmer members) still feel neutral about the statements related to the responsiveness towards community complaints, while the internal respondent cluster (plant staff) feel that the service has been done well so that trust can be generated. Despite the differences, the three groups of respondents had positive perceptions above the value of 3.0-4.0 (neutral - agree).

Other perceptions tested were economic dimension which included six statements related to the economic impact of cheese production and cheese-making wastes utilization (Figure 4). The economic impact of cheese production and cheese-making utilization shows a positive perception above neutral values. The three cluster of respondents agreed with two statements related to increased income generated from cheese production and cheese-making wastes utilization. This is in line with Utama, et al. [14] that mentioned the economic benefit of cheese-making wastes bioconversions into bioethanol and liquid organic fertilizer. The three cluster of respondents also had neutral-agreed perceptions on three statements regarding sales profits, the benefits of community involvement in the use and involvement of farmer members in marketing. Meanwhile, differences in perceptions between internal (plant staff) and external groups (farmer members and influenced communities) occurred in the involvement of farmer members in marketing the by-products resulted by cheese-making wastes of utilization process. However, the external group of respondents still have a neutral perception towards KPBS.

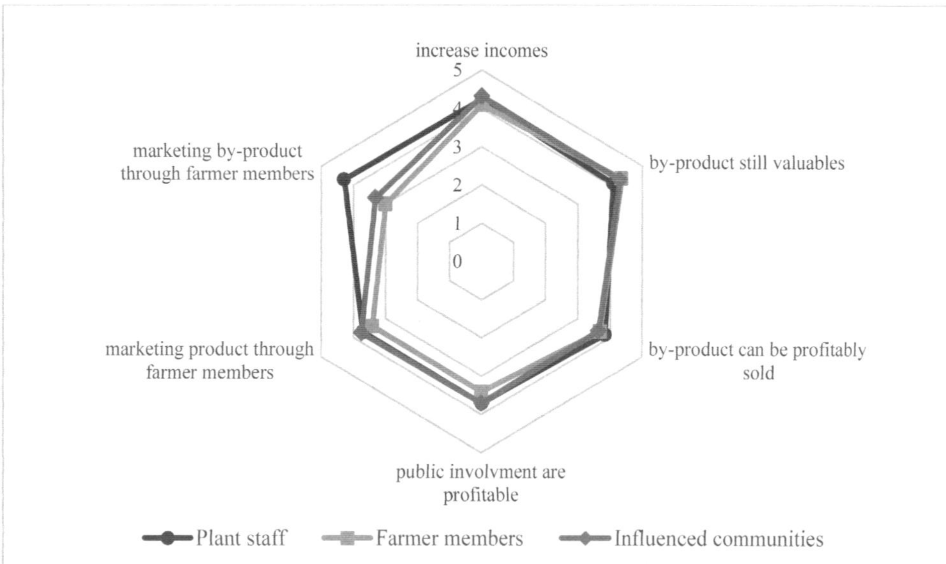

Figure 4. Community perception of economic dimension of cheese-making wastes utilization

Community perceptions of the ecological dimension of cheese production and cheese-making wastes utilization include six questions (Figure 5). The process of cheese production and t cheese-making wastes utilization have resulted in ecological impacts that give rise to diverse perceptions of respondents. Almost all statements of respondents were neutral including statements that did not smell, did not pollute the soil and air. However, the respondent cluster of the community has a disagree-neutral perception regarding the statement of not polluting water. According to Ir. Herry Hidayat statements who is the Production Manager of KPBS Plant, several times the disposal of the cheese-making wastes into the waters caused pollution in surrounding wells belonging to the community around the plant. However, KPBS immediately made compensation related to the pollution caused. This is reflected in the response to the statement on social impacts, related to KPBS 'attention to public complaints and the responsiveness towards community complaints. Cheese-making wastes utilization into bioethanol and liquid organic fertilizer could reduce the wastes disposal to the environment up to 62.1% which can be roles as a solution to the pollution resulted [15]. This is in line with the perceptions of all respondents, especially the internal group (plant staff) who agree with the utilization of cheese-making wastes to results pollution reduction around the KPBS.

Support for the sustainability of the utilization of cheese-making wastes is also determined by the technological knowledge possessed by the community that will be involved (Figure 6). Technological knowledge mastery are important in ensuring sustainable management of natural resources [16].

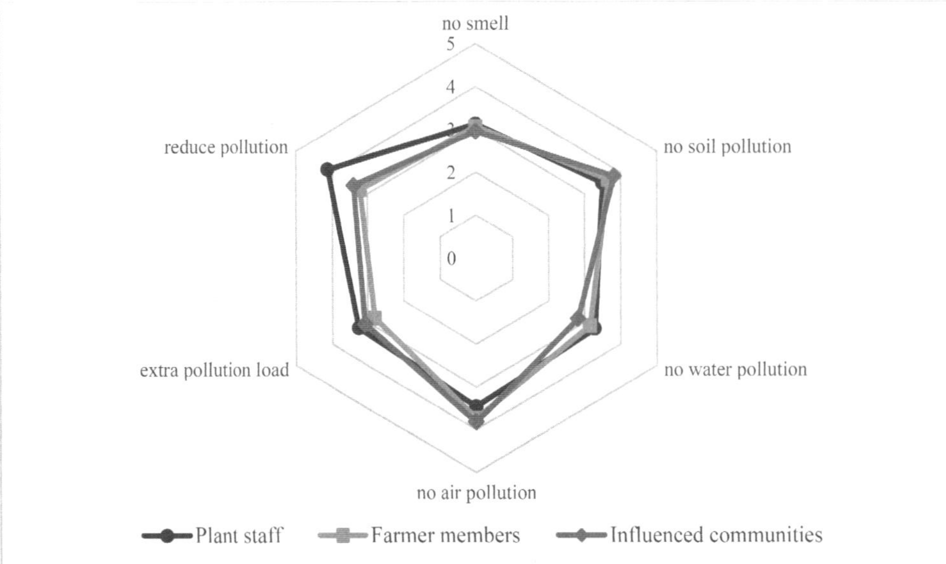

Figure 5. Community perception of ecological dimension of cheese-making wastes utilization

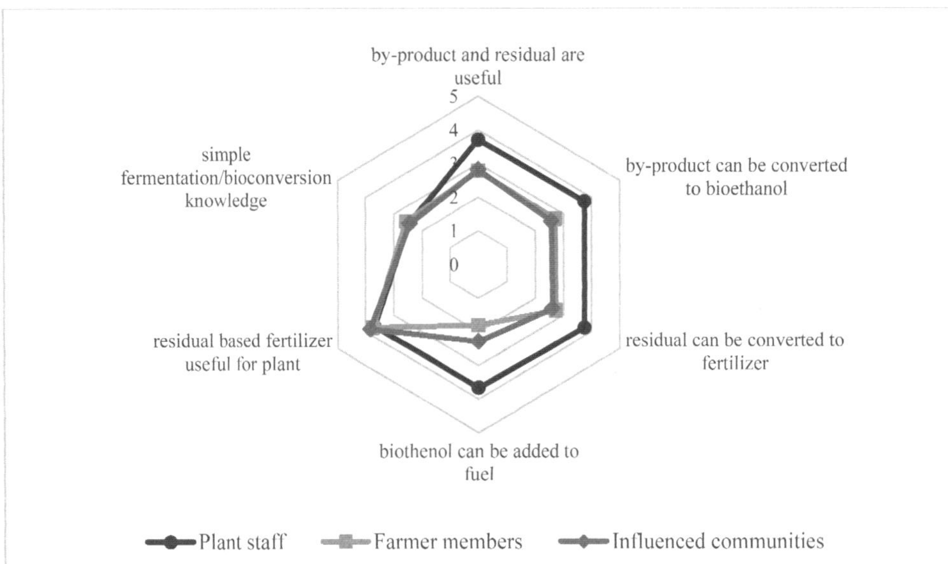

Figure 6. Community perception of technological knowledge of cheese-making wastes utilization

Technological knowledge mastery by respondents, especially external respondents (farmer members and influenced communities) regarding the cheese-making wastes utilization is still categorized as poor. Only internal respondents (plant staff) responded well to the five statements, but the weakness was still in the statement regarding bioconversion knowledge. Meanwhile, external respondents clearly showed a poor response to the five statements except for statements regarding the benefits of by-products utilization. These results indicate the weakness of technological knowledge dimension and mastery by respondents so that it also has an impact on responses to social, economic and environmental aspects.

Knowledge and mastery of technological knowledge, one of which can be obtained through education, therefore the education level of respondents will influence the response produced. This is in line with Hermawan [17] which states that there is a positive correlation between the level of education with perceptions and behavior in ecological preservation. Poor technological knowledge about the utilization of cheese-making wastes and reduction of other residual can be a threat and have a negative impact on the sustainability of the cheese-making process at KPBS.

4. Conclusions

Cheese-making wastes utilization that has been done by KPBS resulted in positives community perception towards social, economic and ecological dimensions. However, still found negative perception and low technological knowledge mastery in the community that could be a threat to the sustainability of cheese production by KPBS. Community empowerment through capacity building related to specific technological knowledge of cheese-making wastes bioconversion or other residual reduction technology could be an alternative to decrease the threat of unsustainable cheese production.

5. Acknowledgments

Authors would like to thank the Student Research Group, KPBS, Vivi Fadilla Sari and Syarah Virgina who helped in the laboratory. This research is funded by the Directorate of Research and Community Services, Directorate General of Research Affirmation and Development, Ministry of Research, Technology and Higher Education of The Republic of Indonesia. This article's publication is partially supported by the United States Agency for International Development (USAID) through the Sustainable Higher Education Research Alliance (SHERA) Program for Universitas Indonesia's Scientific Modeling, Application, Research and Training for City-centered Innovation and Technology (SMART CITY) Project, Grant #AID-497-A-1600004, Sub Grant #IIE-00000078-UI-1. The research was funded by Indonesian Ministry of Research, Technology and Higher Education (Ristekdikti) under the scheme of University Basic Research Excellent 2019 Contract No. NKB-1636/UN2.R3.1/HKP.05.00/2019.

6. References

[1] Gifford R, Steg L and Reser J P 2011 Environmental Psychology *IAAP Handbook of Applied Psychology* ed P R Martin, F M Cheung, M C Knowles, M Kyrios, J B Overmier and J M Prieto (Oxford, UK: Wiley-Blackwell) pp 440–70

[2] Robinson L 2002 Public outrage public trust : a road map for public involvement in waste management decision-making Waste and Recycle 2002 Conference (Perth, Australia)

[3] Balia R L, Kurnani T B A and Utama G L 2018 Selection of Mozzarella Cheese Whey Native Yeasts with Ethanol and Glucose Tolerance Ability *International Journal on Advanced Science, Engineering and Information Technology* **8** 1091-1097–1097

[4] McDevitt J, Langer E and Leckie A 2013 Community Engagement and Environmental Life Cycle Assessment of Kaikōura's Biosolid Reuse Options *Sustainability* **5** 242–55

[5] Roostita R and Fleet G H 1999 Growth of Yeasts Isolated from Cheeses on Organic Acids in the Presence of Sodium Chloride *Food Technology and Biotechnology* **37** 73–9

[6] Zakus J D and Lysack C L 1998 Revisiting community participation *Health Policy Plan* **13** 1–12

[7] Tindana P O, Singh J A, Tracy C S, Upshur R E G, Daar A S, Singer P A, Frohlich J and Lavery J V 2007 Grand Challenges in Global Health: Community Engagement in Research in Developing Countries *PLoS Med* **4**

[8] Clarkson M B E 1995 A Stakeholder Framework for Analyzing and Evaluating Corporate Social Performance *The Academy of Management Review* **20** 92–117

[9] Jones T M 1983 An Integrating Framework for Research in Business and Society: A Step toward the Elusive Paradigm? *The Academy of Management Review* **8** 559–64

[10] Petrini M and Pozzebon M 2010 Integrating sustainability into business practices: learning from Brazilian firms *BAR - Brazilian Administration Review* **7** 362–78

[11] Suyanto B and Sutinah 2005 *Metode Penelitian Sosial [Social Research Methods]* (Jakarta: Kencana)

[12] Indri A, Marina S and Ali M 2015 Public Perceptions of Benefits and Negative Impact of Livestock Waste of Dairy Farm Cattle (Case in The Village of Rancamulya Sumedang West Java) *E-journal Unpad* **4** 1–14

[13] Winahyuningsih P 2010 Pengaruh Kepercayaan dan Kualitas Pelayanan terhadap Kepuasan Konsumen pada Hotel Griptha Kudus *Sosial dan Budaya* **3**

[14] Utama G L, Kurnani T B A, Sunardi S, Cahyandito F and Balia R L 2017 Joint Cost Allocation of Cheese-Making Wastes Bioconversions into Ethanol and Organic Liquid Fertilizer *Bulgarian Journal of Agricultural Science* **23** 1016–20
[15] Utama G L, Kurnani T B A, Sunardi and Balia R L 2017 Reducing cheese-making by-product disposal through ethanol fermentation and the utilization of distillery waste for fertilizer *International Journal of GEOMATE* **13** 103–7
[16] Tenriawaruwaty A, Rukmana D and Salman D 2013 Community Participation in Management of Water Resources in a Sustainable Regency in Bulukumba *E-journal Pascasarjana Unhas* 14
[17] Hermawan Y 2012 Hubungan Antara Tingkat Pendidikan dan Persepsi dengan Perilaku Ibu Rumah Tangga dalam Pemeliharaan Kebersihan Lingkungan *Bumi Lestari Journal of Environment* **5** 1–15

Liquefaction Potential Based on Geology and Geotechnical Data on Sanana Region, Sula Island Regency, North Maluku, Indonesia

U S Pajrin[1], A Mubarak[1], J P Basuki[1], Z Zakaria[2], R I Sophian[2], N Khoirullah[2]

[1] Faculty of Geological Engineering, Universitas Padjadjaran, Bandung, Indonesia
[2] Department of Applied Geology, Faculty of Geological Engineering, Universitas Padjadjaran, Bandung, Indonesia

E-mail: ulfiasari28@gmail.com

Abstract. Soil liquefaction is one of a major problem in infrastructure development. This disaster can occur on sandy soil and high seismicity region. Soil with liquefaction potential will affect as a liquid during earthquakes and has high potential to collapse after the earthquake. The purpose of this study is to identify liquefaction potential in Sanana, Sula Island, North Maluku where many infrastructure developments being carried out. The method used in this research is "simplified procedure" for evaluating liquefaction resistance of soils. This method considers geological and geotechnical characteristics based on Standard Penetration Test. The geological characteristics of Sanana region show that this region has liquefaction potential because it is composed of Holocene alluvium and located in high seismicity region. Furthermore, based on the geotechnical data from the SPT test, this region has liquefaction potential in the sand, silty sand and sandy silt with loose-medium density. The loose-medium properties affect a low seismic resistance to cyclic loads. This will lead to high liquefaction potential in this research area. Therefore, in order to prevent possible disaster caused by liquefaction, this problem needs to be considered in future infrastructure planning and development.

Keywords: Liquefaction, earthquake, SPT, sandy soil

1. Introduction

Liquefaction is one of the seismic hazards caused by earthquakes. During an earthquake, Soil with liquefaction potential will affect as a liquid during earthquakes and resulting in loss of the soil ability to support the building which could lead to building failure. The Sanana region located in Maluku, where this region is at the convergence of three main plates, causing this region to have very high tectonic and seismic complexity. This condition makes this region frequent earthquakes were the source of the earthquake trigger in this area came from the Sula-Sorong active fault activity with the largest history of the earthquake at magnitude 8.1 (depth of 20.0 km) on Sula Island on January 24, 1965 (Figure 1).

IOP Conf. Series: Earth and Environmental Science **396** (2019) 012038 doi:10.1088/1755-1315/396/1/012038

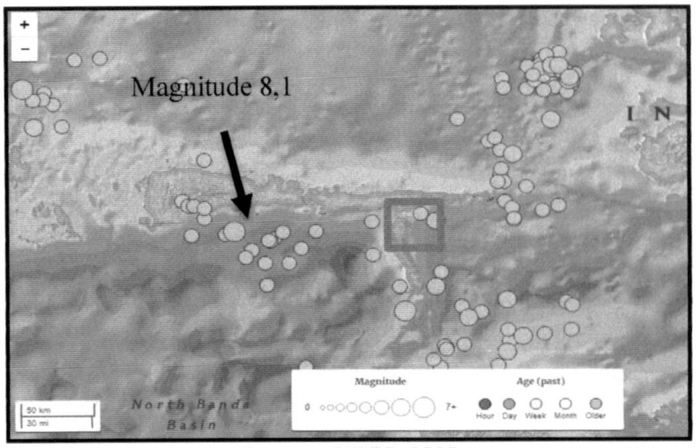

Source : USGS, 2019

Figure 1.
Tectonic Earthquake Research Area in 1919-2019

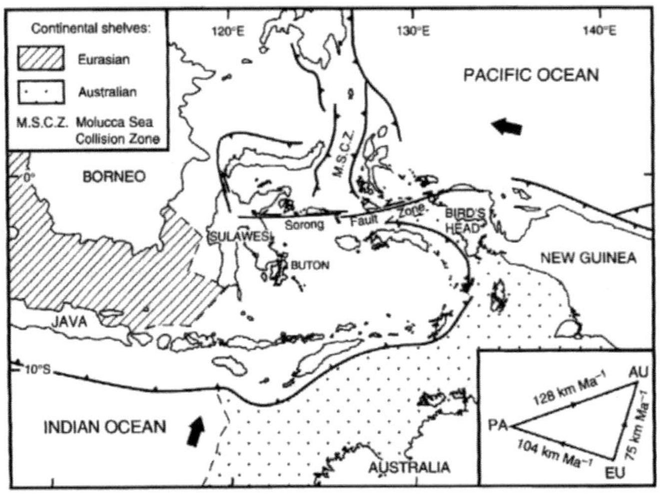

Figure 2. Regional tectonic setting the Sorong fault zone [1]

Sulabesi Island is the result of collision with a subduction system along the eastern boundary of the Sundanese Exposure which produces a tectonic setting of Eastern Indonesia [2,3].The Sorong Fault in the Sula Islands region is separated into two fault systems, each is the North Sula Fault and the South Sula Fault. Tilted movements the Sorong fault to the west along with the Indo-Australian plate movement.

The composition of the regional research is guided by the P.Tabuabu geological map and its surroundings [4,5]. In the Geological Map the Sanana Sheet, Maluku consists of 10 rock formations and 1 surface deposit, ranging from Carbon to Holocene [5].

In addition, In the Sanana Geological Map by [6] (Figure 3), along the coast of the island, the Sanana region consists of alluvium (Qa) which consists of sand, gravel, clusters, and clays where the age of material is Quaternary. The condition of the region which is composed of alluvium deposits such as sand and silt which is susceptible to liquefaction increases the possibility of liquefaction which causes soil failure.

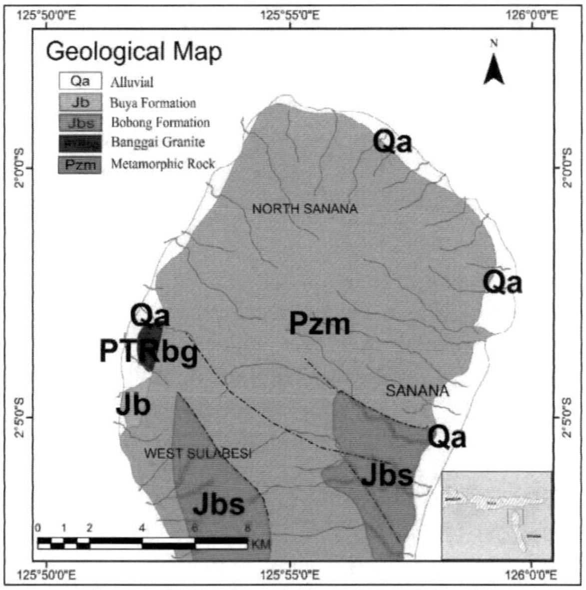

Figure 3. Regional Geological Map [6]

2. Research Methods

Contains research approach, research method, sampling, survey method, etc. The method used in this research is "simplified procedure" for evaluating liquefaction resistance of soils. This method considers geological and geotechnical characteristics based on Standard Penetration Test. The data needed in evaluating liquefaction potential in this study uses geological setting data and is correlated with evaluation of liquefaction resistance.

2.1. Liquefaction Potential Evaluation based on Geological Criteria

Preliminary studies of potential liquefaction were carried out by analyzing the geological conditions of the study area. Factors that cause ground failure susceptibility include sedimentation, sedimentation process, age of deposition, geological history, depth of water table, grain size distribution, density state, depth of burial, ground slope, and nearness of a free face [7]. The geological process will make a region have material with uniform grain distribution and make the material loose so that the soil layer has a high susceptibility to liquefaction. The presence and depth of the water table are one of the important parameters for liquefaction because this event will occur if the material is saturated with water. in addition, the age of the deposit also has an important effect, this is due to the process of lithification in the rock or soil.

Information on geomorphology units can be used as a database for potential liquefaction analysis. This geomorphology unit is categorized into 3, including likely, possible, and not likely of liquefaction (Table 1).

Table 1. Liquefaction Susceptibility of Various Geomorphological Units [8,9]

Rank	Geomorphological Units	Liquefaction potential
A	Present riverbed, old riverbed, swamp, reclaimed land, and interdune low	Liquefaction likely
B	Fan, natural levee, sand dune, flood plain, beach, and other plains	Liquefaction possible
C	Terrace, hill, and mountain	Liquefaction not likely

Table 2. Liquefaction Susceptibility of Various Geomorphological Units [9,10]

Age of Deposit	Depth of Water Table		
	0-3 m	3-10 m	10 m
Latest Holocene	High	Low	Nil
Earlier Holocene	Moderate	Low	Nil
Late Pleistocene	Low	Nil	Nil

2.2. *Evaluation of Liquefaction Potential Based on Geotechnical Data*

In evaluating liquefaction potential, Seed and Idriss developed a "simplified procedure" method in 1972 based on earthquakes that occurred in Alaska and Nigiita in 1964. The criteria for evaluating liquefaction resistance using the N-SPT value are the most commonly used and powerful methods for years.

Basically, analysis of liquefaction potential is looking for 2 variables: Cyclic Stress Ratio (CSR) which is cyclic stress that occurs due to an earthquake divided by effective vertical overburden stresses and Cyclic Resistance Ratio (CRR) is the soil resistance to resist liquefaction. The CRR value obtained from the field test results is one of them is the SPT. The method for evaluating liquefaction potential is by obtaining a safety factor value from the comparison of the CRR value which reflects the strength of the soil against the cyclic load which is usually caused by the earthquake load with CSR. The safety factor used should not be less than one, because if there is less than one liquefaction will occur [11,12]. CSR Value according to [11,12].

$$CSR = \frac{\tau_{av}}{\sigma' v} = 0{,}65\left(\frac{a_{max}}{g}\right)\left(\frac{\sigma_v}{\sigma_v'}\right) rd \quad (1)$$

Relationship between depth z and rd value according to [11,12]

$$\begin{aligned} rd &= 1 - 0{,}00765z & ;z < 9{,}15\ m \quad (2)\\ rd &= 1{,}174 - 0{,}0267z & ;9{,}15 < z < 23\ m\\ rd &= 0{,}744 - 0{,}008z & ;23 < z < 30\ m\\ rd &= 0{,}5 & ;z > 30\ m \end{aligned}$$

In evaluating the CRR value, this value can be calculated by the formula CRR7,5 from Rauch (1998) in [11] with the equation (3):

$$CRR_{7,5} = \left(\frac{1}{34-(N1)_{60}}\right) + \left(\frac{(N1)_{60}}{135}\right) + \left(\frac{50}{[(N1)_{60}+45]^2}\right) - \left(\frac{1}{200}\right) \quad (3)$$

Where the value (N1) 60 can be calculated by the equation Robertson (1998) in [11] provide a correction to get the value (N1)60 :

$$(N1)60 = N_M . C_N . C_E . C_B . C_R . C_S \quad (4)$$

This equation will be valid if (N1) 60 <30, because if (N1) 60 ≥ 30 then the material does not need to be evaluated, and it will be strong to withstand seismic loads because the soil is too dense to experience liquefaction.

To calculate the CRR with an earthquake magnitude other than 7.5, a correction factor called Magnitude Scale Factor (MSF) is needed. Seed (1983) in [11] gives an equation:

$$CRR_{MW} = CRR_{7,5} . MSF . K_\sigma . K_\alpha \quad (5)$$

Where Kα = 1 (assuming a flat surface) and Kσ must be ≤ 1, Pa value = 1atm ≈ 1kpa.

Based on the results of re-evaluation by [11] for the value of MSF for magnitude smaller than 7.5 and magnitude greater than 7.5, as follows:

$$MSF = \left(\frac{10^{2,24}}{M_w.\ 10^{2,56}}\right) \tag{6}$$

Then the calculation of the FS value with the SPT data from the corrected CRR is to use the following equation:

$$FS = \left(\frac{CRR_{Mw}}{CSR}\right) \tag{7}$$

3. Results and Discussion

3.1. Contains findings of the study and research discussion. Liquefaction Potential based on geological criteria

Based on the criteria of geological characteristics, the research area has the potential for liquefaction. The research area is composed of sandy clay, sandy silt, sand, clay sand, and silt, with the dominantly very loose-dense sand. The results of drilling techniques show that the subsurface stratigraphic conditions are Holocene alluvium, which is characterized by the repetition of a sand unit that is dominantly very fine to coarse sand grains, and there are clay and silt mixed with sand and gravel (Figure 4).

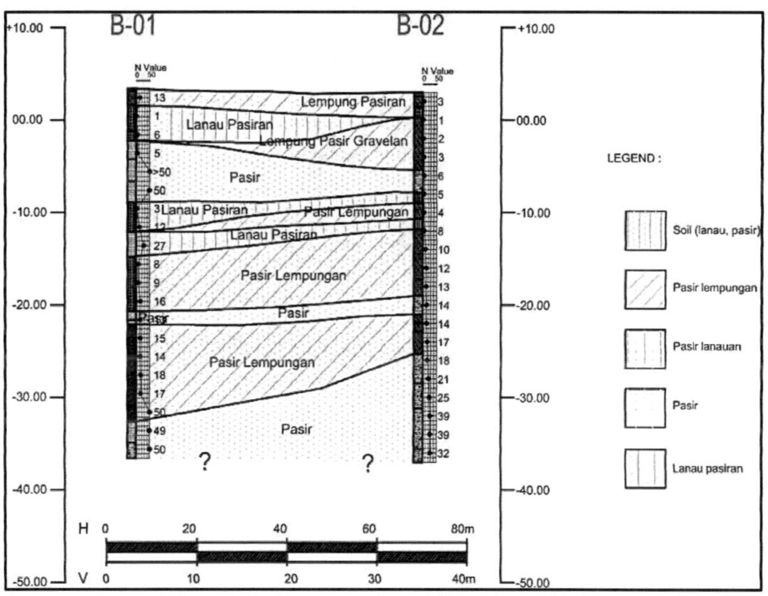

Figure 4. Cross section of B-01 and B-02

The repetition of layers of clay, silt, and sand shows that this area has been carried out repeatedly in the process of sedimentation, either caused by tectonic processes, tides or sedimentation processes during past deposition in this basin. This process is characterized by a fine to coarse sand deposition sequence with the continuity of silt and clay layers that interfingering in several drilling sites. In this study, the distribution of ground water levels was obtained during drilling. Hydrogeological conditions of the Holocene alluvium deposits show shallow depths of MAT from 1.5 to 2 m from the surface. Area research which is generally composed of alluvium deposits consisting of material a layer of fine-coarse sand, silt, and gravel functions as a layer of aquifer due to it permable while silt-clay layers are layers semipermeable. The distribution of layers of loose sand and shallow groundwater can cause liquefaction phenomena. (Figure 5)

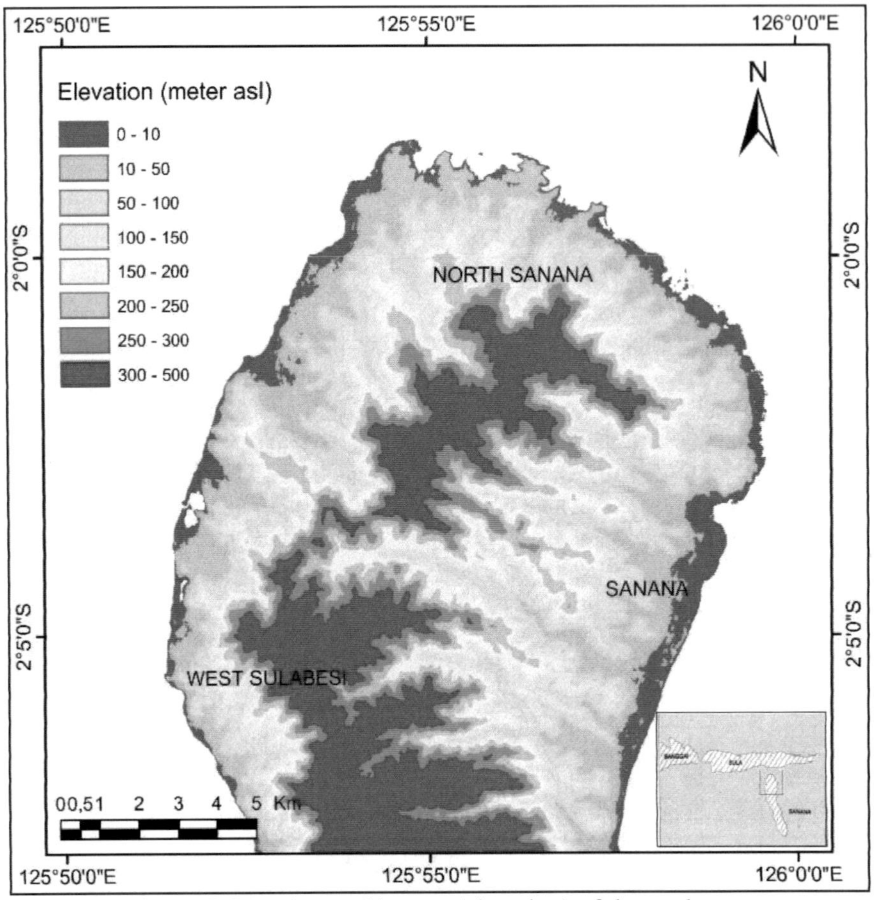

Figure 5. Morphographic map (elevation) of the study area

The results of the geomorphological analysis show that the study area is included in the lowland unit (altitude 0-50 meters above sea level) so that based on Table 1, the study area also has the potential for liquefaction.

3.2. Liquefaction Potential based on geological criteria Geotechnical data

Based on an evaluation of liquefaction potential evaluation from Seed and Idriss (1972) in [11], all SPT sites have liquefaction potential at depths of 3 to 13 m which consisted of silt and very fine-medium sand grains very loose-dense density. This layer has very loose density so that when there is a large earthquake shake, the acceleration of vibration will also be higher (Figure 6 and Figure 7).

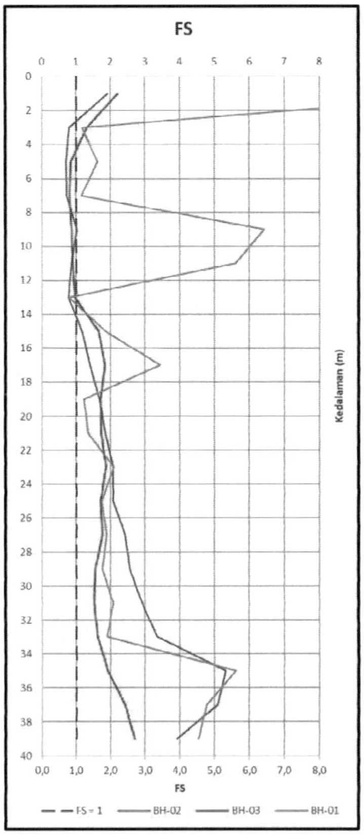

Figure 6. FS graph of BH-01, BH-02, and BH-03

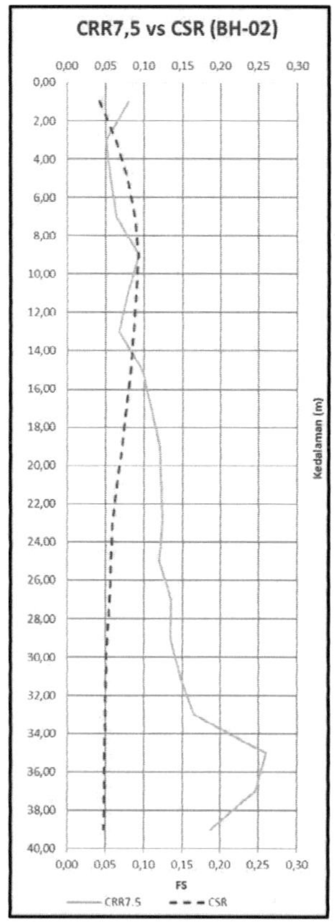

Figure 7. CRR7.5 vs. CSR graph from BH-02

On FK graph (Figure 6) showing FK values <1 indicate that at that depth the soil has the potential for liquefaction because it has a CSR value smaller than the CRR value. The CRR value is large, due to the value of small N-SPT with loose-medium density. The resistance to cyclic stress is shown in the CSR & CRR graph, where at the depth of liquefaction, the soil has a low seismic resistance to stress from the earthquake. FK> 1, indicating that at that depth the soil has a large cyclic resistance ratio compared to the cyclic stress ratio caused by earthquakes.

The results of this study indicate that the area based on the geological characteristics of several geological parameters, has the potential for liquefaction. the position of the sanana region which is in the active fault region of Sorong increases the potential for liquefaction, where in this region frequent earthquakes occur. in addition to that quantitatively using geotechnical data (SPT) with the simplified procedure method, the calculation results show that the area of sanana also has the potential for liquefaction.

Evaluation of liquefaction potential based on geological and geotechnical characteristics shows the consistency of the results. The results of qualitative analysis based on geological characteristics are preliminary studies in evaluating liquefaction potential, then further studies are conducted with quantitative analysis both through soil strength and seismicity that occur so that better and more accurate evaluations are obtained.

4. Conclusion

From the method described above, the results of the analysis of geological conditions, geotechnical conditions and laboratory results to estimate potential liquefaction, it can be concluded that the Sanana region has the potential for liquefaction. Geologically, the Holocene alluvium with low water table depth conditions and the location of the Sanana region in the South Sula-Sorong active fault causes this region to have potential liquefaction. Whereas in the technical geology of the SPT results, the depth of the subsurface layer that has liquefaction potential at depths of 3 to 13 m. Therefore, in order to prevent possible disaster caused by liquefaction, this problem needs to be considered in future infrastructure planning and development.

5. Acknowledgment

Thank you to Dr. Ir. Zufialdi Zakaria MT., And Raden Irvan Shopian S.T., MT., As the supervisors who have guided and provided many inputs in conducting research. In addition to PT. Kwarsa Hexagon which has provided the opportunity to conduct research. This article's publication is partially supported by the United States Agency for International Development (USAID) through the Sustainable Higher Education Research Alliance (SHERA) Program for Universitas Indonesia's Scientific Modeling, Application, Research and Training for City-centered Innovation and Technology (SMART CITY) Project, Grant #AID-497-A-1600004, Sub Grant #IIE-00000078-UI-1. The research was funded by Indonesian Ministry of Research, Technology and Higher Education (Ristekdikti) under the scheme of University Basic Research Excellent 2019 Contract No. NKB-1636/UN2.R3.1/HKP.05.00/2019.

6. References

[1] Charlton T R 1996 Correlation of the Salawati and Tomori Basins, eastern Indonesia: a constraint on left-lateral displacements of the Sorong fault zone *Geol. Soc. London, Spec. Publ.*

[2] Hamilton W 1979 Tectonics of the Indonesian Region. *Geol. Soc. Malaysia, Bull.*

[3] Silver E A and Moore J C 1978 The Molucca sea collision zone, Indonesia *J. Geophys. Res. Solid Earth* **83** 1681–91

[4] Pigram C J and Supandjono S J B 1985 Origin of the Sula platform, eastern Indonesia *Geology* **13** 246–8

[5] Surono and Surana 1993 *Peta Geologi Lembar Sanana, Maluku* (Bandung: Pusat Penelitian dan Pengembangan Geologi)

[6] Surono dan Surana 1993 *Peta Geologi Lembar Sanana, Maluku* (Bandung: Pusat Penelitian dan Pengembangan Geolofi)

[7] Youd T L and Perkins D 1978 Mapping Liquefaction-Induced Ground Failure Potential *J. Geotech. Eng. Div.* **104** 433–46

[8] Iwasaki T, Tokida K I, Tatsuoka F, Watanabe S, Yasuda S and Sato H 1982 Microzonation for soil liquefaction potential using simplified methods *Proceedings of the 3rd international conference on microzonation, Seattle* vol 3 pp 1310–30

[9] Puri N and Jain A 2014 Preliminary investigation for screening of liquefiable areas in Haryana state, India *ISET J. Earthq. Technol* **51** 19–34

[10] Obermeier S F 1996 Use of liquefaction-induced features for paleoseismic analysis—an overview of how seismic liquefaction features can be distinguished from other features and how their regional distribution and properties of source sediment can be used to infer the locatio *Eng. Geol.* **44** 1–76

[11] Youd T L and Idriss I M 2002 Liquefaction Resistance of Soils: Summary Report from the 1996 NCEER and 1998 NCEER/NSF Workshops on Evaluation of Liquefaction Resistance of Soils *J. Geotech. Geoenvironmental Eng.*

[12] Seed H B and Idriss I M 1971 Simplified procedure for evaluating soil liquefaction potential *J. Soil Mech. Found. Div*

The Impact of Go-food on Travel Behaviour for The Fulfillment of Consumption Needs in Higher Education Areas

O R Manullang, N S Limbong

Department of Urban and Regional Planning
Diponegoro University, Indonesia

E-mail: oktomanullang72@gmail.com

Abstract. This study examines the Go-food implications for changing the behavior of people's travels in fulfilling consumption needs, specifically for eating and drinking. This community travel behaviour is seen from two sides, namely: Go-food user request side and provider side (Go-food driver travel intensity). Quantitative methods were applied to analyse Go-food implications towards travel behaviour from provider and user. The results of this study explain that a high level of travel occurs at night and is hostile to the beginning of the driver's travel. It is influenced by the function of the trade area that is located on the surrounding road, especially when weekdays on Sirojudin Street (21%), while the weekend at Ngesrep Timur V Street (20%) and Jatimulyo Street (17%). In addition, the behaviour of the driver's travel goals is spread, which is when weekdays in the Bulusan area (30%), while in the weekend in the Pedalangan area (27%). Fulfillment of this consumption needs affects the change in Go-food user travel behaviour. Before there was Go-food, several factors were considered, namely: distance, time, cost, frequency, and reason for travel, whereas after Go-food, factors that are considered or have a relationship are: frequency factor and goal travel. This is because the Go-food service is used when the community prioritizes its time for productive matters. Moreover, the Go-food facility resulted in the Go-food user making a request to the place where he wanted to eat, although the distance was quite far. This is because cost and distance are not a problem for Go-food consumers.

Keywords: Go-Food, travel behaviour, and higher education

1. Introduction

The area of higher education is usually spread in big cities, one of which is in the city of Semarang. According to the policy of the City Area Section (BWK) contained in the Semarang City RTRW in 2011-2031 [1], the center of higher education in the Tembalang District is Diponegoro University (Undip), Semarang State Polytechnic (Polinnes) and Pandanaran University (Unpand), while in Banyumanik District it has one university, the Health Polytechnic of the Ministry of Health Semarang (Poltekkes). The term "tertiary education area" will be used in this study to describe the area around the four tertiary institutions.

The area of Higher Education is inhabited by the local community, but the majority is dominated by students from outside the City / Regency of Semarang. Judging from the number of students, Diponegoro University is a university that has the most students among the other four, namely around 43,095 students and followed by the Health Ministry of Health Polytechnic Semarang (Poltekkes) around 6,764 students, Semarang State Polytechnic (Polinnes) around 4,961 students, and Pandanaran University around 1,184 students [2]. The total number of students in the Higher Education Area is around 60,000 students who can experience population growth each year. Population growth that occurs can be known based on the number of new students who come from outside the City / Regency of Semarang. New students as migrants have a percentage of 61%, while new students coming from the City / Regency of Semarang are 39% [3]. The addition of new students who come from outside this area will certainly result in an increasing population growth rate each year.

The rate of population growth in this region can trigger an increase in meeting the needs of every individual who is heterogeneous. The community needs in question can be: clothing needs, shelter, consumption, and so forth. But the need that currently dominates is consumption needs with a percentage of business towards trade and services of 64% as the main work in the Higher Education Area [4]. In addition, the trigger for an increase in the fulfillment of consumption needs is seen from the people's habit of consuming 2 to 3 times a day.

The intensity of meeting the needs of sustainable consumption can indirectly result in an increase in the quantity of traffic travel. This is influenced by individual behavior in traveling to the place of consumption needed or referred to as travel behavior [5]. Some examples of trips are: from a place to stay to a traditional market, food stalls, restaurants, or other places. On the other hand, the travel behavior requires transportation facilities, such as walking, private motorbikes, public transportation, and so on. According to data from the Central Statistics Agency (2010) [6] the Higher Education Region has always experienced an increase in motorcycle vehicles each year with an average increase of 289 units of motorcycles per year. When viewed from these data, transportation is highly developed and widely used by the public, namely motorbikes. Therefore, people's travel behavior in meeting consumption needs can have an impact on the trip generation and changes in land use, especially by using private vehicles (motorbikes) [7].

However, at present the fulfillment of community needs has been influenced by the rapid development of technology through online transportation services in the form of Go-jek, Grab and UBER [8]. Began to shift the choice of modes of public transportation to online transportation caused by the comfort and safety obtained from online transportation. Another advantage possessed by online transportation is that it offers convenience for users in ordering via smarthpone without having to come to a motorcycle taxi station [9]. One of the online transportation used through smartphones is the Go-jek application.

The Go-jek application established by Nadiem Makarim ranks first as the most popular online transportation service in Indonesia [9]. This service is also one of the transportation facilities used by the community in the Higher Education area, especially for student migrants from outside the area to facilitate travel. Besides functioning to pick up passengers, the Go-jek service also provides delivery services (Go-food) that function to deliver food / drinks in accordance with the needs of public consumption [10].

Nadiem Makarim said that the message service between Go-food has grown to become the largest consumption service center in the world after China [11]. This development began to emerge during April 2015 with the increasing coverage of Go-food services. In addition, the Go-food service has reached food vendors and food stalls on the outskirts of 50 regencies and cities in Indonesia [11]. This gives benefits to the community because most of the consumption places in this region already exist in Go-food services, so that people can meet their consumption needs easily and varied without traveling. On the other hand, these benefits can also be felt by the community through work as a Go-food driver to increase the income of the people in the Higher Education Area. This is an important reason for researchers to find out about people's travel behavior towards Go-food services in the Higher Education Area.

The existence of this Go-food service is thought to bring changes to people's travel behavior in terms of meeting consumption needs. Travel behavior is measured in several factors, namely: distance traveled (travel distance), travel time (travel time), travel costs (travel cost), travel frequency (travel frequency), and mode selection (mode-choice) [12]. Changes in travel behavior are seen based on two different sides, namely from the demand side of Go-food users and the provider side through Go-food drivers. This is a comparison of travel behavior, both before and after the Go-food service in meeting their daily consumption needs.

Viewed from the side of the Go-food driver, the behavior of the trip to pick up and order has a relationship between spaces from one location to another on an ongoing basis. This is thought to create erratic travel patterns because it adjusts the demand for Go-food users who place orders through the Go-food service. The driver's journey occurs continuously, which is more than 2 or 3 times a trip in 1 day, it is suspected to cause traffic jams. The impact of congestion is usually seen at rush hour or travel time that has a high frequency of travel simultaneously. Then there is the concentration of origin of trip generation in a zone and the imposition of traffic in meeting the consumption needs in the Higher Education Area [13].

Viewed from the perspective of Go-food users, the intended travel behavior aims to look at the supporting factors for changes in people's behavior in meeting consumption needs. In addition to the availability of Go-food service factors, it is also necessary to know how factors change travel behavior towards trip generation generated in the Higher Education Area. This can provide 2 (two) possibilities, namely: the reduction in trip generation or increasing trip generation, depending on the intensity of the trip caused by distance [14]. The statement can be interpreted as the implication of Go-food on people's travel behavior.

Implications of Go-food as an online service can affect lifestyle (life style) or different behaviors [15]. Therefore, it is necessary to analyze how changes in travel behavior that occur after the Go-food service. The behavior of the trip is difficult to know if it does not first pay attention to what factors and impacts that are affected by the existence of Go-food on consumption needs. Based on the description above, this research is important to be carried out to determine changes in travel behavior as an implication of Go-food in meeting consumption needs in the Higher Education Area.

2. Research Methods

The research method used is a quantitative method with descriptive analysis tools and crosstabs analysis. The purpose of this method is to identify the implications of Go-food from the demand and supply side which include: social, economic, demographic conditions, and travel behavior of Go-food users. This study also uses primary collection techniques (questionnaires & observations) and secondary data collection techniques (journals, books, internet sources, articles, etc.). The data collection is divided into 3 times, namely at breakfast, lunch, and dinner.

Data collection method is a very important component in conducting research because data collection determines the success or failure of a study [16]. Data collection techniques used in this study are distinguished based on the source, namely primary data collection techniques and secondary data collection techniques. The data acquisition of this study was conducted for Go-food drivers (supply side) and Go-food users (demand side). The technique used in sampling for Go-food drivers in this study is the non-probability sampling technique using the accidental sampling method. Accidental sampling method is a responsive sampling technique based on coincidences, so anyone who happens to meet with researchers can be used as a response when viewed according to data requirements. Sampling locations are spread evenly within a radius of about 3 kilometers from the Higher Education Area, ie Go-food service users who have data on their travel behavior that they do every day or several days as long as they use the application. The number of Go-food drivers in the Higher Education Region is unknown, so sampling can be calculated using the Lemeshow formula, a total of = 96 ≈ 100 samples is obtained. The sample distribution of Go-food drivers can be seen in detail in the Table 1:

Table 1 Go-food Driver Sampling

Days	Go Food drivers	Quantity of Go Food drivers trips	Category
Monday	15 drivers	88 trips	Weekdays
Tuesday	15 drivers	89 trips	
Wednesday	15 drivers	98 trips	
Thursday	15 drivers	85 trips	
Friday	15 drivers	98 trips	
Saturday	15 drivers	102 trips	Weekend
Sunday	15 drivers	94 trips	
TOTAL	**105** drivers	**654 trips**	

Source: Researcher's Observation Results, 2019

The technique in sampling for Go-food users will be known after observing the results of the recording of Go-food drivers as many as 105 respondents of Go-food drivers. The resulting observations were 654 trips. The population obtained was subject to research using the sampling technique with the Slovin formula. Slovin formula is a formula to calculate the minimum number of samples from a population that is obtained to be smaller but can represent the population as a whole, which is equal to 100 samples of Go-food users. The data to be studied was carried out by a random sampling method that is taking several populations randomly but having the same criteria namely users of Go-food services.

This research has an important objective to learn how to use the large variables that are the main focus of this analysis. This can reflect travel characteristics and land use characteristics that conflict with travel [17]. resulting from Go-food services. By understanding and analyzing these changes, it is hoped that they can provide findings that are consistent with the theory and reality currently being developed, especially in the Higher Education Area.

3. Travel Behaviour From Provider Side (Go-Food Driver Travel)

The existence of Go-jek especially who serves Go-food during weekdays and weekends has a difference. This is because there are different patterns of activity on weekdays and weekends, and can affect the behavior of Go-food driver trips [18]. Therefore, this study will identify the distribution based on two (2) different times, namely working days (Monday to Friday) and weekends (Saturday and Sunday).

3.1. During Weekdays

Go-Food drivers start their travel from the consumption place listed on the Go-food service as per booking. Usually the driver who get the Go-food booking is determined based on the closest distance to the place of consumption needed by the Go-food user community. Therefore, Go-food drivers have a tendency to wait for reservations in places that are close to the restaurants or other consumption places that are already registered on the Go-food service (Figure 1).

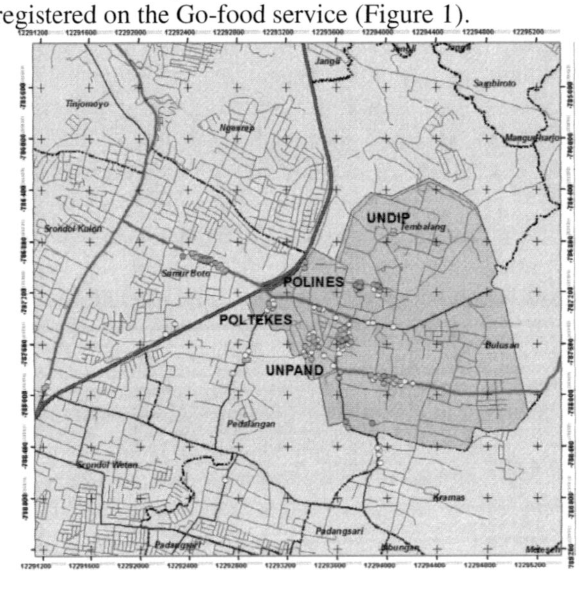

Figure 1. Early distribution map of driver's travel at Weekdays

The distribution of driver travels during weekdays tends to focus on the corridor Sirojudin Street, Ngesrep Timur V Street, Jatimulyo Street, and Banjarsari Street. This centralized distribution pattern occurs because of the large quantity of Go-food driver travels in certain locations (Figure 2).

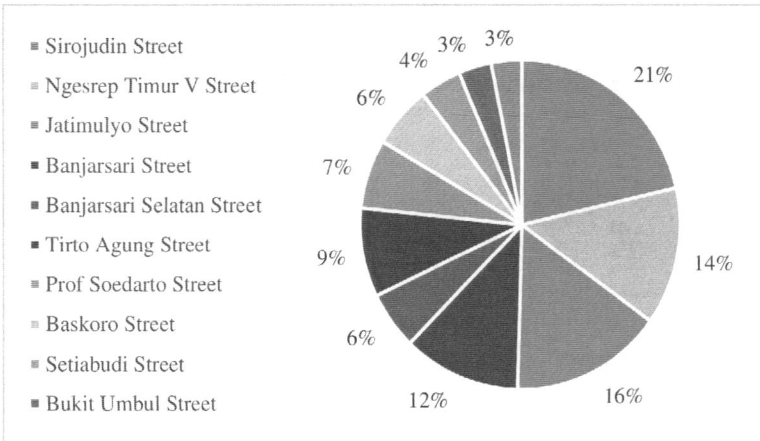

Figure 2. Percentage of Initial Driver Travel During Weekdays

The initial distribution of the travel by the Go-food driver was dominated by the corridor Sirojudin Street is 21% of the total. Location points that have high concentrations include: thuk-thuk thai tea, ternate yaya yellow rice, abahabu kebab, and pong pera-pera-warung. These places have been registered with the Go-food service and have a higher customer quantity compared to other places to eat on weekdays (Figure 3).

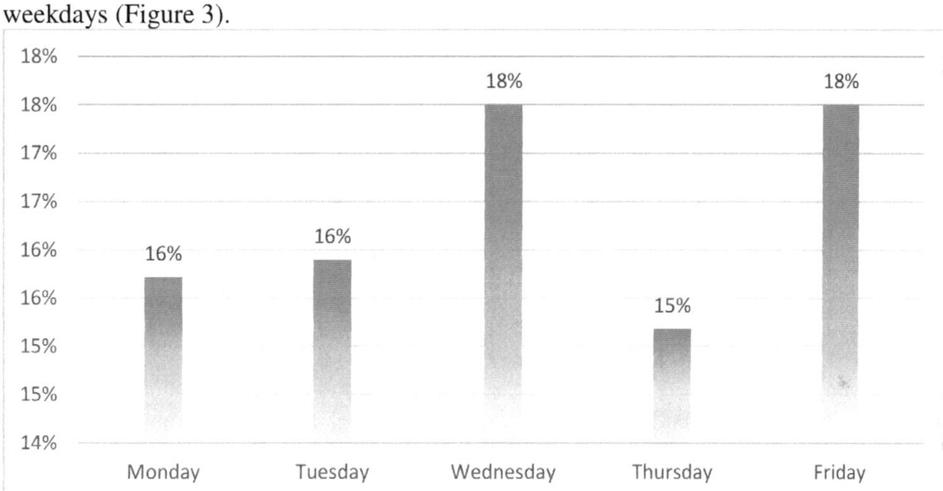

Figure 3. Go-food Driver Travels on Weekdays

The highest frequency is on Wednesday and Friday, compared to other days. Most people in the Higher Education Area do many activities during weekdays, so there is a lack of time to cook on their own. After completing their activities, people prefer to fulfill their food needs without having to cook. The higher the demand for Go-food users, the higher the travel by the driver. The frequency of this travel will increase the number of drivers' movements, a minimum of more than 2 or 3 travels in 1 day. Increased driver travel can contribute to congestion when passing certain road sections at the same time. This congestion is also influenced by the place of eating that is most in demand by Go-food users. Some places of consumption needed during weekdays can be known through Figure 4:

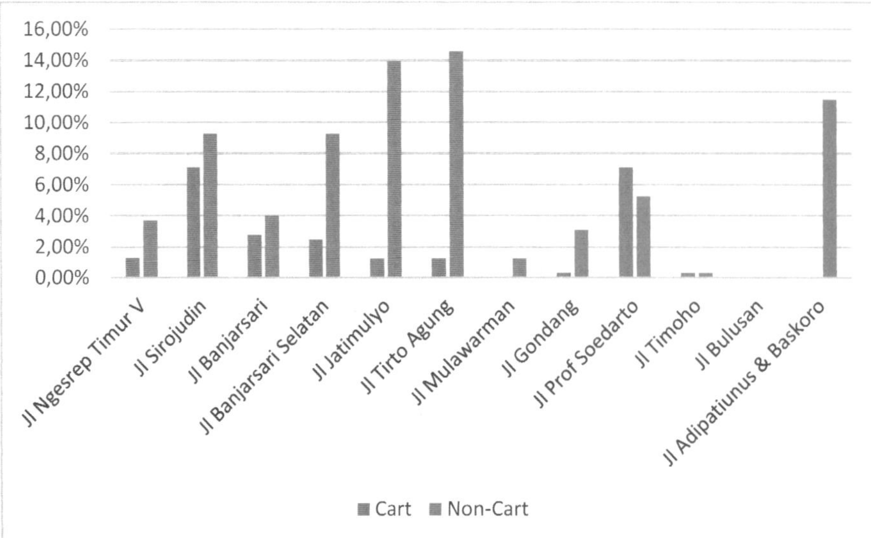

Figure 4. Frequency of Driver's Travel Based on Consumption Place on Weekdays

Based on Figure 4 above, it can be explained that the dominant driver travels to a place to eat on Jl. Tirto Agung, compared to eating places that use carts. Places to eat that have buildings certainly require parking for Go-food drivers, but if seen in the field, parking spaces are only provided by a few eating places.

3,2. During Weekend

Based on the Figure 5, map below, the distribution of origin of driver travels at weekends tends to focus on certain corridors. This centralized distribution pattern occurs because the quantity of Go-food drivers is determined according to the place to eat or drink needed by Go-food users. Every travel made by the Go-food driver depends on the request of Go-food users who are centered on a place to eat or drink .

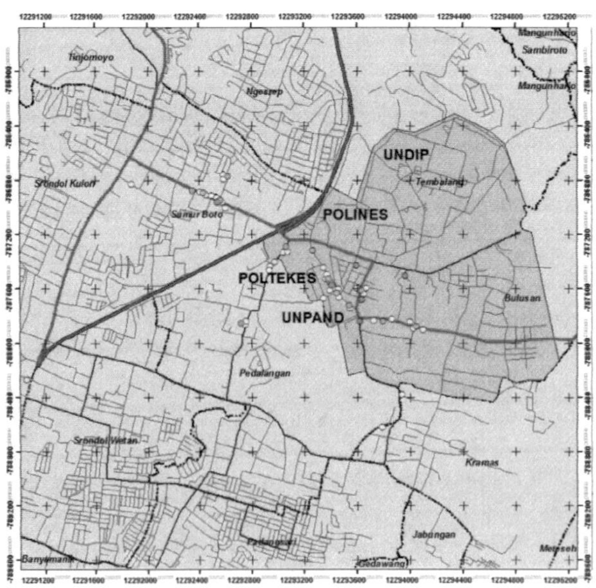

Figure 5. Map of the distribution of the origin of the driver's travel during the Weekend

The following can be seen in Fifure 6 more detail the initial quantity of driver travels based on the corridors in the Higher Education Area:

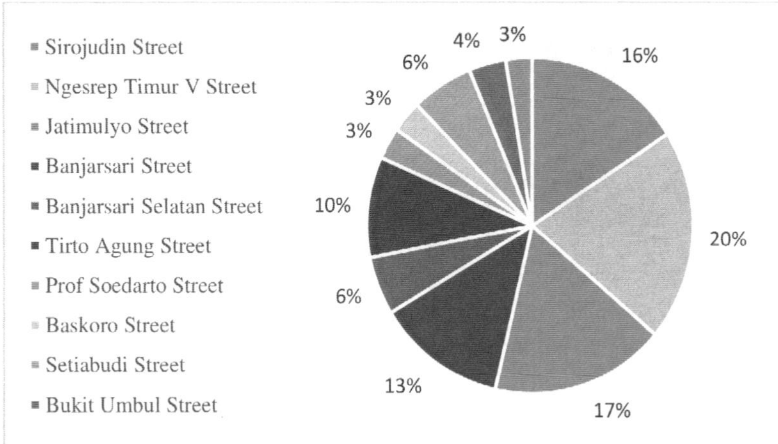

Figure 6. Percentage of Origin of Driver's Travel during Weekend

Based on the Figure 6, it can be seen that the driver's travel during the weekend tends to center on the corridor Jatimulyo Street (17%) and Ngesrep Timur V Street (20%). This centralized distribution pattern occurs because travel generation with high concentrations is in certain locations carried out by Go-food drivers. The location is determined according to where food is needed by Go-food users. From the results of the field survey, along the Jatimulyo Street and Ngesrep Timur V Street is dominated by eating places that are already registered in the Go-food service. The places to eat are: panties pizza, Surabaya noodle hut "cak dul", successful depots, buto express, house of moo, ragil stalls and gemes shops. These eating places are already subscribed to Go-food and have a high quantity of customers on weekends.

The frequency of Go-food driver travels is higher on Saturdays, this can be seen within 2 days on weekends via the Figure 7:

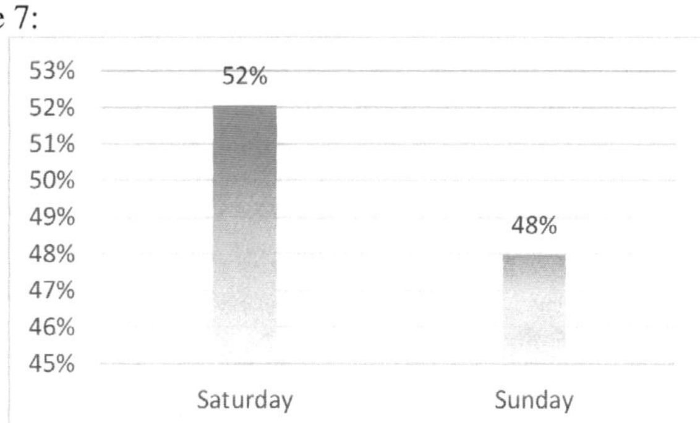

Figure 7. Driver Travel Frequency Go-food at Weekend (Per-day)

Based on the results of the diagram above, the frequency of Go-food driver travels occurs every day in a fluctuating manner. This can be seen from the number of Saturday and Sunday travels that are not much different, namely 67 travels and 63 travels. This is due to the influence of activity behavior, where at the weekend most people do not carry out fulltime work activities and use Go-food to facilitate the fulfillment of their food consumption needs. Thus, the driver will travel to the dining place according to the order of the Go-food user community.

The intended place to eat is a place that has been served by Go-food, both those who have built land for eating places and those who do not have land built (using a cart). Some of the places to eat needed at weekends can be found through the Figure 8:

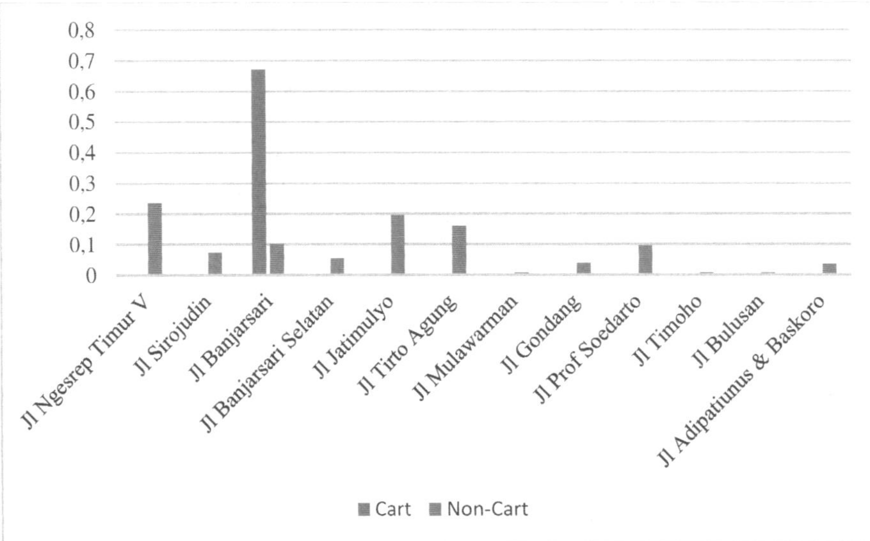

Figure 8. Frequency of Driver Travel by Weekend Consumption Place

Based on the Figure 8, it can be explained that the dominant driver travels to a place of consumption that has built up land on Ngesrep Timur V Street, compared to consumption places that use carts. Consumption sites that have built-up land certainly require parking spaces for Go-food drivers, but if seen in the field, parking lots are only provided by a few places of consumption. This can be one of the considerations related to the provision of parking space for consumption places, so that the arrangement of the space for consumption can be arranged in such a way, especially in places of consumption that have high frequency of demand. The frequency of these requests can be seen based on mealtimes used by the community, as shown in Figure 9:

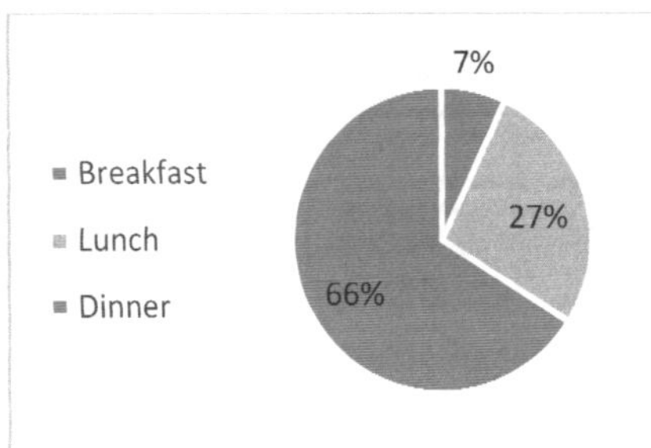

Figure 9. Go-food driver travels during weekends based on mealtimes

Based on the diagram above, it is known that out of 196 most travels were made during dinner as many as 130 movements (66%), while lunch time was 52 travels (27%), and breakfast time was 14 travels (7%). Comparison of this travel explains that the driver's breakfast time is less than the others. The high and low travel done depends on the number of requests for Go-food users who place an order. Go-food driver travels are dominated at night, so that it can cause the concentration of travel in certain corridors at the same time. The distribution of Go-food driver trips is dominated at night, so that it can cause the concentration of travel in certain corridors at the same time.

The distribution of trips over different periods of time has different identification results. This can be seen from the quantity of Go-food drivers' distribution in traveling for 1 week. Therefore, it needs to be seen how the travel distribution spreads during weekdays and weekends in the Table 2:

IOP Conf. Series: Earth and Environmental Science **396** (2019) 012039 doi:10.1088/1755-1315/396/1/012039

Table 2
Comparison of Travel Spreads During Weekdays and Weekend

Paramater	Travel Spreads During weekdays	Travel distribution on weekends	Analysis
Early travel distribution patterns	Centralized distribution pattern with the highest quantity is in the corridor Jl. KH. Sirojudin by 21%.	The distribution pattern is concentrated with the highest quantity in the corridor Jl.Jatimulyo (17%) and Jl.Ngesrep Timur V (20%).	The centralized distribution pattern occurs because the location of consumption places is concentrated in one particular location, with the highest quantity being on Jl. Sirojudin, Jl. Ngesrep Timur V, and Jl. Jatimulyo. The higher the quantity of Go-food driver trips, the higher the density of traffic passing through these roads. This can have an impact on high traffic generation in the corridor.
Distribution pattern of travel destinations	The pattern of travel spread is dominated by the Bulusan area by 30% of the total.	The pattern of travel spread is dominated by the Pedalangan area (27%) which is inhabited by the local community.	Demand for transportation services for passengers is inelastic, whereas demand for transportation services for the transportation of goods is elastic [19]. This statement occurs in meeting the consumption needs of the community through Go-food services that are elastic, in accordance with the needs of each individual. Go-food user dwellings in this region are scattered randomly according to land use that is urban sprawl. This resulted in the point of awakening of the Go-food driver's destination travel spread. Therefore, the availability of Go-food services is an efficient solution in meeting its consumption needs.

4. Changes in Travel Behavior from the Side of Go-food Users

Changes in travel behavior can be seen from the Go-food user profile and its characteristics in 2 (two) different time periods, namely before and after the existence of Go-food.

4.1 Go-food User Profile

Go-food user profile intended in this study is a general description of Go-food users viewed from several variables, namely: distance traveled, travel time, travel costs, travel frequency, reasons for travel, and destination of the travel. Each of these variables will be seen based on 2 (two) different time periods, namely before and after the Go-food service in the Higher Education Area.

a. Travel Distance

The intended travel distance is the distance from the place of origin to the place where the community intended to eat before Go-food. This study will look at three (3) categories of distance, which are near (<500 m), moderate (500 m - 1,000 m) and far (> 1,000 m). Changes in travel behavior based on distance can be seen on the Figure 10:

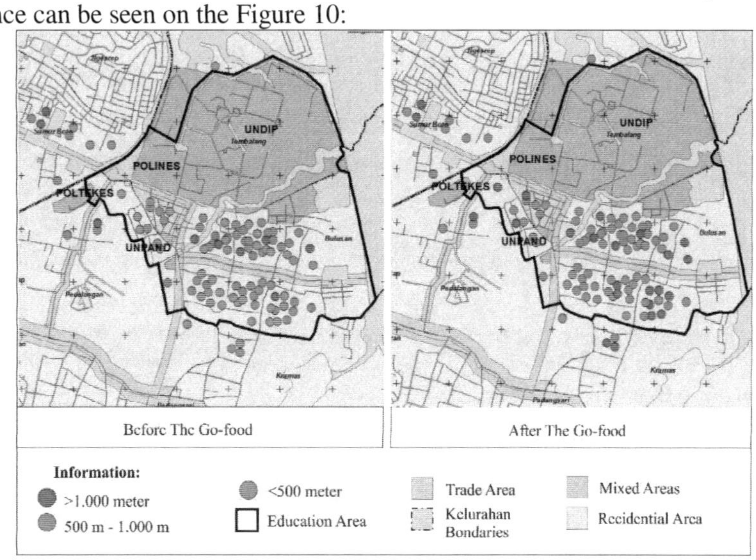

Figure 10. Map of distance traveled before and after the Go-food

Based on the map above, the travel distance of people who dominate their food needs is around 500 m - 1,000 m, both before and after the Go-food. The results of the percentage comparison of the distance traveled can be known more clearly through the Figure 11:

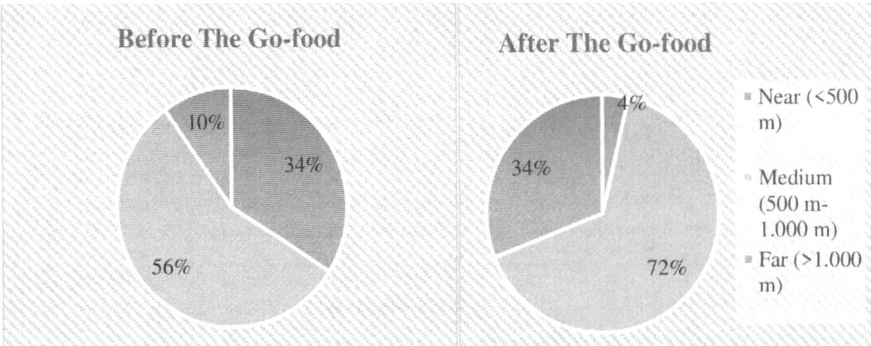

Figure 11. Percentage of distance before and after Go-food

Distance is related to the location of meeting food needs, including food needs that are carried out with Go-food services. Almost all interactions that require travel generate traffic flow, but now these interactions are facilitated by Go-food services. Based on the results of the percentage of distance obtained, there is a change, namely the increase in the distance traveled from before (56%) to after Go-food (72%). This was caused by the ease of Go-food service in reaching places to eat, so that more dominant people chose places to eat as they wished with more distance traveled than before. Therefore, it can be concluded that technology has a major impact on people's travel behavior after Go-food, namely the increasing distance of food ordering done by Go-food users.

b. Travel Time

The travel time of the intended travel is the travel time taken from the place of origin to the place of eating intended by the community before the Go-food. The travel time variable is divided into three (3) categories, namely fast (<15 minutes), medium (15 minutes - 30 minutes) and far (> 30 minutes). Travel behavior based on travel time can be seen on the Figure 12:

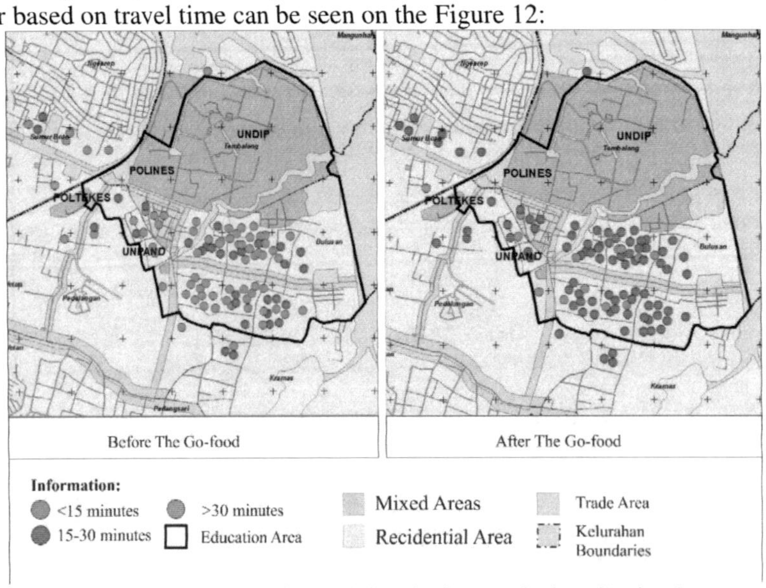

Figure 12. Map of travel time before and after Go-food

The travel behavior map above explains that there are changes in the travel time of the community before and after the Go-food. The travel time before Go-food is dominated by <15 minutes, then changes after Go-food is dominated by about 15-30 minutes to meet its consumption needs. The results of the comparison of before and after travel time can be known more clearly through the Figre 13:

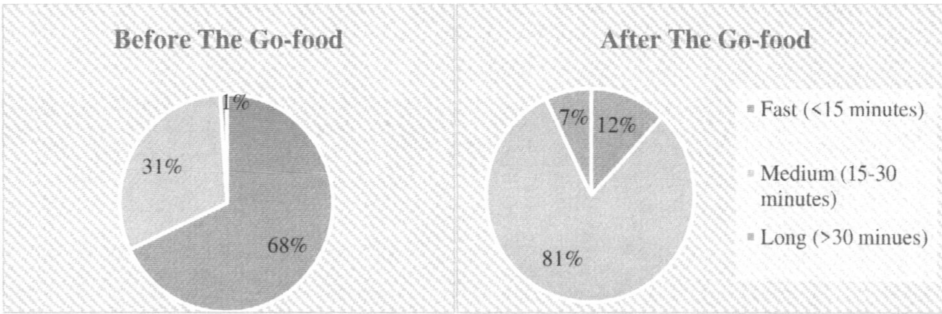

Figure 13. Percentage of travel time before and after Go-food

Travel time is closely related to distance traveled, the further the distance traveled, the greater the required travel time. Based on the results of the study, there was a change in the increase in travel time from fast (<15 minutes) to moderate (15-30 minutes). The travel time needed after the Go-food is longer because the distance to go is further by increasing the waiting time of the driver at the food place. The behavior of the travel can explain that a longer travel time is not an obstacle for Go-food users to meet their food needs.

c. Travel Cost

The intended travel costs are the transportation costs incurred to meet their food needs before the Go-food. Transportation costs are divided into three (3) categories, namely cheap (<Rp. 6,000), medium (Rp. 6,000 - Rp. 12,000) and expensive (> Rp. 12,000). Travel behavior based on travel costs can be seen on the Figure 14:

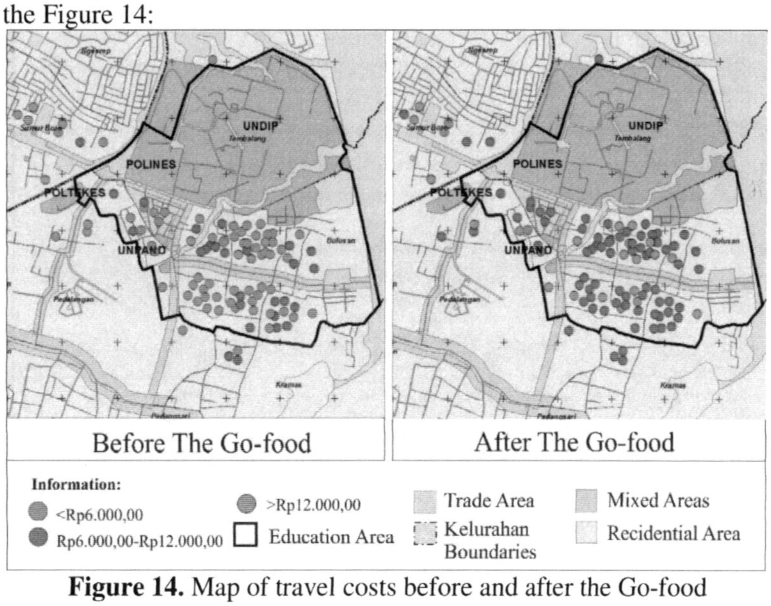

Figure 14. Map of travel costs before and after the Go-food

Based on Figure 14, the travel costs of community travels change between before and after the Go-food. The change in behavior is that before Go-food is dominated by costs of <Rp6,000.00, while after the existence of Go-food, it costs Rp6,000.00 - Rp12,000.00 for travels to meet people's food needs. The results of the comparison of the cost of this travel can be known in more detail through the Figure 15:

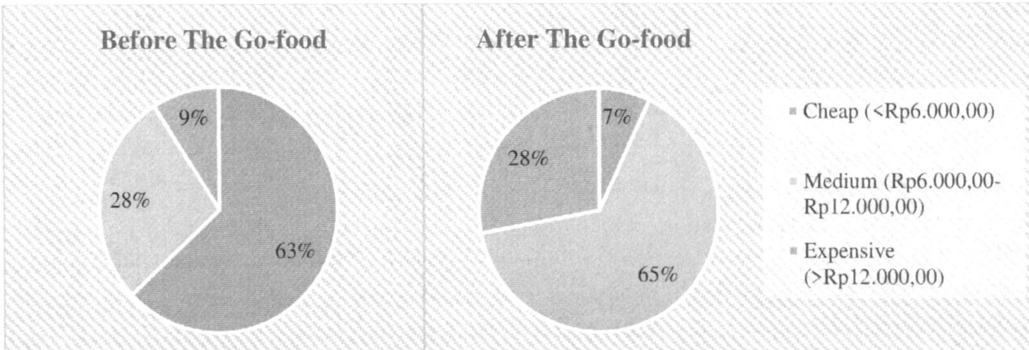

Figure 15. Percentage of user travel costs before and after Go-food

The percentage change in travel costs before and after Go-food has increased, from cheap (<Rp6,000.00) to moderate (Rp6.000.00 - Rp12,000.00). This also affects the increase in distance and travel time after Go-food. Therefore, the costs incurred in accordance with the distance of ordering food needed by Go-food users, as well as additional transportation service costs for Go-food drivers.

d. Travel Frequency

The intended travel frequency is the number of travels taken by the community in meeting their food needs before and after the existence of Go-food services. Travel frequency is divided into three (3) categories, which are rare (<5 times in 1 week), moderate (5-10 times in 1 week) and high (> 10 times in 1 week). Travel behavior based on travel frequency can be seen on the Figure 16:

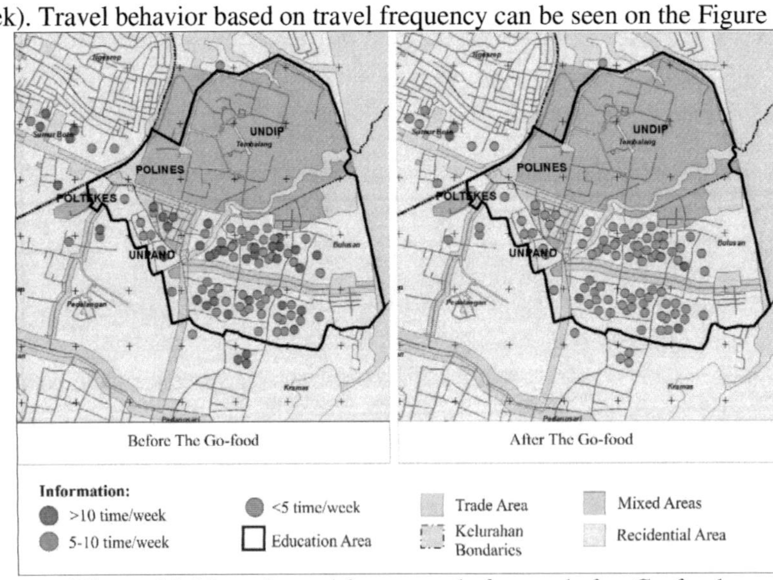

Figure 16. Map of travel frequency before and after Go-food

The map before and after the Go-food above explains that the frequency of people's travels in meeting their food needs is almost balanced from the 3 (three) categories, whereas after the existence of Go-food tends to be dominated by people who do frequency as much as <5 times / week (rarely) The results of the frequency comparison of travels can be known in more detail through the Figure 17:

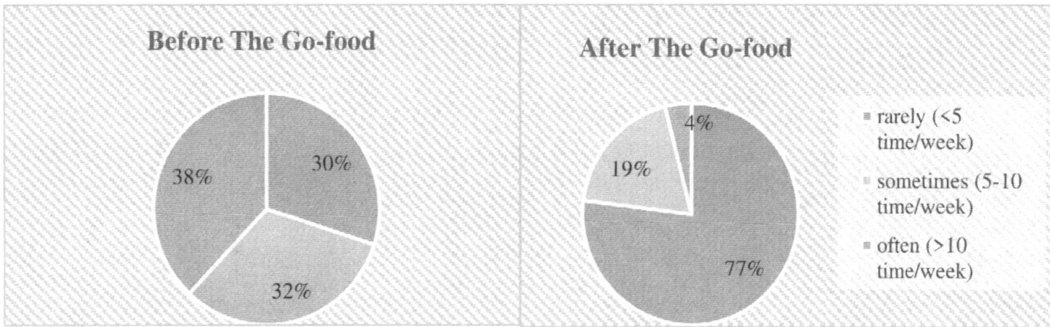

Figure 17. Percentage of travel frequency before and after Go-food

The frequency of travels that are seen to have a change that decreases, i.e. from frequent (> 10 times / week) to rarely (<5 times / week). This is consistent with what was said in Engel's law, namely: if the income of a person rises, then that person will proportionally reduce expenses to obtain daily necessities and replace them with more luxurious or secondary goods.

e. Travel reasons

The reason for the intended travel is the reason for the Go-food user community to travel to the intended dining place to meet their food needs before the Go-food. The reason for this travel is divided into five (5) categories, namely: (1) self-cooking, (2) buying near boarding, (3) bored with food near boarding, (4) amount of busyness, and (5) distance of eating place far away. To find out the user's profile from the reason for the travel, it can be seen on the Figure 18:

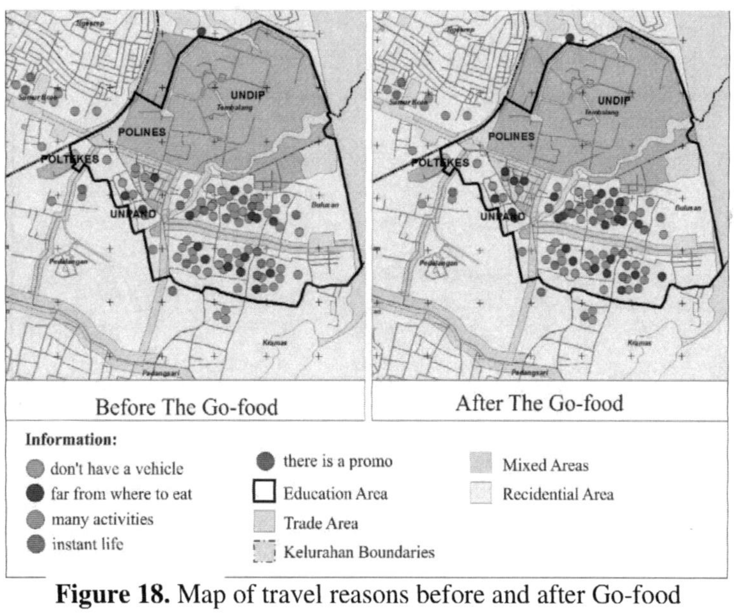

Figure 18. Map of travel reasons before and after Go-food

Based on **Figure 18,** the reasons for community travel before and after Go-food have more reasons for activities than other reasons for meeting food needs. The results of the percentage comparison of the reasons for this travel can be known in more detail through the Figure 19:

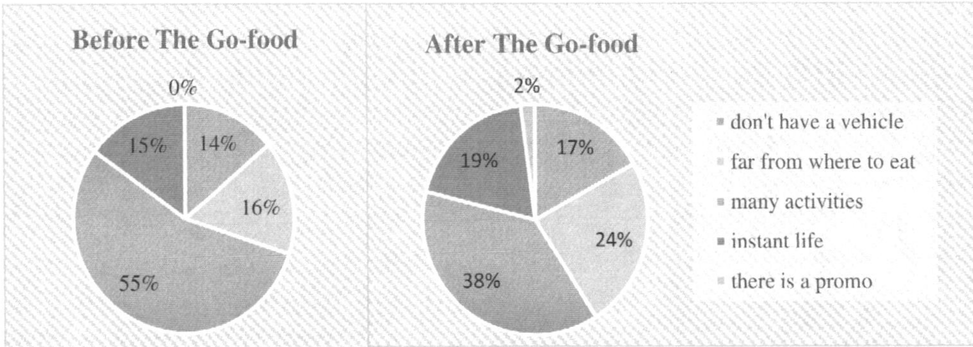

Figure 19. Percentage of travel reasons before and after Go-food

The reason for the travel depends on the behavior of the community, including the behavior of daily activities. Changes in travel reasons that dominated before and after the Go-food were the same but had a difference in percentage values. The percentage of travel reasons that dominate is due to the large number of activities. In addition, the reason for the increased travel is that it is far from where to eat (24%) compared to others. This is related to the increasing travel distance after Go-food to facilitate the fulfillment of food needs, even though the eating place is far from where Go-food users live.

f. Travel Destinations

The purpose of the travel in this study is a place to eat that is often visited by people to meet their food needs before the Go-food. The purpose of the travel is divided into five (5) categories, namely: namely: the nearest market, the nearest food stall, an affordable place, depending on the mood and the place to eat far enough. Travel behavior before Go-food based on the purpose of the travel can be seen on the Figure 20:

Figure 20. Map of travel destinations before and after Go-food

The travel behavior map above explains that the travel destination of the community before and after the Go-food has changed. These changes can be seen from the distribution before the Go-food is dominated by the destination to the nearest place, while after the Go-food is dominated by the destination to a quite far place. The results of the percentage comparison of the purpose of this travel can be seen more clearly through the Figure 21:

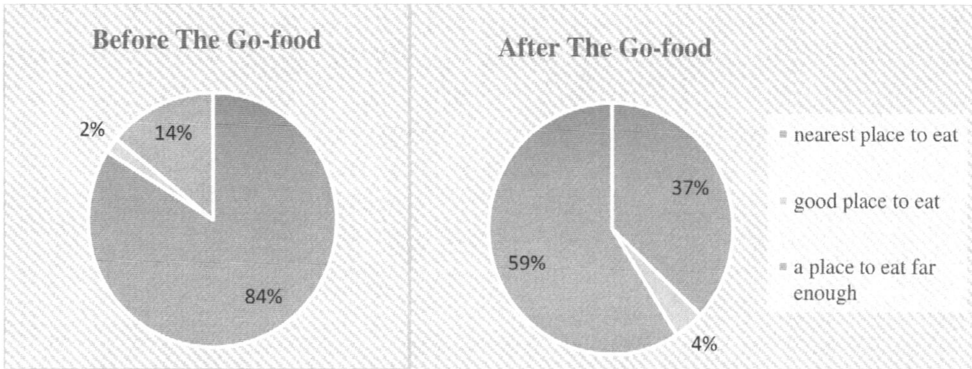

Figure 21. Percentage of user travel destinations before and after Go-food

The purpose of the travel obtained is related to the ease of Go-food service in meeting food needs with distance and near distance. When viewed from the purpose of the travel before and after the travel, there are differences, namely: from the nearest eating place to a place to eat far enough. It can be concluded that the purpose of Go-food users' requests is dominated by places far enough from where they live.

Travel behavior that experiences different changes and tends to influence other factors, is the travel distance factor. Before Go-food, users tended to travel at a distance that was close to where they lived, so that it would also influence the cheap cost factor of travel and time factors that tend to be faster than when using Go-food services. Go-food users are dominated by demand with a distance that is far enough from the place to eat, so that it has an influence on the travel time is quite long and the cost of travel is more expensive than before the Go-food service.

4.2. Characteristics of Travel Behavior

The behavior characteristics of this travel are divided into two (2) different times, namely before and after the Go-food. The aim is to see how changes in the behavior of Go-food users travel, whether to provide significant or reverse changes to the use of Go-food services. This analysis will look at some of the effects of the research variables determined based on the theories from Gliebe and Koppelman in Ettema, et al. (2006) [20] , explains that travel behavior is influenced by social, economic and demographic conditions.

a. Before the Existence of Go-food

Some data obtained to analyze the characteristics of travel behavior before Go-food will be explained in more detail through the results of the crosstab analysis that can be seen in Table 3:

Table 3

Chi Square Analysis Results of Travel Behavior Before the Existence of Go-food

Correlation Coefficient Value (Asymp. Sig. (2-sided))	Mileage	Traveling time	Travel Costs	Travel Frequency	Travel Reasons	Travel Destinations
Age	,949	**,000**	,232	,355	,308	,901
Vehicle Ownership	**,020**	,154	**,017**	,314	,872	**,018**
Income	,242	,931	,186	,780	**,044**	,055

Hypothesis:

Ho: there is no relationship between rows and columns, or between the two variables analyzed
H1: there is a relationship between rows and columns, or between two variables analyzed

This Chi-square test is used to see whether there is a relationship between variables, namely a hypothesis can be obtained from the probability value through the value of Asymp. Sig. (2-sided). If the probability value is> 0.05 then Ho is accepted, but if the probability is <0.05 then H1 is accepted and Ho is rejected. In the case of the crosstab analysis above, there are five (5) relationships that have a probability value of <0.05, namely (Figure 22):

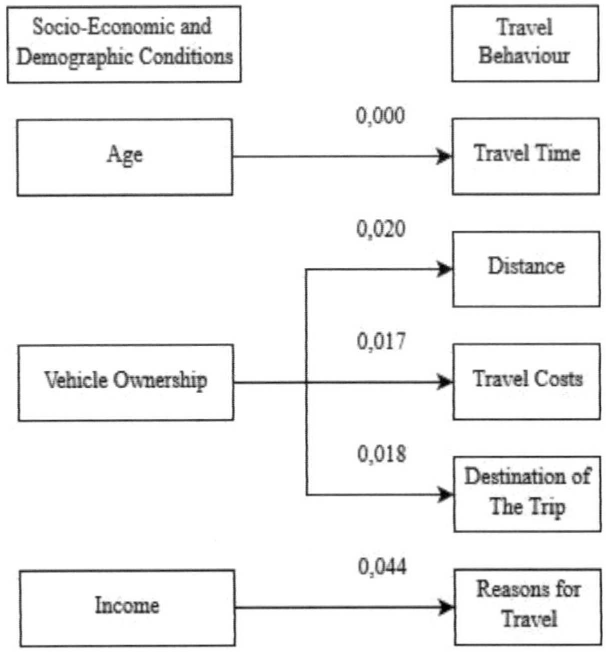

Figure 22. Crosstab Analysis Results That Have a Relationship Before the Existence of Go-food

Based on the results of the relationship of the variables above, there are several things that occur, namely:

1. The trading area is spreading and is increasing in the Higher Education Area, so that it can be reached more closely.
2. The community in the Higher Education Area is dominated by students who need a lot of productive time (activities / activities), so they prefer to meet their consumption needs in the nearest food stalls.
3. The community decides to fulfill their food needs related to distance, travel time, travel costs, travel destinations, and reasons for travel.

b. After the Go-food

Some data obtained to analyze the characteristics of travel behavior after Go-food will be explained in more detail through the results of the crosstab analysis (Table 4):

Table 4

Results of Crosstab Analysis of Travel Behavior After the Go-food

Correlation Coefficient Value	Distance	Travelling Time	Travel Costs	Travel Frequency	Travel Reasons	Travel Destination
Age	,327	,330	,858	**,003**	,673	**,030**
Vehicle Ownership	,557	,936	,959	,251	,104	,588
Income	,987	,947	,200	,629	,600	,399

Hypothesis:

Ho: there is no relationship between rows and columns, or between age and travel distance

H1: there is a relationship between rows and columns, or between age and travel distance.

Chi-square test will see whether there is a relationship between variables that can test the hypothesis by comparing the probability value through the Asymp. sig value. (2-sided). If the probability value is> 0.05 then Ho is accepted, but if the probability is <0.05 then H1 is accepted and Ho is rejected. In the case of the crosstab analysis above, there are two (2) relationships that have a probability value <0.05, namely (Figure 23):

IOP Conf. Series: Earth and Environmental Science **396** (2019) 012039 doi:10.1088/1755-1315/396/1/012039

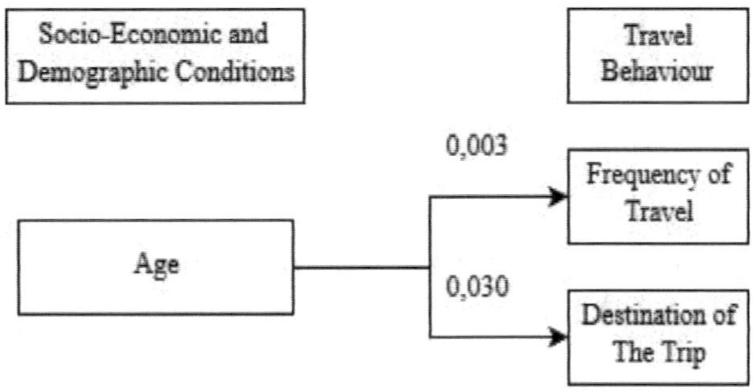

Figure 23. Crosstab Analysis Results That Have a Relationship After the Existence of Go-food

Based on the results of the variable relationship above, there are several things that have happened, namely:

1. Go-food services are used when people prioritize time for productive matters.
2. Go-food users do not think about the distance, cost, time, and reason for the travel.
3. Go-food users make requests to where they want to eat, even though the distance is quite far. This is because cost and distance are not a problem for Go-food users.

Go-food users do not need to leave the house to make their travel requests, because they are already represented by Go-food drivers.

5. Implications of Go-food on Travel Behaviors

The intended implication of Go-food is the impact caused by Go-food services on travel behavior from two (2) sides, namely: demand and supply (provider). Go-food user requests are analyzed through the characteristics of Go-food users, while offers are analyzed through the distribution results and travel frequency of Go-food drivers (Figure 24).

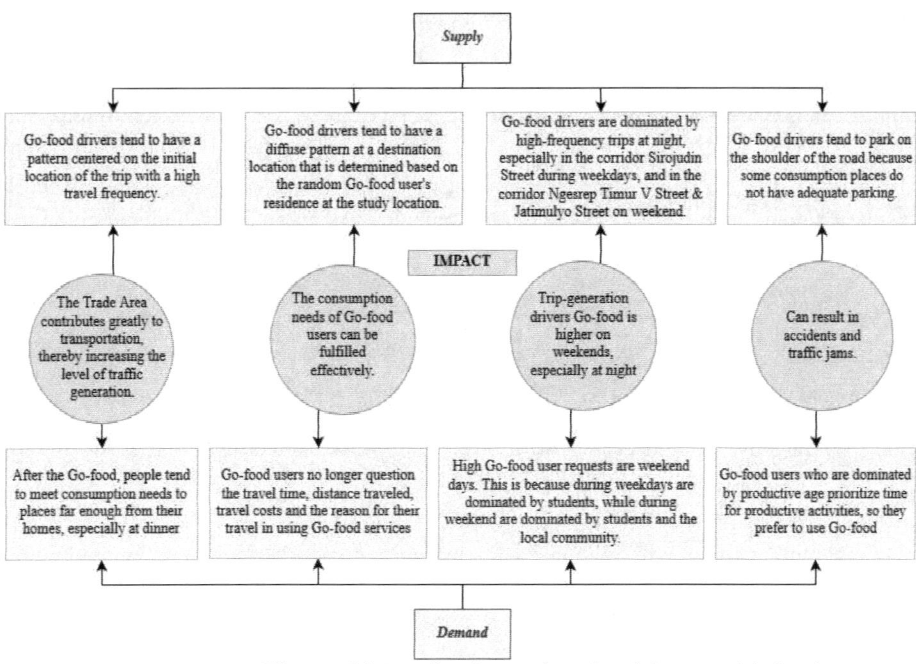

Figure 24. Implications of go-food for travel behavior

6. Conclusion

The initial distribution of the Go-food driver's travel is centralized because the location of the trade is concentrated in one particular location, with the highest quantity when weekdays are on Sirojudin Street (21%), while when weekend is on Jatimulyo Street (17%) and Ngesrep Timur V Street (20%). The higher the quantity carried out, the higher the frequency of the Go-food driver that passes through the road. This results in high traffic generation in the corridor. Concentration of trip generation conducted by Go-food drivers is at dinner time, both on weekdays (62%) and on weekends (66%). This results in high accessibility in the afternoon until the evening.

Meeting the needs of food has an impact on changes in travel behavior of Go-food users. Prior to Go-food, several factors were taken into consideration, namely: distance, time, cost, frequency, and reason for travel, whereas after Go-food, factors that were considered or had a relationship were: the frequency factor and the purpose of the travel. This is because Go-food services are used when the community prioritizes time for productive things. In addition, the ease of Go-food has resulted in Go-food users making requests to the places they want to eat, even though the distance is quite far. According to Levinson (1998) [21], economic aspects will influence travel behavior, including: aspects of income, expenses, work status, household, and type of work. This opinion illustrates the economic condition of Go-food users according to standards and is not a problem for Go-food users.

7. Acknowledgment

This article's publication is partially supported by the United States Agency for International Development (USAID) through the Sustainable Higher Education Research Alliance (SHERA) Program for Universitas Indonesia's Scientific Modeling, Application, Research and Training for City-centered Innovation and Technology (SMART CITY) Project, Grant #AID-497-A-1600004, Sub Grant #IIE-00000078-UI-1. The research was funded by Indonesian Ministry of Research, Technology and Higher Education (Ristekdikti) under the scheme of University Basic Research Excellent 2019 Contract No. NKB-1636/UN2.R3.1/HKP.05.00/2019.

8. References

[1] BAPPEDA 2010 *Rencana Tata Ruang Wilayah Kota Semarang Tahun 2011- 2031* (Bappeda Kota Semarang)

[2] Pangkalan Data Pendidikan 2018 Pangkalan Data Pendidikan Tinggi. Tersedia di

[3] Bapsi U 2016 *Baprsi - Universitas Diponegoro*

[4] Samadikun B P, Sudibyakto S, Setiawan B and Rijanta R 2014 Dampak Perkembangan Kawasan Pendidikan Di Tembalang Semarang Jawa Tengah (the Impact Development of Education Area in Tembalang Semarang Jawa Tengah) *J. Mns. dan Lingkung.* **21** 366–76

[5] Srinivasan S and Bhat C R 2005 Modeling household interactions in daily in-home and out-of-home maintenance activity participation *Transportation (Amst).* **32** 523–44

[6] BPS Kota Semarang 2010 *Kota Semarang dalam Angka 2018* (Semarang, Kota Semarang)

[7] Abbas S 2000 Manajemen Transportasi. Cetakan Pertama *Ed. Kedua. Ghalia Indones. Jakarta*

[8] Septiani R, Handayani P W and Azzahro F 2017 Factors that Affecting Behavioral Intention in Online Transportation Service: Case study of {GO}-{JEK} *Procedia Comput. Sci.* **124** 504–12

[9] Hastuti K R 2019 Gojek Klaim Jadi Aplikasi Transportasi Online Nomor 1 di RI

[10] Go-Jek 2017 PT Go-Jek Indonesia

[11] Haryanto A T 2018 CEO Go-jek Ungkap Alasan Kelulusan Go-food

[12] Kitamura R 2009 Life-style and travel demand *Transportation (Amst).* **36** 679–710

IOP Conf. Series: Earth and Environmental Science **396** (2019) 012039 doi:10.1088/1755-1315/396/1/012039

[13] Wells G R 1975 Comprehensive Transport Planning, Charles Griffin & Company Ltd

[14] Black J 2018 *Urban transport planning: Theory and practice* (Routledge)

[15] Bagley M N and Mokhtarian P L 2002 The impact of residential neighborhood type on travel behavior: A structural equations modeling approach *Ann. Reg. Sci.* **36** 279–97

[16] Bungin B 2005 *Metodologi penelitian kuantitatif* (Jakarta: Kencana)

[17] Sadhu S L N S and Tiwari G 2016 An activity pattern--destination land use choice model of low income households of informal settlements--Case study of Delhi *Transp. Res. Part A Policy Pract.* **85** 265–75

[18] Agarwal A 2004 A comparison of weekend and weekday travel behavior characteristics in urban areas

[19] Nasution M N 2004 Ekonometri Buku Kedua, Edisi Revisi

[20] Ettema D, Schwanen T and Timmermans H 2007 The effect of location, mobility and socio-demographic factors on task and time allocation of households *Transportation (Amst).* **34** 89–105

[21] Levinson D M 1998 Life-Cycle, Money, Space, and the Allocation of Time *Sumitted for possible presentation on Transportation Reseach Board meeting, Washington, DC*

The Hexa-Helix Concept for Supporting Sustainable Regional Development (Case Study: Citatah Area, Padalarang Subdistrict, West Java, Indonesia)

Z Zakaria[*,1,2)], R I Sophian[1)], B Muljana[1,2)], N Gusriani[3)], S Zakaria[4)]

[1]Faculty of Geological Engineering, Universitas Padjadjaran, Jl. Raya Bandung-Sumedang, KM-21, Jatinangor-45363, Indonesia
[2]Postgraduate School, Universitas Padjadjaran, Jl. Dipati Ukur No. 35, Bandung, Jawa Barat, Indonesia
[3]Faculty of Mathematics and Natural Sciences, Universitas Padjadjaran, Jl. Raya Bandung-Sumedang, KM-21, Jatinangor-45363, Indonesia
[4]Faculty of Social and Political Sciences, Universitas Padjadjaran, Jl. Raya Bandung-Sumedang, KM-21, Jatinangor-45363, Indonesia

E-mail:zufialdi.zakaria@unpad.ac.id

Abstract. A regional development requires some analyses in terms of the constraint and the potential. Constraints of a development need to be managed by periodic monitoring, while the potentials need to be maintained or optimally improved. The hexa-helix concept is a development of penta-helix concept with addition in the role of law and regulation. Citatah area is a multipurpose area where there is limestone quarry, factories, tourist attractions, and rock-climbing area as territoral potential. The constraints is found in Citatah are landslides and faults. To support the potential of the region, infrastructures need to be safe from landslide. The stable slopes have been built through terracing and engineered slopes. This research is directed to identify potentials and constraint in development of the Citatah area; evaluate safety factors around landslide prone areas; and inventory the role of hexa-helix agents in the infrastructure of road around the engineeried slope. The result shows that regional potential needs to be managed by utilizing the hexa-helix concept. Besides, the Citatah landslide zone needs to be aware. Environmental management and monitoring are required to avoid failure of development planning in the future. The result of this study will be useful for sustainable regional development in Citatah area.

Keywords: Hexa-helix concept, Sustainable regional development, Citatah Area

1. Introduction

1.1. Background

Sustainable regional development of an area requires comprehensive study of the area concerned. Citatah area requires support for the implementation of sustainable development. In Citatah area there are various potentials that support the local economy through tourism, mining, factories, and education areas, but in that area there are also various constraint [1–3]. One aspect of the disaster that needs to be reviewed is slope stability. In road segments around the Citatah area there are unstable slope. For this reason, an inventory of potential and constraint of the area is indispensable. Potential areas that support development need to be maintained, while constraints that can hamper the development need to be managed by mitigation, stabilization, or limitation. The objectives of this study are: 1) identifying potential and constraint in development of Citatah area; 2) Evaluating Safety Factors around landslide prone areas; 3) Inventorying the role of hexa-helix agents as direction of Environmental Management Planning and Environmental Monitoring Planning on the road infrastructure around the engineered slope.

The research conducted in Citatah area, Padalarang subdistrict, West Bandung Regency. The location can be reached by using a vehicle (motorcycle or car) from Bandung city heading towards Cianjur city. This area is located approximately at KM 23 to KM 27 on Bandung-Cianjur highway, which is an important route for economic facilities and infrastructure, especially as a route that connects Bandung with western cities, such as Cianjur, Sukabumi, Pelabuhanratu, Cisolok, Ciletuh, and others (Figure 1).

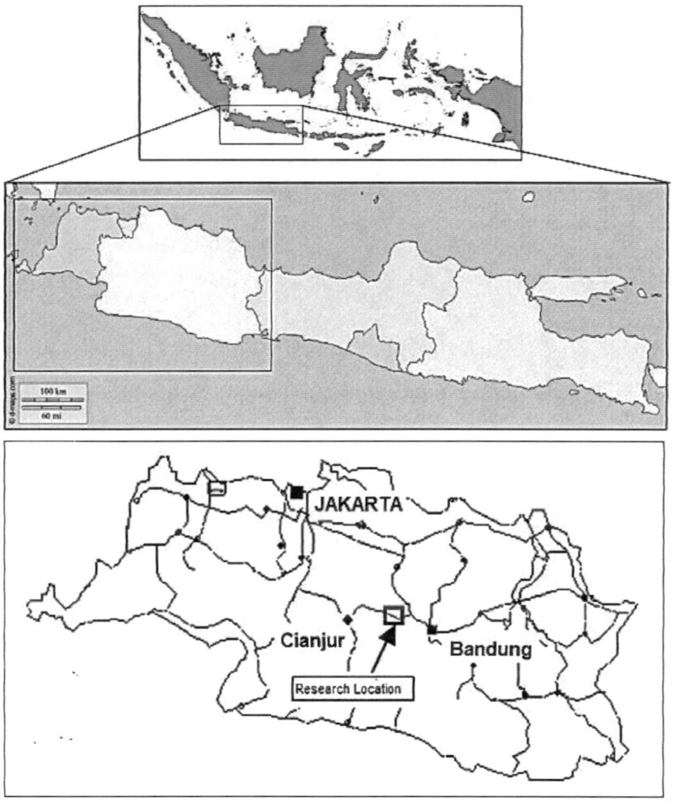

Figure 1. Research location

1.2. Potential and Constraint Characteristics of Citatah Area

In Citatah, there are limestone quarry and limestone processing plant. The location of limestone quarry is in the northern as well as the southern hill. The two hills are separated by Bandung-Cianjur highway. In this area, there are also rock-climbing area for public at Tebing 125 and for soldiers elsewhere; both are in the southern hill. Around the northern hill, there are several hills namely Pasir Pawon and Gunung Masigit. Pasir Pawon has tourist attractions at the top of the hill, i.e. Stone Garden, and at the foot of the hill, i.e. Gua Pawon. Gua Pawon has fairly good economic value for tourism [1]. This condition strongly supports the development of Citatah but requires safe infrastructure so that the potential of the area remains beneficial.

Based on geological condition of Citatah area (Figure 2), there is lithological distribution of Batuasih Formation claystone, Rajamandala Formation limestone, Citarum Formation sandstone, and andesitic intrusion (Figure 2). Position of Batuasih Formation claystone is at the bottom. At the top of it, there is limestone of Rajamandala Formation. Limestone is younger than claystone. Stratigraphic condition of Rajamandala Formation limestone is above the Batuasih Formation claystone. The claystone of Batuasih Formation was revealed due to the geological structure of Cimandiri Fault [2,3]. This condition led to the emergence of geomorphological units of very steep to gently sloping hills. The geomorphological form of the hills gives a negative contribution in the form of landslide-prone zone [3]. Landslide-prone zone is a constraint in regional development. Landslides that occur around the road may cause the Bandung-Cianjur highway as provincial road to be cut off. The last landslide occurred in 1993 [4] which caused the road to be cut off. Landslide-prone areas need slope stability, so it is necessary to create engineered slopes and build new roads towards the hillside.

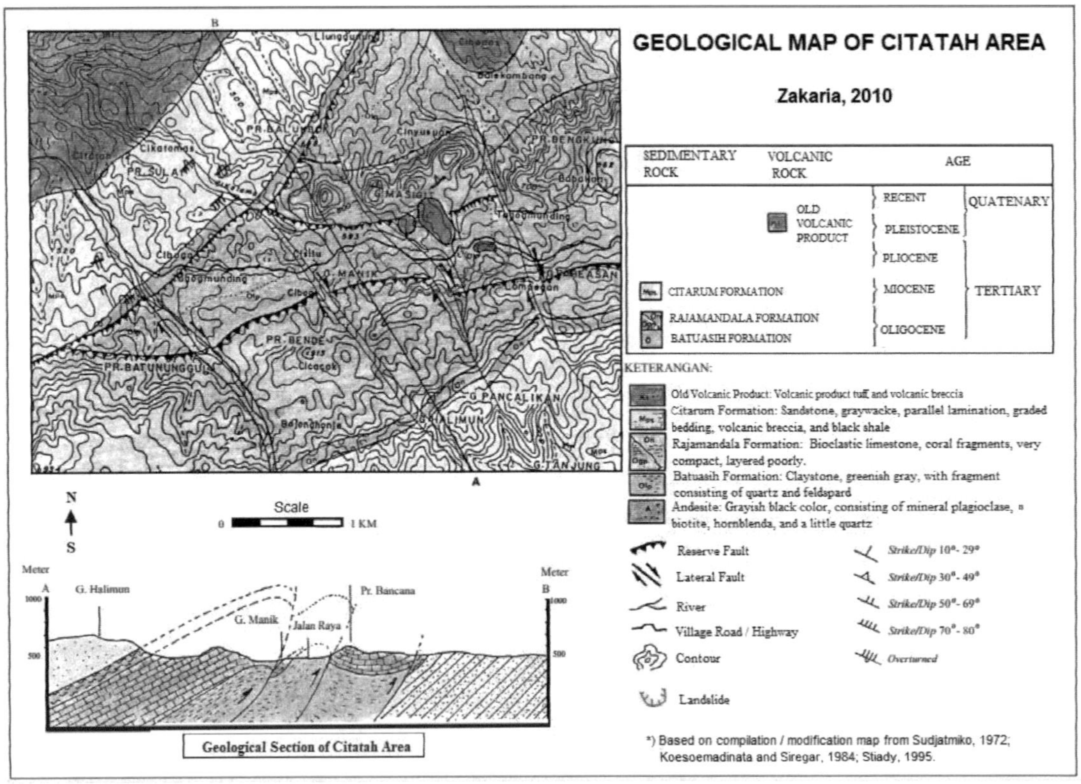

Figure 2. Geological map of Citatah area [3]

2. Methods

2.1. Helix Concept to Support Development

Innovation plays an important role in improving regional development. Realization of innovation requires participation from various parties. This support or participation is often called the helix concept [3]. The concept of helix continues to develop according to changes in society. Several concepts have been developed, from triple-helix, quadruple-helix, penta-helix, to hexa-helix. The triple-helix concept (Figure 3) focuses on the relation of university, industry, and government [5]. Quadruple-helix emphasizes the importance of integrating media-based and culture-based public perspectives [6]. Quadruple-helix complements triple-helix by adding "media-and-culture-based public" and "civil society" as the fourth helix. The quintuple-helix innovation model is even broader and more comprehensive by contextualizing quadruple-helix and by adding the helix (and perspective) of the "natural environment of society" [7]. This last helix developed into the penta-helix (Academic, Business, Community, Government, and Mass Media). Penta-helix also developed into a more complete hexa-helix by adding laws and regulations [8].

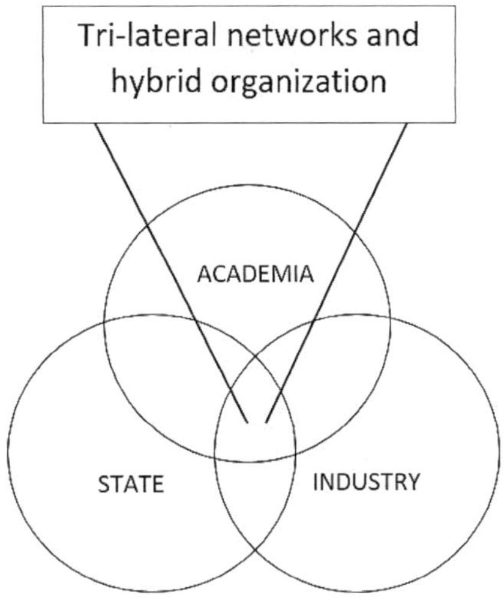

Figure 3. Triple-helix [6]

2.2. *Quadruple-helix and Penta-helix*

In the aspect of environment and disaster, concept like quadruple-helix is expressed in the *Starlet* Model. *Starlet* stands for *Stabilisasi dan Rancangbangun Lereng Terpadu* (Integrated Slope Stabilization and Design) [3] which examines several aspects related to slope stability issues as follows:

- Stabilization and slope construction design with a series of engineering activities
- Engineering geological mapping
- Slope stability analysis, stable slope design simulation, and slope reinforcement
- Environmental management directive accompanied by environmental monitoring

In its implementation, collaboration between helix components is needed by combining four mutually supporting interests, specifically Academics (or education), Business (industry), Community or Society, and Government (Figure 4). This concept is known as the quadruple-helix.

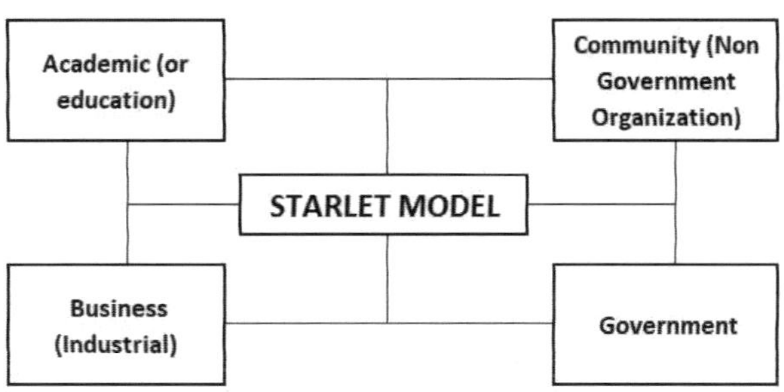

Figure 4. Quadruple-helix in the *Starlet* model

The Quadruple-helix concept developed again into penta-helix (Figure 5). This concept supports many activities such as tourism, economy, and business. The penta-helix model is a relationship between five elements, consisting of government, educational institution, environmental assessment institution, society, and industry; each coordinates well with each other [9]. This concept is used to strengthen and develop alliance collaboration with stakeholder. This concept is a collaboration between Academics (or education), Business (industry)[10], Community, Government, and Media. The concept of penta-helix can also be used in industry , in research and community service [11], and in modification of Participatory Rural Appraisal (PRA) for debriefing the Student Study Service (*Kuliah Kerja Nyata* or KKN) in West Java [12].

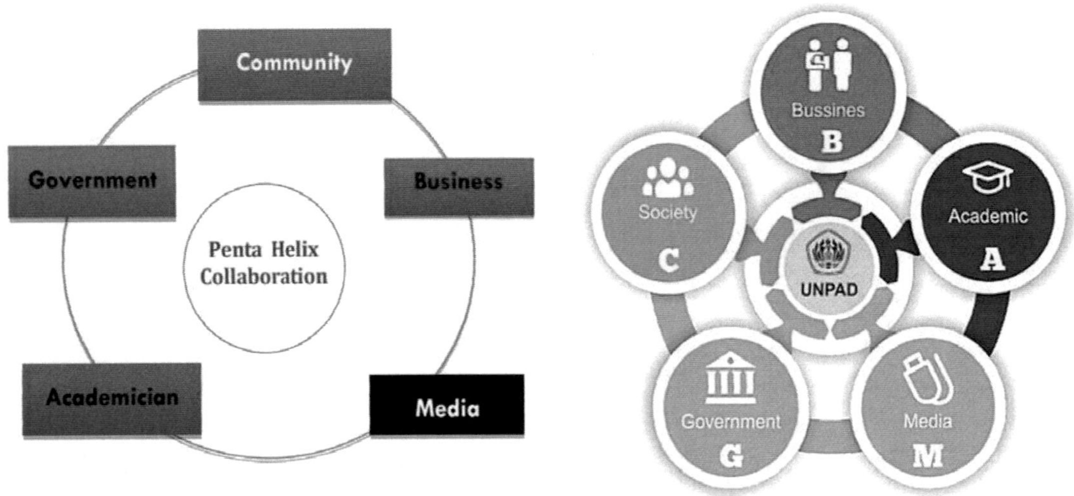

Figure 5. (Left) Penta-helix collaboration in industry [10], (Right) Penta-helix in research activity and community service at Universitas Padjadjaran [11]

2.3. Hexa-helix

Concept of hexa-helix is developed from penta-helix by adding aspects of law and regulation. This concept is used in *Starlet* Model (Model of Integrated Slope Stabilization and Design) for slope stabilization and then modified into *Starlet Perdana* Model (with the addition of soil bearing capacity for the foundation). *Starlet Perdana* Model utilizes hexa-helix concept to be more effective [8]. The six components of hexa-helix are Academics (or education), Business (Industry), Community, Government, Mass Media, and Law & Regulation (Figure 6) with following details:

- Government, consisting of the central government and regional governments
- Entrepreneur, industry, business, service, and trade
- Community, consisting of people from various professions and NGO
- Academics or education, or scientists, including the world of general education and higher education
- Mass media and communication media, including communication devices such as smartphone.
- Laws and regulations, including local government regulations.

The six components share the same views on regional potentials and constraints, as well as disasters that damage infrastructure. The same view of disaster is a must to reduce material losses and avoid casualties. It is required attention, especially for monitoring infrastructure damage.

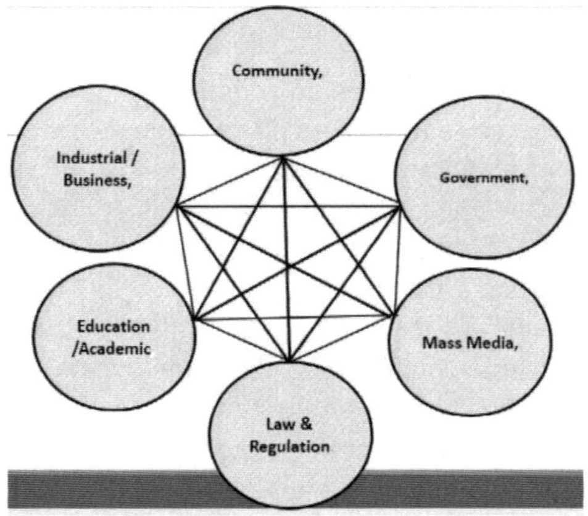

Figure 6. The six hexa-helix components [8]

The six components share the same views on regional potentials and constraints, as well as disasters that damage infrastructure. The same view of disaster is a must to reduce material losses and avoid casualties. It is required attention, especially for monitoring infrastructure damage.

2.4. Slope Safety Factor

The basic principle of Safety Factor (SF or F) calculation is the comparison between driving forces (□) and resisting forces (S) in a slope's body, or expressedd by the formula: F = □ / S (Figure 7). Theoretically, a slope considered to be stable if F> 1, critical if F = 1, and unstable if F <1, the slope is unstable. Based on the experience in slope analysis study, Bowles (1989) in [8] proposed a classification and the meaning of F value (Table 1).

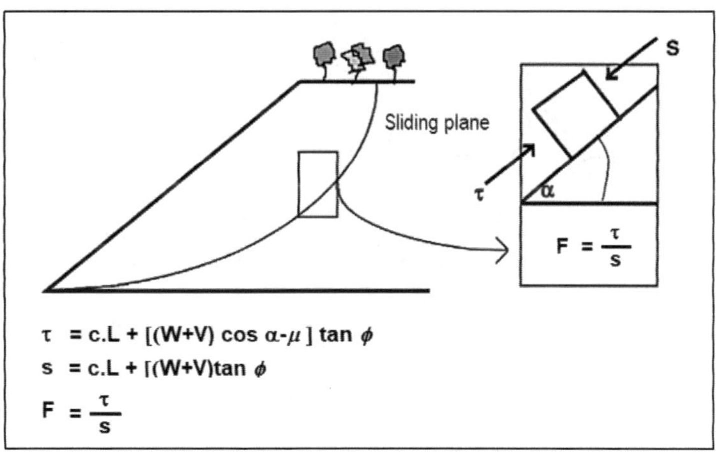

Figure 7. Sketch of a slope section and forces acting on the slope [8]

Table 1. Safety factor and meaning of slope condition Bowles (1989) in [8]

Safety Factor	Slope condition	Meaning
FS < 1.07	Usually occurrence of slope failure	Unstable
1.07 <FS <1.25	Slope failure ever occurred	Critical
FS > 1.25	Slope failure events are rare	Stable

IOP Conf. Series: Earth and Environmental Science **396** (2019) 012040 doi:10.1088/1755-1315/396/1/012040

3. Results and Discussion

3.1. Inventory of Development Potential in Economy, Tourism, and Education

Some activities that support the potential of regional development at Citatah and the surrounding areas (Figure 8) can be described as follows:

- In the southern limestone hills, there are mining activities of non-metallic mineral (limestone) located around Pasir Pabeasan. In the northern hills, some limestone hills have been strived so that they are no longer mined (in Gunung Masigit).
- In the southern part of Bandung-Cianjur highway, limestone processing plant is still active working. In some locations at the southern hill, there are limestone affected by metamorphosis due to high pressure and/or high temperature turning into marble.
- Rock climbing activities by a group of young people and rock climbers are usually done on holidays. The activity site is located in Tebing 125. On another cliff, there is a special site for rock climbing training for soldiers.
- In the northern hills, especially in Pasir Pawon, there are two tourist attractions. At the bottom of Pasir Pawon, a cave namely Gua Pawon is believed to be the home of ancient human. Here, a human skeleton was found. A replica of the skeleton is still stored in one corner of the cave. At the top of Pasir Pawon, there is the Stone Garden tourist area. Here, many stones are naturally arranged forming various configurations. Next to Stone Garden, there is a spot to take geomorphological photo of the Citatah area.
- To support geology education and training, especially Human Resources in the fields of geology, minerals, and coal, a Field Campus has been built for Education and Training Center. This campus belongs to *PPSDM Geominerba* (Center for Development of Geological, Mineral and, Coal Human Resources). It is located in the southern hills of Citatah area.
- To support the potentials mentioned above, it is needed to improve road access from each locations to Bandung-Cianjur highway. Some of the roads that have been built are made of stone, while the others are asphalt roads.

Figure 8. Some potential of development in Citatah area: a) Factory and limestone quarry (top left), b) Gunung Masigit and Pasir Pawon (top right), c) Stone Garden (bottom left), and d) The gate of *PPSDM Geominerba* Field Campus (bottom right)

IOP Conf. Series: Earth and Environmental Science **396** (2019) 012040 doi:10.1088/1755-1315/396/1/012040

3.2. Inventory of Development Constraints

The Citatah area is crossed by the Cimandiri Fault [2,3]. This fault extends almost in the direction of west-east. This fault is believed to be part of the Cimandiri Fault [13], so it is referred as the Eastern Cimandiri Fault. Citatah area has been deformed. An indication of the deformation is the existence of Batuasih Formation and Rajamandala Formation which are known to be Oligocene-aged. Both of these formations came to surface due to strong deformation. Limestone forms hills that extend in the direction of west-east. The Eastern Cimandiri Fault is cut by several dextral strike-slip faults. The indications are found in Pasir Pabeasan and around Pasir Pawon.

In some road segments between KM 23 to KM 26 in Bandung-Cianjur highway, there are landslide-prone areas with some indications of landslides. Landslide occurred in 1993. Slope stabilization was carried out by constructing engineered slopes on 1995. But, now, the engineered slope has changed. Some cracks on the slope terrace have occurred, some have collapsed, and there is bulging at the foot of the slope. Landslide is estimated to be active again. Location of unstable slope at KM 25 in Bandung-Cianjur highway, especially slope which have characterized by an indication of rotational landslides. The indication can be seen in the electric pile that tilt toward the slope (Figure 9). The position of the electric pile changed only in the surrounding area of the engineered slope.

Figure 9. Indication of the land returning to rotational movement. Left image was shot from east to west. Right image was shot from west to east. The sloping part is only around the location of the engineered slope.

3.3. Slope Stability Analysis

In previous studies [3], with data in 1999, the following analysis was obtained mechanical and physical properties i.e. water content, cohesion, internal friction angle, dry unit weight, and wet unit weight. Data are taken at conditions: 1) highest value of water content, 2) lowest value of cohesion, and 3) lowest value of internal friction angle. Dry and wet unit weight are taken at condition of highest water content. The results obtained: Water content (□)= 48.92%, dry unit weight ($\square_{dry}$) = 10.7529

KN/M^3, wet unit weight (γ_{wet}) = 16.1442 KN/M^3, angle of internal friction (ϕ) = 10°, cohesion = 9.3160 KN/M^2.

The results of Safety Factors analysis give the value of FS = 1.249 or close to FS = 1.25 (stable) for slope 45 on terraced slopes with dry slope conditions or unsaturated condition (Figure 10). For wet slopes (saturated condition), the FS value drops to FS = 0.987 or FS < 1.07 or unstable slopes (Figure 11). The condition will be worse if there is an earthquake with seismic load k_h = 0.3 (at Citatah area, according to the National Earthquake Center), the FS value will be even lower by FS = 0.727 in dry conditions (Figure 12). In wet conditions FS = 0.576 (Figure 13). If there is an earthquake with seismic load k_h = 0.3, the slope will collapse.

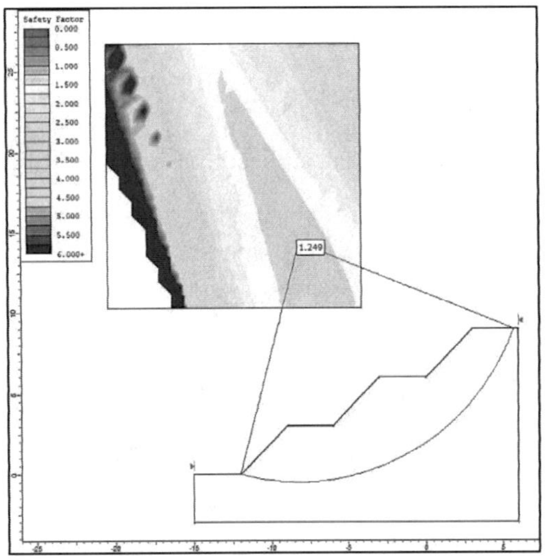

Figure 10. FS value in dry conditions without earthquake effects

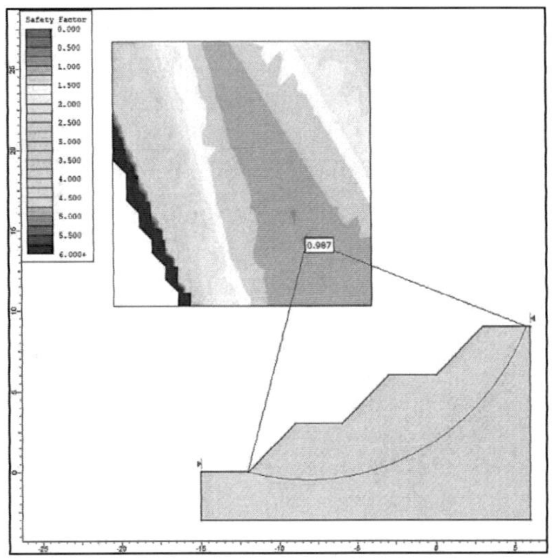

Figure 11, FS value in wet conditions without earthquake effects

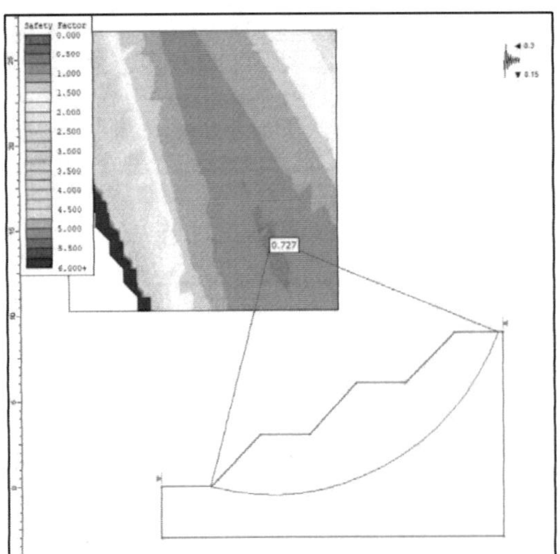

Figure 12. FS value in dry conditions with earthquake effects

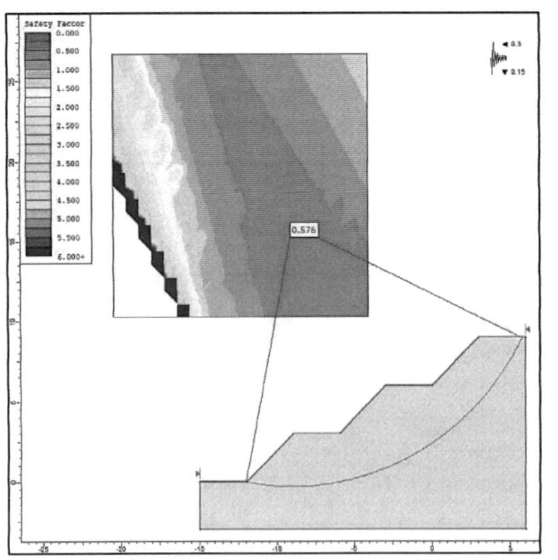

Figure 13. FS value in wet conditions with earthquake effects

Engineered slope has become unstable because there is no monitoring or weak supervision of monitoring. Engineered slopes built in the 1990s began to crack around the 2000s. In 2019, cracks and indications of unstable slopes are increasingly being shown. For example by tilting of electricity pile at the foot of slope.

3.4. The Concept of Hexa-helix to Support Development

To support sustainable development in an area, the concept of hexa-helix can be completely used. This concept needs to be supported by inventory data of potential and constraints in the local area. In sustainable development, the potential can be enlarged, while the constraint can be minimized or limited by mitigation, stabilization, or limitation. The concept of the hexa-helix in *Starlet Perdana* Model (or *Stabilisasi dan Rancangbangun Lereng Terpadu & Perkuatan Dayadukung Tanah*) can be modified to understand the role of six aspects of the project that is being run, tested, or verified; so in Figure 14, *Starlet Perdana* Model focuses on the center position and the other helix components are next to the center of the project. The role of hexa-helix can be reviewed for road infrastructure around engineered slopes, limestone quarries, infrastructure of limestone processing plant, and tourist attractions. In this paper, for example, the concept of hexa-helix for road infrastructure around engineered slope is given at Table 2.

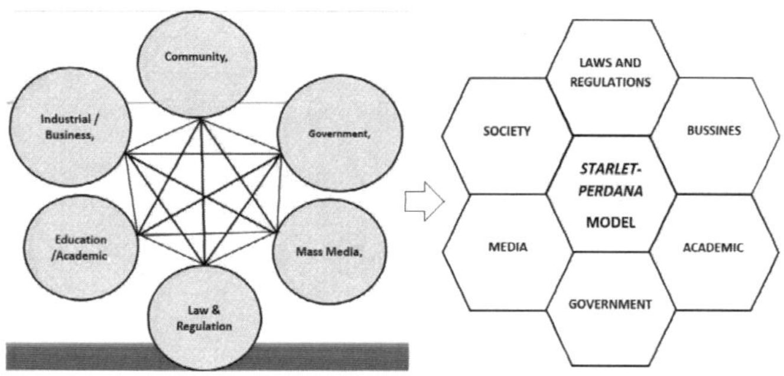

Figure 14. Hexa-helix transformation of *Starlet Perdana* Model

Table 2. Role of hexa-helix in the road infrastructure around the engineered slope

Agent	Management	Monitoring
Academics, scientists, education	Calculate slope safety factor. Design stable slope. Provide direction in the process of management and monitoring. Provide counseling on stable slopes	Get information from the mass media, the public, and government institutions. Monitor directly. Field survey.
Business, entrepreneurship, industry, service, trade	Carry out slope engineering. Build drainage at the foot of the slope.	Get information from the mass media, the public, and government institutions. Monitor slope conditions periodically.
Society, Non-Governmental Organizations (NGO)	Maintain cleanliness of engineered slopes. Get counseling about slope failure warning system.	Monitor the presence/absence of slope change directly.
Central government, local government	Investigate slope stability. Coordinate with scientists and entrepreneurs. Give work permission. Provide counseling on stable slopes	Monitor slope conditions periodically. Get reports from various parties. Field survey
Mass media, communication	Provide information of landslides event and indicating of landslide.	Monitor information of slope change or landslide event directly.
Laws, regulations	Regulation of building construction permits. Regulations regarding the environment. Regulations regarding the Direction of Environmental Management and Monitoring Plan.	Monitor violations of laws and regulations. Supervise the implementation of environmental management and monitoring guidelines

The support or participation of each agent in hexa-helix needs to be considered for the success and sustainability of various projects. This concept can be used for tourism, geopark, mining, infrastructure (road, bridge, foundation, building, engineered slope, factory, office, and other various public facilities), or service. The participation of six agents in hexa-helix can support the success of various projects (Figure 15).

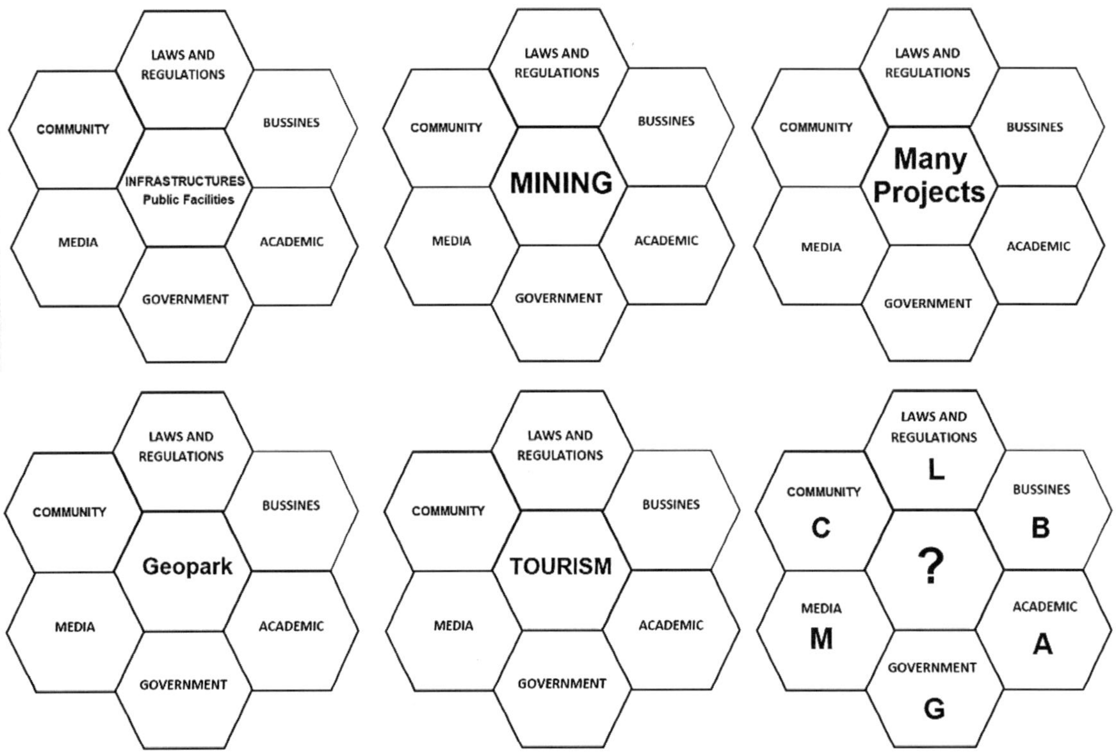

Figure 15. Hexa-helix for various projects

4. Conclusions

- There are several potentials that will support development in Citatah area as well as constraints that will hamper the development. In efforts to support sustainable development, potential aspects need to be maintained and protected from damage, while constraints that will hamper the development need to be minimized (limited by mitigation, stabilization, or limitation).
- Infrastructure of engineered slopes in wet conditions will experience a decline in Safety Factor value. The slope will be unstable. The condition will be worse, if there is an earthquake. The slope will be collapse.
- The concept of hexa-helix may be a choice to address problems comprehensively. The potential and constraint of development will become a shared domain for hexa-helix agents to maintain and monitor together the infrastructure that has been built. Each helix, ABCGML (Academics or education, Business or industry, Community, Government, Media, and Law & Regulation), has its own responsibility in accordance to its function and ability.

5. Acknowledgements

Thanks to the Director of *DRPMI* (*Direktorat Riset, Pengabdian Masyarakat, dan Inovasi* or Directorate of Research, Community Service, and Innovation), Universitas Padjadjaran. This paper is a small part of research which is funded by Universitas Padjadjaran through *HIU* 2019 (*Hibah Internal*

Unpad or Internal Grant of Unpad) Batch-2. This article's publication is partially supported by the United States Agency for International Development (USAID) through the Sustainable Higher Education Research Alliance (SHERA) Program for Universitas Indonesia's Scientific Modeling, Application, Research and Training for City-centered Innovation and Technology (SMART CITY) Project, Grant #AID-497-A-1600004, Sub Grant #IIE-00000078-UI-1. The research was funded by Indonesian Ministry of Research, Technology and Higher Education (Ristekdikti) under the scheme of University Basic Research Excellent 2019 Contract No. NKB-1636/UN2.R3.1/HKP.05.00/2019.

6. References

[1] Pertiwi Y I 2014 *Analisis dampak ekonomi dan strategi manajemen pariwisata untuk Gua Pawon di Wilayah Citatah Karst, Kabupaten Cipatat, Kabupaten Bandung Barat* (Institut Pertanian Bogor)

[2] Z Z 2005 *Bagian timur Patahan Cimandiri dan implikasinya terhadap tanah longsor di Citatah, Padalarang, Jawa Barat* (Majalah Geologi Indonesia)

[3] Zakaria Z 2010 Model Starlet, suatu Usulan untuk Mitigasi Bencana Longsor dengan Pendekatan Genetika Wilayah (Studi Kasus: Longsoran Citatah, Padalarang, Jawa) *Indones. J. Geosci.* **5**

[4] Permana E 2012 *The influence of rainfall intensity and length of time of rain on the landslide in terms of geotechnical aspects Universitas Indonesia*

[5] A L, M R and E S 2009 Difficulties of Collaboration for Innovation, A Study in the Öresund Region, Department of Business Administration, Lund University, Swedish, from

[6] Carayannis E G and Campbell D F J 2009 Mode 3 and Quadruple Helix: toward a 21st century fractal innovation ecosystem *Int. J. Technol. Manag.* **46** 201

[7] Carayannis E G, Barth T D and Campbell D F J 2012 The Quintuple Helix innovation model: global warming as a challenge and driver for innovation *J. Innov. Entrep.* **1** 2

[8] Zakaria Z, Sophian R I and Gusriani N 2019 Environmental Aspect in Infrastructure Planning Using Starlet-Perdana Model *IOP Conf. Ser. Earth Environ. Sci.* **248** 12013

[9] Amrial A, Muhammad A and Muhamad E 2017 Penta helix model: A sustainable development solution through the industrial sector, HISAS 14th Proceedings of Conference *14th Hokkaido Indonesian Student Association Scientific Meeting at Sapporo Japan* pp 152–6

[10] Muhyi H A, Chan A, Sukoco I and Herawaty T 2017 The Penta Helix Collaboration Model in Developing Centers of Flagship Industry in Bandung City

[11] DRPMI UNPAD 2019 *Pedoman Penelitian Internal dan Hibah Pengabdian kepada Masyarakat*

[12] Zakaria Z 2018 Modifikasi Konsep Participatory Rural Apraisal Untuk Pembekalan Kuliah Kerja Nyata Mahasiswa di Jawa Barat, Indonesia *Dharmakarya* **7**

[13] Anugrahadi A 1993 The Biggest Force for East Part of Cimandiri Fault, Bandung Regency, West Java *IAG (Indonesian Geologist Association) Annual Scientific Meeting 22nd* pp 226–40

An Open Market: The Legitimation and Regulation of the Public Realm in Denpasar

G A M Suartika[1], I K Mudra[1] and K E Saputra[1]

[1]Master's Program in Architecture, Udayana University, Bali, Indonesia.

E-mail: ayusuartika@unud.ac.id

Abstract. This paper interrogates the actuality of *'ruang terbuka hijau kota'* (urban green open space - UGOS) in Denpasar Bali. This debate is critical because it is directly linked to two conditions. First, is an ever increasing population, and second, is a diminishing public realm. While the social, economic, and environmental significances of the UGOS in the formation of a liveable city is widely acknowledged, both conditions, in reality, do not go in line with the importance of UGOS to urban living. The paper argues that while UGOS are legitimated by the existing urban planning procedures and reflected in spatial planning, there is no policy directive on securing the deliverance of sustainable UGOS that guarantee the public interest. The following study is supported by relevant case studies that illustrate and demonstrate the above claims, predominantly across Denpasar, the capital city of both Bali Province and Denpasar city. The conclusion to the paper becomes increasingly self-evident as it progresses. It is clear that having new categories of land use will accomplish nothing. Similarly the key is not to re-designate land uses or manipulate existing codes, but to reconstitute how the planning system itself functions in relation to a sustainable public realm and environmental justice.

Keywords: urban infrastructure, green open space, urban planning practices, elimination, conversion.

1. Introduction

The city is the center of accumulation for a diversity of activities. This is similarly a consequence of increasing urban densities in Denpasar. The World Bank notes that by 2025.88% of the world's population will live in urban areas, with 90% populating developing countries [1–3]. The density of the population in urban areas causes the city to continue to develop dynamically, in line with the changing times and demands of people's needs. As a result of global markets and trade agreements urban development is becoming increasingly dynamic. According to Thorns, globalization has been the key to the transformation of cities in the world over the last two decades [2]. Significantly, globalization is not a product but a process. Everything that is currently happening is due to the diverse interactions between economics, society, politics, culture, and ecology. Cities will increasingly be subject to rapid change in each sector [4].

Reflecting the above, cities will be subject to an increasing pressure on urban land use, decreasing provision of urban green open spaces (UGOS), increasing social conflict over ideologies, and the magnification of migration stress, climate change and economic disparity. Invariably, urban densities will continue to escalate, and neo-corporate pressure will be exerted in Bali to increase building heights and relax planning controls in line with free market principles. Whether or not these changes will result in increased economic benefits for all is debatable. History has taught us that wealth accrues to the rich, and it is reduced for the poor.

These changes will also substantially affect the city of Denpasar, Bali. As the provincial capital, Denpasar City provides a reference and barometer for urban development elsewhere on the island, one that is inescapably consequent upon tourism driven economic growth. It is widely recognized that tourism contributes significantly to developing traditions and encourages the expansion of cultural diversity in Bali. Critically, the emergence of various cultural activities is dependent on public space to maintain its prime function, the dominant symbol of a democratic society. According to Ahmad [5], public space has varied functions in pursuit of this principle but will vary in typology and content from one culture to another.

Historically, public space has developed in line with the overall trajectory of civilization, from the archetypal example of the Greek Polis to globalised contemporary society with its parks and gardens, urban squares and pedestrianised precincts. Technological development has also affected human attitudes and behavior through readily accessible international transport, new technologies such as the internet. The erosion of national boundaries in the interests of economic development as in the European Economic Community and international trade agreements and immigration are also deterministic of diversity and influence [6].

On a local scale, articles written by Suartika [7,8], highlighted and delineated the existence of public space in Bali and proposed the classification of public spaces grouped into five categories, namely: (1) open spaces located in each local administrative center; (2) temple open space; (3) market open space; (4) beach open space; and (5) mountain open space. The existence of this public space has an important meaning in protecting the environment, maintaining cultural values and practices, maintaining people's welfare, and achieving economic development goals.

Public space to a large extent designates the character of the city in support of economic activity, social interaction and symbolic representation via monumental architecture. The city's In Bali, economic development is increasing overall, with a concentration in Denpasar. Many investors wish to target the city's public spaces as a profit maximizing exercise unencumbered by existing built form and to a degree, or planning regulation [6]. Since the dominant narrative is that of globalization, it is arguable that public space in the city is threatened by external forces as much as it is on self-generated economic development. According to Carr [9] green open space is one of several typologies of public space.

The above explanation indicates that in the context of the spatial development of vibrant cities, Bali is subject to the same forces, where private interest and the public good collide. But a sustainable built environment depends on citizens needs being met, not only that of capital. So the need for this type of public space is often overlooked or eliminated by physical development which only focuses on the force of capitalist accumulation to pursue material gains, usually with minimal public support or participation. This fact will be the focus of the study, by examining in depth the aspects of the legitimacy and regulation of green open space within the planning system locally. The practices of eliminating green open space in the era of open markets and their consequences for regional spatial characteristics are profound. The results of the study are expected to be useful in supporting regional spatial planning and policies to maintain the public sphere for urban green open spaces continuously and sustainably.

2. Methods

In achieving the focus of discussion within, this study implements an inductive logic and qualitative research approach by case study. Data collection was done by gathering relevant secondary data in the form of statistical data of those provided by related Planning Authority of Denpasar City and other government bodies within the same territorial administration. On top of this, physical observation is also a prominent method for data collection, especially when comes to collecting data in regard to elimination and conversion of urban green open space within Denpasar City administration. Data was analysed by implementing the selected inductive logic.

IOP Conf. Series: Earth and Environmental Science **396** (2019) 012041 doi:10.1088/1755-1315/396/1/012041

3. Legitimation and Regulation of Green Open Space in Urban Planning

Urban planning is an instrument of the state to exert control over land development, and constructs the reality of the state along with that of the law, education, welfare etc. The essential characteristic of urban planning is that of regulation. Who benefits from such regulation and in what manner will depend on an ideological perspective as to whether the state is interpreted as functionalist, materialist, input-output, conservative, etc. Most recognize that all state perspectives represent an effort to organize in an innovative way, the future quality of life in cities and rural areas. The provision of green open space will therefore be dependent on many factors – the relative autonomy of the state in the first place, the power of the private sector to influence state policy while influenced by state neo-corporatism, citizen participation in decision making, the availability of open space and its ownership, the designation of zoning and land use policy, and the state's ability to compulsorily purchase land for public use and access, and the imagination and application of the planning apparatus [3,10–12].

Urban design is also important, and we consider that an urban design approach in this instance must take place as a fundamental planning strategy [13]. According to Shirvani [14], one of the eight physical elements of urban design is green open space. If city design is seen as part of spatial planning on a macro basis, then green open space has an important role and is a basic element in the formation of the physical form of the city. Various types of green open spaces can be distinguished. We may define protected green open space as a wider space or area, either in the form of elongated/lined or clustered areas, where the use is more open/general, and dominated by vegetation that is either cultivated or grows naturally.

Protected green areas consist of natural reserves on land and islands, such as conservation areas including forests, agriculture, areas of special interest such as mangroves, areas under a *banjar* (neighborhood association) control etc. The built-in green open space is a wider space or area, both in the form of elongated/lined or clustered areas, where the use is more open/general, with the land surface dominated by artificial pavements and a small portion of plants. Green open areas/spaces are fostered as an effort to obviate the negative effects of building and, have an ecological function for water absorption, the generation of oxygenation, prevention of air pollution, and protection of flora and fauna [15].

The legitimacy of the regulation of national green open spaces is stipulated in the regulation of legal products in the form of: (1) Law Number 26 year 2007 concerning Spatial Planning [16]; (2) Regulation of Minister of Home Affairs Number 1 year 2007 concerning Green Spatial Planning of Urban Areas [17]; and (3) Regulation of Minister of Public Works Number 05/PRT/M/2008 [18] concerning Guidelines for the Supply and Use of Green Open Space in Urban Areas with the typology of green open spaces as shown in Table 1.

Table 1. Utilization of Public Spaces in 2010 - 2017

	Forms	Function	Structure
Green open Space	Public green open space	Ecological	Ecological Pattern
		Socio-cultural	Planological Pattern GIS
	Private green open space	Aesthetics	
		Economy	

Source: Regulation of the Minister of Home Affairs Number 1 Year 2007 [17]

For the local scale of Bali, the legality of regulating green open spaces is determined through the Regional Regulation of Bali Province Number 16 year 2009 [19] concerning the Bali Province Regional Spatial Plan 2009-2029 [19]. The Regional Regulation of Denpasar City Number 27 year 2011 concerning the Denpasar City Spatial Plan [20] with the composition of the broad public and private green open spaces as presented in Table 2.

Table 2. Areas of Public Spaces in 2010-2017

	Spatial use	District				Denpasar (Ha)	% of Scale
		Denpasar-North	**Denpasar-East**	**Denpasar-West**	**Denpasar-South**		
I	Urban green open public space						
1	Urban forest Ngurah Rai	-	-	-	588.99	588.99	4.61
2	Un-built coastal area	-	16	-	152.5	168.5	1.32
3	Un-built river area	46.11	64.21	45.87	52.34	208.53	1.63
4	Urban forest-green open public space	2.47	9.51	7.13	25.96	45.07	0.35
5	Recreation & sport facilities (90%)	48.65	59.81	36.23	101.44	246.13	1.93
6	Cemetery	3.8	6.4	11	14.2	35.4	0.28
7	Green open public space-agricultural land (paddy field-based ecotourism)	236.88	305.14	12.14	227.62	781.78	6.12
8	Government offices (20%)	8.17	78.4	3.88	3.56	94.01	0.27
9	Education facilities (10%)	1.83	2.03	1.54	2.36	7.76	0.06
10	Health facilities (10%)	0.8	0.33	2.83	1.02	4.98	0.04
11	Ritual facilities (20%)	2.22	3.58	1.98	2.44	10.22	0.08
12	Estuary Dam (10%)	-	-	-	3.3	3.3	0.03
13	Road networks(20%)	43.72	46.82	44.45	65.77	200.76	1.57
14	Ports (10%)	-	-	-	5.2	5.2	0.04
15	Terminal (10%)	0.45	-	0.25	0.16	0.86	0.01
	Total urban green open public space	**395.1**	**592.23**	**167.3**	**1246.86**	**2401.49**	**18.34**
II	Private urban green open public space						
1	Urban green open public space-paddy fields (Wet paddy field)	236.86	305.14	12.14	227.62	781.76	6.12
2	Settlement (20%)	374.42	170.04	291.22	345.26	1180.94	9.24
3	Commerce and services (10%)	35.57	20.89	52.59	61.59	170.64	1.34
4	Zone for tourism (10%)	-	9.43	-	136.74	146.17	1.14
5	Industry and storage (10%)	-	-	-	3.25	3.25	0.03
6	Security (20%)	0.53	1.46	0.94	5.96	8.89	0.07
7	Urban waste area of Suwung (10%)	-	-	-	1	1	0.01
8	Waste recycling area of Suwung (10%)	-	-	-	1.96	1.96	0.02
	TOTAL PRIVATE - URBAN GREEN OPEN PUBLIC SPACE (UGOPS)	**647.38**	**506.96**	**356.89**	**783.38**	**2294.61**	**17.97**
III	**TOTAL UGOS**	**1042.45**	**1039.19**	**524.2**	**2030.25**	**4636.09**	
IV	**Scale (Ha)**	**3112**	**2254**	**2413**	**4999**	**12778**	
V	**% UGOPS**	**33.5**	**46.1**	**21.72**	**40.61**	**36.28**	

Source: Local Government Regulation for Denpasar No. 27 Year 2011 [20]

Table 2 above shows the total green open space in Denpasar City is 36.28% of the total area of the city, consisting of 18.32% public green open space and 17.96% private green open space. Examples of public green open spaces owned by the City of Denpasar and their existence are still maintained can be seen in Figure 1 and Figure 2.

Figure 1. Puputan Badung City Park- I G N M. Agung

Figure 2. City Park of Denpasar
Source: Mudra [27]

In the midst of a very significant existence and position, however, as one of the typologies of the public sphere, urban green open space planning in many cases is made as a matter of fact, with the aim only to fulfill the provisions of the proportion of 30% and 20%. Green open space planning is often not integrated with regional spatial planning as a whole, causing the position and existence of green open spaces to be very weak and vulnerable to manipulation. This is evidenced by the high change in the function of existing urban green open space into a built area.

4. The Elimination of Urban Green Open Space

Spatial planning is very closely related to social reproduction, the physical perpetuation of class structures, the development of capital, and the symbolic representation of power. The 21st century marked an unheard of acceleration in technological development and the creation of the space of flows in the form of electronic communication. The economic sector that promulgated open market activities through the mechanism of free trade, also demanded a new approach in urban planning. This new approach, according to Kriken [21], is broken down into nine principles of urban planning in the 21st century. One of the principles associated with open space planning is to revive natural systems that future life will depend on. This has many implications and we consider the following to have dominant priority. First, the role of planning as a regulator in the interest of the public good will have to overshadow the vocation of big capital to make vast profits from land development. Second, all development must be forced to adopt a triple bottom line approach, where all of the consequences of any development are included, without relying on the public to carry the cost of private inefficiencies. Third, all technologies promoted by big capital such as the smart city idea must be measured against real social gain defined by and including the public, and not some nebulous benefit to urban 'efficiency' to increase corporate market share.

The increasing development of urban economies has caused many investors to target the city's public spaces for development [6]. As often the case, this move aims at establishing new commercial centers that would generate profits. The government also has a tendency to favor this kind of development plan, as it would directly contribute to revenues generated from taxation (tax on land, development, purchase and sale). In this circumstance, public space is obviously not seen as lucrative as commercial facilities. Thus, the conversion of public spaces into developed areas is one of the common occurrences in many developing cities. The loss of city public space is also expressed by Sudikno [22] on the basis of architecture and cities without ethics. In the context of urban development in Bali in general, and in Denpasar especially, the conversion of green open space (considered a type of public space) has shown an uncontrolled pattern.

The tendency in an open market system is for the government (as planning authorities) to support development that supports capital accumulation through competition. This belief is reflected in Suartika [8] which states that, space targeted for conversion is often of those having a strategic role either to protect the environment or to accommodate socio-cultural interactions, such as parks and gardens. This view is grounded by a conception proposed by Keizer et al. [23] emphasizing that, on the one hand, the marginalization of land available for social and environmental interests, and the domination of economic interests in land uses and development, on the other.

In the case of Bali, the conversion of public space into private functions also takes place in many tourist destinations. Sanur, a coastal-based tourist destination located on the eastern side of Denpasar City, is an example of how coastal public spaces are used to accommodate tourism related facilities. In fact, public space located in front of hotels and restaurants, in this case, has been treated as the extension of premises' private territories. Thus they can be utilized to extend their businesses (Figure 3 and Figure 4).

Figure 3. Private Tourism Facilities in Public Beach (Sanur Beach)
Source: Mudra [27]

Figure 4. Private Tourism Facilities in Public Beach (Sindu Beach)
Source: Mudra [27]

The existence of green open space in Denpasar City has also experienced a similar disturbance. Green open spaces dedicated as urban lungs that absorb pollutants and carbon dioxide flying in the air that urban dwellers breathe in, are also in unsafe position. Many of them have been developed for housing units, shop-houses, commercial areas, restaurants, etc. (Figure 5 and Figure 6).

Figure 5. Elimination of open green area at Sidakarya- Denpasar
Source: Mudra [27]

Figure 6. Elimination of green area at Jalan Suradira, Denpasar
Source: Mudra [27]

5. Urban Green Open Public Space and Regulation

Every policy decided by the government is the result of synergy, compromise or even competition between various ideas, theories, ideologies, and interests [24]. The existence of various interests as a result of urbanization on a very broad scale suggests that, similar conditions also occur on a more micro scale, especially in planning and utilization of public spaces. In line with this, Aminah [25] also emphasizes that the practice of urban spatial planning is not merely an arrangement of economic and political interests. In this relationship, negotiations between actors determine the political practice in structuring a space, both for the public interest where all members of the community can fully use the space, and for commercial interests in which the space can only be accessed by certain groups of people [22,26].

Spatial planning policy in the context of allocating public space including urban green open space, should be done by considering harmony, balance and protection functions as mandated by the Indonesian Regulation Number 26 year 2007 [16] concerning Spatial Planning. This policy however does not give any detailed explanation as to how this objective to be achieved. This absence has opened a gate for subjective and various perceptions as to how this regulation should be interpreted before it can be enforced and sanctioned. These perceptions sometime can be so differing and confusing to those who have to conform. In relating this circumstance to the practice of converting and eliminating urban public spaces, one can see that the disappearance of such a space is an expected encounter. We will not be able to talk about law enforcement when the law itself is not yet sufficient to be enforced. This is the root of the problem.

In each regional spatial planning activity, the government has an obligation to place the interests of the community as the main consideration. Likewise, the planning of the public space, the social and cultural activities of the people that take place in it must be preserved by preparing a set of regulations as a controlling instrument. This needs to be supported by positioning public spaces as an important part of maintaining cultural continuity and the balance of spatial patterns based on the philosophical local wisdom of *Tri Hita Karana* (a harmonious relationship between beliefs system, human, and environment). It is unfortunate that in the midst of free market competition, the practices of public space is not yet given a prime attention.

6. Conclusion

The three sub-topics of the study above has explained that public space and green open space as one of its typologies are legitimated and implemented at national, regional and local levels. In many cases, the emergence of violations of the provisions of regional regulations, indicates a problem in planning green open space in Denpasar city. The problem is becoming increasingly complex in the midst of the 21st Century open market era, where capital power represented by the expansion of the tourism industry is based on economic benefits, believed to be the trigger for the emergence of practices to convert public space and eliminate green open spaces into built-up areas. This poses a real threat to the preservation of social and cultural activities of the community that take place in public space, which in fact should be the basis for the realization of the distinctive character of the regional sphere. Clearly, the planning authorities need to pay a more serious attention to the prolonged existence of the public space. In doing so, the attempt may be started with the conduct of an extensive research on public space needs that can lead to the development of a firm foundation for policy making process. By so doing, it is expected that the instigation of a planning policy oriented toward the provision of adequate public space is not far away from reach.

7. Acknowledgement

This article was presented at the 2nd International Conference on Smart City INNOVATION (ICSCI) 2019, jointly held by Universitas Indonesia and Universitas Diponegoro. ICSCI conferences have been supported by the United States Agency for International Development (USAID) through the Sustainable Higher Education Research Alliance (SHERA) Project for Universitas Indonesia's Scientific Modelling, Application, Research, and Training for City-centered Innovation and Technology (SMART CITY) Center for Collaborative Research, administered through Grant #AID-497-A-1600004, Sub Grant #IIE-00000078-UI-1.

8. References

[1] The World Bank 2010 *Cities and Climate Change: an Urgent Agenda* (Washington DC)
[2] Thorns D C 2017 *The transformation of cities: urban theory and urban life* (Macmillan International Higher Education)
[3] Sutriadi R 2018 *Perencanaan Kota Abad 21: Inovasi dan Tujuan Pembangunan Berkelanjutan* (Bandung: ITB Press)
[4] Coopers P-W 2005 Cities of the future. Global competition Local leadership *PriceWatershouse*

Coopers

[5] Ahmad A 2002 *Re-desain Jakarta: tata kota tata kita, 2020* (Kotakita Press)

[6] Darmawan E 2005 *Analisa Ruang Publik Arsitektur Kota* (Badan Penerbit Universitas Diponegoro)

[7] Suartika G A M 2013 Open Space Traditions, Development and Modernity: The Case of Bali *Mudra J. Art Cult.*

[8] Suartika G A M 2010 Substansi Budaya Dalam Kebijakan Tata Ruang Di Bali *J. Hum.* **22** 313–26

[9] Carr S, Stephen C, Francis M, Rivlin L G and Stone A M 1992 *Public space* (Cambridge University Press)

[10] Tarigan R 2012 Perencanaan Pembangunan Wilayah: ed *Revisi. Jakarta PT. Bumi Aksara*

[11] Yomtov N 2015 *Urban Planner* (Ann Arbor: Cherry Lake Pub)

[12] Watson D J 2016 *Local Government Management: Current Issues and Best Practices* (Routledge)

[13] Cuthbert A R 2008 *The form of cities: Political economy and urban design* (John Wiley & Sons)

[14] Shirvani H 1985 *The urban design process* (Van Nostrand Reinhold Company)

[15] Hakim R 2000 *Analisis Kebijakan Pengelolaan Ruang Terbuka Hijau Kota DKI Jakarta* phdthesis (Tesis Institut Teknologi bandung. Jakarta)

[16] Sekretariat Negara Republik Indonesia 2007 *Undang-Undang Republik Indonesia Nomor 26 Tahun 2007 tentang Penataan Ruang*

[17] Kementerian Dalam Negeri 2007 *Peraturan Menteri Dalam Negeri Nomor 1 Tahun 2007 tentang Penataan Ruang Terbuka Hijau Kawasan Perkotaan*

[18] Direktorat Jenderal Penataan Ruang Departemen Pekerjaan Umum 2008 *Peraturan Menteri Pekerjaan Umum Nomor 05/PRT/M/2008 tentang Pedoman Penyediaan dan Pemanfaatan Ruang Terbuka Hijau di Kawasan Perkotaan*

[19] Pemerintah Provinsi Bali 2009 *Peraturan Daerah Provinsi Bali Nomor 16 Tahun 2009 tentang Rencana Tata Ruang Wilayah Provinsi Bali Tahun 2009-2029*

[20] Pemerintah Kota Denpasar 2011 *Peraturan Daerah Kota Denpasar Nomor 27 Tahun 2011 tentang Rencana Tata Ruang Wilayah Kota Denpasar Tahun 2011-2031*

[21] Kriken J L, Enquist P J and Rapaport R 2010 *City building: Nine planning principles for the twenty-first century* (Princeton Architectural Press)

[22] Sudikno A 2016 Sejarah Arsitektur Perkotaan dalam Etika dan Urban Politik *Seminar City of Tomorrow Arsitektur, Urbanisme Berkelanjutan dan Warisan Budaya*

[23] Kaiser E J, Godschalk D R and Chapin F S 1995 *Urban land use planning* vol 4(University of Illinois press Urbana, IL)

[24] Yunus H S 2006 Struktur Tata Ruang Kota. cetakan ke IV *Yogyakarta: Pustaka Pelajar*

[25] Aminah S 2016 Konflik dan Kontestasi Penataan Ruang Kota Surabaya *{MASYARAKAT} J. Sosiol.* **20**

[26] Zahnd M 1999 *Strategi Arsitektur 2 Perancangan Sistem Kota Secara Terpadu, Teori Perancangan Kota dan Penerapannya* vol 2(Kanisius)

[27] Mudra I K 2018 Public Space Elimination Practice in the Development of Tourism Industry in Bali *Proceeding the First International Conference on Cultural Communication and Space and the 9th International Conference on International Socie* (Denpasar)

The spectrum of city park service scope: case study of Lapangan Puputan Badung and Lapangan Puputan Margarana Denpasar-Bali

N Kohdrata[1], G A M Suartika[2], A A K Krisnandika[1], L S Yusiana[1], I M A Dharmadiatmika[1]

[1] Landscape Architecture Study Program, Faculty of Agriculture, Udayana University, Bali, Indonesia
[2] Architecture Study Program, Faculty of Engineering, Udayana University, Bali, Indonesia

E-mail: naniek_kohdrata@unud.ac.id

Abstract. This research aims to find the spectrum of a park service scope from two city parks in Denpasar, a densely populated city. The city parks in study are Lapangan Puputan Badung and Lapangan Puputan Margarana as city public facilities of Denpasar – Bali that is said to improve citizen quality of life and environment as well. However, this research focuses to see how city parks in Bali serve their users. This study uses a survey research method. Sample of respondents are chosen using accidental sampling technique with questionnaire as media of data collection. Data series include, users' age, frequency of visit, activities, and domicile. Research findings are analysed spatially and descriptively as well. Result shows that most visitors are young people who spend time mostly for exercising and other recreational activities. It is also interesting to see that both city park users are regular visitors. They visit the parks at least 1 to 2 time per week and spend 1 to 2 hours in there. The parks users mostly come from densely populated areas. That shows the importance of city parks as complement to a city. However, as a city that going to smart city, ICT has not been recognized much. As conclusion, the spectrum of city parks service scope of Denpasar still ranging in traditional functions of public green open space. A city park serves a man-made nature-like place for people to get refresh within an urban landscape setting.

Keywords: city park, green open space, park service scope, liveable city

1. Introduction

Denpasar is a capital city for Bali province. The city population 914,300 within area of 127.78 km^2 that gives about 7,155 people/km^2 [1]. Compare with Surabaya as the capital city for East Java province with area of 350 km^2 has population density around 8,274 people/km^2 [2]. Surabaya has about 21 notable city parks according to data that has been published at the city official website. Meanwhile for Denpasar, there are only three prominent city parks that are well known among its citizens. Nonetheless, the focus of this paper is not on the quantity of city parks own by a city but on how city parks can serve the city dwellers in term of social and recreational functions. As capital city of Bali province, Denpasar has similar challenges as other urbanized cities. One of the challenges is providing good living environment for the people. One issue that has been lingered around urban setting is competition use of space. A city like Denpasar has been thrived into urban landscape to accommodate its status as the capital city of Bali since 1958. Less green open spaces have been a major problem of the city whilst the need of space for housing has been increased high. A research by Riana *et al* [3] shows most of land conversion in Denpasar city is turned into residential use. About two-third of converted land was green open space which in this case is farm land. The reason behind land use conversion mostly because a need for housing. The increasing housing demand in Denpasar corresponds with data from a report of Denpasar city that points out the increasing percentage of smaller occupational space per person in a household between 2015 and 2016 [4]. As the consequences, people have less a private green open space if not have none. This premise is supported by Nuarsa [5] in his research using remote sensing imagery and spatial computer program to identify green open space in dwelling areas. His research finding supports an issue commonly face by a city that private gardens have been a luxury for most city dwellers since price of land in the city is raising along with the increasing demand for housing. Nonetheless, open green spaces are still needed in a city to maintain balance between hardscape structures and natural environment. Indonesia government mandates every city to have at least 30% of green open space within its area. At this point, a city park has important role in providing green open space for city dwellers to compensate the lack of private green space in their property.

As Eckbo [6] stated about parks,*"... spaces of limited use and flexible form, developed with a minimum of construction and a maximum of natural unprocessed materials.... They are spaces of limited intensity and nonspecific use"*. City parks give opportunity for everyone to have time, do positive activities, have contact with nature, and socialize with others in a "garden-like" setting. That notion gets along with Fisher [7] arguments about city park functions when he discusses Central Park as a public space where culture, ecology, economic, and politics matters happen among the users and stakeholders. A city park cannot be seen as merely a rural escape for a city landscape as imagined by Olmsted when he designed the Central Park in the heart of busy New York City. Nowadays, a city park is functioning more in social aspect where citizens can go, socialize, and still have contact not only with other persons but also with plants. Suartika [8] denotes the importance of city parks as public space to accommodate people needs in a city with growing population and decreasing green space. With that in mind, it is interesting see how the Denpasar citizens make use of their public spaces. Moreover, since Denpasar government has promoted the city as a smart city, is there any internet of things that relate to city parks? This study is to find out how a park serves its users by recognizing people activities and connection to the park, instead of looking at the aesthetic value and physical quality aspects of a park. Do city parks in Denpasar serve users in tradisional way as Olmsted picturing a city park function? Is there any other functions of a city park? How those activities and people-park's connection get along with recent information technology fast invention.

2. Method

This research uses survey research method to collect data from both city park users by using questionnaire as a tool. The data set is taken from 100 respondents from each park. Respondents are visitor of the park by the time questionnaire was given. Samples are taken from visitors on weekdays and weekends as well. The minimum age of a respondent is 15 years old, assuming that he or she has independence to travel without supervision. Fifty questionnaires are distributed in workdays and other fifty during weekends for each park. Total of 200 respondents had contributed to this research data.

Data is analyzed descriptively to give more opportunities to portray park users as the subject of the study. Garret Eckbo proclaimed his view about garden design with people as his stand point [9][10]. His acknowledgement on three dimensional design is to ensure a person enjoying a garden when in it. This premise become the basis of analyzing quantitative data of the study. A descriptive explanation of people responses are used to explain park's physical condition instead of the opposite way. Subjects' reaction will be interpreted using physical features within the perimeter of both parks and the demographic of Denpasar city as well. Sztompka [11] stated that human societies is constantly change and it is unavoidable. It happens at all levels of societies, macro – mezzo – micro, through the time that result in what people named tradition. Sztompka stressing on continuity of process for a tradition to survive. It can happen through two types of mechanisms, first of material or physical and the second is ideal or psychological mechanism. This study tries see and formulate in what extent city parks in Denpasar have served citizens. Two city parks in study are Lapangan Puputan Margarana and Lapangan Puputan Badung. Another matter to be addressed is the information and communication technology (ICT) within Sztompka framework dynamic tradition. Since Denpasar has proclaimed itself as a smart city, it is also necessary to see how internet things have been or can be incorporated in city parks. Government policies, installment of hardware or software, and internet base applications will be used as indicators for ICT involvement in a city park.

3. Results and Discussions

3.1. Age of respondents

Result shows that more than half of respondents are teenagers and young-adults that is around 80 % (table 1). Middle-aged adult and adult-aged shares the rest portion of the population. This indicate that city parks still popular among young people in Denpasar. It also a good sign to improve green open space, especially city parks by government to give young people a place to use. Surprisingly, the major category of park users shown in both parks in Denpasar parallel with user age characteristic in Franklin Park - Boston designed by Olmsted [12].

Table 1. Age of Park Users

Age	Lap. Puputan Margarana (%)	Lap. Puputan Badung (%)
16 – 25	47%	47%
26 – 35	38%	27%
36 – 45	14%	16%
46 – 55	1%	4%
56 – 65	0	2%
> 66	0	4%

A research conducted by Susanti and Natalia [13] on Indonesian young people preference in public space denotes that the millenial generation choice of public space for outdoor is city park as for the opposite indoor is shopping mall. Activities that stand out from the research are taking pictures, strolling, and eating. In this case, taking pictures activity is still in correlation with an intention of uploading those pictures in social media. Accordingly, data of respondents showing that two third of Denpasar city parks visitors are young adult which is a good indication for anybody who concern with future generation of being asocial.

However, there is a pessimistic view on young people attachment to a gadget. In the era of high technology and easy internet access, gadgets have been a significant tool for communication, business, entertainment, source of information, and others as far as human can imagine. People can socialize with others through their gadgets, for texting, talking, and even video calling. The attachment of a man to gadgets is common. Abdel-Aziz *et al.* [14] state the down-side of social media "The activities and entertainment in physical environment are diminished, because social engagement can easily be achieved through far-distance connections in social media network. ...Facebook and others such networks ...[make] people live in one place but carry out most of their social engagement in other places."

Taking those pessimistic and optimistic view of young people and gadgets into account, the millennial attachment to digital technology can be managed at some point. A city park can be designed into to drive the youngster to green open public space by providing a convenient Wi-Fi connection. At the past, Olmsted draws people to city parks by promoting having fresh air for healthy life and having some relaxing time after working days full of stress in daily urban life demand. Today, Olmsted might say "Come to city park for free Wi-Fi connection". A city park as a tool in a public realm to connect people and also to some extent can drive interaction between people and nature. Lapangan Puputan

Margarana and Lapangan Puputan Badung still be promising spaces for creating a livable urban landscape.

3.2. Frequency and Time Spent on visit

For Denpasar citizens, it seems that visiting city park for 1 – 2 times per week is best suit for them (table 2). It is interesting to know that both parks users visit the place in regular basis. More than a quarter from total respondents show an intensive use of park by visiting 3 – 4 times and some other choose more than 5 times visits per week. Would that be an indication that there is a positive engagement – whether active or passive engagement – of park users with the environment?

Table 2. Frequency of Park Visit

Times of visit per week	Lap. Puputan Margarana (%)	Lap. Puputan Badung (%)
1 – 2	71%	64%
3 – 4	17%	21%
>5	12%	15%

The frequency of visit indicates that users feel comfortable with the environment. Feeling comfortable in a place can be physical or psychological. It means, despite some design comment about the park that says need to be upgrade [15], those parks have been successful in providing public open space for Denpasar citizens. The frequency of visit also shows that park visitors are park 'users". As define by Francis [16] "Users are those who frequent public places and rely on them for passive and active engagement." It indicates that engagement is important for a park to be function as it is.

One to two hours per visit seems an ideal time spend for visitors of both parks (see Table 3). The time spent confirms with the United Nations report on young people leisure time across the world, which for youth in Asia tend to have leisure time less than two hours compare to their fellow in Europe and United States [17]. It also corresponds with the use of park as a public recreation space. Different groups, type of uses, and time are expected to take place a city park.

Table 3. Time Spend in Park

Time spent per visit (hour)	Lap. Puputan Margarana (%)	Lap. Puputan Badung (%)
1 – 2	71%	85%
3 – 4	28%	15%
>5	1%	0%

The frequency and hours of time spend in park indicate users feeling in using parks. There are several reasons for a person to feel comfortable in a certain environment. Carmona, *et al* [18] stated that "*Comfort is a prerequisite of a successful public spaces. The length of time people stay in a public space is a function and an indicator of its comfort.*" I believe that the enjoyment of staying within park landscape is individual feeling. Carr, *et al.* assert that comfort is the basic need [19]. This construction of thinking gives us a way to find similarity features of the comfort. It shall be recognized at some extend that government and citizens have come to "agreement" in terms of fulfilling one of park function goal that is for public. Government as the service provider of public open space has successfully creating a space and Denpasar citizen have been perceptive users.

The data also tells a unique use of both parks in Denpasar that show majority of the users stay in parks between 1 – 2 hours. Compare to respondents in a research conducted by Mark Francis in community park and gardens in Sacramento, where time spend by park users is less than one hour while gardens users is between 1 – 2 hours [20]. Garden users spend more time in the place because they have

IOP Conf. Series: Earth and Environmental Science **396** (2019) 012042 doi:10.1088/1755-1315/396/1/012042

gardening activities to do, such as planting, weeding, and so on. It seems that users at Lapangan Puputan Margarana and Lapangan Puputan Badung have "stronger motive" when they come to visit the city parks. The parks are not just *a place to go*, instead it is *a place to go to do something.*

3.3. Activities in park

Findings for activities aspects shows an interesting data. Users of Lapangan Puputan Margarana is dominated by citizens who do sports. Meanwhile in Lapangan Puputan Badung, users for physical exercise and recreation activities almost at comparable number (Table 4). This response is coherent with the size of the city parks. Lapangan Puputan Margarana has a size of four times larger than Lapangan Puputan Badung. It also has more lawns than the last (Figure 1). Therefore, it can support more variety of physical exercises, such as football, basketball, volley, wood-ball, baseball, yoga, jogging, and some other sport that use grass field. Even occasionally, you might see some people do juggling and agility sports. Lapangan Puputan Badung does not have enough space those variety of sport, less space can be divide and share among group of people to do physical activities there. However, in this park, chess is the most prominent sport and well-known among citizen. It is a place for chess players to find partners to play.

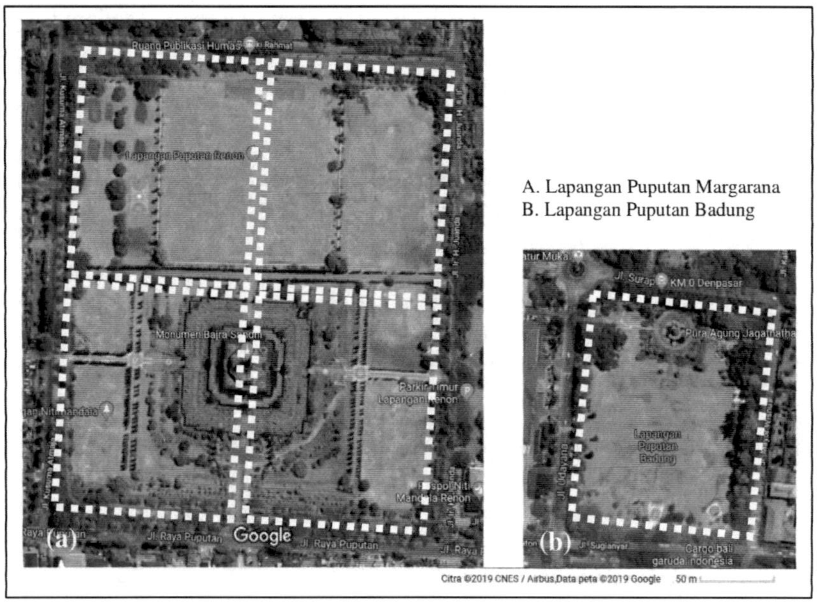

Figure 1. Size difference between two city parks
Source: Google Map with Modification

Visitors at Puputan Badung city park seem to consider their activities as recreational rather than a serious physical exercise. Maybe it is due to the limitation of space to use for some sports that required certain size of area, thus people who do the sports more likely to consider it as recreation rather than doing serious sports. Social interaction as an activity appears in Lapangan Puputan Badung and none at Lapangan Puputan Margarana. Although the number only 16%, it is interesting to pursue in future research about the topic on social interaction in a city park.

Table 4. Activities in Park

Activity	Lap. Puputan Margarana (%)	Lap. Puputan Badung (%)
physical exercise	71%	36%
recreation	28%	44%
social interaction	-	16%
others	3%	4%

Activities in city parks can be categories in two types that are passive and actives engagements [6][16][19]. The passive type such as watching people or wildlife, reading, enjoying scenes, and chess playing. While, jogging, walking, playing football, playing kites, etc. will be consider as active type of people engagement in a park. The data shows that Lapangan Puputan Margarana can contain more active elements compare to the other one. This correspond to the size of the park which is five times bigger the other. Thus, more group sports activities may take place without conflicting. To name some sports activities that take place in Lapangan Puputan Margarana are football, basket-ball, volley, baseball, wood-ball, jogging, walking, and yoga. Even in almost the same time, two or three groups play football in the grass-field area or a group of people doing yoga and next of them there is a group of people doing gymnastic. Meanwhile, with more limited space and almost similar users enthusiastic in using park, Lapangan Puputan Badung can only provide less space for group sports. Most physical exercises people do in it are jogging, walking, and some small group working on static exercise bikes and another set of exercise equipment had been provided by park management. Other facility for active engagement is a set of children play equipment in the children play ground area, a giant-chess play set, and an area with tile finishing that suit for in-line skate or skateboarding. Interestingly, 16% users in Lapangan Puputan Badung said social interaction is their reason to visit the park. Although not many seatings has been provided within the park, but it can always be seen people suit themselves by sitting on grass-field, planter boxes short wall, and steps.

Parks users seem adjust themselves and their activities to the park size. They create their own "territory" in terms of area and time when using city parks. Territoriality is crucial in defining types of activity engagement. Sharing space in a park can be seen a learning process that can lead to more tolerance and solidarity among citizen. Shaftoe [21] affirms public spaces as one of places that give opportunities for social learning. Accessibility to social arena and possibility to meet people with different background – cultures, norms, ages, behaviors, educations, and so on are valuable things to learn of. The learning point of view may contradictory to more skeptical view to see activity-territory as potential social conflict. The level of space contested can be micro - mezzo [22][7] means users of park "claiming" space for their group activity. Competition in micro level of space happen among groups or people with same interest to claim their arena. At mezzo level, non-users group crash into supposed to be a park arena and claim it for other uses that is semi-public. A city is a macro level of the contested ground of public spaces with other semi-public and private as the competitors [23]. Considering a possibility for a city park to be a contested ground, city authority can partake in creating condition that promote democratic uses of a city park. At micro level, park design can be a solution, at mezzo there can be park use regulation by government agency, and at macro level it can be city ordinance that protect the existence of public green open space.

3.4. Domicile area to park

Considering both parks location is relatively close to each other – about 3 km – and their position are considerably at the "center" area of Denpasar city, result shows a tendency that park users come from area with proximity to a place. See figure 2 for users domicile composition to each parks and figure 3 to a get sense of each park position relative to sub-districts. Users domicile for Lapangan Puputan Margarana show a normal disposition base on park location relative to its prospect users. It corresponds with Eckbo given figure for a community or district park service radius between 1.5 – 2.5 km and with acreage around 101,000 m^2 [6]. However, for the case of Lapangan Puputan Badung, which is located at the center of three sub-districts – West Denpasar, North Denpasar, and East Denpasar – the park users' composition shows an interesting phenomenon. People from South Denpasar sub-district – which is quite far – also go to that park and even take almost the same portion as citizens from West Denpasar. Distance from the closest South Denpasar area to Lapangan Puputan Badung is about 2 km and the size of park only about 35,000 m^2 which is a size for neighborhood park. But interestingly, the park serves its users like a community or district park. This phenomena can be justified using Mark Francis notion of accessibility as one of public-space dimensions [16]. That park complies with three types of access that are convenient direct physical access, wide-opened social access, and see through visual access. Lapangan Puputan Badung is quite easy to access from every side of its perimeter because there is no distinct entrance gate or designated openings to enter. Compare to Lapangan Puputan Margarana which is surrounded by fence at north side part and planting area on three other sides. People can enter only from certain designated entrances. For second type of access, both parks show same welcoming atmosphere to any social class in Denpasar. The third type access that is visual access, Lapangan Puputan Badung give more sense of secure, since there is almost no visual barrier the park. Shrubs are maintained pretty low $\pm$60 cm that allow people to see inside the park. It is also smaller in dimension, when a person stands at one side of the park, he/she can see to another side clearly. Meanwhile Puputan Margarana has fence barrier, though it is quite see through. Other visual barrier is the undulating park surface and extensive park size that makes people visually not secure [16][24].

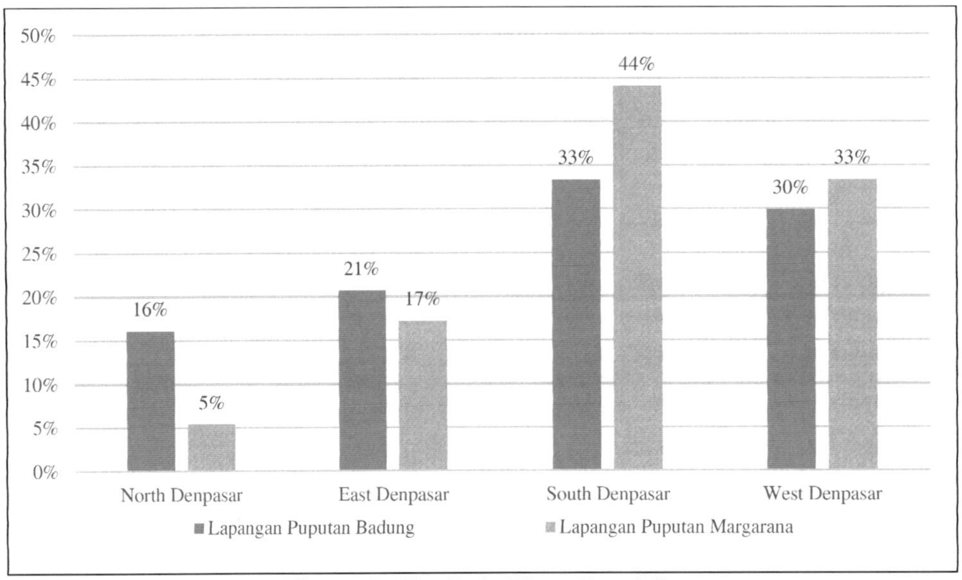

Figure 2. City Parks Users Domicile

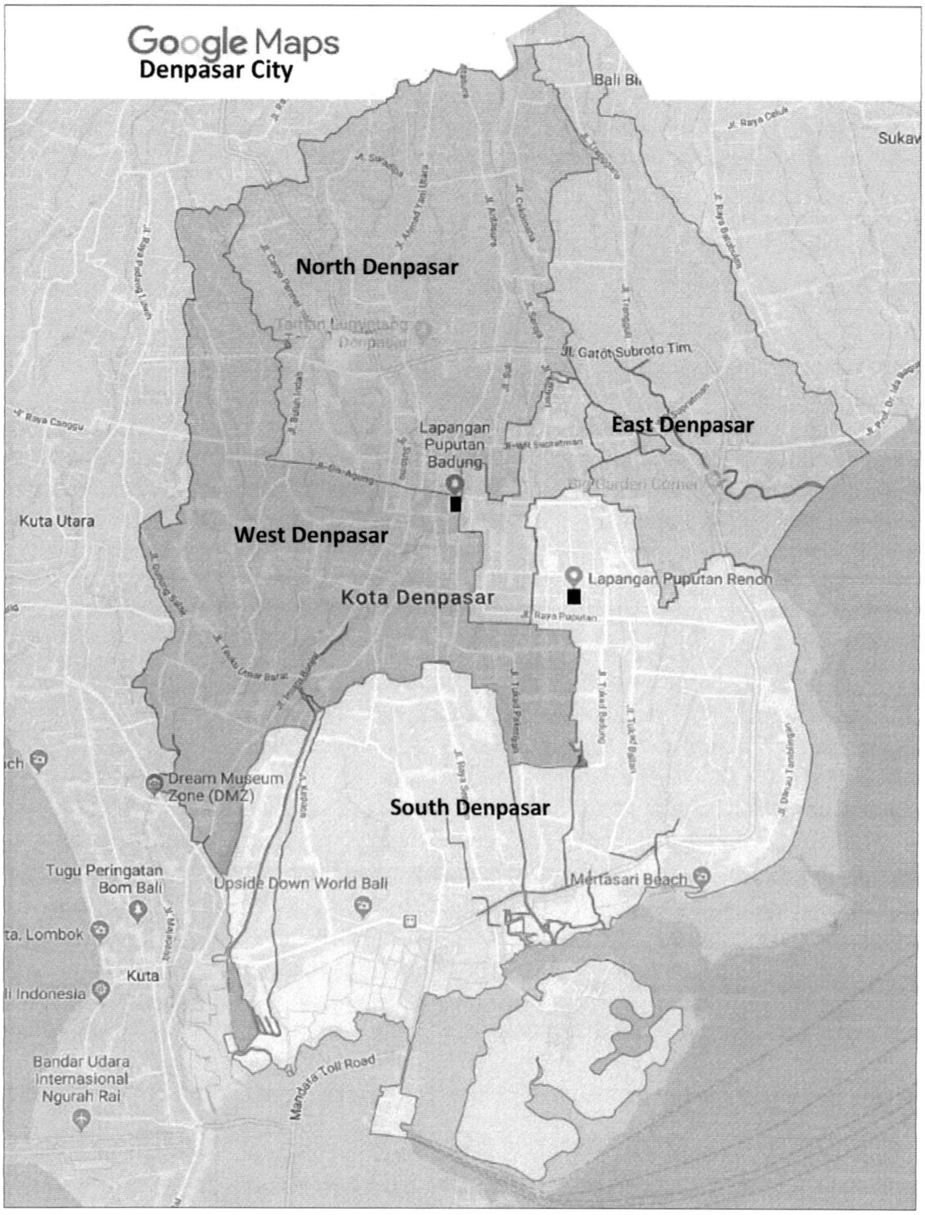

Figure 3. City Parks Location at Denpasar City
Source: Google Map with Modification

The data shows that most users come from highly populated area (see Table 5). It corresponds with built space coverage in each of sub-districts area. Users to Lapangan Puputan Margarana dan Lapangan Puputan Badung mostly come from South Denpasar sub-district and West Denpasar sub-district with higher number of built space and less green open space. Compare with less users from sub-district of East Denpasar which still have about 40% green open space for that area, which mostly is categorized as urban rice field. On the other hand, citizen from North Denpasar sub-district seems not using much on both parks in this study. Albeit the population density is relatively high and three-quarter of the area is built space. It might be due to the presence of Lapangan Lumintang, another city park that located in that sub-district. The location is only about 3 km to the north of Lapangan Puputan Badung, so it seems that people from North Denpasar have been serve by that city park.

Table 5. Population Density, Percentage of Built Space and Green Open Space

Domicile Area (sub-district)	Population Density (persons/sq.km)	Green Open Space (%)	Built Space (%)
North Denpasar	4028	24%	76%
East Denpasar	3116	38%	62%
South Denpasar	5855	21%	79%
West Denpasar	5291	10%	90%

Eckbo [6] recommends space dimension for recreational use is every 4,000 m^2 to serve 100 to 300 persons, which gives about 40 m^2 to – 13 m^2 per person. Based on that formula, assuming that both parks serve the sub-district of West and South Denpasar, it gives a number of 15 m^2 per person. This calculation is based on an assumption that both parks is located quite close to each other then they can serve the users need of green public open spaces from two sub-districts. J.O Simonds asserted that no community can satisfy all resident's needs, so does Denpasar city cannot fulfill all citizen's needs. But an optimum habitat can be created to invoke sense of belonging, sharing, and contributing. A place that users can feel safe, comfort, and sense of well-being [25].

3.5. The spectrum of services

At this point the study result shows that Lapangan Puputan Margarana and Lapangan Puputan Badung have been functioning as public area to citizens of Denpasar. By serving and functioning as such, city parks of Denpasar play significant role to create a live-able urban environment. Both city parks provide connection to nature for urban citizens. This study in not talking about natural rural like setting that Olmsted tried to bring back to urban setting in city of New York or Ebenezer Howards notion of "Garden City" that tries to create an idealized town surrounded by lush urban forest, parks, and agriculture land. McHarg proposition in his book *Design with Nature* works better with growing condition of today's Denpasar. The nature of Denpasar city is urban that is area with more built-space. Man-made structures, such as buildings, paved streets, signage, and so on are natural for a city. As consequences, even the green open spaces are also "man-made". It means that human has significant influence in creating it. Thus, the "naturalness" of a city park is shaped by people who use it. City people may like trees and shrubs that neatly trim or grass field that nicely mowed for them to enjoy visually and functionally as well. Their connection to living creatures is much well-ordered. This study denotes that Denpasar citizens appreciate green public open space such as a city park. It is not merely about a green public open space with natural grow of plants, but about a space where people can do fun things for themselves or with group, interaction with others, or even just silently enjoying the presence of others in a broad sense – human, plants, birds, and so on. This is supported by Sztompka theory of "Species Being" which recognizes relation between human with nature [11]. The nature of Lapangan Puputan Margarana and Lapangan Puputan Badungare are "designed" by users of each city parks. According to findings of this study, physical exercise and recreation are strong "colors" for both parks that works interchangeably.

As green public space, city parks in Denpasar serve well mostly for active engagement of their users. Interestingly, citizens of Denpasar have tendency to use city parks as physical exercise area. Sport activities have been triumph follow by recreation and socialization. Area size seems to have influence the way a park is used. Considering in both parks, sports have been popular choice of activity by respondents. In Lapangan Puputan Margarana where space is available more, physical exercises such as football, basketball, volleyball, baseball, wood-ball, running, jogging, martial arts and yoga are commonly seen sports in the park. It can fit more groups with sports activities that require "big space" at the same time within acreage space of four times than Lapangan Puputan Badung. Since the last has less space to accommodate users, especially sport lovers, therefore chess is the most popular sport in that park. Almost at any time, people can see some tables with two persons facing each other in front of checker board and surrounded by some enthusiast onlookers. Other sport that popular is jogging that has been accommodate by jogging track around the city park. Other group sports such as football and volleyball, need to squeeze itself with other recreational and socializing uses of the park. Recreational uses, such as children flying kites, ball playing, tag games, are type of active engagements of users that have most potential to conflict with group sports. However, observing the condition at the park, football players tend to have more flexibility in their playing space. It means they play football more for fun instead of "real" sport. They are willing to share space with children who play kites, kicking balls, and the "indefinite walking game" of toddlers. They imaginatively "marking" much smaller area of football field for themselves. The children and their parents at the same time also will not cross into area of football players. At this point, sense of tolerance has been built among park users. Each person or group define their arena, working on their position, and enjoying what is he/she or are they do. To the context of Denpasar, city parks can be an arena to learn to live with others in a "short" time when many interests are involved.

The outdoor and freedom to access characteristic has been added values for city people in Denpasar. With growing population and decreasing space available for housing, it pushes people to be efficient in building their house. A person will carefully be dividing his land for rooms he needs and there will be priority consideration for what to build or not. The value of importance of a garden as private open space is the last after bedrooms, bathroom(s), kitchen, and living room. Therefore, city parks can complement the needs of open space for the citizens. Spending some leisure time in a park can fill in the needs of doing things during free time. A range of time when people are not required to do their paid work or household jobs, or to be in school [17]. Realizing that majority of city parks users are the young adults, there is a positive connection of green public open space and young people. The groups of young people now correlate with people who familiar with technology. They work, play, and doing most of things using help or support of technology. Nowadays, the internet of things has been got into every single life of human beings. This group of young people are the most familiar with it. They good with it, they live in it.

In regards to Denpasar as a smart city, there is not yet any ICT application find in those two city parks of study. Most of facilities from both parks are still considered "traditional" and common park elements, such as strolling path, benches, children play equipment, grass area for multi-purpose sports and other recreational activities. Providing free Wi-Fi access can be appealing for young citizens. Meanwhile, having digital media façade can benefit in educating, informing, and engaging park users of things. It might lead from a passive engagement, e.g. listening/absorbing information from display, to a fruitful exciting discussion. So, it is a necessary thing to connect city park as one of urban life element with information and communication technology. The internet through social-medias and other possible channel can be designed to draw more people to city parks. Informing events and other interesting activities that can be done in a park is a way to keep city folks grow with vibrant but humanized environment. Active or passive engagements in recreation and socialization can be the easiest way to connect with information and communication technology things.

The spectra of city parks services in Denpasar is illustrated in diagram (a) at Figure 4. It is strongly indicated that city parks still about relation and interaction between people and nature. The people itself are surrounded by ICT, which has been common need for nowadays globalized world. To keep Denpasar citizens as "species being", I suggest to shift the ICT boundary as independent element by itself (see illustration (b) - Figure 4).

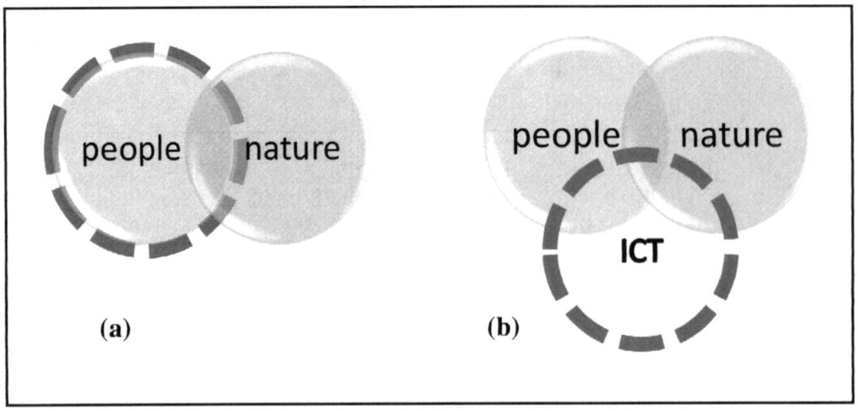

Figure 4. Diagrammatic Service Spectra – Now and Future

ICT as an independent element will be easier to adjust itself accordingly based on a city condition. It can move itself close to human needs or nature needs. When it is embedded with people, it becomes personal and serve specifically to a person it attached to. But when ICT be independent, it can be designed accordingly to the needs of a city, whether it designated for group or personal. A tradition of using a city park by integrating ICT shall be created and promoted to adjust with technology invention and urban people life-habits.

4. Conclusions

This study shows Denpasar city parks users of Lapangan Puputan Margarana and Lapangan Puputan Badung still stay in "traditional" uses of a city park. The green public open space is seen as an area where citizens can do active and passive engagements, such as sports, playing games, pleasure walk, and watching people. Both city parks also serve as good place of natural man-made creation of urban setting for incorporating trees, shrubs, and ground covers in a manageable way. They do serve as an outdoor space for citizens to enjoy. Working as a place to get refresh, healthy, and to acquire social skill and matureness. It is also important to note that young people of Denpasar have contributed to the vibrant of parks uses. This is an asset to the city for having healthy community and in turns will affect to good physical and psychological of urban life. However, government has not been recognized the need of integrating ICT within a park setting. The discourse within groups of authority is still dominated by disengaged people from their gadget. The worry of people-gadget attachment and asocial community have been an issue in promoting activities for a city park. Instead of worrying about it, I suggest a new perspective and approach that is integrating ICT in citypark. Free Wi-Fi, on-line social media, interactive apps, *etc.* can be utilized to draw people to parks. Then, let the site, condition, situation, and the park users decide their own level of enjoyment in creating connection between people to people or people to nature with the support of ICT.

5. Acknowledgements

The publication of this article has been supported by the United States Agency for International Development (USAID) through the Sustainable Higher Education Research Alliance (SHERA) Program for Universitas Indonesia's Scientific Modeling Application, Research and Training for City-centered Innovation and Technology (SMART CITY) Project, GRANT #AID-497-A-1600004, Sub Grant #IIE-00000078-UI-1.

6. References

[1] Anon 2018 Denpasar Municipality in Figures 2018
[2] Anon 2018 Surabaya Municipality in Figures 2018

[3] Riana I N, Widiastuti and Primayatna I B G 2014 Kajian Alih Fungsi Ruang Terbuka Hijau Kota Denpasar *SPACE* **1** 85–98
[4] Suarta I G and Putra I G A 2016 *Indikator Kesejahteraan Rakyat Kota Denpasar 2016* (Denpasar, Bali)
[5] Nuarsa I W 2013 Penggunaan teknologi penginderaan jauh dan sistem informasi geografis untuk menghitung persentase ruang terbuka hijau di daerah permukiman kota denpasar *J. Bumi Lestari* **13** 9–15
[6] Eckbo G 1964 *Urban Landscape Design* (New York: McGraw-Hill Book Company)
[7] Fisher C 2011 Nature in the City : Urban Environmental History and Central Park **25** 27–31
[8] Suartika G A M 2013 Lost in Translation: The Balinese Vernacular Open Space Tradition *Vernacular Transformations: Architecture, Place, and Tradition* ed G M A Suartika (Denpasar, Bali: Pustaka Larasan) p 314
[9] Treib M 1994 Axioms for a Modern Landscape Architecture *Modern Landscape Architecture: A Critical Review* ed M Treib (Cambridge, MA: MIT Press) p 306
[10] Streatfield D C 2009 Introduction to reprinted Landscape for Living *Landscape for Living* (Amherts: University of Massachusetts Press and Library of American Landscape History) p 274
[11] Sztompka P 1994 *The Sociology of Social Change* (Oxford UK and Cambridge USA: Blackwell)
[12] Hayward J 1989 Urban Parks: research, planning, and social change *Public Places and Spaces* ed I (University of U Altman and E H (University of A Zube (New York: Plenum Press) pp 193–215
[13] Susanti A and Natalia T W 2018 Public space strategic planning based on Z generation preferences *IOP Conf. Series: Mater. Sci. Eng. 407 012076* (IOP Publishing) pp 1–6
[14] Abdel-aziz A A, Abdel-salam H and El-sayad Z 2016 The role of ICTs in creating the new social public place of the digital era *Alexandria Eng. J.* **55** 487–93
[15] Utaminingtyas A, Astiningsih A A M and Mayun I A 2012 Studi Hubungan Vandalisme dengan Setting Taman Lapangan Puputan Badung I Gusti Ngurah Made Agung Denpasar - Bali *E-Journal Agroekoteknologi Trop.* **1** 144–53
[16] Francis M 1989 Control as a Dimension Public-Space Quality *Public Places and Spaces* ed I Altman and E H Zube (New York: Plenum Press) pp 147–72
[17] Nations U 2003 *World Youth Report 2003* (New York)
[18] Carmona M, Heath T, Oc T and Tiesdell S 2003 *Public Places - Urban Spaces* (Burlington MA: Architectural Press)
[19] Carr S, Francis M, Rivlin L G and Stone A M 2007 Needs in public space *Urban Design Reader* ed M Carmona and S Tiedell (Burlingkton. MA, USA: Architectural Press) pp 230–40
[20] Francis M 1987 Some Different Meanings Attached to a City Park and Community Gardens *Landsc. J.* **6** 101–12
[21] Shaftoe H 2008 *Convivial urban spaces: creating effective public places* (London UK: Earthscan)
[22] Sasongko P D 2002 *Kajian perubahan fungsi taman kota di kota semarang* (Universitas Diponegoro)
[23] Aminah S 2015 Konflik dan Kontestasi Penataan Ruang Kota Surabaya *Masy. J. Sosiol.* **20** 59–79
[24] Lawson B 2001 *The Language of Space* (Boston: Architectural Press)
[25] Simonds J O 1994 *Garden Cities 21: creating a livable urban environment* (New York: McGraw-Hill, Inc.)

AUTHOR INDEX